AN INTRODUCTION TO THINKING LIKE A SOCIOLOGIST

RECENT SOCIOLOGY TITLES FROM W. W. NORTON

TO LEARN MORE ABOUT NORTON SOCIOLOGY, PLEASE VISIT:
WWNORTON.COM/SOC

Core Sixth Edition

You May Ask Yourself

AN INTRODUCTION TO THINKING
LIKE A SOCIOLOGIST

Core Sixth Edition

You May Ask Yourself

AN INTRODUCTION TO THINKING LIKE A SOCIOLOGIST

Dalton Conley

PRINCETON UNIVERSITY

 W. W. NORTON
NEW YORK | LONDON

W. W. Norton & Company has been independent since its founding in 1923, when William Warder Norton and Mary D. Herter Norton first published lectures delivered at the People's Institute, the adult education division of New York City's Cooper Union. The firm soon expanded its program beyond the Institute, publishing books by celebrated academics from America and abroad. By mid-century, the two major pillars of Norton's publishing program—trade books and college texts— were firmly established. In the 1950s, the Norton family transferred control of the company to its employees, and today—with a staff of four hundred and a comparable number of trade, college, and professional titles published each year—W. W. Norton & Company stands as the largest and oldest publishing house owned wholly by its employees.

Editor: Justin Cahill
Project Editor: Michael Fauver
Assistant Editors: Erika Nakagawa, Rachel Taylor
Managing Editor, College: Marian Johnson
Managing Editor, College Digital Media: Kim Yi
Production Manager: Eric Pier-Hocking
Media Editor: Eileen Connell
Associate Media Editor: Ariel Eaton
Media Editorial Assistant: Samuel Tang
Marketing Director, Sociology: Julia Hall
Book Designer: Kiss Me I'm Polish LLC, New York
Design Director: Rubina Yeh
Director of College Permissions: Megan Schindel
Photo Editor: Travis Carr
College Permissions Specialist: Bethany Salminen
Composition: Brad Walrod/Kenoza Type, Inc.
Manufacturing: Transcontinental Printing

ISBN: 978-0-393-67418-7 (pbk.)

Library of Congress Cataloging-in-Publication Data
Names: Conley, Dalton, 1969– author.
Title: You may ask yourself : an introduction to thinking like a sociologist/
    Dalton Conley, Princeton University.
Description: Sixth edition. | New York : W.W. Norton, [2019] | Includes bibliographical
    references and index.
Identifiers: LCCN 2018048480 | ISBN 9780393674170 (pbk.)
Subjects: LCSH: Sociology—Methodology. | Sociology—Study and teaching.
Classification: LCC HM511 .C664 2019 | DDC 301.01—dc23 LC record available at
    https://lccn.loc.gov/2018048480

W. W. Norton & Company, Inc., 500 Fifth Avenue, New York, NY 10110
wwnorton.com
W. W. Norton & Company, Ltd., 15 Carlisle Street, London W1D 3BS

1 2 3 4 5 6 7 8 9 0

Brief Contents

Contents

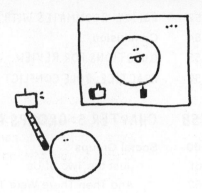

120 CHAPTER 4: SOCIALIZATION AND THE CONSTRUCTION OF REALITY

290 CHAPTER 8: GENDER

390   CHAPTER 10: FAMILY

Preface

I came to sociology by accident, so to speak. During the 1980s, there were no sociology courses at the high-school level, so I entered college with only the vaguest notion of what sociology—or even social science—was. Instead, I headed straight for the pre-med courses. But there was no such thing as a pre-med major, so I ended up specializing in the now defunct "humanities field major." This un-major major was really the result of my becoming a junior and realizing that I was not any closer to a declared field of study than I had been when arriving two years earlier. So I scanned a list of all the electives I had taken until then—philosophy of aesthetics, history of technology, and so on—and marched right into my advisor's office, declaring that it had always been my lifelong dream to study "art and technology in the twentieth century." I wrote this up convincingly enough, apparently, because the college allowed me to write a senior thesis about how the evolution of Warner Brothers' cartoon characters—from the stuttering, insecure Porky Pig to the militant Daffy Duck to the cool, collected, and confident Bugs Bunny—reflected the self-image of the United States on the world stage during the Depression, World War II, and the postwar period, respectively. Little did I know, I was already becoming a sociologist.

After college, I worked as a journalist but then decided that I wanted to continue my schooling. I was drawn to the critical stance and reflexivity that I had learned in my humanities classes, but I knew that I didn't want to devote my life to arcane texts. What I wanted to do was take those skills—that critical stance—and apply them to everyday life, to the here and now. I also was rather skeptical of the methods that humanists used. What texts they chose to analyze always seemed so arbitrary. I wanted to systematize the inquiry a bit more; I found myself trying to apply the scientific method that I had gotten a taste of in my biology classes. But I didn't want to do science in a lab. I wanted to be out in the proverbial real world. So when I flipped through a course catalog with these latent preferences somewhere in the back of my head, my finger landed on the sociology courses.

Once I became a card-carrying sociologist, the very first course I taught was Introduction to Sociology. I had big shoes to fill in teaching this course at Yale. Kai Erikson, the world-renowned author of *Wayward Puritans* and

Everything in Its Path and the son of psychologist Erik Erikson, was stepping down from his popular course, The Human Universe, and I, a first-year assistant professor, was expected to replace him.

I had a lot of sociology to learn. After all, graduate training in sociology is spotty at best. And there is no single theory of society to study in the same way that one might learn, for example, the biochemistry of DNA transcription and translation as the central dogma of molecular biology. We talk about the sociological imagination as an organizing principle. But even that is almost a poetic notion, not so easily articulated. Think of sociology as more like driving a car than learning calculus. You can read the manual all you want, but that isn't going to teach you how to do it. Only by seeing sociology in action and then trying it yourself will you eventually say, "Hey, I've got the hang of this!" The great Chinese philosopher Confucius said about learning: "By three methods we may learn wisdom: First, by reflection, which is noblest; second, by imitation, which is easiest; and third by experience, which is the bitterest." Hopefully you can skip the bitterness, but you get the general idea. For example, by trying to fix a local problem through appealing to your elected officials, you might better grasp sociological theories of the state.

Hence the title of this book. In *You May Ask Yourself,* I show readers how sociologists question what most others take for granted about society, and I give readers opportunities to apply sociological ways of thinking to their own experiences. I've tried to jettison the arcane academic debates that become the guiding light of so many intro books in favor of a series of contemporary empirical (gold) nuggets that show off sociology (and empirical social science more generally) in its finest hour. Most students who take an introductory sociology class in college will not end up being sociology majors, let alone professional sociologists. Yet I aim to speak to both the aspiring major and the student who is merely fulfilling a requirement. So rather than having pages filled with statistics and theories that will go out of date rather quickly, *You May Ask Yourself* tries to instill in the reader a way of thinking—a scientific approach to human affairs that is portable, one that students will find useful when they study anything else, whether history or medicine.

To achieve this ambitious goal, I tried to write a book that was as "un-textbook"-like as possible, while covering all the material that a student in sociology needs to know. In this vein, each chapter is organized around a motivating paradox, meant to serve as the first chilling line of a mystery novel that motivates the reader to read on to find out (or rather, figure out, because this book is not about spoon-feeding facts) the nugget, the debate, the fundamentally new way of looking at the world that illuminates the paradox. Along with a paradox, each chapter begins with a profile of a relevant person who speaks to the core theme of the chapter. These range from myself to Angelina Jolie to a guy who wore a rainbow-colored clown wig to try and get media attention to share his Christian message. In addition, to show the usefulness of sociological knowledge in shaping the world around us, each chapter also culminates in a Policy discussion and a Practice activity, which has been reimagined for the Sixth Edition.

WHAT'S NEW IN THE SIXTH EDITION

Higher education is in rapid transition, with online instruction expanding in traditional institutions, in the expanding for-profit sector, and in the new open-courseware movement. The industry is still very much in flux, with Massive Open Online Courses (MOOCs) failing to displace traditional class-room education (yet). With these changes, textbooks must also reinvent and reorient themselves. Students now expect, I believe, an entire multimedia experience when they purchase a textbook.

To that end, the Third Edition included animations of the associated chapter Paradoxes. (For the Sixth Edition, we updated the gender animation to match the new Paradox for that chapter.) For the Fourth Edition, in addition to a new round of interviews with sociologists, we filmed Sociology on the Street assignment videos. To illustrate a "breaching experiment," for example, I went on camera to perform one myself. It has been years since I had been as nervous speaking on camera as I was the day I walked—barefoot but dressed in a suit—into W. W. Norton's conference room filled with unsuspecting volunteers and proceeded to clip my toenails while I explained the plan for the day and we surreptitiously filmed their (surprisingly unflinching) response.

In the Fifth Edition, we brought the streets into the classroom. Along with new Q&A videos with professional sociologists, we added videos (and text) from folks outside the ivory tower who are doing sociology in their work. For instance, I spoke with journalist and author Jennifer Senior, who wrote the best-selling book *All Joy and No Fun: The Paradox of Modern Parenting*— an obviously sociological domain. We also heard from Zephyr Teachout, an insurgent candidate for governor of New York State who ran on an anti-corruption platform. Other guests included a former FBI agent and a Wall Street fund manager, among others. These interviews help students understand the real-world relevance of sociology and reflect the applied turn in the field.

Further reflecting the increasing emphasis on applications within the discipline, the major new feature in this Sixth Edition is the revamping of the fourth "P" (the first three being the Paradox, Person, and Policy). Rather than just answering review questions, these new Practice activities send students out into the proverbial "streets" (sometimes just metaphorically), where they get to learn by doing—whether that's discovering the true price of unpaid labor in our personal economies, analyzing the structural forces that contribute to one's own carbon emissions, better managing competing roles and statuses in one's life, or, failing that, figuring out how to completely disappear in today's totally connected society.

The other major change to the Sixth Edition is an overhaul of the Gender chapter. Perhaps no domain of social life in US society has changed more dramatically in the past few years than that of gender. As a result, no chapter was more outdated than Chapter 8. In the revamped version I—with the extensive help of experts in the fields of sex, gender, and sexuality—really tried to dig into the concept of gender, turning it inside out, in the service of conveying an understanding of gender and the sex—gender system as something processual and fluid. The new Person for the chapter embodied this shift. Elliot Jackson was someone I "met" on a website I'm addicted to, Quora.com. It's a forum where people post questions and answers to them, and I was browsing responses to the question, "Have you ever reconsidered being transgender?" I was so taken by Elliot's first-person story about navigating the bathroom in his high school that I reached out to ask if we could reprint it in the book. Much to my delight, he agreed. An aspiring young writer, Elliot Jackson is a fantastic chronicler of the trans experience and much else besides. I urge you all to follow him on Quora, like I did, if you are taken by his narrative in the chapter.

In addition to these new features, we revised every chapter in the book to include updated data, research, and examples. Here are some of the highlights:

WHAT'S NEW BY CHAPTER

CHAPTER 1: THE SOCIOLOGICAL IMAGINATION: AN INTRODUCTION

The discussion of the merits of a college degree includes updated data on the cost of college and earnings by degree holders. A new table illustrates the concept of overcredentialism, comparing the percentage of bachelor's degree holders and high-school graduates in various professions from 1970 to 2015. In the new Practice feature, "Seeing Sociologically," students differentiate between natural laws and social norms.

CHAPTER 2: METHODS

Students often struggle with differentiating a theory from an idea, so at reviewer request, I've added two new key terms, **scientific method** and **theory**, to this chapter. A redesigned figure on the research process makes clearer how theory and hypothesis differ. The new Practice feature invites students to think about how sociological methods may be useful in their future careers.

CHAPTER 3: CULTURE AND MEDIA

In the section on Ideology, I explain how the 2016 presidential election has proven that our notions of democratic ideology are remarkably resilient despite recent issues threatening our confidence in democratic institutions, such as fake news. I've added a new discussion about Elijah Anderson's notion of "code switching." The chapter notes the increasing role of computer algorithms in cultural production, including news articles written by artificial intelligence and algorithms that limit information on social media, creating the so-called "online echo chamber." The section on Advertising and Children now considers Google's expansion into classrooms with its low-cost Chromebooks and suite of education software. In the new Practice feature, "Subculture Wars," students investigate subcultures and think about how they reinterpret mainstream cultural memes.

CHAPTER 4: SOCIALIZATION AND THE CONSTRUCTION OF REALITY

The discussion of the Turing Test has been updated. In the section on how families influence socialization, new findings have been added about how daughters make parents more politically conservative, especially about sexuality. To complement updates in Chapter 8: Gender, the section on Gender Roles now defines the idea of the gender binary and includes new information about male and female behavior in the workplace, including sexual harassment. In the new Practice activity, "Role Conflict and Role Strain," students map out their potentially conflicting roles and statuses, from roommate to waitress.

CHAPTER 5: GROUPS AND NETWORKS

In a new chapter-opening vignette, students learn about the mysterious Satoshi Nakamoto, the founder of bitcoin, as a preview of the power of social networks. The discussion of the strength of weak ties is newly illustrated by the example of multilevel marketing schemes. Again to complement the newly revised Gender chapter, the section on the Social Structure of Teenage Sex was updated to include Lisa Wade's recent work on hook-up culture. In the new Practice feature, "How to Disappear," students make a plan for getting off the grid—and think critically about their embeddedness in social life and institutions.

CHAPTER 6: SOCIAL CONTROL AND DEVIANCE

At reviewer request, the discussion of Durkheim's theories of suicide has been condensed. A feature on the Stanford Prison Experiment and Abu Ghraib acknowledges recent controversy surrounding Zimbardo's original methods and findings. Opioid use as an "epidemic" has been added as an example of labeling theory. Figures and data throughout the chapter have been updated with the most recently available information about crime and homicide rates, prison population and demographics, and executions. In the new Practice feature, I make a list of laws I break on a daily basis and invite students to follow suit—and to think critically about what kinds of people are prosecuted for these small infractions.

CHAPTER 7: STRATIFICATION

The word **stratification** has been added as a key term with a corresponding marginal definition. In the discussion of status hierarchy systems, I show how statuses can obscure differences within a particular status group, such as professors—pointing out the wide differences in income and job security between adjunct and tenure-track faculty. The chapter includes new data on how much CEOs of America's largest companies make compared to the average worker, and throughout the chapter, updated data includes the distribution of net wealth, the poverty line, and outlook on future prospects. In the new Practice feature, students research the most and least expensive versions of a particular good or service in their area—such a $5,000 toothbrush—and think about how these extremes can serve as an indicator for class stratification.

CHAPTER 8: GENDER

Thoroughly revised based on extensive reviewer feedback, this chapter now begins with a personal narrative from Elliot, a trans boy who is harassed for using a restroom at his high school, and whose story I follow throughout the chapter. I frame the revised chapter with a brand-new Paradox and corresponding animation: "How do we investigate inequality between men and women without reinforcing binary thinking about gender?" The chapter includes updated research throughout, including Jane Ward's work on men who have sex with men, Georgiann Davis's research about the intersex community, and Tristan Bridges and C. J. Pascoe's notion of "hybrid masculinities." In addition to more material on intersectionality, LGBTQIA people, and the changing world of campus sexual life, the chapter better addresses topics of recent student concern, especially sexism and sexual harassment in the workplace, including a new Policy feature discussing possible solutions to support the #metoo movement. The Practice feature helps students understand the idea of "mansplaining"—and measure it empirically in a meeting or classroom. The chapter ends where it began: Elliot, whose story opens the chapter, reflects on what gender equality looks like to him.

CHAPTER 9: RACE

The chapter now discusses the recent spike in anti-Islamic hate crimes during Trump's campaign and administration, as well as federal assistance for Native Americans, the controversy behind Elizabeth Warren's ancestry, and Native American political activism related to the Dakota Access Pipeline and gerrymandering in Colorado. The discussion of Latinos now considers the deportation of Cubans amid disintegrating relations between the United States and Cuba. I show how the Asian American "model minority" myth has been held up by white supremacist groups. There is new coverage of white nationalist backlash against the "browning of America," and other "market dominant minorities" worldwide, as well as a discussion of Rachel Dolezal and racial passing. The chapter includes a new key term and marginal definition of Eduardo Bonilla-Silva's theory of color-blind racism. New discussions of employment discrimination in France against those with Arab and North African heritage, and of marijuana policy and policing among whites and African Americans. I draw a new comparison between recent proposals for reparations for black slavery and the 1988 reparations for Japanese Americans after internment. In the new Practice feature, students use American Fact Finder to investigate the racial and ethnic diversity of their communities and compare their findings to national averages.

CHAPTER 10: FAMILY

A new section deals with the unique challenges facing immigrant families, especially in light of recent US policy. At reviewer request, I now use the phrase "traditional family" to reflect current sociological thinking, carefully differentiating it from the modern "nuclear family." Again to coincide with the updated Gender chapter, the material about gender and household labor has been rewritten, including a new discussion about how same-sex couples divide housework, and there is a new discussion of transgender parents and their acceptance in society. Data and figures have been updated throughout, including adoption and fertility rates, household demographics, instances of multiple births, divorce, housework and child care, and household earnings. In the new Practice feature, student calculate the value of work done by their parents or caretakers—and think critically about the social division of unpaid labor in their family.

ACKNOWLEDGMENTS

You May Ask Yourself originated in the Introduction to Sociology course that I have taught on and off since the mid-1990s at New York University, Yale University, and Columbia University. However, the process of writing it made me feel as if I were learning to be a sociologist all over again. For example, I never taught religion, methodology, or the sociology of education. But instructors who reviewed the manuscript requested that these topics be covered, so with the assistance of an army of graduate students who really ought to be recognized as coauthors, I got to work. The experience was invaluable, and in a way, I finally feel like a card-carrying sociologist, having acquired at last a bird's-eye view of my colleagues' work. I consider it a great honor to be able to put my little spin (or filter) on the field in this way, to be able not just to influence the few hundred intro students I teach each year, but to excite (I hope) and instill the enthusiasm I didn't get to experience until graduate school in students who may be just a few months out of high school (if that).

I mentioned that the graduate students who helped me create this book were really more like coauthors, ghost writers, or perhaps law clerks. Law clerks do much of the writing of legal opinions for judges, but only a judge's name graces a decision. I asked Norton to allow more coauthors, but they declined—perhaps understandably, given how long such a list would be—so I will take this opportunity to thank my students and hope that you are still reading this preface.

The original transcription of my lectures that formed the basis of this text was completed by Carse Ramos, who also worked on assembling the glossary and drafted some parts of various chapters, such as sections in the economic sociology chapter, as well as some text in the chapters on authority and deviance. She also served as an all-around editor. Ashley Mears did the heavy lifting on the race, gender, family, and religion chapters. Amy LeClair took the lead on methods, culture, groups and networks, socialization, and health. Jennifer Heerwig cobbled together the chapter on authority and the state and deviance (a nice combo), while her officemate Brian McCabe whipped up the chapter on science, technology, and the environment and the one on social movements. Melissa Velez wrote the first draft of the education chapter (and a fine one at that). Michael McCarthy did the same for the stratification chapter. Devyani Prabhat helped revise the social movements chapter. My administrative assistant, Amelia Branigan, served as fact-checker, editor, and box drafter while running a department, taking the GREs, and writing and submitting her own graduate applications. When Amelia had to decamp for Northwestern University to pursue her own doctorate, Lauren Marten took over the job of chasing down obscure references, fact-checking, and proofreading. Alexandre Frenette drafted the questions and activities in the practice sections at the end of each chapter.

For the Second Edition, much of the work to integrate the interview transcripts and update material based on reviewer feedback fell to a great extent on the shoulders of Laura Norén, a fantastic New York University graduate student who has worked on topics as far ranging as public toilets (with my colleague Harvey Molotch) to how symphonies and designers collaborate (as part of her dissertation). I hope Laura will find her crash-course overview of sociology useful at some point in what promises to be a productive and exciting scholarly career.

When it was time to begin the Third Edition, the updating of all the statistics, fact-checking, and so on that is the bread and butter of a revision fell upon the capable shoulders of Emi Nakazato, who though trained as a social worker in graduate school, adeptly pivoted to that field's cousin, sociology.

For the Fourth and Fifth Editions, Laura Norén returned as the research assistant. With her prior experience she picked up the task ably without dropping a beat. Finally, for the Sixth Edition, I turned to then-graduate student Thomas Laidley, who did a more-than-thorough job of not only updating facts and figures but in questioning them as well.

In addition to the students who have worked with me on the book, I need to give shouts out to all the top-notch scholars who found time in their busy schedules to sit down with me and do on-camera interviews: Julia Adams, Andy Bichlbaum, danah boyd, Andrew Cherlin, Nitsan Chorev, Susan Crawford, Adam Davidson, Matthew Desmond, Stephen Duncombe, Mitchell Duncier, Paula England, John Evans, Michael Gaddis, David Grusky, Fadi Haddad, Michael Hout, Jennifer Jacquet, Shamus Khan, Annette Lareau, Jennifer Lee, Ka Liu, Douglas McAdam, Amos Mac, Ashley Mears, Steven Morgan, Alondra Nelson, Devah Pager, Nathan Palmer, C. J. Pascoe, Frances Fox Piven, Allison Pugh, Adeel Qalbani, Marc Ramirez, Asha Rangappa, Jen'nan Read, Victor Rios, Jeffrey Sachs, Jennifer Senior, Mario Luis Small, Zephyr Teachout, Duncan Watts, and Robb Willer.

For the interview videos, the filmmaking, editing, and postproduction were done by Erica Rothman at Nightlight Productions with the assistance of Jim Haverkamp, Kevin Wells, Saul Rouda, Dimitriy Khavin, and Arkadiy Ugorskiy. This was no easy task, because we wanted a bunch of cuts ranging from 30-second sound bites to television-show-length segments of 22 minutes. Although a bunch of interviews with academic social scientists on topics ranging from estimating the effects of Catholic schools on student outcomes to the political economy of global trade to the social contagion of autism are not likely to win any Emmys or rock the Nielsens (with the possible exception of the one on college sex), it was certainly one of the most exciting highlights in my sociological career to host this makeshift talk show on such a wide range of interesting topics. (If only more of our public discourse would dig into issues in the way that we did in these interviews, our society and governance would be in better shape—if I do say so myself!)

When I began work on the Sixth Edition's overhaul of the Gender chapter, I knew I could use some expert help. I'm especially grateful for the reviewers—consultants, really—who gave thorough feedback on nearly every word of the chapter: Kristen Barber (Southern Illinois University, Carbondale) and Carla A. Pfeffer (University of South Carolina). Their comments were generous, thoughtful, and constructive. I'm also grateful for the instructors who reviewed a draft of the revised chapter—Max A. Greenberg (Boston University), Alexandra Hendley (Murray State University), Lauren Jade Martin (Penn State Berks), and Naomi McCool (Chaffey College)—and to Jennifer Haskin (Arizona State University) and Chris Grayson (St. Philip's College), who reviewed the final product. Finally, of course, thanks to Elliot Jackson for sharing his story with us.

I also relied on a number of scholars who generously read chapters of this book and offered valuable feedback, criticisms, and suggestions:

REVIEWERS FOR THE SIXTH EDITION

David Arizmendi, South Texas College, Pecan Campus

Diana Ayers-Darling, Mohawk Valley Community College

Lilika A. Belet, University of Colorado, Colorado Springs / Pikes Peak Community College

Gayle Gordon Bouzard, Texas State University

Gina Carreno-Lukasik, Florida Atlantic University

Nathaniel G. Chapman, Arkansas Tech University

Ann-Renee Clark, Florida International University

Tuesday L. Cooper, Manchester Community College

Cheryl DeFlavis, Hillsborough Community College

Kathy Dolan, Georgia State University

Mark G. Eckel, McHenry County College

Sean French, Manchester Community College

Skylar Gremillion, Louisiana State University

Karen E. Hettinger, Delaware Technical Community College, Owens Campus

Dale Hoffman, Folsom Lake College

Liddy Hope, Elgin Community College / Purdue University Northwestern / Kankakee Community College

Emily Horowitz, St. Francis College

Carrie Hough, Florida Atlantic University

Wesley James, University of Memphis

Isabella Kasselstrand, California State University, Bakersfield

David G. LoConto, New Mexico State University

Gregory Lukasik, Florida Atlantic University

Ryan MacDonald, Hillsborough Community College

Naomi McCool, Chaffey College

Jericho McElroy, Arkansas Tech University

Michael D. McKain, Delaware Technical Community College

Kelly Mosel-Talavera, Texas State University

Romney Norwood, Georgia State University Perimeter College

Jenna Rawlins, American River College

stef m. shuster, Appalachian State University

Karl Smith, Delaware Technical Community College

Julie Song, Chaffey College

Anna Sorensen, SUNY Potsdam

Rachel Sparkman, Virginia Commonwealth University

Santos Torres Jr., American River College

Stephen Vrla, Michigan State University

Tim Wadsworth, University of Colorado Boulder

Lauren Wilson, California State University, Chico

As you can see, it took a village to raise this child. But that's not all. At Norton, I need to thank, first and foremost, Justin Cahill, the editor into whose lap this project landed (after having passed through the hands of Steve Dunn, Melea Seward, and most notably Karl Bakeman, who got promoted onward and upward). Justin deserves great credit for bringing a fresh set of eyes and a powerful brain to help me sociologically question my own assumptions about the book, which, in turn, led to this edition's overhaul. In addition, I am grateful to assistant editors Erika Nakagawa and Rachel Taylor, project editors Diane Cipollone and Michael Fauver, and production manager Eric Pier-Hocking, who handled every stage of the manuscript and managed to keep the innumerable pieces of the book moving through production. Agnieszka Gasparska and her team at Kiss Me I'm Polish are responsible for the terrific new book design, including the reimagined Practice features—as well as the new Paradox video animation to accompany the revised Gender chapter. I also must thank Norton's sociology marketing manager Julia Hall and the social science sales specialists Jonathan Mason and Julie Sindel. Much of You May Ask Yourself's success is due to their boundless energy and enthusiasm. Finally, I owe a special thanks to Eileen Connell, Ariel Eaton, Sam Tang, and Alice Garrard. They are responsible for putting together all of the video and electronic resources that accompany You May Ask Yourself. When it comes to developing new digital products to help instructors teach in the classroom or teach online, they are the most creative and resourceful folks working in college publishing today.

CORRELATION WITH PSYCHOLOGICAL, SOCIAL, AND BIOLOGICAL FOUNDATIONS OF BEHAVIOR SECTION OF THE MCAT®

In 2015, the Association of American Medical Colleges revised the Medical College Admissions Test (MCAT) to include fundamental concepts from sociology. To help students prepare for the test, here is a correlation guide for *You May Ask Yourself*, Sixth Edition.

FOUNDATIONAL CONCEPT 7

Biological, psychological, and sociocultural factors influence behavior and behavior change.

FOUNDATIONAL CONCEPT 8

Psychological, sociocultural, and biological factors influence the way we think about ourselves and others as well as how we interact with others.

CHAPTER	HEADING/DESCRIPTION	PAGE
3	Ethnocentrism	84
3	Cultural relativism	88
4	Me, Myself, and I: Development of the Self and the Other	125
4	Agents of Socialization	129
4	Social Interaction	136
4	Dramaturgical Theory	144
5	Social Groups	160
5	From Groups to Networks	170
5	Network Analysis in Practice	183
5	Organizations	188
6	Stigma	221
9	Minority–Majority Group Relations	364
9	Prejudice, Discrimination, and the New Racism	375
9	Institutional Racism	380

FOUNDATIONAL CONCEPT 10

Social stratification and access to resources influence well-being.

You May Ask Yourself

Core Sixth Edition

AN INTRODUCTION TO THINKING LIKE A SOCIOLOGIST

1

A SUCCESSFUL SOCIOLOGIST
MAKES THE FAMILIAR STRANGE.

The Sociological Imagination: An Introduction

If you want to understand sociology, why don't we start with you. Why are you taking this class and reading this textbook? It's as good a place to start as any—after all, sociology is the study of human society, and there is the sociology of sports, of religion, of music, of medicine, even a sociology of sociologists. So why not start, by way of example, with the sociology of an introduction to sociology?

For example, why are you bent over this page? Take a moment to write down the reasons. Maybe you have heard of sociology and want to learn about it. Maybe you are merely following the suggestion of a parent, guidance counselor, or academic adviser. The course syllabus probably indicates that for the first week of class, you are required to read this chapter. So there are at least two good reasons to be reading this introduction to sociology text.

Let's take the first response, "I want to educate myself about sociology." That's a fairly good reason, but may I then ask why you are taking the class rather than simply reading the book on your own? Furthermore, assuming that you're paying tuition, why are you doing so? If you really are here for the education, let me suggest an alternative: Grab one of the course schedules at your college, decide which courses to take, and just show up! Most introductory classes are so large that nobody notices if an extra student attends. If it is a smaller, more advanced seminar, ask the professor if you

SOCIOLOGY

the study of human society.

can audit it. I have never known a faculty member who checks that all class attendees are legitimate students at the college—in fact, we're happy when students *do* show up to class. An auditor, someone who is there for the sake of pure learning, and who won't be grade grubbing or submitting papers to be marked, is pure gold to any professor interested in imparting knowledge for learning's sake.

You know the rest of the drill: Do all the reading (you can usually access the required texts for free at the library), do your homework, and participate in class discussion. About the only thing you won't get at the end of the course is a grade. So give yourself one. As a matter of fact, once you have compiled enough credits and written a senior thesis, award yourself a diploma. Why not? You will probably have received a better education than most students—certainly better than I did in college.

But what are you going to do with a homemade diploma? You are not here just to learn; you wish to obtain an actual college degree. Why exactly do you want a college degree? Students typically answer that they have to get one in order to earn more money. Others may say that they need credentials to get the job they want. And some students are in college because they don't know what else to do. Whatever your answer, the fact that you asked yourself a question about something you may have previously taken for granted is the first step in thinking like a sociologist. "Thinking like a sociologist" means applying analytical tools to something you have always done without much conscious thought—like opening this book or taking this class. It requires you to reconsider your assumptions about society and question what you have taken for granted in order to better understand the world around you. In other words, thinking like a sociologist means *making the familiar strange*.

This chapter introduces you to the sociological approach to the world. Specifically, you will learn about the *sociological imagination,* a term coined by C. Wright Mills. We'll return to the question "Why go to college?" and apply our sociological imaginations to it. You will also learn what a social institution is. The chapter concludes by looking at the sociology of sociology—that is, the history of sociology and where it fits within the social sciences.

The Sociological Imagination

SOCIOLOGICAL IMAGINATION

the ability to connect the most basic, intimate aspects of an individual's life to seemingly impersonal and remote historical forces.

More than 50 years ago, sociologist C. Wright Mills argued that in the effort to think critically about the social world around us, we need to use our sociological imagination, the ability to see the connections between our personal experience and the larger forces of history. This is just what we are doing when we question this textbook, this course, and college in general.

In *The Sociological Imagination* (1959), Mills describes it this way: "The first fruit of this imagination—and the first lesson of the social science that embodies it—is the idea that the individual can understand his own experience and gauge his own fate only by locating himself within his period, that he can know his own chances in life only by becoming aware of those of all individuals in his circumstances. In many ways it is a terrible lesson; in many ways a magnificent one." The terrible part of the lesson is to make our own lives ordinary—that is, to see our intensely personal, private experience of life as typical of the period and place in which we live. This can also serve as a source of comfort, however, helping us realize we are not alone in our experiences, whether they involve our alienation from the increasingly dog-eat-dog capitalism of modern America, the peculiar combination of intimacy and dissociation that we may experience on the internet, or the ways that nationality or geography affect our life choices. The sociological imagination does not just leave us hanging with these feelings of recognition, however. Mills writes that it also "enables [us] to take into account how individuals, in the welter of their daily experience, often become falsely conscious of their social positions." The sociological imagination thus allows us to see the veneer of social life for what it is and to step outside the "trap" of rapid historical change in order to comprehend what is occurring in our world and the social foundations that may be shifting right under our feet. As Mills wrote after World War II, a time of enormous political, social, and technological change, "The sociological imagination enables us to grasp history and biography and the relations between the two within society. That is its task and its promise. To recognize this task and this promise is the mark of the classic social analyst."

Sociologist C. Wright Mills smoking his pipe in his office at Columbia University. How does Mills's concept of the sociological imagination help us make the familiar strange?

Mills offered his readers a way to stop and take stock of their lives in light of all that had happened in the previous decade. Of course, we almost always feel that social change is fairly rapid and continually getting ahead of us. Think of the 1960s or even today, with the rise of the internet and global terror threats. In retrospect, we consider the 1950s, the decade when Mills wrote his seminal work, to be a relatively placid time, when Americans experienced some relief from the change and strife of World War II and the Great Depression. But Mills believed the profound sense of alienation experienced by many during the postwar period was a result of the change that had immediately preceded it.

Another way to think about the sociological imagination is to ask

HOW TO BE A SOCIOLOGIST ACCORDING TO QUENTIN TARANTINO: A SCENE FROM *PULP FICTION*

Have you ever been to a foreign country, noticed how many little things were different, and wondered why? Have you ever been to a church of a different denomination—or a different religion altogether—from your own? Or have you been a fish out of water in some other way? The only guy attending a social event for women, perhaps? Or the only person from out of state in your dorm? If you have experienced that fish-out-of-water feeling, then you have, however briefly, engaged your sociological imagination. By shifting your social environment enough to be in a position where you are not able to take everything for granted, you are forced to see the connections between particular historical paths taken (and not taken) and how you live your daily life. You may, for instance, wonder why there are bidets in most European bathrooms and not in American ones. Or why people waiting in lines in the Middle East typically stand closer to each other than they do in Europe or America. Or why, in some rural Chinese societies, many generations of a family sleep in the same bed. If you are able to resist your initial impulses toward xenophobia (feelings that may result from the discomfort of facing a different reality), then you are halfway to understanding other people's lifestyles as no more or less sensible than your own. Once you have truly adopted the sociological imagination, you can start questioning the links between your personal experience and the particulars of a given society without ever leaving home.

In the following excerpt of dialogue from Quentin Tarantino's 1994 film *Pulp Fiction*, the character Vincent tells Jules about the "little differences" between life in the United States and life in Europe.

VINCENT: It's the little differences. A lotta the same shit we got here, they got there, but there they're a little different.

JULES: Example?

VINCENT: Well, in Amsterdam, you can buy beer in a movie theater. And I don't mean in a paper cup either. They give you a glass of beer, like in a bar. In Paris, you

Vincent Vega (John Travolta) describes his visit to a McDonald's in Amsterdam to Jules Winnfield (Samuel L. Jackson).

can buy beer at McDonald's. Also, you know what they call a Quarter Pounder with Cheese in Paris?

JULES: They don't call it a Quarter Pounder with Cheese?

VINCENT: No, they got the metric system there, they wouldn't know what the fuck a Quarter Pounder is.

JULES: What'd they call it?

VINCENT: Royale with Cheese.

[. . .]

VINCENT: [Y]ou know what they put on french fries in Holland instead of ketchup?

JULES: What?

VINCENT: Mayonnaise.
           [. . .] And I don't mean a little bit on the side of the plate, they fuckin' drown 'em in it.

JULES: Uuccch!

Your job as a sociologist is to get into the mind-set that mayonnaise on french fries, though it might seem disgusting at first, is not strange after all, certainly no more so than ketchup.

ourselves what we take to be natural that actually isn't. For example, let's return to the question "Why go to college?" Sociologists and economists have shown that the financial benefits of education—particularly higher education—appear to be increasing. They refer to this as the "returns to schooling." In today's economy, the median (i.e., typical) annual income for a high-school graduate is $35,984; for those with a bachelor's degree, it is $60,112 (2016 data; Bureau of Labor Statistics, 2017d). That $24,128 annual advantage seems like a good deal, but is it really? Let's shift gears and do a little math.

WHAT ARE THE TRUE COSTS AND RETURNS OF COLLEGE?

Now that you are thinking like a sociologist, let's compare the true cost of going to college for four or five years to calling the whole thing off and taking a full-time job right after high school. First, there is the tuition to consider. Let's assume for the sake of argument you are paying $9,970 per year for tuition (College Board, 2018). That's a lot less than what most private four-year colleges cost, but about average for in-state tuition at a state school. (Community colleges, by contrast, are usually much cheaper, especially because they tend to be commuter schools whose students live off-campus, but they typically do not offer a four-year bachelor's degree.)

In making the decision to attend college, you are agreeing to pay $9,970 this year, about $10,300 next year, 3.3 percent more the following year, and another 3.3 percent on top of that amount in your senior year to cover tuition hikes and inflation. The $9,970 you have to pay right now is what hurts the most, because costs in the future are worth less than expenses today. Money in the future is worth less than money in hand for several reasons. The first is inflation. We all know that money is not what it used to be. In fact, taking into account the standard inflation rate—as measured by the government's Consumer Price Index—it took $17.75 in 2015 to equal the buying power of a single dollar back in 1940 (Bureau of Labor Statistics, 2018a). The second reason money today is worth more than money tomorrow is that we could invest the money today to make more tomorrow.

Using a standard formula to adjust for inflation and bring future amounts into current dollars, we can determine that paying out $9,970 this year and the higher amounts over the next three years is equivalent to paying $41,900 in one lump sum today; this would be the direct cost of attending college. Indirect costs—so-called opportunity costs—exist as well, such as the costs associated with the amount of time you are devoting to school. Taking into account the typical wage for a high-school graduate, not counting differences by gender, age, or level of experience, we can calculate that if you worked full time instead of going to college, you would make $30,000 this year. Thus we find that the present value of the total wages lost over the next four years by choosing full-time school over full-time work is about $144,000. Add these opportunity costs to the direct costs of tuition, and we get $185,900.

Next we need to calculate the "returns to schooling." For the sake of simplicity, we will mostly ignore the fact that the differences between high-school graduates and college graduates change over time—given years of experience and the ups and downs of the economy. We will regard the $60,112 annual earnings figure for recent college graduates as fixed for the first 10 years past college graduation. We will use a higher estimate for annual earnings after that, to take into account the fact that mid-career workers make more. But remember, those who start working right out of high school begin earning about 5 years earlier than those who spend that time in college. The average time it takes to complete a bachelor's degree at a public university is 5.2 years so we are rounding down to 5 (Shapiro et al., 2016). Assuming you attend college for 5 years and retire at age 65, you will have worked 42 years (high-school grads will be in the workforce for 47 years because they get a 5-year head start). When we compare your college-degree-holding lifetime earnings to the lifetime earnings of someone who has only a high-school education, we find that with a college degree you will make about $1,500,000 more than someone who went straight to work after high school (Figure 1.1). (To simplify, we are conveniently ignoring the fact that future money is inherently worth less than present money and that some college degrees, like those in engineering, lead to higher-paying jobs than others.) On top of this substantial financial return to schooling, research also shows that earning a high-school and/or college degree makes us healthier and less likely to engage in unhealthy behaviors like smoking, even after accounting for other factors like income (Heckman et al., 2016).

But wait a minute: How do we know for sure that college really mattered in the equation? Individuals who finish college might earn more because they actually learned something and obtained a degree, or—a big OR—they might earn more regardless of the college experience because people who stay in school (1) are innately smarter, (2) know how to work the system, (3) come from wealthier families, (4) can delay gratification, (5) are more efficient at managing their time, or (6) all of the above—take your pick. In other words, Yale graduates might not have needed to go to college to earn higher wages; they might have been successful anyway.

Maybe, then, the success stories of Mark Zuckerberg, Steve Jobs, Lady Gaga, and other college dropouts don't cut against the grain so sharply after all. Maybe they were the savvy ones: Convinced of their ability to make it on their own, thanks to the social cues they received (including the fact that they had been admitted to college), they decided that they wouldn't wait four years to try to achieve success. They opted to just go for it right then and there. College's "value added," they might have concluded, was marginal at best.

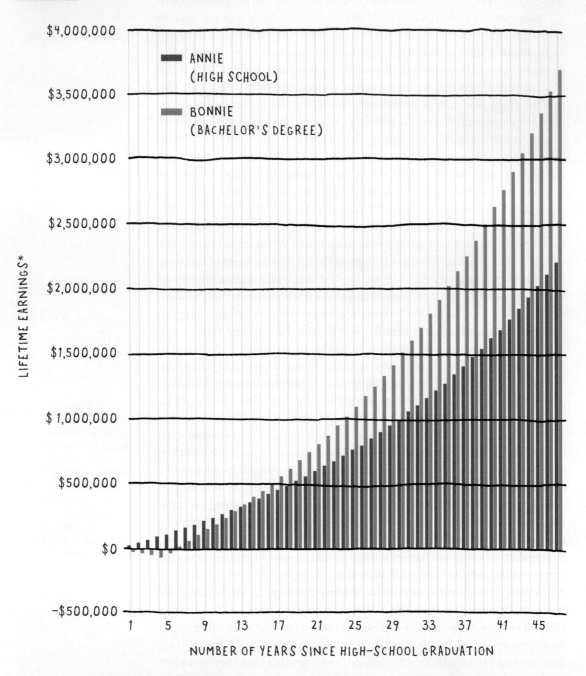

FIGURE 1.1 Returns to Schooling

LIFETIME EARNINGS*

- ANNIE (HIGH SCHOOL)
- BONNIE (BACHELOR'S DEGREE)

NUMBER OF YEARS SINCE HIGH-SCHOOL GRADUATION

*This set of hypothetical women—Annie and Bonnie—live in a world that is not quite like reality. We did not flatten Annie's trajectory to account for the fact that high-school diploma holders are more likely to experience periods of forced part-time work and/or unemployment. We also assumed the same rate of income increase over time (i.e., raises) for these two, although high-school diploma holders are more likely to experience wage stagnation than college diploma holders.

SOURCE: Carnevale et al., 2014.

GETTING THAT "PIECE OF PAPER"

Even if college turns out to matter in the end, does it make a difference because of the learning that takes place there or because of our credentialist society that it aids and abets? The answer to this question has enormous implications for what education means in our society. Imagine, for example, a society in which people become doctors not by doing well on the SATs, going to college, taking premed courses, acing the MCATs, and then spending more time in the classroom. Instead, the route to becoming a doctor—among the most prestigious and highly paid occupations in our society—starts with emptying bedpans as a nurse's aide and working your way up through the ranks of registered nurse, apprentice physician, and so forth; finally, after years of on-the-job training, you achieve the title of doctor. Social theorist Randall Collins has proposed just such a medical education system in the controversial *The Credential Society: A Historical Sociology of Education and Stratification* (1979), which argues that the expansion of higher education has merely resulted in a ratcheting up of credentialism and expenditures on formal education rather than reflecting any true societal need for more formal education or opening up opportunity to more people.

Two famous college dropouts. Facebook CEO Mark Zuckerberg (left) attended Harvard but dropped out before graduating. Oprah Winfrey (right) left Tennessee State University as a sophomore to begin a career in media.

College bulletin boards are covered with advertisements like this one promoting websites that generate diplomas. Why are these fake diplomas not worth it?

If Collins is correct and credentials are what matter most, then isn't there a cheaper, faster way to get them? In fact, all you need are $29.95 and a little guts, and you can receive a diploma from one of the many online sites that promise either legitimate degrees from nonaccredited colleges or a faux college diploma from any school of your choosing. Thus why not save four years and lots of money and obtain your credentials immediately?

Obviously, universities have incentives to prevent such websites from undermining their exclusive authority over degree-conferring ability. They rely on a number of other social institutions, ranging from copyright law to magazines that publish rankings to protect their status. Despite universities' interest in protecting their reputations, I had never had a university employer verify my education claims when applying to teach until 2015 when I moved to Princeton. Every other employer, including NYU, accepted my résumé without calling my graduate or undergraduate universities. NYU does check to make sure student applicants have completed high school, however. Are universities too lazy to care? Probably not.

There are strong informal mechanisms by which universities protect their status. First, there is the university's alumni network. Potential employers rarely call a university's registrar to make sure you graduated, but they will expect you to talk a bit about your college experience. If your interviewer is an alumnus/alumna or otherwise familiar with the institution, you might also be expected to talk about what dorm you lived in, reminisce about a particularly dramatic homecoming game, or gripe about an especially unreasonable professor. If you slip up on any of this information, suspicions will grow, and then people might call to check on your graduation status. Perhaps there are some good reasons not to opt for that $29.95 degree and to pay the costs of college after all.

On a more serious note, the role of credentialism in our society means that getting in—especially to a wealthy school with plenty of need-based aid available—can make a huge difference for lower-income students. I sat down with Asha Rangappa, the dean of admissions at Yale Law School (and former FBI agent), who explained that by the time students are old enough to apply to law school, middle-class and upper-class students have already been granted all sorts of opportunities that make them appear to be stronger

TABLE 1.1 Overcredentialed? Workers with Bachelor's Degrees in 1970 and 2015

JOB TITLE	PERCENTAGE OF WORKFORCE, AGE 25-64, WITH BACHELOR'S DEGREE		
	1970	2015	% CHANGE
Bartenders	3%	20%	567%
Photographers	10%	50%	400%
Electricians	2%	7%	250%
Accountants and auditors	42%	78%	86%
Actors, directors, and producers	41%	74%	80%
Writers and authors	47%	84%	79%
Military	24%	37%	54%
Dental hygienists	29%	36%	24%
Primary-school teachers	84%	94%	12%
Doctors	96%	99%	3%
Preschool teachers	53%	40%	−9%
Human-resource clerks	40%	29%	−28%
Machine programmers	27%	7%	−74%

A recent *Economist* article found that since the 1970s, more American workers in most professions have earned a bachelor's degree. In half of these professions, the authors found that real wages have actually fallen. Do these examples support Collins's argument about overcredentialism? Have these jobs gotten more demanding or technologically complex over time or are there just more people doing them with a diploma?

SOURCES: University of Minnesota IPUMS; *The Economist* (2018a).

candidates even if they do not have better moral character or a stronger apti-tude for law than their less affluent counterparts in the same applicant pool:

> I read anywhere from three thousand, four thousand applications a year, and I do a kind of character and personality assessment. I decide who gets in. . . . I think that there's a meritocracy at the point where I'm doing it, but I think accessing the good opportunities that allow you to take advantage of the meritocracy is limited. I think that's the problem, I think that's what somebody like me in my position works very hard to correct for. When twenty-two to twenty-five years of someone's life are behind them, it is too

DIGITAL.WWNORTON.COM/YOUMAYASK6CORE

To see my interview with Asha Rangappa, go to
digital.wwnorton.com/youmayask6core

late to correct the disparity in access that really needed to have been corrected from like zero to five years, zero to ten years.

Do I think that people who have had access to more resources and opportunities and money are going to do better in the admissions process? Yeah. Because they're just going to have the richer background. And [admissions counselors] have to be able to be in a position where [we] can afford to account for that and take the risk. (Conley, 2015a)

What's more, law schools' rankings in magazines like *US News and World Report* are impacted by the average LSAT scores of their incoming classes. Rangappa's school, Yale, is so elite that it is able to maintain its ranking even if its average LSAT score drops a little. But for other schools, a drop in LSAT scores will cause their ranking to drop, leading to a decline in high-quality applicants and a further drop in rankings. From Rangappa's perspective:

> I've now read over ten years, over thirty thousand admission files. To me, the LSAT is one number, and I can look at the rest of the file. There may very well be somebody who has a crappy LSAT score, but . . . I can tell in the totality of the application that the applicant is going to be a better person at the school. I have the luxury of taking that person because it's Yale Law School and the way the US News formula is created, we're not going to suffer a consequence if our median LSAT drops one point.

Rangappa's account shows that elite educational institutions perform a balancing act whereby they often seek to broaden the population who gains from the opportunities and status they provide while at the same time maintaining their own rank in the hierarchy of similar institutions. That is, even when such institutions do want to level the playing field, they themselves are trapped in a highly competitive environment that does not allow them to fully counteract preexisting inequalities. Social institutions thus have a tendency to reinforce existing social structures and the inequalities therein. College (and graduate school) is no exception.

What Is a Social Institution?

The university, then, is more than just a printing press that churns out diplomas, and, for that matter, it does not merely impart formal knowledge. It fulfills a variety of roles and provides links to many other societal institutions. For example, a college is an institution that acts as a gatekeeper to what are considered legitimate forms of educational advantage by certifying what is legitimate knowledge. It is an institution that segregates great swaths of the population by age. (You won't find a more age-segregated environment than a four year college; it even beats a retirement home in having the smallest amount of age variation in its client population.) A college is a proprietary brand that is marketed on sweatshirts and mugs and through televised sporting events. Last but not least, it is an informal set of stories told within a social network of students, faculty, administrators, alumni, and other relevant individuals.

This last part of the definition is key to understanding one of sociology's most important concepts, the social institution. A social institution is a complex group of interdependent positions that, together, perform a social role and reproduce themselves over time. A way to think of these social positions is as a set of stories we tell ourselves; social relations are a network of ties; and the social role is a grand narrative that unifies these stories within the network. In order to think sociologically about social institutions, you need to think of them not as monolithic, uniform, stable entities—things that "just are"—but as institutions constructed within a dense network of other social institutions and meanings. Sound confusing? Bear with me as I provide an example: I teach at Princeton University. What exactly is the social institution known as Princeton? It is not the collection of buildings I frequent. It certainly cannot be the people who work there, or even the students, because they change over time, shifting in and out through recruitment and retirement, admission and graduation. We might thus conclude that a social institution is just a name. However, an institution can change its name and still retain its social identity. Duke University was once called Normal College and then Trinity College, yet it remains the same institution.

Of course, all such transitions involving a change of name, location, mission, and so on require a great deal of effort and agreement among interested parties. In some cases, changes in personnel, function, or location may be too much for a social institution to sustain, causing it to die out and be replaced by something that is considered new. Sometimes institutions even try to rupture their identity intentionally. Tobacco company Philip Morris had received such bad press as a cigarette manufacturer for so long that it

SOCIAL INSTITUTION

a complex group of interdependent positions that, together, perform a social role and reproduce themselves over time; also defined in a narrow sense as any institution in a society that works to shape the behavior of the groups or people within it.

Tobacco company Philip
Morris changed its name
to Altria at a stockholders'
meeting in January 2003.

changed its name to Altria, hoping to start fresh and shake off the negative connotations of its previous embodiment. For that effort to succeed,
the narrative of Philip Morris circulating in social networks had to die out
without being connected to Altria.

This grand narrative that constitutes social identity is nothing more
than the sum of individual stories told between pairs of individuals. Think
about your relationship to your parents. You have a particular story that you
tell if asked to describe your relationship with your mother. She also has a
story. Your story may change slightly, depending on whom you are talking
to; you may add some details or leave out others. Your other relatives have
stories about your mother and her relationship to you. So do her friends and
yours. Anyone who knows her contributes to her social identity. The sum
total of stories about your mom is the grand narrative of who she is.

All of this may seem like a fairly flimsy notion of how things operate
in the social world, but even though any social identity boils down to a set
of stories within a social network, that narrative is still hearty and robust.
Imagine what it would take to change an identity. Let's say your mom is 50
years old. You want to make her 40 instead. You would not only have to convince her to refer constantly to herself as 10 years younger; you would also
have to get your other relatives and her friends to abide by this change. And
it wouldn't stop there. You'd have to change official documents as well—her
driver's license, passport, and so on. This is not so easily done. Even though
your mother's identity (in this case her age, although the same logic can be
extended to her name, ethnicity, and many other aspects of her identity) could
be described as nothing more than an understanding among her, everyone
who knows her, and the formal authorities, the matter is a fairly complicated
one. If that sounds hard, just think about trying to change the identity of a
major institution such as your university. You'd have to convince the board of

directors, alumni, faculty, students, and everyone else who has a relationship to the school of the need for a change. Altering an identity is fairly difficult, even though it is ultimately nothing more than an idea.

I mentioned that if you wanted to change your mother's age, name, or race, you'd have to convince not only her friends but also the formal authorities, which are social institutions with their own logics and inertias. Let's take the example of college once again. Other than the informal ties of people who have a relationship to the grand narrative of a particular college, a number of social structures exist that make colleges, which are themselves made up of a series of stories within social networks, possible:

1. The *legal system* enforces copyright law, making fake diplomas illegitimate.

2. The *primary and secondary educational system* (i.e., K–12 schooling) prepares students both academically and culturally for college as well as acting as an extended screening and sorting mechanism to help determine who goes to college and to which one.

3. The *Educational Testing Service* and *ACT* are private companies that have a duopoly on the standardized tests that screen for college admission.

4 The *wage labor market* encompasses the entire economy that allows your teacher to be paid, not to mention the administrators, staff, and other outside contractors who maintain the intellectual, fiscal, and physical infrastructure of the school you attend.

5. *English*, although not the official language of the United States (there isn't one!), is the language in which instruction takes place at the majority of US colleges. Language itself is a social phenomenon; some would argue that it is the basis for all of social life. But a given language is a particular outcome of political boundaries and historical struggles among various populations in the world. It is often said that the only difference between a language and a dialect is that a language has an army to back it up. In other words, deciding whether a spoken tongue is a language or merely a dialect is a question of power and legitimacy.

Trying to understand social institutions such as the legal system, the labor market, or language itself is at the heart of sociological inquiry. Although social institutions shape every aspect of our behavior, they are not monolithic. In fact, every day we construct and change social institutions through ordinary interactions and the meanings we ascribe to them. By becoming aware of the intersections between social institutions and your life, you are already thinking like a sociologist.

The Sociology of Sociology

Now that we have an idea of how sociologists approach their analysis of the world, let's turn that lens, the sociological imagination, to sociology itself. As a formal field, sociology is a relatively young discipline. Numerous fields of inquiry exist, such as molecular genetics, radio astronomy, and computer science, that could not emerge until a certain technology was invented. Sociology might seem to fall outside this category, but to study society, we need not only a curious mind and a certain willingness but also the specific frame of reference—the lens—of the sociological imagination.

TWO CENTURIES OF SOCIOLOGY

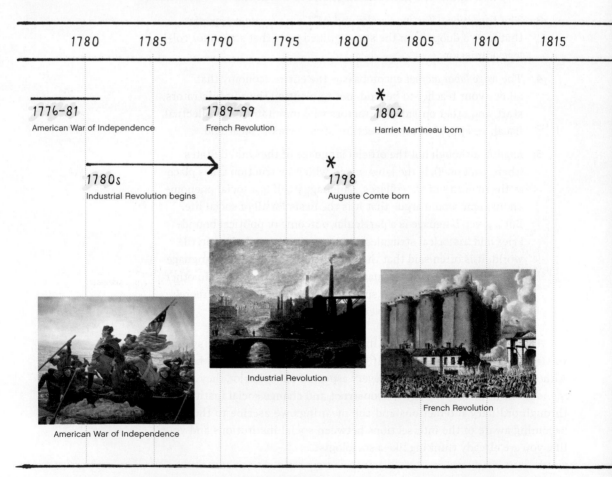

| 1780 | 1785 | 1790 | 1795 | 1800 | 1805 | 1810 | 1815 |

1776–81
American War of Independence

1789–99
French Revolution

1802
Harriet Martineau born

1780s
Industrial Revolution begins

1798
Auguste Comte born

Industrial Revolution

American War of Independence

French Revolution

The sociological imagination is a technology of sorts, a technology that could have developed only during a certain time. That time was, arguably, the nineteenth century, when French scholar Auguste Comte (1798–1857) invented what he called "social physics" or "positivism."

AUGUSTE COMTE AND THE CREATION OF SOCIOLOGY

According to Comte, positivism arose out of a need to make moral sense of the social order in a time of declining religious authority. Comte claimed that a secular basis for morality did indeed exist—that is, we could determine right and wrong without reference to higher powers or other religious concepts. And that was the job of the sociologist: to develop a secular morality. Comte further argued that human society had gone through three historical,

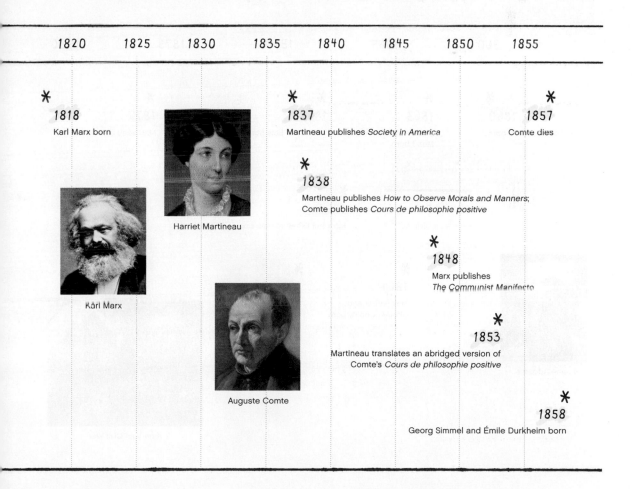

| 1820 | 1825 | 1830 | 1835 | 1840 | 1845 | 1850 | 1855 |

*
1818
Karl Marx born

*
1837
Martineau publishes *Society in America*

*
1857
Comte dies

Harriet Martineau

*
1838
Martineau publishes *How to Observe Morals and Manners*; Comte publishes *Cours de philosophie positive*

Karl Marx

*
1848
Marx publishes
The Communist Manifesto

*
1853
Martineau translates an abridged version of Comte's *Cours de philosophie positive*

Auguste Comte

*
1858
Georg Simmel and Émile Durkheim born

epistemological stages. In the first, which he referred to as the theological stage, society seemed to be the result of divine will. If you wanted to understand why kings ruled, why Europe used a feudal and guild system of labor, or why colonialism took root, the answer was that it was God's plan. To better understand God's plan and thus comprehend the logic of social life, scholars of the theological period might consult the Bible or other ecclesiastical texts. During stage two, the metaphysical stage according to Comte, Enlightenment thinkers such as Jean-Jacques Rousseau, John Stuart Mill, and Thomas Hobbes saw humankind's behavior as governed by natural, biological instincts. To understand the nature of society—why things were the way they were—we needed to strip away the layers of society to better comprehend how our basic drives and natural instincts governed

TWO CENTURIES OF SOCIOLOGY

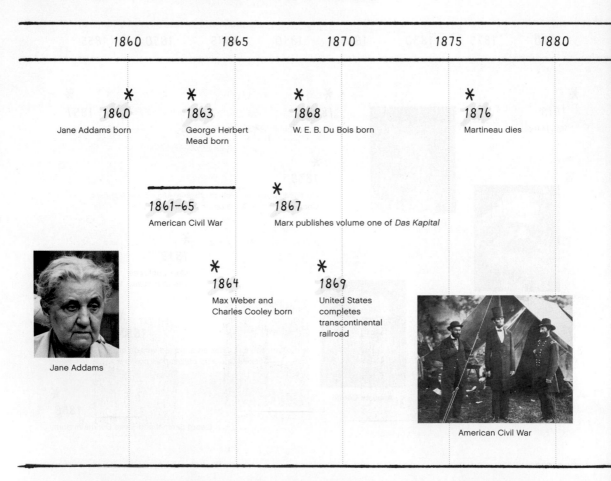

| 1860 | 1865 | 1870 | 1875 | 1880 |

✳ 1860
Jane Addams born

✳ 1863
George Herbert Mead born

✳ 1868
W. E. B. Du Bois born

✳ 1876
Martineau dies

1861–65
American Civil War

✳ 1867
Marx publishes volume one of *Das Kapital*

✳ 1864
Max Weber and Charles Cooley born

✳ 1869
United States completes transcontinental railroad

Jane Addams

American Civil War

and established the foundation for the surrounding world. Comte called the third and final stage of historical development the scientific stage. In this era, he claimed, we would develop a social physics of sorts in order to identify the scientific laws that govern human behavior. The analogy here is not theology or biology but rather physics. Comte was convinced we could understand how social institutions worked (and didn't work), how we relate to one another (whether on an individual or group level), and the overall structure of societies if we merely ascertained their "equations" or underlying logic. Needless to say, most sociologists today are not so optimistic.

Harriet Martineau Harriet Martineau (1802—1876), an English social theorist, was the first to translate Comte into English. In fact, Comte assigned

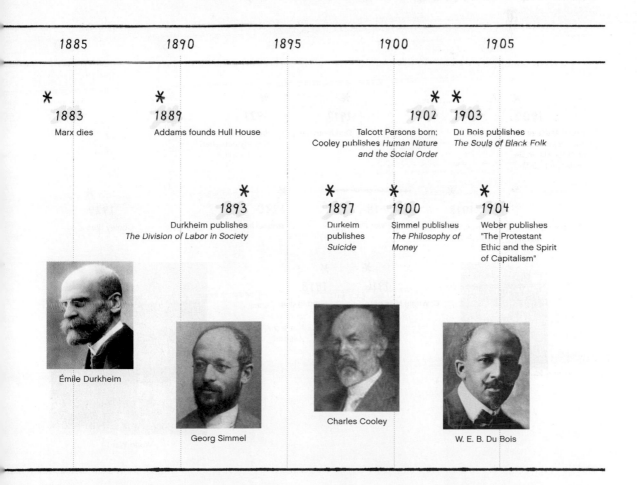

1885	1890	1895	1900	1905

✳
1883
Marx dies

✳
1889
Addams founds Hull House

✳
1902
Talcott Parsons born;
Cooley publishes *Human Nature
and the Social Order*

✳
1903
Du Bois publishes
The Souls of Black Folk

✳
1893
Durkheim publishes
The Division of Labor in Society

✳
1897
Durkeim
publishes
Suicide

✳
1900
Simmel publishes
*The Philosophy of
Money*

✳
1904
Weber publishes
"The Protestant
Ethic and the Spirit
of Capitalism"

Émile Durkheim

Georg Simmel

Charles Cooley

W. E. B. Du Bois

her translations to his students, claiming that they were better than the original. She also wrote important works of her own, including *Society in America* (1837), in which she describes our nation's physical and social aspects. She addressed topics ranging from the way we educate children (which, she attests, affords parents too much control and doesn't ensure quality) to the relationship between the federal and state governments. She was also the author of the first methods book in the area of sociology, *How to Observe Morals and Manners* (1838), in which she took on the institution of marriage, claiming that it was based on an assumption of the inferiority of women. This critique, among other writings, suggests that Martineau should be considered one of the earliest feminist social scientists writing in the English language.

TWO CENTURIES OF SOCIOLOGY

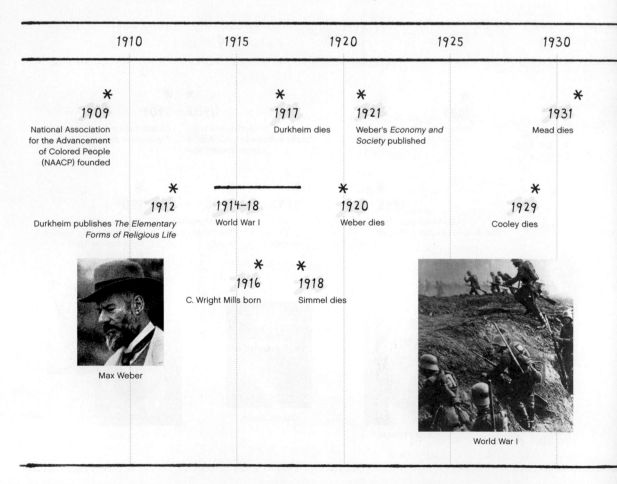

1910	1915	1920	1925	1930

* 1909
National Association for the Advancement of Colored People (NAACP) founded

* 1917
Durkheim dies

* 1921
Weber's *Economy and Society* published

* 1931
Mead dies

* 1912
Durkheim publishes *The Elementary Forms of Religious Life*

1914–18
World War I

* 1920
Weber dies

* 1929
Cooley dies

* 1916
C. Wright Mills born

* 1918
Simmel dies

Max Weber

World War I

CLASSICAL SOCIOLOGICAL THEORY

Although Comte and Martineau preceded them, Karl Marx, Max Weber, and Émile Durkheim are often credited as the founding fathers of the sociological discipline. Some would add a fourth classical sociological theorist, Georg Simmel, to the triumvirate. A brief overview of each of their paradigms follows. We will return to the work of these thinkers throughout the book.

Karl Marx Karl Marx (1818–1883) is probably the most famous of the three early sociologists; from his surname the term *Marxism* (an ideological alternative to capitalism) derives, and his writings provided the theoretical basis for Communism. When Marx was a young man, he edited a newspaper that

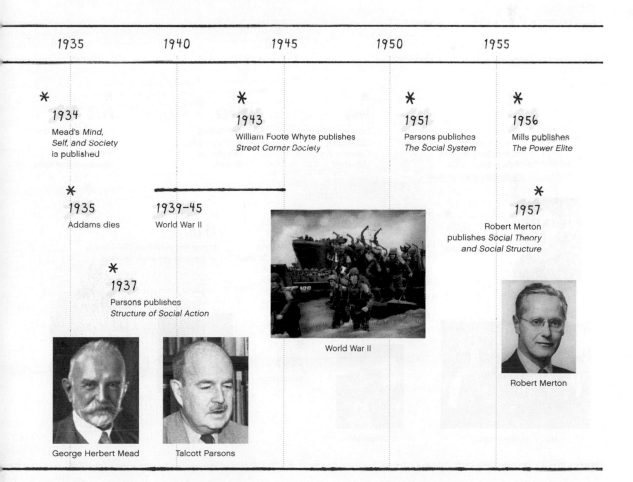

1935   1940   1945   1950   1955

1934
Mead's *Mind,
Self, and Society*
is published

1943
William Foote Whyte publishes
Street Corner Society

1951
Parsons publishes
The Social System

1956
Mills publishes
The Power Elite

1935
Addams dies

1939–45
World War II

1957
Robert Merton
publishes *Social Theory
and Social Structure*

1937
Parsons publishes
Structure of Social Action

World War II

George Herbert Mead

Talcott Parsons

Robert Merton

was suppressed by the Prussian government for its radicalism. Forced into exile, Marx settled in London, where he wrote his most important works. Marx was essentially a historian, but he did more than just chronicle events. He elaborated a theory of what drives history, now called historical materialism. Marx believed that it was primarily the conflicts between classes that drove social change throughout history. Marx saw history as an account of man's struggle to gain control of and later dominate his natural environment. However, at a certain point—with the Industrial Revolution and the emergence of modern capitalism—the very tools and processes that humans embraced to survive and to manage their surroundings came to dominate humans. Instead of using technology to master the natural world, people became slaves to industrial technology in order to make a living. In Marx's version of history, each economic system, whether small-scale farming or

TWO CENTURIES OF SOCIOLOGY

| 1960 | 1965 | 1970 | 1975 | 1980 |

* 1959
Erving Goffman publishes *The Presentation of Self in Everyday Life*; Mills publishes *The Sociological Imagination*

* 1966
Equality of Educational Opportunity (the Coleman Report) is published

* 1972
Ann Oakley publishes *Sex, Gender, and Society*

* 1978
William Julius Wilson publishes *The Declining Significance of Race*

Erving Goffman

* 1962
Mills dies

* 1963
Du Bois dies; the March on Washington; Betty Friedan publishes *The Feminine Mystique*

* 1967
Elliot Liebow publishes *Tally's Corner*; Peter M. Blau and Otis Dudley Duncan publish *The American Occupational Structure*

* 1973
Daniel Bell publishes *The Coming of the Post-Industrial Society*

March on Washington

C. Wright Mills

* 1969
Woodstock

Ann Oakley

* 1977
Paul Willis publishes *Learning to Labor*

* 1979
Parsons dies

factory capitalism, had its own fault lines of conflict. In the current epoch, that fault line divided society into a small number of capitalists and a large number of workers (the proletariat) whose interests were opposed. This political struggle, along with escalating crises within the economic system itself, would produce social change through a Communist revolution. In the ensuing Communist society, private property would be abolished and the resulting ideology governing the new economy would be "from each according to his abilities, to each according to his needs" (Marx & Engels, 1848/1998). We will explore Marx's theories in depth in Chapter 7 on social stratification.

Max Weber If Karl Marx brought the material world back into history, which had been thought of as mostly idea driven until then, Max Weber

1985   1990   1995   2000   2005   2010   2015

* 1984
Pierre Bourdieu publishes *Distinction*; Anthony Giddens publishes *The Constitution of Society*

Anthony Giddens

* 1989
Arlie Hochschild publishes *The Second Shift: Working Parents and the Revolution at Home*

* 1991
James Coleman publishes *Foundations of Social Theory*

* 1993
Douglas S. Massey and Nancy A. Denton publish *American Apartheid*

William Julius Wilson

Arlie Hochschild

James Coleman

* 1999
Duncan Watts publishes *Small Worlds*; Mary Pattillo publishes *Black Picket Fences*

Duncan Watts

Mary Pattillo

* 2016
Matthew Desmond publishes *Evicted: Poverty and Profit in the American City*

Matthew Desmond

(1864–1920), writing shortly after Marx, is said to have brought ideas back into history. Weber and others believed Marx went too far in seeing culture, ideas, religion, and the like as merely an effect and not a cause of how societies evolve. Specifically, Weber criticized Marx for his exclusive focus on the economy and social class, advocating sociological analysis to accommodate the multiple influences of culture, economics, and politics. Weber is most famous for his two-volume work *Economy and Society* (published posthumously in 1922) as well as for a lengthy essay titled "The Protestant Ethic and the Spirit of Capitalism" (1904/2003). In the latter, he argued that the religious transformation that occurred during the Protestant Reformation in the sixteenth and early seventeenth centuries laid the groundwork for modern capitalism by upending the medieval ethic of virtuous poverty and replacing it with an ideology that saw riches as a sign of divine providence. *Economy and Society* also provided the theories of authority, rationality, the state (i.e., government), and status and a host of other concepts that sociologists still use today.

One of Weber's most important contributions was the concept of *Verstehen* ("understanding" in German). By emphasizing *Verstehen*, Weber was suggesting that sociologists approach social behavior from the perspective of those engaging in it. In other words, to truly understand why people act the way they do, a sociologist must understand the meanings people attach to their actions. Weber's emphasis on subjectivity is the foundation of interpretive sociology, the study of social meaning.

Émile Durkheim Across the Rhine in France, the work of Émile Durkheim (1858–1917) focused on themes similar to those studied by his German colleagues. He wished to understand how society holds together and how modern capitalism and industrialization have transformed the ways people relate to one another. Durkheim's sociological writing began with *The Division of Labor in Society* (1893/1997). The division of labor refers to the degree to which jobs are specialized. A society of hunter-gatherers or small-scale farmers has a low division of labor (each household essentially carries out the same tasks to survive); today the United States has a high degree of division of labor, with many highly specialized occupations. What made the substance of Durkheim's sociology of work, in addition to economics, was the fact that the division of labor didn't just affect work and productivity but had social and moral consequences as well. Specifically, the division of labor in a given society helps determine its form of social solidarity—that is, the way social cohesion among individuals is maintained. Durkheim followed this work with *Suicide* (1897/1951), in which he shows how this individual act is, in reality, conditioned by social forces: the degree to which we are integrated into group life (or not) and the degree to which our lives follow routines. Durkheim argues that one of the main social forces leading to suicide is the sense of normlessness resulting from drastic changes in

---

VERSTEHEN

German for "understanding." The concept of *Verstehen* comes from Max Weber and is the basis of interpretive sociology in which researchers imagine themselves experiencing the life positions of the social actors they want to understand rather than treating those people as objects to be examined.

living conditions or arrangements, which he calls anomie. He also wrote about the methods of social science as well as religion in *The Elementary Forms of Religious Life* (1917/1995). Although the concept originated with Comte, Durkheim is often considered the founding practitioner of positivist sociology, a strain within sociology that believes the social world can be described and predicted by certain observable relationships.

Georg Simmel Until recently, Georg Simmel (1858–1918) has historically received less credit as one of the founders of sociology. In a series of important lectures and essays, Simmel established what we today refer to as formal sociology—that is, a sociology of pure numbers. For example, among the issues he addressed were the fundamental differences between a group of two and a group of three or more (independent of the reasons for the group or who belongs to it). His work was influential in the development of urban sociology and cultural sociology, and his work with small-group interactions served as an intellectual precedent for later sociologists who came to study microinteractions. He provided formal definitions for small and large groups, a party, a stranger, and the poor. (These are antecedents of network theory, which emerged in the latter half of the twentieth century.)

AMERICAN SOCIOLOGY

Throughout the history of sociology, the pendulum has tended to swing back and forth between a focus on big, sweeping theories and on more focused empirical research. The emergence of American sociology was characterized by the latter, applied perspective and was best embodied by what came to be referred to as the Chicago School, named for many of its proponents' affiliation with the University of Chicago. If the Chicago School had a basic premise, it was that humans' behaviors and personalities are shaped by their social and physical environments, a concept known as social ecology.

Chicago, which had grown from a midsize city of 109,260 in 1860 to a major metropolis by the beginning of the twentieth century, when these scholars were writing, served as the main laboratory for the Chicago School's studies. Chicago proved to be fertile ground for studying urbanism and its many discontents. Immigration, race and ethnicity, politics, and family life all became topics of study, primarily through a community-based approach (i.e., interviewing people and spending time with them). Robert Park (1864–1944), for example, exhorted scholars to "go and get the seat of [their] pants dirty in real research." This was a time of rapid growth in urban America thanks to a high rate of foreign immigration as well as the Great Migration of African Americans from the rural South to the urban North. The researchers of the Chicago School were concerned with how race and ethnic divisions played out in cities: how Polish peasants and African

a sense of aimlessness or despair that arises when we can no longer reasonably expect life to be predictable; too little social regulation; normlessness.

POSITIVIST SOCIOLOGY

a strain within sociology that believes the social world can be described and predicted by certain observable relationships (akin to a social physics).

American sharecroppers adapted to life in a new, industrialized world, or how the anonymity of the city itself contributed to creativity and freedom on the one hand and to the breakdown of traditional communities and higher rates of social problems on the other. For example, in the classic Chicago School essay "Urbanism as a Way of Life" (1938), Louis Wirth—himself an immigrant from a small village in Germany—described how the city broke down traditional forms of social solidarity while promoting tolerance, rationality (which led to scientific advances), and individual freedom. Much of the work was what would today be called cultural sociology. For example, in their studies of ethnicity, Park and others challenged the notion inherited from Europe that ethnicity was about bloodlines and instead showed ethnicity "in practice" to be more about the maintenance of cultural practices passed down through generations. Likewise, the stages of immigrant assimilation into American society (contact, then competition, and finally assimilation), which are today regarded as common knowledge and indeed part of our national ideology, were first described by Park.

If there was a theoretical paradigm that undergirded much of the research of the Chicago School, it would be the theory of the "social self" that emerged from the work of the social psychologists Charles Horton Cooley (1864–1929) and George Herbert Mead (1863–1931). Cooley and Mead adapted the Chicago School's theme of how the social environment shapes the individual to incorporate some of the key ideas of the pragmatist school of philosophy (which argues that inquiry and truth cannot be understood outside their environment—i.e., that environment affects meaning). Cooley, who taught at the University of Michigan, is best known for the concept of the "looking-glass self." He argued that the self emerges from an interactive social process. We envision how others perceive us; then we gauge the responses of other individuals to our presentation of self. By refining our vision of how others perceive us, we develop a self-concept that is in constant interaction with the surrounding social world. Much of Cooley's work described the important role that group dynamics played in this process. (See Chapter 5 on groups and networks.)

In his book *Mind, Self, and Society* (1934), George Herbert Mead, a social psychologist and philosopher at the University of Chicago, described how the "self" itself (i.e., the perception of consciousness as an object) develops over the course of childhood as the individual learns to take the point of view of specific others in specific contexts (such as games) and eventually internalizes what Mead calls the "generalized other"—our view of the views of society as a whole that transcends individuals or particular situations. (Mead's theories are discussed in depth in Chapter 4 on socialization.) Key to both Cooley's and Mead's work is the notion that meaning emerges through social interaction. This theoretical paradigm is perhaps best summarized by another Chicago scholar, W. I. Thomas, who stated that "if men define situations as real they are real in their consequences" (Thomas &

Thomas, 1928, p. 572). This statement was an important precursor to the notion of the social construction of reality.

W. E. B. Du Bois However, even as the Chicago School questioned essentialist notions of race and ethnicity (and even the self), the community of scholars was still dominated by white men. The most important black sociologist of the time, and the first African American to receive a PhD from Harvard, W. E. B. Du Bois (1868–1963) failed to gain the renown he deserved. The first sociologist to undertake ethnography in the African American community, Du Bois made manifold contributions to scholarship and social causes. He developed the concept of double consciousness, a mechanism by which African Americans constantly maintain two behavioral scripts. The first is the script that any American would have for moving through the world; the second is the script that takes the external opinions of an often racially prejudiced onlooker into consideration. The double consciousness is a "sense of always looking at one's self through the eyes of others, of measuring one's soul by the tape of a world that looks on in amused contempt and pity" (1903, p. 2). Without a double consciousness, a person shopping for groceries moves through the store trying to remember everything on the list, maybe taste-testing the grapes, impatiently scolding children begging for the latest sugary treat, or snacking on some cookies before paying for them at the register. With a double consciousness, an African American shopping for groceries is aware that he or she might be watched carefully by store security and makes an effort to get in and out quickly. He or she does not linger in back corners out of the gaze of shopkeepers and remembers not to reach into a pocket lest this motion be perceived as evidence of

DOUBLE CONSCIOUSNESS

a concept conceived by W. E. B. Du Bois to describe the two behavioral scripts, one for moving through the world and the other incorporating the external opinions of prejudiced onlookers, which are constantly maintained by African Americans.

W. E. B. Du Bois (second from right) at the office of the NAACP's *Crisis* magazine.

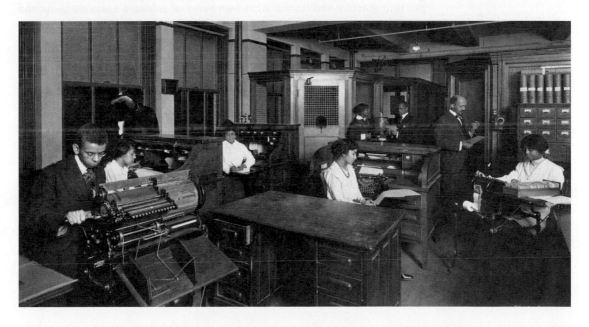

shoplifting. Snacking on a bag of chips before reaching the register or sampling a tasty morsel from the bulk bins is totally out of the question. Those operating with a double consciousness risk conforming so closely to others' perceptions that they are fully constrained to the behaviors predicted of them. Du Bois was also interested in criminology, using Durkheim's theory of anomie to explain crime rates among African Americans. Specifically, Du Bois theorized that the breakdown of norms resulting from the sudden and newfound freedom of former slaves caused high crime rates among blacks (at least in the South). He also analyzed the social stratification among Philadelphia's black population and argued that such class inequality was necessary for progress in the black community. African Americans, he argued, would be led by what he coined "the talented tenth," an elite of highly educated professionals. In addition to being a major academic sociologist, Du Bois worked to advance a civil rights agenda in the United States. To this end, he co-founded the National Association for the Advancement of Colored People (NAACP) in 1909.

Jane Addams Women, much like African Americans, didn't always receive the respect they deserved among sociologists. Jane Addams (1860–1935), for example, was considered a marginal member of the Chicago School, yet many of the movement's thinkers drew some of their insights from her applied work. In Chicago, Addams founded Hull House, the first American settlement house that attempted to link the ideas of the university to the poor through a full-service community center, staffed by students and professionals, which offered educational services and aid and promoted sports and the arts. It was at Hull House that the ideas of the Chicago School were put into practice and tested. Although many of Addams's observations and experiences at Hull House were influential in the development of the Chicago School's theories and Addams herself was a prolific author on both the substance and methodology of community studies, she was regarded as a social worker by the majority of her contemporaries. This label, which she rejected, partly resulted from the applied nature of her work, but undoubtedly gender also played a role in her marginalization: Many of the men of the Chicago School also engaged in social activism yet retained their academic prestige.

Talcott Parsons Although American sociology was born in a tradition of community studies that avoided grand theory and drew its insights from the careful observation of people in their environments, it was largely characterized by the concept of functionalism for much of the twentieth century. Drawing on the ideas of Durkheim and best embodied by the work of Talcott Parsons (1902–1979), functionalism derived its name from the notion that the best way to analyze society was to identify the roles that different aspects or phenomena play. These functions may be manifest (explicit) or

FUNCTIONALISM

the theory that various social institutions and processes in society exist to serve some important (or necessary) function to keep society running.

latent (hidden). This lens is really just an extension of a nineteenth-century theory called *organicism*, the notion that society is like a living organism, each part of which serves an important role in keeping society together. The state or government was seen to be the brain, industry was the muscular system, media and mass communications were the nervous system, and so on.

Twentieth-century sociologists had moved beyond such simplistic biological metaphors, yet the essential notion that social institutions were present for a reason persisted. Hence analysis by Parsons and others sought to describe how the various parts of the whole were integrated with, but articulated against, one another. Almost every social phenomenon was subjected to functionalist analysis: What is the function of schooling? The health care system? Even crime and the Mafia were seen to play a role in a functioning society. For example, functionalists view social inequality as a "device by which societies ensure that the most important positions are conscientiously filled by the most qualified persons" (Davis & Moore, 1944).

Although associated with mid-twentieth-century sociology, the functionalist impulse originated in the nineteenth century, most notably in the work of Durkheim, who in 1893 (1972) wrote, "For, if there is nothing which either unduly hinders or favours the chances of those competing for occupations, it is inevitable that only those who are most capable at each type of activity will move into it. The only factor which then determines the manner in which work is divided is the diversity of capacities." Functionalism was still being applied into the late twentieth century by Richard J. Herrnstein and Charles Murray, who argued in *The Bell Curve* that "no one decreed that occupations should sort us out by our cognitive abilities, and no one enforces the process. It goes on beneath the surface, guided by its own invisible hand" (1994, p. 52).

MODERN SOCIOLOGICAL THEORIES

Conflict Theory Even though functionalism continues to reappear in many guises, its fundamental assertions have not gone unchallenged. Sociologists such as C. Wright Mills, writing from 1948 through 1962, criticized Parsons and functionalist theory for reinforcing the status quo and the dominant economic system with its class structures and inequalities instead of challenging how such systems evolved and offering alternatives. Functionalism also took a beating in the turbulent 1960s, when its place was usurped by a number of theories frequently subsumed under the label Marxist theory or conflict theory. Whereas functionalists painted a picture of social harmony as the well-oiled parts of a societal machine working together (with some friction and the occasional breakdown), conflict theorists viewed society from exactly the opposite perspective. Drawing on the ideas of Marx,

CONFLICT THEORY

the idea that conflict between competing interests is the basic, animating force of social change and society in general.

↑

Female textile workers struggle with a national guardsman during a 1929 strike in Gastonia, North Carolina. How might a conflict theorist interpret labor unrest?

the theory—as expressed by Ralph Dahrendorf, Lewis Coser, and others—stated that conflict among competing interests is the basic, animating force of any society. Competition, not consensus, is the essential nature, and this conflict at all levels of analysis (from the individual to the family to the tribe to the nation-state), in turn, drives social change. And such social change occurs only through revolution and war, not evolution or baby steps.

According to conflict theorists, inequality exists as a result of political struggles among different groups (classes) in a particular society. Although functionalists theorize that inequality is a necessary and beneficial aspect of society, conflict theorists argue that it is unfair and exists at the expense of less powerful groups. Thus functionalism and conflict theory take extreme (if opposing) positions on the fundamental nature of society. Today most sociologists see societies as demonstrating characteristics of both consensus and conflict and believe that social change does result from both revolution and evolution.

Feminist Theory Emerging from the women's movement of the 1960s and 1970s, feminist theory shares many ideas with Marxist theory—in particular, the Marxist emphasis on conflict and political reform. Feminism is not one idea but a catchall term for many theories. What they all have in common is an emphasis on women's experiences and a belief that sociology and society in general subordinate women. Feminist theorists emphasize equality between men and women and want to see women's lives and experiences represented in sociological studies. Early feminist theory focused on defining concepts such as sex and gender, and on challenging conventional wisdom by questioning the meanings usually assigned to these concepts. In *Sex, Gender, and Society* (1972), sociologist Ann Oakley argued that much of what we attribute to biological sex differences can be traced to behaviors that are learned and internalized through socialization (see Chapter 8 on gender).

In addition to defining sex and gender, much feminist research focuses on inequalities based on gender categories. Feminist theorists have studied women's experiences at home and in the workplace. They have also researched gender inequality in social institutions such as schools, the family, and the government. In each case, feminist sociologists remain interested in how power relationships are defined, shaped, and reproduced on the basis of gender differences.

Symbolic Interactionism A strain of thought that developed in the 1960s was symbolic interactionism, which eschewed big theories of society (macrosociology) and instead focused on how face-to-face interactions create the social world (microsociology). Exemplified most notably by the work of Herbert Blumer, one of George Herbert Mead's students, this paradigm operates on the basic premise of a cycle of meaning—namely, the idea that people act in response to the meaning that signs and social signals hold for them (e.g., a red light means stop). By acting on perceptions of the social world in this way and regarding these meanings as *sui generis* (i.e., appearing to be self-constituting rather than flimsily constructed by ourselves or others), we then collectively make their meaning so. In other words, John G. Roberts Jr. is chief justice of the US Supreme Court because we all act as if he is and believe that it is an objective fact he is (even if that belief is the creation of our somewhat arbitrary collective thinking). We then act in ways that reify, or make consequential, this consensus and arrive at our calculations based on the supposed objectivity of this "fact."

The groundwork for symbolic interactionism was laid by Erving Goffman's dramaturgical theory of social interaction. Goffman had used the language of theater to describe the social facade we create through devices such as tact, gestures, front-stage (versus backstage) behavior, props, and scripts. For example, in *The Presentation of Self in Everyday Life* (1959), Goffman explored how our everyday personal encounters shape and reinforce our notions about class and social status. According to Goffman, we make judgments about class and social status based on how people speak, what they wear, and the other tiny details of how they present themselves to others, and at the same time, they rely on the same information from our everyday interactions with them to classify us too.

Postmodernism If symbolic interactionism emphasizes the meanings negotiated through the interaction of individuals, postmodernism can perhaps be summarized succinctly as the notion that these shared meanings have eroded. A red light, for instance, may have multiple meanings to different groups or individuals in society. There is no longer one version of history that is correct. Everything is interpretable within this framework; even "facts" are up for debate. It's as if everyone has become a symbolic interactionist and decided that seemingly objective phenomena are social constructions, so that all organizing narratives break down. Postmodernists may not feel compelled to act on these shared meanings as seemingly objective, because the meanings aren't, in fact, objective. The term

SYMBOLIC INTERACTIONISM

a micro-level theory in which shared meanings, orientations, and assumptions form the basic motivations behind people's actions.

POSTMODERNISM

a condition characterized by a questioning of the notion of progress and history, the replacement of narrative within pastiche, and multiple, perhaps even conflicting, identities resulting from disjointed affiliations.

SOCIAL CONSTRUCTION

an entity that exists because people behave as if it exists and whose existence is perpetuated as people and social institutions act in accordance with the widely agreed-on formal rules or informal norms of behavior associated with that entity.

Las Vegas, the ultimate postmodern city, borrows from various regions, times, and cultures to shape its constantly changing landscape.

MIDRANGE THEORY

a theory that attempts to predict how certain social institutions tend to function.

itself derives from the idea that the grand narratives of history are over (hence, "after" modernism, *postmodernism*).

Midrange Theory Although many sociologists have taken to the postmodernist project of deconstructing social phenomena (i.e., showing how they are created arbitrarily by social actors with varying degrees of power), most sociologists have returned, as the pendulum swings yet again, to what sociologist Robert Merton called for in the middle of the last century: midrange theory. Midrange theory is neither macrosociology (it doesn't try to explain all of society) nor microsociology. Rather, midrange theory attempts to predict how certain social institutions tend to function. For example, a midrange theorist might develop a theory of democracy (under what political or demographic conditions does it arise?), a theory of the household (when do households expand to include extended kin or nonkin, and when do they contract to the nuclear family unit or the individual?), or a theory explaining the relationship between the educational system and the labor market. The key to midrange theory is that it generates falsifiable hypotheses—predictions that can be tested by analyzing the real world. (Hypothesis generation is covered in Chapter 2.)

Sociology and Its Cousins

We have already noted that a sociology of sports exists, as can a sociology of music, of organizations, of economies, of science, and even of sociology itself. What then, if anything, distinguishes sociology from other disciplines? Certainly overlap with other fields does occur, but at the same time, there is something distinctive about sociology as a discipline that transcends the sociological imagination discussed earlier. All of sociology boils down to comparisons across cases of some form or another, and perhaps the best way to conceptualize its role in the landscape of knowledge is to compare it with other fields.

HISTORY

Let's start with history. History is concerned with the *idiographic* (from the Greek *idio*, "unique," and *graphic*, "depicting"), meaning that historians have traditionally been concerned with explaining unique cases. Why did Adolf Hitler rise to power? What conditions led to the Haitian slave revolt 200 years ago (the first such rebellion against European chattel slavery)? How did the Counter-Reformation affect the practices of lay Catholics in France? What was the impact of the railroad on civilization's sense of time? How did adolescence arise as a meaningful stage of life? Historians' research questions center on the notion that by understanding the particularity of certain past events, individual people, or intellectual concepts, we can better understand the world in which we live.

Italian dictator Benito Mussolini (on the left) and Nazi leader Adolf Hitler at a 1937 rally in Munich. How do different disciplines provide various tools to analyze the rise of fascism under these leaders?

Sometimes historians use comparative frameworks to situate their analysis—for example, comparing Hitler's rise to Mussolini's rise in order to examine why the Third Reich pursued a genocidal agenda, whereas the Italian fascists had no such agenda and, in fact, somewhat resisted cooperating with Germany's deportation of Jews. Another strategy historians often use, although it is sometimes frowned on, is the notion of the counterfactual: What would have happened had Hitler been killed rather than wounded in World War I? Would World War II have been inevitable even if the victors in World War I had not pursued such a punitive reparations policy after defeating Germany?

The preceding description is, of course, an oversimplification of a diverse field. History runs the gamut from "great man" theories (a focus on figures like Hitler) to people's histories (a focus on the lives of anonymous, disempowered people in various epochs, or on groups traditionally given short shrift in historical scholarship, such as women, African Americans, and subaltern colonials) to historiography (in essence, metahistory examining the intellectual assumptions and constraints on knowledge entailed by the subjects and methods historians choose) in which historians rely on archival material and primary resources.

Sociology, by contrast, is generally concerned not with the uniqueness of phenomena but rather with commonalities that can be abstracted across cases. Whereas the unique case is the staple of the historian, the comparative method is the staple of sociologists. Historical comparative sociologist Julia Adams explained what makes her work different from what a historian would do: "If you define the difference according to whether one is engaged in primary archival research or not, I would . . . also be an historian as well as a sociologist," but the big difference is that historical sociologists are "consciously theoretical" and "very keen to explain and illuminate" historical patterns (Conley, 2013a). Whether looking at contemporary American life, the formation of city-states in medieval Europe, or the origins of unequal economic development thousands of years ago, sociologists formulate hypotheses and theorems about how social life works or worked.

Instead of inquiring why Hitler rose to power, the sociologist might ask what common element allowed fascism to arise during the early to mid-twentieth century in Germany, Italy, Spain, and Japan and not in other countries such as France, Great Britain, or the Scandinavian nations. Of course, sociologists recognize that there is no one-size-fits-all hypothesis that will explain all these cases perfectly, but the exercise of considering competing explanations is illuminating in and of itself. Instead of asking what specific conditions led to the Haitian slave revolt in 1791, the sociologist might ask under what general conditions uprisings have occurred among indentured populations. Instead of asking how the Counter-Reformation affected the practices of lay Catholics in one region of Europe,

To see my interview with Julia Adams, go to
digital.wwnorton.com/youmayask6core

the sociologist might ask what aspects of the conditions in various regions of Europe made the reaction to the Catholic Church's reforms different. This is a subtle but important difference in focus. Instead of examining the impact of the railroad on our sense of time, the sociologist might compare the railroad to other forms of transportation that both collapsed distances and opened up the possibility of more frequent and longer migrations. And finally, instead of asking how adolescence arose as a meaningful stage of life, the sociologist might compare various societies that have distinct roles for individuals aged 13 to 20 to examine how those labor markets, educational systems, family arrangements, life expectancies, and so on lead to different experiences for that age group.

Whether sociologists are comparing two tribes that died out 400 years ago, two countries today, siblings within the same family, or even changes in the same person over the course of his or her life, they are always seeking a variation in some outcome that can be explained by variation in some input. This holds true, I would argue, whether the methods are interviews, research using historical archives, community-based observation studies of participants, or statistical analyses of data from the US census. Sociologists are always at least implicitly drawing comparisons to identify abstractable patterns.

ANTHROPOLOGY

The field of anthropology is split between physical anthropologists, who resemble biologists more than sociologists, and cultural anthropologists, who study human relations similarly to the way sociologists do. Traditionally, the distinction was that sociologists studied "us" (Western society and culture), whereas anthropologists studied "them" (other societies or cultures). This distinction was helpful in the early to mid-twentieth century, when anthropologist Margaret Mead studied rites of passage in Samoa and sociologists interviewed Chicago residents, but it is less salient today. Sociologists increasingly study non-Western social relations, and anthropologists do not hesitate to tackle domestic social issues.

Take two recent examples that confound this division from either side of the metaphorical aisle: Caitlin Zaloom is a cultural anthropologist whose "tribe" (she likes to joke) is the group of commodities traders who work on the floor of the Chicago Mercantile Exchange. Natasha Dow Schüll, another anthropologist, has studied gamblers in Las Vegas and how they lose themselves and their money in the machine zone. Meanwhile, recent sociological scholarship includes Stephen Morgan's study of African status attainment and patronage relations. Another project, spearheaded by business school professor Doug Guthrie in 2006, investigated the way informal social connections facilitate business in China. Given the era of globalization we have entered, it is appropriate that division on the basis of "us" and "them" has

How does anthropologist Natasha Dow Schüll's research on slot-machine gamblers challenge the traditional boundaries between anthropology and sociology?

diminished. Many scholars now question the legitimacy of such a division in the first place, asking whether it served a colonial agenda of dividing up the world and reproducing social relations of domination and exclusion—accomplishing on the intellectual front what European imperialism did on the military, political, and economic fronts.

What then distinguishes sociology from cultural anthropology? Nothing, some would argue. However, although certain aspects of sociology are almost indistinguishable from those of cultural anthropology, sociology as a whole has a wider array of methods to answer questions, such as experimentation and statistical data analysis. Sociology also tends more toward comparative case study, whereas anthropology is more like history in its focus on particular circumstances. This does not necessarily imply that sociology is superior; a wider range of methods can be a weakness as well as a strength because it can lead to irreconcilable differences within the field. For example, demographic and ethnographic studies of the family may employ different definitions of what a household is. This makes meaningful dialogue between the two subfields challenging.

THE PSYCHOLOGICAL AND BIOLOGICAL SCIENCES

Social, developmental, and cognitive psychology often address many of the same questions that sociologists do: How do people react to stereotypes? What explains racial differences in educational performance? How do individuals respond to authority in various circumstances? Ultimately, however, psychologists focus on the individual, whereas sociologists focus on the supra-individual (above or beyond the individual) level. In other words, psychologists focus on the individual to explain the phenomenon under consideration, examining how urges, drives, instincts, and the mind can account for human behavior, whereas sociologists examine group-level dynamics and social structures.

Biology, especially evolutionary biology, increasingly attempts to explain phenomena that once would have seemed the exclusive dominion of social scientists. There are now biological (evolutionary) theories of many aspects of gender relations—even rape and high-heeled shoes (Posner, 1992). Medical science has claimed to identify genes that explain some aspects of social behavior, such as aggressiveness, shyness, and even thrill seeking. Increasingly, social differences have been medicalized through diagnoses such as attention-deficit hyperactivity disorder (ADHD) and disorders on the autism spectrum. The distinction between these areas of biology and the social sciences lies not so much in the topic of study (or even the scientific methods in some cases), but rather in the underlying variation or causal mechanisms with which the disciplines are concerned. Sociology addresses supra-individual-level dynamics that affect our behavior; psychology addresses individual-level dynamics; and biology typically deals with the intra-individual-level factors (those within the individual) that affect our lives, such as biochemistry, genetic makeup, and cellular activity. So if a group of biologists attempted to explain differences in culture across continents, they would typically analyze something like local ecological effects or the distribution of genes across subpopulations.

ECONOMICS AND POLITICAL SCIENCE

The quantitative side of sociology shares many methodological and substantive features with economics and political science. Economics traditionally has focused on market exchange relations (or, simply, money). More recently, however, economics has expanded to include social realms such as culture, religion, and the family—the traditional stomping ground of sociologists. What distinguishes economics from sociology in these contexts is the underlying view of human behavior. Economics assumes that people are rational utility maximizers: They are out to get the best deal for themselves. Sociology, on the other hand, has a more open view of human motivation that includes selfishness, altruism, and simple irrationality. (New branches of economics are moving, however, into the realm of the irrational.) Another difference is methodological: Economics is a fundamentally quantitative discipline, meaning that it is based on numerical data.

Similarly, political science is almost a subsector of sociology that focuses on only one aspect of social relations—power. Of course, power relations take many forms. Political scientists study state relations, legal structure, and the nature of civic life. Like sociologists, political scientists deploy a variety of methods, ranging from historical case studies to abstract statistical models. Increasingly, political science has adopted the rational actor model implicit in economics in an attempt to explain everything from how lobbyists influence legislators to the recruitment of suicide bombers by terrorist groups.

Having explained all of these distinctions, we should keep in mind that disciplinary boundaries are in a constant state of flux. For example, Stanford English professor Franco Moretti argues that books should be counted, mapped, and graphed. In his research, Moretti statistically analyzes thousands of books by thematic and linguistic patterns. Economist Steven Levitt has explored how teachers teach to the test (and sometimes cheat) and how stereotypically African American names may or may not disadvantage their holders. Historian Zvi Ben-Dor Benite has assembled a global, comparative history of how different societies execute criminals. It all sounds like sociology to me! For now, let's just say that significant overlap exists between various arenas of scholarship, so that any divisions are simultaneously meaningful and arbitrary.

Divisions within Sociology

Even if sociologists tend to leverage comparisons of some sort, significant fault lines still persist within sociology. Often, the major division is perceived to exist between those who deal in numbers (statistical or quantitative researchers) and those who deal in words (qualitative sociologists). Another split exists between theorists and empiricists. These are false dichotomies, however; they merely act as shorthand for deeper intellectual divisions for which they are poor proxies. A much more significant cleavage exists between interpretive and positivist sociology. Positivist sociology is born from the mission of Comte—that mission being to reveal the "social facts" (to use the term Durkheim later coined) that affect, if not govern, social life. It is akin to uncovering the laws of "social physics," although most sociologists today would shun Comte's phrase because it implies an overly deterministic sense of unwavering, time-transcendent laws.

To this end, the standard practice is to form a theory about how the social world works—for instance, that members of minorities have a high degree of group solidarity. The next task is to generate a hypothesis that derives from this theory, perhaps that minority groups should demonstrate a lower level of intragroup violence than majority groups. Next, we make predictions based on our hypotheses. Both the hypotheses and predictions have to be falsifiable by an empirical, or experimental, test; in this case, it might involve examining homicide rates among different groups in a given society or in multiple societies. And last comes the acceptance or rejection of the hypothesis and the revision (or extension) of the theory (in the face of contradictory or confirming evidence). These scientific methods are the same as in any basic science. For that reason, positivism is often called the "normal science" model of sociology.

Normal science stands in contrast to interpretive sociology, which is much more concerned with the meaning of social phenomena to individuals (remember Weber's *Verstehen*). Rather than make a prediction about homicide rates, the interpretive sociologist will likely seek to understand the experience of solidarity among minority groups in various contexts. An interpretive sociologist might object to the notion that we can make worthwhile predictions about human behavior—or more precisely, might question whether such an endeavor is worth the time and effort. It is a sociology premised on the idea that situation matters so much that the search for social facts that transcend time and place may be futile. Why measure the number of friends we have by the number of people we see face-to-face every day, given the existence of the internet and how it has redefined the meaning of social interaction so completely?

MICROSOCIOLOGY AND MACROSOCIOLOGY

A similar cleavage involves the distinction between *microsociology* and *macrosociology*. Microsociology seeks to understand local interactional contexts—for example, why people stare at the numbers in an elevator and are reluctant to make eye contact in this setting. Microsociologists focus on face-to-face encounters and the types of interactions between individuals. They rely on data gathered through participant observations and other qualitative methodologies (for more on these methods, see Chapter 2).

Macrosociology is generally concerned with social dynamics at a higher level of analysis—across the breadth of a society (or at least a swath of it). A macrosociologist might investigate immigration policy or gender norms or how the educational system interacts with the labor market. Statistical analysis is the most typical manifestation of this kind of research, but by no means the only one. Macrosociologists also use qualitative methods such as historical comparison and in-depth interviewing. They may also resort to large-scale experimentation. That said, a perfect overlap does not exist between methodological divisions and level of analysis. For example, microsociologists might use an experimental method such as varying the context of an elevator to see how people react. Or they might use statistical methods such as conversation analysis, which analyzes turn-taking, pausing, and other quantifiable aspects of social interaction in localized settings.

MICROSOCIOLOGY

a branch of sociology that seeks to understand local interactional contexts; its methods of choice are ethnographic, generally including participant observation and in-depth interviews.

MACROSOCIOLOGY

a branch of sociology generally concerned with social dynamics at a higher level of analysis— that is, across the breadth of a society.

Conclusion

The bottom line is that anything goes; as long as you use your sociological imagination, you will be asking important questions and seeking the best way to answer them. As you read the subsequent chapters, keep in mind that a sociologist "makes the familiar strange." The first six chapters introduce the methodological and theoretical tools that you need in order to think like a sociologist. Chapters 7 through 10 ask you to study the inequalities and differences that divide people in our society.

QUESTIONS FOR REVIEW

1. Some people accuse sociologists of observing conditions that are obvious. How does looking at sociology as "making the familiar strange" help counter this claim? How does sociology differ from simple commonsense reasoning?

2. What is the sociological imagination, and how do history and personal biography affect it? If a sociologist studies the challenges experienced by students earning a college degree, how could the lessons gained be described as "terrible" as well as "magnificent"?

3. What is a social institution, and how does it relate to social identity? Choose a sports team or another social institution to illustrate your answer.

4. A sociologist studies the way a group of fast-food restaurant employees do their work. From what you read in this chapter, how would Max Weber and Émile Durkheim differ in their study of these workers?

5. Compare functionalism and conflict theory. How would the two differ in their understanding of inequality?

6. You tell a friend that you're taking a class in sociology. There's a chance she knows about sociology and is quite jealous. There's also a chance she's confusing sociology with the other social sciences. How would you describe sociology? How does sociology differ from history and psychology?

7. Sociology, like any discipline, features some divisions. What are some of the cleavages in the field, and why might these be described as false dichotomies?

8. Why do people go to college, and how does Randall Collins's book *The Credential Society* make the familiar reality of college education seem strange?

SEEING SOCIOLOGICALLY

TRY IT!

Much of what we accept as "natural" in our daily lives are actually agreed-on social norms. Identify all the mistakes here. For example, notice that the shopping cart has square wheels. Or that the moon and the sun are out at the same time. Which are violating laws of nature, and which are sociological?

Now do it for real: Go out and about today with your sociological notebook and note how many "rules" of social life you had been taking for granted.

SOCIAL NORM	WHERE YOU SAW IT
LINING UP AT THE CASHIER	GROCERY STORE
STRANGER NODDING HELLO	ELEVATOR AT WORK
STAYING QUIET IN A STUDY SPACE	CAMPUS LIBRARY

THINK ABOUT IT

What was the most surprising aspect of daily life that you now realize is social in origin rather than natural? Is this social norm unique to your community or does it apply pretty much anywhere? Is it a recent development (like smartphone etiquette) or has it been around since humans tamed fire? What is a specific annoying social norm you wish would go away, and why?

SOCIOLOGY ON THE STREET

The neighborhood where you grow up exerts a significant effect on the rest of your life. How did your house, neighbors, street, and town influence you? Watch the Sociology on the Street video to find out more: **digital.wwnorton.com/youmayask6core**.

WANT MORE PRACTICE?

Complete the InQuizitive activity for this chapter at digital.wwnorton.com /youmayask6core

PARADOX

2

IF WE SUCCESSFULLY ANSWER ONE QUESTION, IT ONLY SPAWNS OTHERS. THERE IS NO MOMENT WHEN A SOCIAL SCIENTIST'S WORK IS DONE.

Methods

danah boyd (yes, all lowercase) grew up in Lancaster, Pennsylvania, during a period in which the region was trying to adapt to the decline of the farming economy by reinventing itself as an industrial center. The bad news for central Pennsylvania was that factory employment in the United States had already had its heyday, so by the time Lancaster tried to invigorate its economy with manufacturing, the odds were stacked against it. Though Lancaster would later find its economic footing, other towns in the region would not be as lucky.

Raised by a single mother, danah saw her own financial fortunes ebb and flow with the transformation of the American economic landscape. For example, her mother supported herself by running a local franchise of the direct-mail coupon company Valpak. Of course, direct mail has also now been largely displaced by online marketing.

danah's own intellectual trajectory reflects all the subtleties of this background. Though she was taught some computer programming as a young child in school, her interest in this new technology really took off thanks to her brother's hogging of the home phone line. Back in the days of dial-up connections, her brother would take over the family phone line to communicate with other users on various online bulletin boards. Little is more annoying to a teenage girl, danah says, than a younger brother bogarting the telephone, but when she realized that there were actually other people whose computers were deciphering the clicks, beeps, and screeches emanating from her line, that's when the die was cast. Computing was social. And it was cool.

At Brown University danah combined her varied interests by concocting a hybrid concentration: computational gender studies. Her senior thesis research involved studying how sex hormones affect depth cue prioritization in virtual reality spaces by investigating the changes experienced by transgender people undergoing hormone therapy. Evidently, the retina of the eye

DIGITAL.WWNORTON.COM/YOUMAYASK6CORE

To see my interview with danah boyd, go to
digital.wwnorton.com/youmayask6core

is the place in the body with the most sex hormone receptors outside of our reproductive system (who knew?), so how we see the world—quite literally—is shaped by sex.

She continued her cross-disciplinary studies at MIT's Media Lab, where she showed how to graph the social networks of individuals without ever seeing their electronic communications but rather only observing those of their friends. It was during a gap year between completing her master's degree and starting her PhD studies at the University of California that she bluffed her way into conducting her first real independent qualitative study. Her subject was early Friendster adopters. (Friendster was among the first social network sites, followed later by Myspace and ultimately Facebook.)

As an early adopter herself, she managed to gain incredible access to both Friendster and Myspace. Although many researchers study youth culture ethnographically by talking to kids and hanging out at malls and other spaces where teenagers congregate, and others perform "content analysis" of online interactions such as Twitter feeds, blogs, and so on, few have combined the two so deftly as danah. Access to online social networks allows her to draw samples of users to contact both online and offline in order to make sure that her interviews and participant observation is not only deep but also broad. She can cover not only different groups based on race, class, gender, and sexuality but also different cliques that are not so easily categorized by demographers. This dual online–offline approach makes her work more generalizable than the typical approach that solely studies online data or offline socializing.

The face-to-face ethnography that she does is critical to understanding the social lives of teens. This is because teens often speak in code online to avoid being understood by their parents or other adult figures who may be observing them. At other times, what teens say face-to-face is belied by what danah observes online. For example, although some teens talk about racial integration and cross-race friendships, online she observes very segregated social networks. Understanding who is left out of social networks can be just as important as mapping who is included in them.

As more and more social life takes place across platforms, we need more researchers like danah who carefully stitch together the domains of our fragmented lives and help us make sense of them.

As scientists, we follow the scientific method: That is, we observe the world, form a theory about an aspect of it, generate hypotheses (testable predictions from that theory) about our subject matter, and set up an experiment or systematic observations to test those hypotheses. After conducting our analysis, we accept or reject our hypotheses and revise our theory, if necessary. Rinse and repeat. This has been the approach of science since the Enlightenment in the 17th century, and it has worked remarkably well.

In everyday speech we use the word *theory* as a synonym for "idea" or "hunch." But in science, a theory is a systematic, generalized model of how some aspect of the world works. It is more abstract and general than a specific hypothesis, and in fact, may generate multiple, testable hypotheses. A theory articulates a system of relationships between facts and suggests causes and effects emerging out of those relationships. One scientific theory you may be familiar with is Darwin's theory of evolution by natural selection. Now fundamental to the way evolutionary biologists see the world, this theory wasn't just an idea that popped into Darwin's head. It was the product of many years of hypothesis, revision, and further hypothesis: the scientific method at work.

As social scientists, we sociologists have a set of standard approaches that we follow in investigating our questions. We call these rules research methods. They're the tools we use to describe, explore, and explain various social phenomena in an ethical fashion. There are two general categories of methods for gathering sociological data: quantitative and qualitative.

Quantitative methods seek to obtain information about the social world that is already in or can be converted to numeric form. This methodology then uses statistical analysis to describe the social world that those data represent. Some of this analysis attempts to mimic the scientific method of using treatment and control (or placebo) groups to determine how changes in one factor affect another social outcome, while factoring out every other simultaneous event. Such information is often acquired through surveys but may also include data collected by other means, ranging from sampling bank records to weighing people on a scale to hanging out with teens at the mall.

Qualitative methods, of which there are many, attempt to collect information about the social world that cannot be readily converted to numeric form. The information gathered with this approach is often used to document the meanings that actions engender in social participants or to describe the mechanisms by which social processes occur. Qualitative data are collected in a host of ways, from spending time with people and recording what they say and do (participant observation) to interviewing them in an open-ended manner to reviewing archives.

SCIENTIFIC METHOD

a procedure involving the formulation, testing, and modification of hypotheses based on systematic observation, measurement, and/or experiments.

THEORY

an abstracted, systematic model of how some aspect of the world works.

RESEARCH METHODS

approaches that social scientists use for investigating the answers to questions.

QUANTITATIVE METHODS

methods that seek to obtain information about the social world that is already in or can be converted to numeric form.

QUALITATIVE METHODS

methods that attempt to collect information about the social world that cannot be readily converted to numeric form.

Both quantitative and qualitative research approaches provide ways to establish a causal relationship between social elements. Researchers using quantitative approaches, by eliminating all other possibilities through their study's design, hope to state with some certainty that one condition causes another. Qualitative methodology describes social processes in such detail as to rule out competing possibilities.

This chapter gives examples of sociological research conducted using different methods, starting with the various theoretical viewpoints from which social scientists approach research. We'll then examine some techniques used by researchers to tell causal stories and give examples of specific studies that have employed these methods. Finally, we'll talk about the ways that social research can be used for ends other than filling textbooks and keeping sociologists busy.

Research 101

The general goal of sociology is to allow us to see how our individual lives are intimately related to (and, in turn, affect) the social forces that exist beyond us. Good sociological research begins with a puzzle or paradox and asks, "What causes such and such a thing to happen?" Once you pick a question to investigate, there are two ways to approach research: *deductively* and *inductively*. A deductive approach starts with a theory, forms a hypothesis, makes empirical observations, and then analyzes the data to confirm, reject, or modify the original theory. Conversely, an inductive approach starts with empirical observations and then works to form a theory. These different approaches are represented in the research cycle shown in Figure 2.1.

DEDUCTIVE APPROACH

a research approach that starts with a theory, forms a hypothesis, makes empirical observations, and then analyzes the data to confirm, reject, or modify the original theory.

INDUCTIVE APPROACH

a research approach that starts with empirical observations and then works to form a theory.

FIGURE 2.1 The Research Cycle

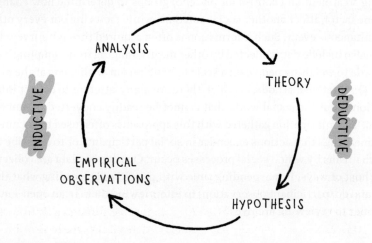

CAUSALITY VERSUS CORRELATION

Regardless of which method we use, social research is about telling a story. The goal is to recount the story as completely as possible so we're fairly certain it can't be told any other way. Let's take an example of the relationship between income and health. We know that a correlation (or association) exists between income and health—that is, they tend to vary together. For example, people with higher levels of income tend to enjoy better overall health. But to say that two things are correlated is very different from stating that one causes the other. In fact, there are three possible causal stories about the relationship between income and health. We might reasonably assert that bad health causes you to have a lower income—because when you get sick you can't work, you lose your job, and so forth. If we drew a diagram of such a scenario, it would look like this:

CORRELATION OR ASSOCIATION

when two variables tend to track each other positively or negatively.

POORER HEALTH $\longrightarrow$ LOWER INCOME

However, we could just as easily tell the opposite story—that higher income leads to better health because you can afford better doctors; you have access to fresh, healthful foods in your upscale neighborhood; and there's a gym at the office. The diagram of this story would look like this:

POORER HEALTH $\longleftarrow$ LOWER INCOME

Finally, we could conclude that a third factor causes both income and health to vary in the same direction. For the sake of argument, we will call this factor "reckless tendencies"—a love of fast cars, wine, and late nights. Such shortsighted behavior could negatively affect our health (especially the wine). And it could also affect our income. Maybe we are unable to get to work on time or are spending too much money on those fast cars instead of investing it in the stock market. In that case, the causal diagram would look like this:

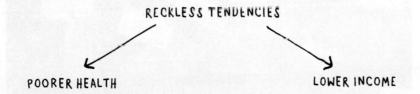

In this scenario, if we merely observe health and income, it may appear as if one causes the other, but the truth of the matter is that they are not related in the slightest apart from being connected through a third factor.

How can we arrive at any conclusions with respect to the health–income correlation? We can't randomly assign people to different jobs at various pay levels and see what happens, nor can we independently affect people's health and (ethically) observe what happens to their income, and it is certainly difficult to curb or instill reckless tendencies on a random basis. Thus we can't really rule out any of the arrows, but we can confirm some of them. We can rule out many third factors by measuring them and then comparing individuals who are similar in that particular respect (like education level) but differ in other key regards (say, income). Other factors may not be so easy to measure, and in these cases we might look for a natural experiment—that is, an event or change in the real world that affects the factor we believe causes an outcome but does not affect the outcome in any way other than through that factor. For example, some researchers have used lottery winnings as a natural experiment, comparing the health of winners who won a significant sum with those who won only a token amount. The assumption, based on how lotteries work, is that the amount won is not determined by the winner's health, but that subsequent changes in health may well be driven by the money won.

Although few certain answers exist in social science, we can safely conclude that low income does contribute to poor health, at least to some extent in specific contexts. (We can also be fairly sure that bad health has a negative effect on income.) However, our case would be stronger if we knew exactly what effect low income has on health. In other words, we don't know the causal mechanism. Is it that low-wage jobs are stressful? (It is well known that certain types of chronic stress are bad for you.) If so, does such stress cause these workers' behavior to change, perhaps by increasing their consumption of fast food, smoking rates, or alcohol intake? Or is there a more direct, psychobiological pathway—say, the stress of a verbally abusive boss that can cause higher corticosteroid levels in the bloodstream? Or is it all

How did studying lottery winners help sociologists understand the relationship between wealth and health?

of these factors and more? The more dots we can connect, the stronger our causal story becomes and the better prepared we are to intervene.

Remember, it is very difficult, especially in social science, to assert causality—that change in one factor causes a change in another. It's much easier to say two things are correlated, which just means that we observe change in both. For example, as race varies (across individuals), so does average life expectancy. Likewise, as nutrition changes across or within populations, so does average height, but can we say that better nutrition causes some populations to be taller? Maybe, but maybe not. Let's further examine this.

To establish causality, three factors are needed: correlation, time order, and ruling out alternative explanations. We've already covered correlation. We notice variations in nutrition across countries and simultaneously observe different average heights across the same countries that tend to correspond statistically to those differences in nutrition. Now we need to establish time order. Have people in country A always been taller than those in country B (a bad sign for the "better nutrition causes height" case)? Or did changes in nutrition occur before increases (or decreases) in height? We can imagine a situation where a drought, flood, frost, or some other environmental factor destroyed a main food source in country B, leading to dramatic changes in people's diet and altering average heights in that nation. Finally, we have to rule out alternative explanations for the variations observed in both nutrition and height. Is a third factor responsible for changes in both? The groundwater supply perhaps? Groundwater supply could lead to better nutrition through higher crop yields (which turns out not to matter for height, let's say), but it could also lead to cleaner drinking water and thus less infection (which turns out to matter for height, again for the sake of argument). If this were the case, then the relationship between nutrition and height would be termed spurious or false, whereas the relationship between infection and height might be described as a "true" causal relationship. Figure 2.2 illustrates this possibility.

CAUSALITY

the notion that a change in one factor results in a corresponding change in another.

FIGURE 2.2 The Charge of Spuriousness

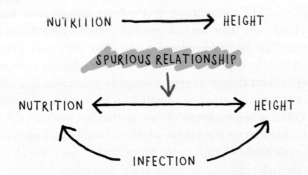

Why is the correlation between nutrition and height spurious?

The Problem of Reverse Causality Reverse causality is just what it sounds like: You think A is causing B when, in fact, B is causing A. Let's look at the relationship between income and health. We know that people who are sick often have less income. But which one is causing the other to happen? Is it that when you're sick, you miss lots of school, don't receive as much education, must take off more time from work, are passed over for promotions, and ultimately remain stuck in a lower-level job and therefore have a lower income than your comparatively healthy neighbor? Or is it that when you're employed in a lower-paying job, you may experience more on-the-job stress and more stress as a result of worrying about money, which has a negative impact on your health by putting you at higher risk for cardiovascular disease? On top of that, you can't afford good health care, a gym membership, or fresh, nutritious food. The problem of reverse causality is why it's so important to establish time order. If a person's income drops only after they get sick, we can be surer that it was sickness that led to the decline in income.

However, we should also note that time order is no guarantee by itself: People may alter their current behavior based on future expectations. Perhaps, for example, I choose to save less today because I assume that my children will become rich adults on their own and support me in my old age. Strict reliance on time order—observing current savings behavior and children's income 30 years later—would lead me to the wrong conclusion: all the things I bought for my kids that put me into debt caused my children's success later in life, when it was actually the other way around. Their future success caused me to spend wantonly, assuming that they would bail me out once they got their six-figure salaries. One way to fix this, of course, would be to directly measure my expectations for my children's future income to resolve the matter of time order.

VARIABLES

In research we talk a great deal about variables. Simply put, you should always have one dependent variable, which is the outcome you are trying to explain, and one or more independent variable, which is the measured factor that you believe has a causal impact on the dependent variable. Because it's possible to have more than one independent variable, we will call the most important one the *key independent variable*. The difference between the independent and the dependent variables is that change in your dependent variable depends on change in your independent variable. Knowing which variable is which is important for complying with mandates for establishing causality. Often, when we establish correlation but can't do the same for causality, it's because we don't know which variable is causing change in the other—we can't establish time order, for example, so we don't know which variable is the independent and which is the dependent.

In high-school science class, you may have learned that a hypothesis is an educated guess. In social research, we use the term *hypothesis* to refer to a proposed relationship between two variables, usually with a stated direction. The *direction* of the relationship refers to whether your variables move in the same direction (positive) or in opposite directions (negative).

Let's take some examples. We know that income is positively related to education: As people's education increases, usually so too does their income. Overt prejudice, on the other hand, is negatively related to education: As people's educational levels increase, generally their levels of expressed prejudice decrease.

HYPOTHESIS TESTING

Are you starting to see how the pieces fit together in the design of a research project? Perhaps we have a special interest in one concept, say, poverty. Poverty is a broad concept, so we need to specify what we mean by poverty in this particular study. The process of assigning a precise definition for measuring a concept being examined in a particular study is called operationalization. When you read a study, it's important to understand how the author is operationalizing his or her concept. If I do a study on poverty among Americans who fall below the official US poverty line, and someone else completes a study that examines poverty using the United Nations' definition of it—namely, subsistence on less than $1.25 per day—we're discussing two very different concepts of poverty. As the old adage says, we're comparing apples and oranges. Once I decide how I'm defining poverty, I can begin to consider all the variables related to my concept. In the case of poverty, we might take a look at education, employment status, race, or gender.

It's time to make some decisions. First of all, is poverty my dependent or independent variable? Am I thinking about poverty as the cause of something else (poor health) or its result (lack of formal education)? Let's say I want to examine the factors that cause poverty, and I'm especially interested in the effect of parental education on children's poverty levels, because theory tells me that a link exists. Assuming that I've defined educational level (Number of years in school? Grades or degrees completed? Scores on a certain test? Prestige of any college attended?), now I'm ready to pose my research question: What effect do parents' educational levels have on children's chances of living in poverty as adults? And I can form a hypothesis:

> Hypothesis: The lower the educational level of the parents, the greater the chance that their children will be poor as adults.

Also for each hypothesis, an equal and opposite alternative hypothesis exists:

HYPOTHESIS

a proposed relationship between two variables, usually with a stated direction.

Alternative hypothesis: There is a positive relationship between parental education and children's likelihood of living in poverty as adults.

Parental education is my key independent variable, but I also believe that race and family structure may affect how my independent variable matters. In this example, race and family structure would be moderating variables—that is, they affect the relationship between my independent and my dependent variables. (Children's education or test scores in this example would be mediating variables that are positioned between the independent and dependent variables but do not interact with either to affect the relationship between them.)

I am not quite ready to test my hypotheses, however. First, I need to tell stories—that is, causal stories about why I would expect the hypotheses to be true. In support of the hypotheses, I might say that parents who are more educated have acquired more confidence and skills for succeeding in our economy and that they are then more likely to pass on some of this knowledge and positive outlook to their kids at home. In support of my alternative hypothesis, parents who have spent a great deal of money on their own education may have less left over to help their children. Establishing the groundwork for a reasonably "fair fight" between main and alternative hypotheses is important so we do not spend time discovering trivialities that are already well known (e.g., low-income individuals tend to be poor). So how good were my guesses or hypotheses?

VALIDITY, RELIABILITY, AND GENERALIZABILITY

VALIDITY

the extent to which an instrument measures what it is intended to measure.

RELIABILITY

the likelihood of obtaining consistent results using the same measure.

Validity, reliability, and generalizability are simple but important concepts. To say a measure has validity means that it measures what you intend it to. So if you step on a scale and it measures your height, it's not valid. Likewise, if I ask you how happy you are with your life in general, and you tell me how happy you are with school in particular, at this exact moment my question isn't a valid measure of your life satisfaction. Reliability refers to how likely you are to obtain the same result using the same measure the next time. A scale that's off by 10 pounds might not be totally valid—it will not give me my actual weight—but the scale is reliable if every time I step on it, it reads exactly 10 pounds less than my true weight. Likewise, a clock that's five minutes fast is reliable but not valid. But if I ask you to rate your overall satisfaction with your life so far, and you honestly tell me you would give yourself a 10 out of 10, and then a week from now I ask you to rate your overall satisfaction with life so far, and you honestly tell me you'd give yourself a 7 out of 10 (because that first time you had just found $10 on the floor), then that measure (my question) is valid but not reliable. Ideally, we'd like our measures to be both valid and reliable, but sometimes we have to make trade-offs between the two. Keep this in mind as we discuss the various methods of data collection.

Finally, generalizability is the extent to which we can claim that our findings inform us about a group larger than the one we studied. Can we generalize our findings to a larger population? And how do we determine whether we can?

ROLE OF THE RESEARCHER

Experimenter Effects As if social research weren't hard enough already, there are also "white coat" effects—that is, the effects that researchers have on the very processes and relationships they are studying by virtue of being there. Often, subjects change their behavior, consciously or not, just because they are part of a study. Have you ever been in a classroom when the teacher is being observed? It might prove to be the best class the teacher has ever taught, even if she didn't mean to put on the charm for the observer.

When we do qualitative fieldwork (interviews, ethnography, or participant observation) we talk about reflexivity, which means analyzing and critically considering the white coat effects you may be inspiring with your research process. What is your relationship to your research subjects? Frequently, research focuses on groups that are disadvantaged relative to the researcher in one way or another. Researchers might have more money, more education, or more resources in general. How does that shape the interactions between researcher and study participants, and ultimately, the findings?

Sociologist Shamus Khan, now a faculty member at Columbia University, attended an exclusive, private boarding school called St. Paul's. Khan has used his alumnus status to gain access to the institution in order to study how elites are groomed and trained (and whether kids on scholarships end up doing just as well as everyone else). He notes that the degree of privilege in his upbringing is not "so rare" among sociologists working in elite sociology departments like his. He admits, "a lot of sociologists...obscure their class backgrounds. They tend [to act] as if they were middle class, but if you looked at the backgrounds of a lot of the people at elite departments, they would

DIGITAL.WWNORTON.COM/YOUMAYASK6CORE

To see my interview with Shamus Khan, go to
digital.wwnorton.com/youmayask6core

be from relatively advantaged class backgrounds" (Conley, 2014b). You can learn more from Khan by watching my interview with him, in which he talks about what it was like to go back and research the imbuing of social status at St. Paul's. For now, think about how researchers' class backgrounds might (unintentionally) influence what they choose to study and how they present their findings.

In a completely different setting, urban ethnographer Mitchell Duneier spent five years hanging out with booksellers in Manhattan's Greenwich Village. He wanted to understand how these street vendors and their groups of friends, many of them homeless, functioned in the community. During the course of his research, Duneier became friends with many of his participants, which was apparent when he and I talked about how he made sure he was not exploiting them. Duneier's firm belief in the researcher's responsibility to his or her participants is not always easy for him. He told me, "in the process of doing my research, before I publish my work, I make an effort to try to show the research that I'm going to publish to the people who are depicted in it. . . . Doing this is a very stressful process. Sometimes you have to read people things that are very unflattering to them. And for me, it has taken a lot of courage to sit in a room with someone, with a manuscript, and to read those things to them. Sometimes I can be shaking when I do it because I'm so afraid of the way people are going to respond. But I feel that if I am going to be writing something about people, and I'm going to be putting it out there for the public to read, whether it's with their name on it or not, then they deserve the basic respect of hearing it from me in advance." Even after the private reading and scrutiny of the work for its most important audience—the subject—Duneier notes that the researcher must show ongoing respect for subjects:

> One of the most basic things that one can do, is to try really hard, to make sure that people don't feel used by you, after you leave the field. It's very easy given the constraints of our lives and the academy, and the great pressures we are under within our own universities, running departments, engaging in teaching, supervising students in research, it's very possible that given those demands that when we leave the field site that we can walk away from our subjects' lives and not stay in touch with them. Not ever send them any acknowledgment of the time that they spent with us. Not ever send them the results of the research. Especially if we end up moving to other parts of the country from where they live, and it's hard to stay in touch with them. It's very easy for them to get the impression that we have just walked away, benefited at their expense. They have given us their emotions, we have appropriated their lives for our sociological purposes, and then have gone on with ours. (Conley, 2009a)

As researchers, we're supposed to remain objective, but even if you want to (and some people may not want to remain impartial in certain situations), it's not always possible. One day an incident occurred between the police and the street vendors when Duneier was there (with his tape recorder running in his shirt pocket, unbeknownst to the police). He defended his friends to the officers. Because he was a white, well-spoken, and highly educated professional, his interactions with the police probably differed significantly from those between African American street vendors and the police. Duneier could speak his mind with a bit less fear of arrest and with the knowledge that he could afford a competent lawyer to defend himself. How did Duneier's presence change the interaction that transpired? Most social scientists would argue that once subjects become accustomed to the researcher's presence, they again behave as normal, but we don't have any real way to determine this. When we're engaged in qualitative research, we may find ourselves in situations where we have to choose whether objectivity and distance are more important than standing up for what we believe is right. At these times, we need to take a step back and think about our own role as both researcher and participant, because it is our perception and experience of events that eventually become the data from which we

How is the white coat effect in play here? Sociologist Mitch Duneier (center), who studied sidewalk booksellers for his book *Sidewalk* (1999), talks with a police officer. To see an interview with Duneier, go to digital.wwnorton.com /youmayask6core.

make our claims. Duneier acknowledges that ethnographers are not perfect observer-reporters whose presence has absolutely no effect on subjects' attitudes and behavior. The ethnographer, like any other scientist, has a responsibility to readers to make his or her data collection methods public. Furthermore, readers "have to have a sense for how the conclusions were drawn...by [researchers] making the lens through which the reality is being refracted apparent. That means telling them something about who we are, not only our race, not only our class or our gender, but a number of other fundamental things."

Power: In the Eyes of the Researcher, We're Not All Equal Along these lines, it is worth asking the following: What role does power play in research? As social researchers, we're not supposed to make value judgments; we should put aside our personal biases, strive for neutrality, and remain impartial and objective. The truth, however, is that we make judgments all the time, beginning at the most basic level of deciding what to study. What does the field in general deem worthy of scholarly attention? What topics am I sufficiently interested in to spend 2, 5, or 10 years, or my entire career, studying? What research do grant-making institutions regard as important enough to fund? What does the social scientific community more broadly view as problematic or interesting and in need of explanation?

Historically, sociology, like most sciences, has been male dominated. But it's also a discipline founded on the idea of making the natural seem unnatural, so it's a good place from which change can percolate. Following the second-wave feminist movement of the late 1960s, a growing stream of thought within sociology sought to turn a critical, feminist, lens on the discipline itself. Because research ultimately forms the foundation of our work, methods became a key site of debate, and thus the concept of feminist methodology was born. What do feminist research methods look like? First, it's important to understand that there is no one feminist research method, just as there is no single school of feminism. Feminist researchers use the same techniques for gathering data as other sociologists, but they employ those techniques in ways that differ significantly from traditional methods. As Sandra Harding (1987) explains it, feminist researchers

> listen carefully to how women informants think about their lives and men's lives, and critically to how traditional social scientists conceptualize women's and men's lives. They observe behaviors of women and men that traditional social scientists have not thought significant. They seek examples of newly recognized patterns in historical data. (p. 2)

The feminist part doesn't lie in the method per se, or necessarily in having women as subjects. Rather, Harding proposes three ways to make research distinctly feminist. First, treat women's experiences as legitimate empirical

FEMINIST METHODOLOGY

a set of systems or methods that treat women's experiences as legitimate empirical and theoretical resources, that promote social science *for* women (think public sociology, but for a specific half of the public), and that take into account the researcher as much as the overt subject matter.

and theoretical resources. Second, engage in social science that may bring about policy changes to help improve women's lives. Third, take into account the researcher as much as the overt subject matter. When we enter a research situation, an imbalance of power usually exists between the researcher and the research subjects, and we need to take this power dimension seriously. The point of adopting feminist methods isn't to exclude men or male perspectives: It's not *instead of*; it's *in addition to*. It means taking all subjects seriously rather than privileging one type of data, experience, or worldview over another.

CHOOSING YOUR METHOD

In Chapter 1, I described the differences between positivist and interpretive sociology. As distinct as they are in their focus, they also lend themselves to different methodological approaches to research. Because positivists are concerned with the factors that influence social life, they tend to rely more heavily on quantitative measures. If, however, you're more concerned with the meanings actors attach to their behavior, as interpretive sociologists are, then you'll likely be drawn to more qualitative measures.

Ultimately, the distinction between quantitative and qualitative methods is a false dichotomy: The most important thing is to determine what you want to learn and then contemplate the best possible way to collect the empirical data that would answer your question—that is, deploy whatever tool or set of tools is called for by the present research problem. That's why getting the research question right is so important to the entire endeavor. Once the question is precisely operationalized, the method to answer the question should be obvious. If the question still could be approached in several ways, then you probably haven't refined it enough. Figure 2.3 gives an overview of the entire process.

DATA COLLECTION

Remember that social science research is largely about collecting empirical evidence to generate or test empirical claims. So how do we go about collecting the evidence needed to support our claims? Let's use case studies—that is, particular examples of good research—and see what these researchers wanted to know, how they obtained their data, and what they found.

Participant Observation Neighborhoods rise and fall and (sometimes) rise again. Often these changes in fortune are accompanied by racially tinged dynamics: white flight in one direction and, when the tides turn again, (white) gentrification. But sometimes neighborhoods are reshaped by class dynamics within racial groups. With the rise of a black professional class since the civil rights era, gentrification—the movement of higher-income individuals into

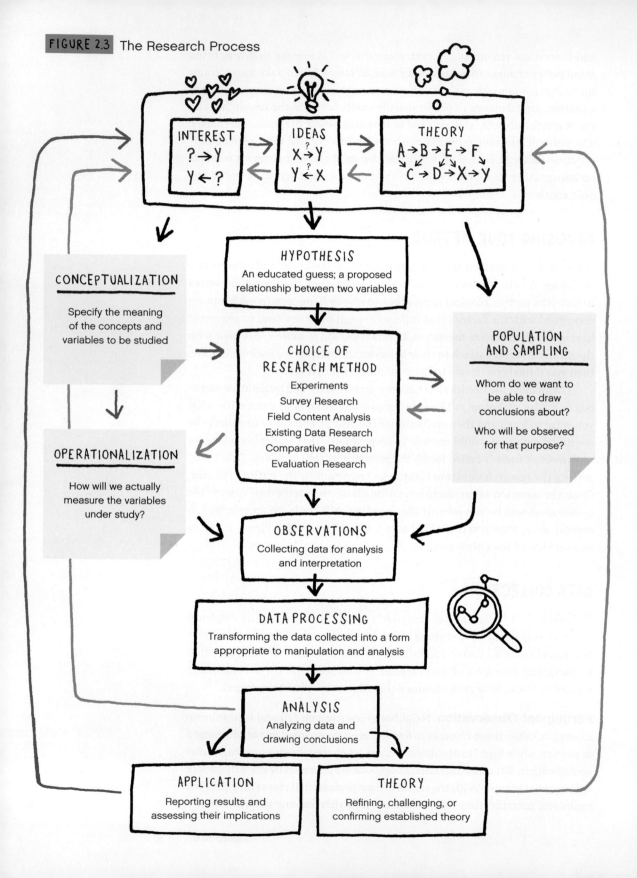

FIGURE 2.3 The Research Process

INTEREST
? → Y
Y ← ?

IDEAS
X → Y
Y ← X

THEORY
A → B → E → F
C → D → X → Y

HYPOTHESIS
An educated guess; a proposed relationship between two variables

CONCEPTUALIZATION
Specify the meaning of the concepts and variables to be studied

CHOICE OF RESEARCH METHOD
Experiments
Survey Research
Field Content Analysis
Existing Data Research
Comparative Research
Evaluation Research

POPULATION AND SAMPLING
Whom do we want to be able to draw conclusions about?
Who will be observed for that purpose?

OPERATIONALIZATION
How will we actually measure the variables under study?

OBSERVATIONS
Collecting data for analysis and interpretation

DATA PROCESSING
Transforming the data collected into a form appropriate to manipulation and analysis

ANALYSIS
Analyzing data and drawing conclusions

APPLICATION
Reporting results and assessing their implications

THEORY
Refining, challenging, or confirming established theory

once-distressed inner-city neighborhoods—can be an entirely black affair. But does that create the same tensions as cross-race urban transitions? Or does it provide its own set of unique problems?

To answer this (and other) questions, sociologist Mary Pattillo (2007) studied the Chicago neighborhood of North Kenwood–Oakland, using the method called participant observation. This approach aims to uncover the meanings people give to their own social actions (and those of others) by observing their behavior in practice, in contrast to just asking them about it after the fact. This strategy is predicated on the notion that surveys, interviews, and other approaches are more easily "managed" by the respondents. That is, studying racial attitudes by surveying respondents, for instance, may lead to subjects telling the researcher what they think is the "right" answer and what the researcher wants to hear. But by hanging out in a given community over a long period, integrating into its daily fabric, and letting the dramas of everyday life unfold with minimal intrusion, participant observers are more likely to capture what folks actually think, feel, and do rather than what they would like to think they do.

This type of sociology typically involves a significant time investment because the participant observer must gain access to a given community, learn its local norms and logic of behavior, and then watch social dynamics unfold. Often it can take years to do an ethnographic study of this nature. In Pattillo's case, she moved into the community right as other professional blacks were as well. She attended local community board meetings, went to block parties, socialized with neighbors—basically she hung out like every-one else. But the difference was that she was engaging her sociological imagination and took extensive field notes about what was unfolding around her.

What she found was that unlike in typical cross-race situations of gentrification, the middle-class blacks who moved in and revitalized the area occupied a "middleman" status. They offered social resources to their lower-class neighbors—helping them navigate city politics and bureaucracy, for example. But the role came with costs and conflicts as well: Residents of different classes disagreed about the proper use of space, such as whether the front porch was a proper hangout or where barbecuing should take place (in the front yard or backyard). There were also more consequential struggles, such as those over local schools. In the end, Pattillo's combination of historical analysis and time spent "hanging out" in the neighborhood painted a rich portrait of class and race as it plays out in Chicago's South Side.

Interviews Want to know how and why someone does something? Why not just ask them to tell you all about it? Interviews are one common form of gathering qualitative data. For her book *Money, Morals, and Manners* (1992), sociologist Michèle Lamont interviewed upper-middle-class men in France and the United States about their tastes. She chose the men in her sample based on their social status—they were employed as managers,

PARTICIPANT OBSERVATION

a qualitative research method that seeks to uncover the meanings people give their social actions by observing their behavior in practice.

professionals, and entrepreneurs—arguing that these people hold enormous power in their jobs and communities, and consequently their tastes are influential in shaping the culture around them. Lamont (1992) conducted more than 160 interviews, trying to determine how the people in her sample defined what it means to be a "worthy person" and analyzing "the relative importance attached to religion, honesty, low moral standards, cosmopolitanism, high culture, money, [and] power." The comparative aspect of her research design allowed her to identify some of the cultural differences between American and French tastes. For example, she ascertained that the French men valued art more than their American counterparts, whereas the Americans cared more about money than their French counterparts.

By using unstructured, open-ended interviews, Lamont allowed her subjects to go off on tangents, to vent, and to share intimacies that might not appear at first glance to be related to the study. But she also did probe—that is, she pushed subjects past their initial, comfortable answers on somewhat delicate, controversial issues. Knowing how and when to probe and when to back off is part of the art of interviewing that results from practice. Other researchers may rely on semistructured or structured interviews—that is, interviews in which the researchers have more than just a set of topics to cover in no preset order; rather, the researchers develop a specific set of questions to address with all respondents in a relatively fixed sequence. If an interview becomes very structured, it falls into the next category: survey research.

Survey Research Chances are you've filled out a survey at some point. One customarily receives them from the manufacturers and retailers of electronics and from restaurants and hotels. Surveys are an ordered series of questions intended to elicit information from respondents, and they can be powerful methods of data collection. Surveys may be done anonymously and distributed widely, so you can reach a much larger sample than if you relied solely on interviews. At the same time, however, you have to pay attention to your response rate. Out of all the surveys you distributed, how many were actually completed and returned to you? Lately, we have been bombarded with more and more surveys soliciting our opinions about everything from what soap we prefer to how to stop global warming. It has become increasingly difficult for researchers to get their surveys answered amid the din of our information society, and response rates, in general, continue to fall.

Why does this matter? If those who answered your survey or tore it up were truly random, then the only concern would be the cost in time and money to obtain, say, 200 completed surveys. But, as it turns out, who responds and who doesn't is not random. As a researcher, you need to consider the ways that selection bias can enter your sample. Are the people who

SURVEY

an ordered series of questions intended to elicit information from respondents.

completed the survey different in some significant way from the people who didn't complete it? If your survey is about the sacrifices respondents would be willing to make to slow global warming and the only folks who bother to respond are environmentalists, your results will likely indicate that the population is willing to sacrifice far more than it actually would. In fact, the nonrespondents were not even willing to sacrifice the 10 minutes to take the survey. Surveys can also be done in person or over the phone. This method of survey design differs from interviews in that a set questionnaire exists. Surveys are generally converted into quantitative data for statistical analysis—everything from simple estimates (How many gay policemen are there in America?) to comparisons of averages across groups (What proportion of gay policemen support abortion rights, and what proportion of retired female plumbers do?) to complex techniques such as multiple regression, where one measured factor (such as education level) is held constant, or statistically removed from the picture, to pin down the effect of another factor (such as total family income) on, say, reported levels of happiness.

The General Social Survey (GSS) run by the National Opinion Research Center of the University of Chicago is one of the premier surveys in the United States. Since 1972, the GSS has asked respondents a battery of questions about their social and demographic characteristics and their opinions on a wide range of subjects. Conducted every other year, the GSS includes some new questions, but many are the same from one survey to the next. This consistency has allowed researchers to track American attitudes about a range of important issues, from race relations to abortion politics to beliefs about sexual orientation, and to see how the beliefs of different demographic subgroups have converged or diverged over four decades. The GSS is an example of a repeated cross-sectional survey. That is, it samples a new group of approximately 2,000 Americans in each yearly survey wave. Each sample should represent a cross-section of the US population of that particular survey year.

A cross-sectional study stands in contrast to a panel survey, also known as a longitudinal study, which tracks the same individuals, households, or other social units over time. One such survey, the Panel Study of Income Dynamics (PSID), run by the Institute for Social Research at the University of Michigan, has followed 5,000 American families each year since 1968. (Recently, the PSID had to trim back to every other year because of budget cutbacks.) It even tracks family members who have split off and formed their own households and families. In this way, the survey has taken on the structure of a family tree. The PSID has contributed to important research on questions about how families transition in and out of poverty, what predicts if marriages will last, and how much economic mobility exists in the United States across generations—just to name a few of the topics that PSID analysis has illuminated.

SAMPLES:
THEY'RE NOT JUST THE FREE TASTES
AT THE SUPERMARKET

POPULATION

an entire group of individual persons, objects, or items from which samples may be drawn.

SAMPLE

the subset of the population from which you are actually collecting data.

People use the word *sample* in many different ways, but in social research it has a very specific and important meaning. You are always studying a population. It could be the entire US population, gay fathers, public schools in the rural South, science textbooks, gangs, Fortune 500 companies, or middle-class, Caucasian, single mothers. Most of the time it's too time-consuming and expensive to collect information about the entire population you want to study, so instead you focus on a sample. Your sample, then, is the subset of the population from which you are actually collecting data. (If you do collect information on the entire population, it's called a *census*.)

How you go about collecting your sample is probably one of the most important steps of your research. Let's say I want to study attitudes toward underage drinking in the United States and hand out a survey to your sociology class. Based on the findings of that survey, I make claims about how the entire US population feels about underage drinking: They're in favor of it. Would you believe my claims? I hope not! Your sociology class is probably not

Volunteer Phyllis Evans (center) questions a homeless man about his living situation and encourages him to seek help while conducting a survey with team members in New York City.

representative of the US population as a whole. Age would be the most important factor, but differences in socioeconomic status, education, race, and the like would exist. In other words, the results I would obtain from a survey of a college sociology class would not be generalizable to the US population and probably not to college students as a whole—maybe not even to students next door in organic chemistry might have very different thoughts about underage drinking). I would be "speaking beyond my data."

Although the issues of generalizability are always at play, they become particularly acute when social scientists use case studies. A case study, often used in qualitative research, is an in-depth look at a specific phenomenon in a particular social setting. If we wanted to understand the interaction among parents, teachers, and administrators in the American public school system, we might do a case study of your high school. How representative of all US high schools do you think your school is? Does your town have a higher or lower average household income than the United States as a whole? Is the PTA particularly vocal? Is yours a regional high school whose students travel long distances to attend? All of these factors—these variables—are important, and if your town isn't typical (statistically speaking), we'd question the usefulness of the findings. This is perhaps the main drawback of the case study method. The findings have very low generalizability. One benefit, however, is that we typically obtain very detailed information. So there is often a trade-off between breadth (i.e., generalizability) and depth (i.e., amount of information and nuanced detail). A case study can serve as a useful starting point for exploring new topics. For example, researchers often use case studies to develop hypotheses and to generate and refine survey questions that the researchers will then administer to a much larger sample. Likewise, qualitative case studies are sometimes used to try to understand causal mechanisms that have been indicated in large-scale survey studies.

↑

Census taker talking with Charles F. Piper as he works on his car.

CASE STUDY

an intensive investigation of one particular unit of analysis in order to describe it or uncover its mechanisms.

Historical Methods, Comparative Research, and Content Analysis

HISTORICAL METHODS

research that collects data from written reports, newspaper articles, journals, transcripts, television programs, diaries, artwork, and other artifacts that date back to the period under study.

Children of a plantation sharecropper preparing food on a woodstove in a sparsely furnished shack in 1936. How did Jill Quadagno use historical methods to analyze the ways in which people like these children were excluded from the benefits of the New Deal?

How do we study the past? We can't interview or survey dead people, and we certainly can't observe institutions or social settings that no longer exist. What researchers employing historical methods do is collect data from written reports, newspaper articles, journals, transcripts, television programs, diaries, artwork, and other artifacts that date back to the period they want to study. Researchers often study social movements using historical methods, because the full import of the movement may not be apparent until after it has ended.

How did America end up with a relatively weak welfare state and a high tolerance for inequality, particularly in the context of race, compared with other industrialized democracies? To answer this question, sociologist Jill Quadagno (1996) went back to the archives to research what had been said officially (in regulations and other government documents) and unofficially (in the press) about the passage and implementation of the New Deal in the 1930s and the War on Poverty in the 1960s.

Quadagno took into account many different explanations: timing (the United States industrialized early, before the adequate development of protective political institutions), institutions (the United States has one of the most fragmented political systems in the developed world, making it comparatively difficult to marshal large-scale government programs), and "American exceptionalism" (the notion that our culture, lacking a history of feudalism, was uniquely individualistic and nonpaternalistic).

Finally, the issue on which Quadagno focused her research was the looming shadow of racism in America. According to Quadagno, in order to prevent blacks from participating fully in the American social contract, authority devolved from the federal government to state and local authorities, which could then exclude blacks overtly or covertly. The result was a much weaker safety net and one that, for a long time, excluded minorities disproportionately. For example, to ensure that congressional committees controlled by racist Southern Democrats passed Social Security, President Franklin D. Roosevelt had to agree to exclude agricultural and domestic workers when the system was established in 1935. This exception was made purposely to exclude African Americans, who were disproportionately employed in these two sectors. Thus by conducting historical, archival research, Quadagno and others have been able to show the relevance of race in explaining the particularities of the social safety net in the United States.

Whereas the example just discussed focuses on one case, sometimes sociologists compare two or more historical societies; we call this "comparative historical" research. For example, Rogers Brubaker (1992) compared the conceptions of citizenship and nationhood in France and Germany. Comparative research is a methodology by which a researcher compares two or more entities with the intent of learning more about the factors that differ between them. By examining official documents and important texts written over a period of many years leading up to, during, and after the formation of the German and French states, Brubaker showed that their historical circumstances led to very different visions of citizenship in each nation. France was formed from a loosely knit group of powerful duchies and principalities. There was no preexisting French nation or nationality before the creation of the French government. The idea of nationhood—that is, of French identity—had to be forged by the state itself, and this led to a very inclusive notion of citizenship. Germany, by contrast, grew out of an already well-refined, tribal sense of Prussian nationality. Thus Germany's citizenship policy was based on excluding others rather than including them.

The general approach to comparative research is to find cases that match on many potentially relevant dimensions but vary on just one, allowing researchers to observe the effect of that particular dimension. Although all social science research makes inferences based on implicit or explicit comparison, comparative research usually refers to cross-national studies. For example, in studying the effects of gun ownership rates on deaths from firearms, it would be better to compare the United States (which has one of the most hands-off approaches to gun regulations and a high degree of gun ownership) with Australia (which was very much like the United States with respect to gun culture—as well as other aspects of culture—but which changed its policy drastically in 1996 in response to a massacre) than to compare Yemen (which, like the United States, has heavily armed populations) with Sweden (which doesn't), because the latter two countries are so different from each other geographically and culturally.

One distinct subtype of historical methods research, content analysis, is a systematic analysis of the content in written or recorded material. Race scholar Ann Morning (2004) used content analysis to investigate depictions and discussions of race in American textbooks across academic disciplines over time. Morning analyzed both manifest and latent content on race in a

Greek miners seeking work in the German Ruhr Basin in 1960 after West Germany began a guest-worker program. What did Rogers Brubaker's comparative research about European immigration policies reveal about definitions of citizenship?

COMPARATIVE RESEARCH

a methodology by which two or more entities (such as countries), which are similar in many dimensions but differ on one in question, are compared to learn about the dimension that differs between them.

CONTENT ANALYSIS

a systematic analysis of the content rather than the structure of a communication, such as a written work, speech, or film.

sample of 92 high-school textbooks published in the United States between 1952 and 2002 in the fields of biology, anthropology, psychology, sociology, world culture, and world geography. *Manifest content* refers to what we can observe; Morning's study included overt discussions and definitions of race and images of different races. *Latent content* refers to what is implied but not stated outright; Morning looked for sections of the texts where race was directly implied, even if the word *race* wasn't used. She chose her sample from all high-school textbooks published in the United States between 1952 and 2002.

Ultimately, Morning's analysis disputed earlier findings that biology textbooks no longer discuss race. Her findings showed that only social sciences texts employed constructivist approaches (i.e., the belief that race is a social construct), whereas biology books reinforced essentialist conceptualizations (i.e., the belief that race is innate and genetic). However, contrary to her original hypotheses, social sciences textbooks also used biological components in their definitions of race, and only the fields of anthropology and sociology critiqued the traditional concept of race. Why does all this matter? As Morning points out, if textbooks aren't changing, how are students supposed to learn new concepts or viewpoints? Anthropology and sociology aren't widely taught at the high-school level, whereas biology is mandatory in most public school systems.

Fans line up outside a bookstore in Tel Aviv, Israel, in anticipation of the release of the final book in the Harry Potter series by J. K. Rowling. Using Duncan Watts's ideas, how can we explain the massive success of this cultural product?

Experimentation Because social scientists deal with people, the controlled environment of a laboratory-based experiment is not always an option. For example, I dream of randomly assigning the students in one of my classes to married or single life in order to examine marital status on some dimension. Assuming seating is random, I'd just draw a line straight down the middle of the lecture hall; all the students on one side would have to marry, while those on the other side would have to remain single. Think of the possibilities! For better or worse, however, I'm not allowed to perform such experiments.

Some sociologists do use experimental methods, however. For instance, sociologist Duncan Watts conducted an experiment to find out the extent to which taste in cultural products like music and movies is determined by what other people think. Watts speculated: "When we look at cultural markets it's very clear that the successful ones are many times more successful than average, so you see things like Harry Potter books have sold probably about 400 million copies around the world. . . . What was puzzling to us about this fact, is that nobody seems to be able to predict which movies, books, albums, etc., are going to be these massive hits" (Conley, 2009b). Watts hypothesized that it was not just the innate cultural quality of a product but the luck of catching on through peer-to-peer influence that determined success in film, music, or publishing. Along with his co-researchers Matthew Salganik and Peter Dodds, Watts tested his hypothesis by creating an online site called Music Lab where they posted

To learn more about Duncan Watts's Music Lab experiment, you can watch my interview with him at digital.wwnorton.com/youmayask6core

mp3s from unknown bands (Salganik et al., 2006). Visitors to the site could download songs and rate how much they liked the music. Some of the 14,341 downloaders saw how many times these previously unheard-of songs from unknown bands had been downloaded by the previous downloaders, some were unable to see download traffic, and still others could see the song rankings but they had been altered, unbeknownst to the subjects. The researchers found that the best songs in one condition rarely did poorly in another condition and the worst rarely did well, but that the songs in the middle could be heavily influenced by how they were rated (if, that is, the downloaders could see those ratings). I asked Watts if his Music Lab experiment could explain other social phenomena. He explained that we can see the same kinds of social influence in the recent housing bubble and recession. According to Watts, "If you look at the recent financial crisis many things went wrong, right? But the big things that went wrong on the way up were this belief that housing prices could not go down, and on the way down a real collapse of our trust in the financial system. Both of them are driven by people thinking whatever they think because of what other people think. These are not fundamentally financial phenomena or economic phenomena even, they're really social phenomena" (Conley, 2009b).

EXPERIMENTAL METHODS

methods that seek to alter the social landscape in a very specific way for a given sample of individuals and then track what results that change yields; they often involve comparisons to a control group that did not experience such an intervention.

Ethics of Social Research

At the beginning of this chapter, I mentioned the contributions feminists made to research methods with their emphasis on examining relationships between power and the process of knowledge generation. Today, we have more codified standards that must be met by all researchers. Many professional associations have their own ethical standards; doctors, lawyers, journalists, psychologists, and sociologists all do, just to name a few. Colleges and universities, too, often have guidelines for research conducted with humans (as well as with animals, particularly vertebrates). As a professional sociologist, I am beholden to the ethical guidelines established by my peers and by the American Sociological Association. As a professor, I am also responsible to my home institution. And as a researcher, I am ultimately responsible to my research subjects. I work with already-collected statistical information (or secondary data) for the most part, which makes the process a little easier, but that doesn't mean I don't have to pay careful attention to the ethical standards of my discipline.

A few golden rules exist in research. The first is "Do no harm." This may seem obvious; you don't want to cause physical harm to your subjects. But what about psychological or emotional harm? What if you want to interview men and women on their attitudes toward abortion, and a respondent becomes very upset because he or she cares deeply about the subject? The initial charge not to do harm gets complicated quickly. Often, we try to design research projects so that subjects will encounter no more risk than that associated with everyday life.

The second rule is regarding informed consent. Subjects have a right to know they are part of a study, what they are expected to do, and how the results will be used. If you're interviewing people or asking them to complete a survey, this makes sense. But how far do you take the rule of informed consent in participant observation? Generally, you have to obtain permission to be at your chosen site, but do you remind every person you bump into that you're doing research? What's more, sometimes mild forms of deception are necessary for the sake of research. In Duncan Watts's research about the way a song's rating influenced the way a new listener would rate it, he could not tell his subjects that the ratings they saw were fake. This small amount of deception is usually considered ethically acceptable, but because deception is a slippery slope, panels of academic peers review all research designs to ensure that subjects are protected. If researchers deceive subjects, the deception must be absolutely necessary to the study, and above all else, the subjects better be safe.

The third rule is to ensure voluntary participation, which usually goes hand in hand with informed consent. People have a right to decide if they want to participate in your study. They are allowed to drop out at any point with no penalty. If you're interviewing someone who doesn't want to answer a question, he or she can skip questions; if the interviewee wants to stop for whatever reason or no reason, that's his or her prerogative. Ethically, the researcher cannot badger respondents into participating in the study or completing it once they have started. There are also certain protected populations—minors, prisoners and other institutionalized individuals, pregnant women and their unborn fetuses, people with disabilities—whom you need additional approval to study. As danah boyd's research with teens shows, it's not impossible to study these populations; it just requires additional effort and caution.

Find out if your college or university has an institutional review board and what the requirements are to gain approval for a research project before you start your budding career as a sociologist. And then you are all set to question everything and make the familiar strange. Good luck!

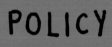

POLICY

 # THE POLITICAL BATTLE OVER STATISTICAL SAMPLING

You might think that sampling is an arcane and unsexy concern of nerdy researchers trying to make sure their results apply to the population as a whole. In most cases, you would be right. But for almost 40 years, the notion of statistical sampling was a red-hot political issue. At one point, some Republican members of Congress talked about shutting down the government over it.

It all started in 1937, when the Census Bureau used statistical sampling methods to survey unemployment. Of course, knowing the rate of unemployment was of utmost importance during the economic crisis of the Great Depression. The idea of sampling a portion of the population was then deployed in the 1940 Census, when enumerators asked a random subset of the population (approximately 5 percent) additional questions. They wanted to get more detailed information about what Americans' lives were like without the cost and burden of asking everyone. This was the

The proposed readjustments of the US Census are an example of a political issue associated with population sampling.

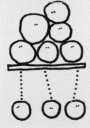

origin of the so-called census long form (which was eliminated in 2010 after being replaced by the yearly American Community Survey).

So far, so good.

The problems started in 1980 with post-enumeration sampling. Official census counts are based on a mail-in survey and in-person follow-up to addresses that do not respond. In 1980, census officials randomly selected (i.e., sampled) certain blocks and went back and redid the counts there to see if some people were missed or double-counted. In fact, the post-enumeration count showed that the US population was indeed undercounted, by about 1.2 percent. The more troubling aspect of the undercount was that certain groups were more likely to be missed than others. For example, African Americans were undercounted at a rate that was 3.7 percent higher than other races.

This result prompted a fight. It started when the City of Detroit sued to require adjustment of the overall census by the purported undercounts (and overcounts) revealed by the sampling. Soon other plaintiffs seeking the same redress also sued, including, most notably, the City and State of New York. This kind of "reweighting" is a common,

relatively uncontroversial adjustment done by researchers using survey studies. But when the apportionment of congressional seats is at stake—not to mention all of the federal dollars that are divided up by population size—then sampling-based adjustments become *very* political.

The Commerce Department (in which the Census Bureau is based) decided not to readjust the "raw" census figures, and the Supreme Court ultimately upheld this decision. The same statistical question had another trip through the courts after the 1990 count, and then again in preparation for the 2000 Census. Groups that tend to be undercounted also tend to vote for Democratic candidates, so it did not come as a surprise that the Clinton administration hoped to use the sampling-adjusted count following the 2000 Census. Meanwhile, in Congress, the issue caused the usually routine budget appropriation for census operations to stall until a compromise two-track planning approach was adopted. Ultimately, the Supreme Court ruled that not only did the Census Bureau not have to adjust its counts to better reflect the actual population numbers and composition but it could not legally do so even if the Clinton administration wanted the adjustment made.

Regardless of the political implications, many statistical and social scientists were unhappy with this outcome because they knew the adjusted counts were more accurate.

Even today, the issue refuses to die. The Court's decision left some wiggle room with respect to adjusting census numbers for the purposes of more accurately distributing the population within states (but not between them). This possibility caused the issue to surface once again during the confirmation hearing of sociologist Robert Groves as President Barack Obama's first Census Bureau director.

Senator Susan Collins from Maine asked Groves, "Will you advocate for the statistical adjustment or use of sampling during the 2010 Census?"

It was, after all, a fair question, because Groves had dissented during George H. W. Bush's presidency from the administration's decision to ignore post-enumeration sampling.

"No, Senator," the University of Michigan professor now responded.

Senator Collins pushed him on the issue with respect to the plans for the 2020 Census.

"I have no plans to do that for 2020," was Groves's response (US Department of Commerce, 2009).

After a hint of doubt, Groves ended up being confirmed and served for three years. Today the post-enumeration sampling approach is used, but only to improve methods for reaching those undercounted populations in future census counts. Even that limited use faces an uncertain future. A year into his administration, President Trump has yet to formally nominate a census director; however, it is likely that whomever he picks will seek to roll back post-enumeration adjustments to any sample-based estimates made by the Census Bureau.

Conclusion

Sociology is a field that deploys a variety of methodologies from survey research to participant observation to historical approaches. Therefore, we sociologists often feel that we have to defend our very identity as scientists. Indeed, even some sociologists would argue that sociology is not a science. I would assert, however, that sociology is among the most difficult sciences of all. Sociology is a science in which you can't complete the controlled experiments that are the staple of most bench science. Perhaps zoology and paleontology are other examples of fields in which the scientist is called on to piece together observational data without the ability to run experiments. Nonetheless, sociologists also must face the task of imputing causal processes, not just describing or classifying the world.

How does one assess causality with only observational data to go by, especially when there are multiple factors to analyze—factors that may all interact with one another? And add to that this complication: Reality changes as you study it and by virtue of the fact that you study it. Our basic units of analysis, such as the family, and our conceptual frameworks, such as race and class, are always shifting as we study them. On top of that, the fact remains that many of the topics we study, such as gender and sexuality, race and class, family life, politics, and so on, are, by design, the most politically charged and most personally sensitive topics in our society. This doesn't make research easy. So what we sociologists are trying to do in this difficult field is to inch our way toward causality.

QUESTIONS FOR REVIEW

1. What is the difference between causality and correlation? Use the example from the beginning of the chapter, on the link between health and income, to illustrate this difference.

2. Describe one of the studies discussed in this chapter, its methodology (e.g., interviews), and its general findings. Then imagine how an additional study using a different methodology (e.g., comparative research) might build on these findings and generate new questions.

3. A sociologist observes the work-seeking habits of welfare recipients. After weeks of observation, trends emerge and the researcher forms a theory about the behaviors of this group. Is the sociologist in this example using a deductive or inductive approach? How would the sociologist study this phenomenon using the other approach?

4. Participant observation research is often long, painstaking, and personally demanding for the sociologist. Why bother with this data collection method? Use the example of Mary Pattillo's research to support your answer.

5. Surveys are complicated to design, costly to administer, and potentially suffer from response bias with respect to who answers them. Why use this data collection method? Draw on the case of the General Social Survey to support your answer.

6. Why do sociologists have to run their projects by institutional review boards? What are the "golden rules" sociologists should keep in mind when conducting research?

PRACTICE

SOCIOLOGY, WHAT IS IT GOOD FOR?

You may think this chapter (indeed, this book) won't have much application to your life after the semester is over. But think again! The research methods described in this chapter—the tools of a social scientist—are the same ones used in many different jobs and careers.

TRY IT!

Let's say you're a brand-new marketing assistant at a cable company. You are given this draft of a survey for customers. Don't forget your research question: You are hoping to figure out why an increasing number of consumers are dropping their cable subscription service, i.e., "cutting the cord." Putting your sociology cap on, see if you can identify and fix some issues:

Age: _____ **Gender:** _____ **Race/ethnicity:** _____

Do you currently subscribe to cable? ☐ Yes ☐ No

If yes:

Have you considered "cutting the cord"?

☐ Yes ☐ No

Do you agree that cable programming is superior to free video available on the internet?

☐ Yes ☐ No

If no:

Do you not subscribe to cable because you can't afford it?

☐ Yes ☐ No ☐ Refuse to answer

Do you watch cable programming illegally?

☐ Yes ☐ No ☐ Refuse to answer

THINK ABOUT IT

Customer surveys are just one area of the economy that uses sociological tools. Using the pairings below, describe how each sociological method might be useful in the career it's matched with.

SOCIOLOGICAL SKILL

Goffman's front stage
(dramaturgical analysis)

Survey

Participant observation

Mills's method of difference

Social network analysis

Experimental methods

CAREER

Policy analyst for education department

Human resources manager

FAA crash site investigator

Real estate agent

Marketer for pharmaceutical company

Family therapist or social worker

SOCIOLOGY ON THE STREET

There are many ways to research a sociological issue. How might your choice of research methods, subjects, and even your perspective alter your results? Watch the Sociology on the Street video to find out more: **digital.wwnorton.com/youmayask6core**.

WANT MORE PRACTICE?

Complete the InQuizitive activity for this chapter at digital.wwnorton.com /youmayask6core

PARADOX

3

DO MASS MEDIA CREATE SOCIAL NORMS
OR MERELY REFLECT THEM?
CULTURE IS LIKE TWO MIRRORS
FACING EACH OTHER:
IT SIMULTANEOUSLY REFLECTS AND
CREATES THE WORLD WE LIVE IN.

SOCIAL
NORMS

MASS
MEDIA

Culture and Media

The year: 1977. The place: a Portland Trailblazers basketball game. The star was not on the court but in the audience—Rollen Stewart, aka Rockin' Rollen, aka Rainbow Man. Caught on camera at the game in a rainbow-colored clown's wig, Rockin' Rollen began his 15 minutes of fame. His plan: First create a celebrity image; then, cash in on that notoriety by landing roles on TV or in the movies. After that first appearance in Portland, Rollen became a fixture at sports events, from Super Bowls to the 1984 Olympics in Sarajevo. He was featured in a beer commercial, was invited to glamorous parties, and lived the Hollywood high life.

After a few years of the extravagant lifestyle, however, he found it "shallow and unhappy." Watching a television program called *Today in Bible Prophecy* following a Super Bowl game, he experienced a religious epiphany and became a born-again Christian. Retaining his colorful wig, he decided to use major sporting events to garner media attention and share a Christian message. He sold everything he owned, lived in his car, and used his money for tickets to key sporting events. He wore shirts and signs quoting John 3:16: "For God so loved the world that he gave his one and only Son, that whoever believes in him shall not perish but have eternal life." He would even carry a small TV set with him into the stands so that he could best position himself to be captured on the cameras, much to the chagrin and annoyance of TV stations. Some members of the media threatened him, so frustrated were they by his co-optation of their precious airtime to spread a religious message for free.

In the mid-1980s Rollen came to believe that Judgment Day was nearing. He decided to focus on the message, discarding the wig so it wouldn't

be a distraction. To get out the final message, he decided to use negative press. He set off a series of stink bombs, mainly in Christian bookstores, and soon the FBI was hunting for him. He claimed that he was trying to direct people's attention to religious scholars. In a dangerous and emotional media-hungry escalation, he held a maid hostage in a hotel room for nine hours, posting religious messages in the hotel window for the media. In the end, the former Rainbow Man admitted that he wasn't receiving the press he wanted—he had lost control of the final presentation. He received three life sentences in prison.

What do we make of the story of Rollen Stewart? We might describe him as a crazy zealot. Alternatively, we might describe him as a media-savvy cultural icon. It's this second definition we're interested in, for the sake of this chapter, at least. What does it mean to be media savvy? What does it mean to be a cultural icon? Before we can answer these questions, we need to back up and take a look at two key concepts: culture and media.

How did Rockin' Rollen manipulate the media to spread his own message?

Definitions of Culture

Culture is a vague term we use to rationalize many behaviors and describe all sorts of peoples and patterns. We talk about a culture of poverty in the United States. We hear about corporate cultures and subcultures, culture wars, the clash of cultures, culture shock, and even cultural conflicts on a global scale. Culture is casually used as shorthand for many things, ranging in meaning from innate biological tendencies to social institutions, and everything in between.

CULTURE

the sum of the social categories and concepts we embrace in addition to beliefs, behaviors (except instinctual ones), and practices; everything but the natural environment around us.

CULTURE = HUMAN − NATURE

We might say that culture is the sum of the social categories and concepts we recognize in addition to our beliefs, behaviors (except the instinctual ones), and practices. In other words, culture is everything but nature.

The last sentence captures exactly how culture has been defined through the ages—in opposition to nature. The word *culture* derives from the Latin verb *colere* ("to cultivate or till"), suggesting the refinement of crops to meet human needs. (We still use *culture* as a verb in a similar sense, as when we culture bacteria in a petri dish.) The more common meaning of *culture* as a noun developed from the same kind of human control and domination over nature. We could say that culture began when humans started acting as the architects of nature by growing crops rather than hunting and gathering, hence the terms *agriculture* and *aquaculture* (growing fish and other aquatic organisms for human consumption). Dating back centuries, the term *culture* has referred to the distinction between what is natural—what comes directly from the earth and follows the laws of physics—and what is modified or created by humans and follows (or breaks) the laws of the state. That said, culture is both the technology by which humans have come to dominate nature and the belief systems, ideologies, and symbolic representations that constitute human existence.

In the fifteenth century, when European nations organized expeditions to extend commerce and establish colonies in North America, Africa, and Asia, Western peoples confronted non-Western natives. The beliefs and behaviors of these peoples served as a foil to European culture. Today, we recognize that culture is always relative. We cannot talk about culture without reference to the global world, but the definitions, practices, and concepts that we use in this chapter largely emanate from a Western viewpoint. It may also be easier to identify cultural elements when they are different from our own. The challenge in this chapter will be to take what we see as natural and view it as a product of culture. We'll also explore the media and the role they play in the birth and dissemination of culture.

CULTURE = (SUPERIOR) MAN − (INFERIOR) MAN

As colonialism led to increased interaction with non-Westerners, Europeans came to recognize that much of what they took for granted as natural was not. Alternative ways of living existed, as manifested in a variety of living arrangements and marital rules, different styles of dress (or lack thereof), other ways of building cities, and other kinds of foodstuffs. Take, for example, the fact that American architecture of the nineteenth and twentieth centuries tended to use the three major Greek styles of columns: Doric, Ionic, and Corinthian. This was merely a product of tradition: There's nothing natural about these columnar styles that made them the dominant choices in the American architecture of those centuries. And if you traveled to non-Western cities, you would encounter examples of other columnar styles—for instance, the massive fluted columns in the Blue Mosque in Istanbul, Turkey, or the Toltec columns in Tula, Mexico. Such differences may not seem so striking to us now, but think about what it was like for Westerners who rarely came into contact with non-Westerners.

Clockwise from top left: US Capitol, Washington, D.C.; Lincoln Memorial, Washington, D.C.; Blue Mosque, Istanbul, Turkey; and Toltec columns, Tula, Mexico

ETHNOCENTRISM

the belief that one's own culture or group is superior to others, and the tendency to view all other cultures from the perspective of one's own.

In the wake of these colonial encounters with the New World, philosophers began to define culture in contrast not just to nature but also to what other peoples did, realizing that the way they performed tasks or lived life was a historical product of specific cultural influences. People started to contemplate why their traditions, beliefs, styles of art, and other ways of living arose. There was nothing inevitable about them, and valid alternative approaches to interacting with the world existed. Coming face-to-face with these alternative ways of living caused Westerners to question the culture that they had so far taken for granted. Philosophers such as Jean-Jacques Rousseau idealized non-Western "savages" in contrast to corrupt and debased Europeans; others declared their own culture superior, going so far as to claim that non-Western peoples didn't have culture. Ethnocentrism is a term that encapsulates the sense of taken-for-granted superiority in the context of cultural practices and attitudes. It is both the belief that one's own culture or group is superior to others and the tendency to view all other cultures from the perspective of one's own. At the time, some even believed that non-Westerners did not have souls and weren't human, and this notion was used to justify slavery, violence, and oppression. Such claims obviously weren't true (and we owe a lot to anthropologists for disproving them), but the long history of racism with which we still struggle does have some of its roots in these ideas.

CULTURE = MAN – MACHINE

Not long after European empires began to emerge and spread, new technologies and forms of business upended European definitions of culture at home. Beginning in the eighteenth and nineteenth centuries, goods that were previously expensive and handcrafted began to be mass-produced and priced within the reach of the average European. New industries and a growing middle class of merchants and industrialists started to transform the political and social climate of Europe, particularly in Great Britain. In response to these rapid social changes, the poet and cultural critic Matthew Arnold (1822–1888) redefined culture as the pursuit of perfection and broad knowledge of the

A sixteenth-century Aztec's drawing of the conquistador Hernán Cortés. Why did Western definitions of culture change during the Age of Exploration?

world in contrast to narrow self-centeredness and material gain. Intellectual refinement was the "pursuit of our natural perfection by means of getting to know, on all the matters which most concern us, the best which has been thought and said in the world." Arnold's definition of culture in his book *Culture and Anarchy* (1869) elevates it beyond dull, middle-class institutions such as religion, liberal economics, and political bureaucracy.

Arnold's definition of culture was an extension of views advanced in Plato's *Republic*, which argues that culture is an ideal, standing in opposition to the real world. Plato argues that God is the ideal form of anything. A carpenter, for example, tries to construct a material embodiment of that ideal form. He starts with a vision, the divine vision of what a chair or table should look like, and he works his hardest to bring that vision to fruition. Of course, it can never be perfect; it can never approximate the platonic ideal.

The artist's job, in contrast, is to *represent* the ideal within the realm of the real. In fact, there's a long history of artists attempting to represent the ideal female in sculpture and painting, but in reality, no woman could ever exist as a flawless object, content to be gazed upon. Jean Auguste Dominique Ingres's *La Grande Odalisque* has an unrealistically long spine, allowing her to appear smooth, supple, and gracefully elegant as she shows us her backside but turns her face to meet the viewer's gaze with a hint of a smile. In this

conception of art—and this understanding of culture—there is a single, best example of any element in the world, from the ideal woman to the ideal form of government to the ideal citizen, which humanity ought to emulate. Furthermore, we see that the ideal woman (and meal and family and governmental structure) is a fluid notion, changing from one place to another and across time periods. Can an ideal be discovered, as Plato believed, or is it constructed?

Material versus Nonmaterial Culture

NONMATERIAL CULTURE

values, beliefs, behaviors, and social norms.

MATERIAL CULTURE

everything that is a part of our constructed, physical environment, including technology.

CULTURAL LAG

the time gap between the appearance of a new technology and the words and practices that give it meaning.

Today, we tend to think that everything is a component of culture. Culture is a way of life created by humans, whatever is not natural. We can divide culture into nonmaterial culture, which includes values, beliefs, behaviors, and social norms, and material culture, which is everything that is a part of our constructed, physical environment, including technology. Well-known monuments, such as the Statue of Liberty or Mount Rushmore, are part of our culture, but so are modern furniture, books, movies, food, magazines, cars, and fashion. Of course, a relationship exists between nonmaterial culture and material culture, and that can take many forms. When someone conjures up a concept like a portable computer, such an invention flows directly from an idea into a material good. Other times, however, it is technology that generates ideas and concepts, values and beliefs. Before phones with cameras and apps such as Instagram, the word *selfie* did not exist, and before selfies there were no selfie sticks. When it takes time for culture to catch up with technological innovations, there is a cultural lag.

LANGUAGE, MEANING, AND CONCEPTS

Another way to contemplate culture is the following: It is what feels normal or natural to us but is, in fact, socially produced, like saying "Bless you" when someone—even a stranger—sneezes. Another way to put it is this: Culture is what we do not notice at home but would spot in a foreign context (although remember, the sociologist's job is to notice these things at home too). In France, no one says "Santé" when a stranger sneezes in the grocery store. When meeting someone for the first time, if you are holding packages in your right hand, have you instead extended your left in greeting? In Saudi Arabia, this action would be construed as highly disrespectful. Have you ever taken public transportation during rush hour? Do the passengers waiting to board generally move aside to let people off the train first? In Moscow, nobody steps aside; in Japan, people wait so long that they practically miss the train! Even your sociology class probably has a different culture—language, meanings, symbols—from a biology or dance class.

Another way to think about culture is that it is a way of organizing our experience. Take our symbols, for example. What does a red light mean? It could mean that an alarm is sounding. It could mean that something is X-rated (the "red-light" district in Amsterdam, for instance). It could also mean "stop." There is nothing inherent about the meaning of the red light. It is embedded within our larger culture and therefore is part of a web of meanings. You cannot change one without affecting the others. Moving from one culture to another can induce feelings of culture shock—that is, confusion and anxiety caused by not knowing what words, signs, and other symbols mean. People who move fluidly from one cultural setting to another learn to code switch by swapping out one set of meanings, values, and/or languages on the fly. Elijah Anderson (1999) and others have pointed out how many minority groups—such as African Americans—learn to code switch in their daily lives by going back and forth between standard English and African American English as they move between predominantly white social contexts (perhaps their workplace) to environments in which they form the majority (such as at home or in church).

Language is an important part of culture. According to the Sapir-Whorf Hypothesis in linguistics, the language we speak directly influences (and reflects) the way we think about and experience the world. On a more concrete level, if you speak another language, you understand how certain meanings can become lost in translation—you can't always say exactly what you want. Many English words have been adopted in other languages, often as slang, such as *hamburger* and *le weekend* in French. Some staunch traditionalists are opposed to such borrowing because they regard it as a threat to their culture. How many words do we have for college? How central does that make higher education to our society? Many words describe the state of being intoxicated. What does that say about our culture?

CULTURE SHOCK

doubt, confusion, or anxiety arising from immersion in an unfamiliar culture.

CODE SWITCH

to flip fluidly between two or more languages and sets of cultural norms to fit different cultural contexts.

There are no inherent meanings behind a red light; its symbolism varies depending on context.

Concepts such as race, gender, class, and inequality are part of our culture as well. If you try to explain the American understanding of racial differences to someone from another country, you might get frustrated because it may not resonate with him or her. That's because meanings are embedded in a wider sense of cultural understanding; you cannot just extract concepts from their context and assume that their meanings will retain a life of their own. In some cases, when opposing concepts come into contact, one will necessarily usurp the other. For example, when European colonization first hit the Americas, Native Americans believed that owning land was similar to the way Americans now feel about owning air—a resource that was very difficult to put a price on and best understood as a collective responsibility. From a real estate perspective, the Europeans must have been very excited. ("All this land and *nobody* owns it?") The issue was not a language barrier: Native Americans had a social order that had nothing to do with assigning ownership to pieces of the earth. Acting on their concepts of ownership, Europeans thus began the process of displacing native peoples from their homelands and attacking them when they resisted.

IDEOLOGY

IDEOLOGY

a system of concepts and relationships; an understanding of cause and effect.

Nonmaterial culture, in its most abstract guise, takes the form of ideology. Ideology is a system of concepts and relationships, an understanding of cause and effect. For example, generally on airplanes you're not allowed to use the toilets in the first-class cabin if you have a coach-class ticket. Why not? It's not as if the lavatories in first class are that much better. What's the big deal? We subscribe to an ideology that the purchase of an airline ticket at the coach, business, or first-class fare brings with it certain service expectations—that an expensive first-class ticket entitles a passenger to priority access to the lavatory, more leg room, and greater amenities, such as warm face towels. The ideology is embedded within an entire series of suppositions and, if you cast aside some of them, they will no longer hold together as a whole. If everyone flying coach started to hang out in first class, chatting with the flight attendants and using the first-class toilets, the system of class stratification (in airplanes at least) would break down. People would not be willing to pay extra for a first-class ticket; more airlines might go bankrupt, and the industry itself would erode.

Even science and religion, which may seem like polar opposites, are both ideological frameworks. People once believed that the sun circled around the earth, and then, in the late fifteenth and early sixteenth centuries, along came Copernicus, Kepler, and Galileo, and this system of beliefs was turned inside out. The earth no longer lay at the center of the universe but orbited the sun. This understanding represented a major shift in ideology, and it was not an easy one to make. In a geocentric universe, humans living on earth stand at its center, and this idea corresponds to Christian notions that humans are the lords of the earth and the chosen children of God. However, when we view the earth as a rock orbiting the sun, just like seven other planets and countless subplanetary bodies, we may feel significantly less special and have to adjust our notion of humanity's special role in the universe. People invest a lot in their belief systems, and those who go against the status quo and question the prevailing ideology may be severely punished, as was Galileo.

More recently, the 2016 presidential election had the potential to shatter the ideology of democracy—that is, the belief that the candidate who receives the most votes ascends to power. The winner of the popular vote, Hillary Clinton, was in fact defeated by Donald Trump, who accumulated the requisite number of Electoral College votes. This was the second time in the last five elections that the winner lost the popular vote, challenging the idea that the United States is a true democracy. That faith was further tested by the proliferation of fake news during the election, Russian meddling through social media, and leaks of stolen e-mails. At the time of this writing, President Trump and Congress both were at or near record low approval ratings; many norms of how we have practiced government have shattered (such as the need for 60 votes to end a filibustered nominee to the Supreme Court or the tradition that presidential candidates release their taxes to the public before the election); and yet the institutions of government and the democratic ideology behind them have soldiered on without collapsing.

Of course, on occasion ideologies do shatter. The fall of the former Soviet Union, for example, marked not just a transition in government but the shattering of a particular brand of Communist ideology. Similarly, when apartheid was abolished in South Africa, more than just a few laws changed; a total reorganization of ideas, beliefs, and social relations followed. Often, ideological change comes more slowly. The fight for women's rights, including equal pay, is ongoing even today, but women won the right to vote way back in 1920.

STUDYING CULTURE

In the United States, the scholarly study of culture began in the field of anthropology. Franz Boas founded the first PhD program in anthropology at

CULTURAL RELATIVISM

taking into account the differences across cultures without passing judgment or assigning value.

Columbia University in the early 1930s and developed the concept of cultural relativity. Ruth Benedict, following Boas, her teacher and mentor, coined the term *cultural relativism* in her book *Patterns of Culture* (1934). Cultural relativism means taking into account the differences across cultures without passing judgment or assigning value. For example, in the United States you are expected to look someone in the eye when you talk to him or her, but in China this is considered rude, and you generally divert your gaze as a sign of respect. Neither practice is inherently right or wrong. By employing the concept of cultural relativism, we can understand difference for the sake of increasing our knowledge about the world. Cultural relativism is also important for businesses that operate on a global scale.

But what should one's position be when local traditions conflict with universally recognized human rights? For example, should Western businesspeople condone the cutting of the clitoris in young girls as a local cultural practice, something to be respected, as they go about their business in parts of Africa? There are, of course, limits to cultural relativism. In some countries, it is both legal and socially acceptable for a man to beat his wife. Should we accept that wife beating is part of the local culture and therefore conclude that we are not in a position to judge those involved? In the United States, some Jehovah's Witnesses reject blood transfusions because they believe blood is sacred and not for "consumption" by Christians. If parents refuse a potentially lifesaving surgery for their child because it will require a transfusion, do we respect their right to religious freedom or arrest them for neglect? Where we draw the lines is a difficult matter to decide and sparks a great deal of political debate on topics such as domestic violence, female genital mutilation, and medical practices versus religious beliefs in treating the critically ill.

CULTURAL SCRIPTS

modes of behavior and understanding that are not universal or natural.

Margaret Mead with two Samoan women, 1926.

Margaret Mead, Benedict's student, further developed ideas about cultural relativity when she wrote *Coming of Age in Samoa* (1928), which has become part of the canon of anthropology and cultural studies. Based on her ethnographic fieldwork among a small group of Samoans, she concluded that women there did not experience the same emotional and psychological turmoil as their American counterparts in the transition from adolescence to adulthood. She found that young women engaged in and enjoyed casual sex before they married and reared children. The book, published in 1928, caused an uproar in the United States and eventually contributed to the feminist movement. The validity of Mead's findings has been disputed, but her work continues to be a landmark of early anthropology for introducing the idea that cultural scripts, modes of behavior and understanding that are not universal or natural, shape our notions of gender. This concept stands in opposition to the belief that such ideas derive from biological programming.

Cockfighting and Symbolic Culture Clifford Geertz, another American anthropologist, was well known for his studies of and writings

A cockfight in Bali, Indonesia. How are roosters central to Bali's symbolic culture?

on culture, one of which concerned the meaning of cockfighting in Bali. Cockfighting involves placing two roosters together in a cockpit, a ring especially designed for the event, and watching them fight. The meaning of cockfighting varies, however. Some see it as one of the basest forms of cruelty to animals. Others attach religious and spiritual meaning to the event. In the Balinese village where Geertz lived, cockfighting was primarily a vehicle for gambling, but it was also an important cultural event. Very few men are allowed to referee these fights, and their decisions are treated with more regard than the law. People bet a lot of money, often forming teams that pool their resources. Bettors profess themselves to be "cock crazy."

The owners of the prized cocks expend an enormous amount of time caring for them. They feed them special diets, bathe them with herbs and flowers, and insert hot peppers in their anuses to give them "spirit." The cock takes on larger symbolic meaning within Balinese society. According to Geertz (1973),

> The language of everyday moralism is shot through, on the male side of it, with roosterish imagery. *Sabung*, the word for cock, is used metaphorically to mean "hero," "warrior," "champion," "man of parts," "political candidate," "bachelor," "dandy," "lady-killer," or "tough guy."...Court trials, wars, political contests, inheritance disputes, and street arguments are all compared to cockfights. Even the very island itself is perceived from its shape as a small, proud cock, poised, neck extended, back taut, tail raised, in eternal challenge to large, feckless, shapeless Java. (pp. 412, 454)

In the United States, cocks do not have much symbolic meaning. A more central metaphor in American society is baseball. According to anthropologist Bradd Shore (1998), baseball's function in the United States is similar to cockfighting's function in Bali. We call baseball America's favorite pastime and regularly use baseball metaphors in our daily conversations. If your parents inquire about how you're doing in school this semester, you might say you're "batting a thousand." If you ask your friend how it went the other night at a party when she spoke to the smart guy from your sociology class, and she says she "struck out," you know not to ask, "So when's your first date?" We could learn a lot about American culture by studying baseball, how people watch the game, and what symbolic meanings they attach to it, just as Geertz learned such things about the Balinese by using cockfights as the center of his analysis.

In *The Interpretation of Cultures* (1973), perhaps his most famous book, Geertz wrote, "Culture is a system of inherited conceptions expressed in symbolic forms by means of which people communicate, perpetuate, and develop their knowledge about and attitudes toward life" (p. 89). He was trying to get away from a monolithic definition of culture. So for some, culture is watching players hit a small, hard ball into a field and run around a diamond; for others, it's squatting down in the dust beside a ring and watching two roosters brawl. One pastime isn't inherently better than the other. They're both interesting in their own right, and by understanding the significance of these events for the local people, we can better understand their lives.

SUBCULTURE

SUBCULTURE
the distinct cultural values and behavioral patterns of a particular group in society; a group united by sets of concepts, values, symbols, and shared meaning specific to the members of that group distinctive enough to distinguish it from others within the same culture or society.

Like culture, subculture as a concept can be a moving target: It's hard to lock into one specific definition of the term. Historically, subcultures have been defined as groups united by sets of concepts, values, symbols, and shared meaning specific to the members of that group. Accordingly, they frequently are seen as vulgar or deviant and are often marginalized. Part of the original impetus behind subculture studies was to gain a deeper understanding of individuals and groups who traditionally have been dismissed as weirdos at best and deviants at worst.

For example, many music genres have affiliated subcultures: hip-hop, hardcore, punk, Christian rock. High-school cliques may verge on subcultures—the jocks, the band kids, the geeks—although these groups don't really go against the dominant society, because athleticism, musical talent, and intelligence are fairly conventional values. But what about the group of kids who dress in black and wear heavy eyeliner? Maybe teachers simply see them as moody teenagers with a penchant for dark fashion and extreme makeup just seeking to annoy the adults in their life, but perhaps their style of self-presentation means more to them.

Goths in Germany (left) and Japan. What characteristics of goth culture make it a subculture?

Goth culture has its roots in the United Kingdom of the 1980s. It emerged as an offshoot of post-punk music. Typified by a distinctive style of dress—namely, black clothing with a Victorian flair—and a general affinity for gothic and death rock, goth culture has evolved over the last three decades, with many internal subdivisions. Some goths are more drawn to magical or religious aspects of the subculture, whereas others focus mainly on the music. Even the term *goth* has different meanings to people within the subculture: Some see it as derogatory; some appropriate it for their own personal meaning. An internal struggle has grown over who has the right to claim and define the label.

What makes today's goths a subculture? They are not just a random group of people in black listening to music (classical musicians usually wear black when they perform, but we don't consider them gothic). Certain words and phrases are unique to goth communities, such as *baby bat* (young goth poseur) and *weekend goth* (someone who dresses up and enters the subculture only on the weekends). Goths in Germany may look very different from those in the United Kingdom, and norms even differ among US cities, so each is a distinct branch of the subculture. Yet as a whole, goths do have their own shared symbols, especially with regard to fashion, so they are visible as a subculture. Not all subcultures, however, adopt characteristic dress or other easily identifiable features. For example, black men on the down low (secretly seeking sex with men) by definition do not want to be identifiable.

CULTURAL EFFECTS: GIVE AND TAKE

How does culture affect us? As we've discussed, culture may be embedded in ideologies. Our understanding of the world is based on the various ideologies we embrace, and ideologies tend to be culturally specific. Culture also affects us by shaping our values, our moral beliefs. The concept of equal opportunity is a good example of this. The majority of us have been taught that everybody should have an equal shot at the "American dream": going to college, obtaining a job, and becoming economically self-sufficient. This

VALUES

moral beliefs.

is a relatively recent cultural conception. In England 600 years ago, there was no such culture of equal opportunity but rather a feudal system. If you had asked someone in the government of that period, a social elite, if he (and it would definitely have been a man) believed that everyone should have equal opportunity, he would probably have rejected such a claim. He would likely have insisted that the elite, the nobles, should have more rights, privileges, and opportunities than everybody else. Class mobility simply wasn't a concept that existed. Similarly, the way that the concept of equal opportunity is expressed in the contemporary United States has a particularly American flavor. We have a very individualistic culture, meaning that we hold dear the idea that everyone should have the opportunity to advance, but we believe that people should do it on their own—"pull yourself up by your bootstraps," as we say. Americans hold tightly to the rags-to-riches dream of triumph over adversity, of coming from nothing and becoming a success despite hardship. The problem with this cultural trope is that, as sociologists like to point out, the larger, structural, macro-level forces—general social stratification, racial segregation, sexism, differential access to health care, and education—keep the concept of equal opportunity more fiction than reality. As a culture, Americans tend to suffer a bit from historical amnesia. Slavery was abolished in 1863, and legal segregation lasted until the 1960s. Historically speaking, this is not very long ago. Our notion of equal opportunity often fails to take into account the very unequal starting positions from which people set out to achieve their goals.

If values are abstract cultural beliefs, norms are how values are put into play. We value hygiene in our society, so it is a norm that you wash your hands after going to the bathroom. When you are little, your parents and teachers must remind you, because chances are you haven't yet fully internalized this norm. Once you're an adult, however, you are in charge of your own actions, yet others may still remind you of this norm by giving you a dirty look if you walk from a stall straight past the sinks and out the door of a public restroom. In an office, people might gossip about the guy who doesn't wash his hands—they are shaming him for not following this norm. When you arrive at college, you enter a new culture with different norms and values, and you must adjust to that new environment. If you're attending a school with a major emphasis on partying, you might be reading this textbook secretly because the cultural norm in that kind of environment is not to buy the assigned books or even go to class on a regular basis. If you are enrolled at a community college, commuting may be the main cultural practice. You go there for class and then leave shortly thereafter.

REFLECTION THEORY

Culture affects us. It's transmitted to us through different processes, with socialization—our internalization of society's values, beliefs, and norms—

NORMS

how values tell us to behave.

SOCIALIZATION

the process by which individuals internalize the values, beliefs, and norms of a given society and learn to function as members of that society.

being the main one. But how do we affect culture? Let's start with reflection theory, which states that culture is a projection of social structures and relationships into the public sphere, a screen onto which the film of the underlying reality or social structures of our society is shown. For example, some people claim that there is too much violence in song lyrics, particularly in rap music. How do hip-hop artists often respond? "I live in a violent world, and I'm like a reporter. I'm telling it like it is; so if you want to fix that, then fix the problems of violence in my community. I'm just the messenger." They are invoking reflection theory.

A different version of reflection theory derives from the Marxist tradition, which says that cultural objects reflect the material labor and relationships of production that went into them. Earlier in this chapter, we discussed the distinction between material and nonmaterial culture. Karl Marx asserted that it is a one-way street—from technology and the means of production to belief systems and ideologies. According to Marx's view of reflection theory, our norms, values, sanctions, ideologies, laws, and even language are outgrowths of the technology and economic means and modes of production. Likewise, for Marx, ideology has a very specific definition: culture that justifies given relations in production.

By way of example, consider the creation of limited liability partnerships. The concept of limited liability emerged in the nineteenth century during the Industrial Revolution, when new technologies enabled the growth of factories and long-distance travel. At the same time, European countries such as Great Britain were colonizing regions all over the world and establishing large global trade networks. As merchants and factory owners tried to expand, they needed more capital from investors. To attract the most money, they came up with the idea of limited liability partnerships. Limited liability means that when you invest money in a publicly traded corporation, you are not responsible for its debt (you can't lose more than what you paid for your shares) or its actions (unless you are on the board of directors). So if you have stock in a cereal company and it goes bankrupt, the farmers who supply the grain can't hit you up for unpaid bills. Likewise, if several small children choke on the prize included in their boxes of cereal, you cannot be held personally responsible for this tragedy. All you lose is the money you invested in the shares. Even if you didn't know what limited liability meant before you read this paragraph, the legal concept is something we take for granted as part of our common understanding of how capitalism works. Historically, however, it arose from a choice made in England in a specific context. Marx would argue that the combination of factory labor and global trade relations between England and its colonies necessitated and inevitably led to these kinds of legal structures.

Like most theories, reflection theory has its limitations. It does not explain why some cultural products have staying power, whereas others fall by the wayside. Why is it that *The Cuckoo's Calling* (2013) by Robert Galbraith

REFLECTION THEORY

the idea that culture is a projection of social structures and relationships into the public sphere, a screen onto which the film of the underlying reality or social structures of a society is projected.

How has the cultural significance of Shakespeare's plays changed over the last 400 years? Compare the poster for an 1884 performance of Macbeth (left) with the poster for the 2015 movie starring Michael Fassbender and Marion Cotillard (right).

had terrible sales until it was leaked that the actual author was J. K. Rowling, who wrote the Harry Potter series? Conversely, Wolfgang Amadeus Mozart was a popular composer in his own day. Why is he still so popular today? Clearly, the relations of production and underlying social structures—indeed, Western society as a whole—have drastically changed since the late eighteenth century, yet the appeal of Mozart's music is as strong as ever. If culture is just a reflection of the state of society in a given epoch, then Mozart should have fallen completely out of favor and we would no longer be interested in him or his music, but this doesn't seem to be the case.

Likewise, if reflection theory is true, why do some products change their meaning over time? Shakespeare is a good example. In nineteenth-century America, Shakespeare was the poet of the people, the playwright for the common man. Scenes from various Shakespearean plays were performed during the intermissions of other events. Conversely, during the intermissions for a full-length Shakespearean play, carnival-like entertainment for the masses called spectacles would be presented. Shakespeare was the most widely performed playwright in England's former colonies, but not revered in terms of high culture. This is not quite what we associate Shakespeare with today, correct? It's just the opposite, in fact: His work is now considered high art. In this instance, the same product changed its meaning over time, and reflection theory doesn't help us understand that change.

Most important, reflection theory has been rejected largely because it is unidirectional—that is, it basically buys the rappers' defense that culture has no impact on society. Do we really believe that the media have no impact on the way we live or think? Do we really believe that ideologies have no effect on the choices we make? No. Most people now understand that an interactive process exists between culture and social structure. Most would agree that culture has an impact on society and it is not just a unidirectional phenomenon.

Media

Among the most pervasive and visible forms of culture in modern societies are those produced by the mass media. We might define media as any formats or vehicles that carry, present, or communicate information. This definition would, of course, include newspapers, periodicals, magazines, books, pamphlets, and posters. But it would also include wax tablets, sky writing, web pages, and the children's game of telephone. We'll first discuss the history of the media and then tackle theory and empirical studies.

FROM THE TOWN CRIER TO THE FACEBOOK WALL: A BRIEF HISTORY

When we talk about the media, we're generally talking about the mass media. The first form of mass media was the book. Before the invention of the printing press, the media did exist—the town crier brought news, and royal messengers traveled by horseback, every now and then hopping off to read a scroll—but they did not exactly reach the masses. People passed along most information by word of mouth. After the 1440s, when Johannes Gutenberg developed movable type for the printing press, text could be printed much more easily. Books and periodicals were produced and circulated at much greater rates and began to reach mass audiences. Since that time the terms *media* and *mass media* have become virtually synonymous.

Innovations in mass media include the invention of the printing press and movable type in the fifteenth century, the creation of moving pictures at the turn of the twentieth century, and the adoption of the scrolling ticker by today's 24-hour news channels.

The innovations didn't stop there, however. In the 1880s along came another invention: the moving picture or silent film. Its quality was not the best at first, but it improved over time. In the 1920s sound was added to films. For many, this represented an improvement over the radio, which had come along about the same time as the silent film. Television was invented in the 1930s, although this technology didn't make its way into most American homes until after World War II. During the postwar period, new forms of media technology quickly hit the market, and the demand for media exploded: glossy magazines; color televisions; blockbuster movies; Betamax videos, then VHS videos, then DVDs; vinyl records, then 8-track tapes, then cassette tapes, then CDs—and once the Internet came along, the sky was the limit! In 2016, 73 percent of the American population had broadband internet at home, with access clustered among the younger, wealthier, and better educated, while more than 10 percent used their smartphones in place of high-speed internet service (Pew Research Center, 2017a).

You're well aware of all the forms the media come in, but let's stop for a minute and contemplate the impact certain forms, such as television, have had on society. Again, televisions didn't become household items in the United States until after World War II. From 1950, the year President Harry S. Truman first sent military advisers to South Vietnam, to 1964, when Congress approved the Gulf of Tonkin Resolution calling for victory by any means necessary, the share of American households with television sets increased from 9 to 92 percent. During the Vietnam War, the American public witnessed military conflict in a way they never had before, and these

In 1963, for the first time, televisions beamed images such as this photo of police officers attacking a student in Birmingham, Alabama. How did television influence the reaction to events such as the civil rights movement?

images helped fuel the antiwar movement. Likewise, television played a large role in the civil rights movement of the 1950s and 1960s. It was one thing to hear of discrimination secondhand, but quite another to sit with your family in the living room and watch images of police setting attack dogs on peaceful protesters and turning fire hoses on little African American girls dressed up in their Sunday best.

HEGEMONY: THE MOTHER OF ALL MEDIA TERMS

All of us might be willing to agree that, on some level, the media both reflect culture and work to produce the very culture they represent. How does this dynamic work? Antonio Gramsci, an Italian political theorist and activist, came up with the concept of hegemony to describe just that. Gramsci, a Marxist, was imprisoned by the Fascists in the 1920s and 1930s; while in jail, he attempted to explain why the working-class revolution Marx had predicted never came to pass. He published his findings in his "prison notebooks" of 1929–35 (Gramsci, 1971). In this vein, then, hegemony "refers to a historical process in which a dominant group exercises 'moral and intellectual leadership' throughout society by winning the voluntary 'consent' of popular masses" (Kim, 2001). This concept of hegemony stands in contrast to another of Gramsci's ideas, *domination*. If domination means getting people to do what you want through the use of force, hegemony means getting them to go along with the status quo because it seems like the best course or the natural order of things. Although domination generally involves an action by the state (such as the Fascist leaders who imprisoned those who disagreed with them), "hegemony takes place in the realm of private institutions...such as families, churches, trade unions, and the media" (Kim, 2001). For example, if free-market capitalism is the hegemonic economic ideology of a given society, then the state does not have to explicitly work to inculcate that set of principles into its citizenry. Rather, private institutions, such as families, do most of the heavy lifting in this regard. Ever wonder why children receive an allowance for taking out the trash and doing other household chores? Gramsci might argue that this is the way capitalist free-market ideology is instilled in the individual within the private realm of the family.

The concept of hegemony is important for understanding the impact of the media. It also raises questions about the tension between structure and agency. Are people molded by the culture in which they live, or do they actively participate in shaping the world around them? As we discussed earlier in the chapter, it's not an either/or question. Later in the chapter we'll talk about some of the debates between structure and agency in regard to the media, look at some examples of hegemony in practice, and discuss the possibilities for countercultural resistance.

HEGEMONY

a condition by which a dominant group uses its power to elicit the voluntary "consent" of the masses

The Media Life Cycle

We live in a media-saturated society, but one of the most exciting aspects of studying the media is that it allows us to explore the tensions and contradictions created when large social forces conflict with individual identity and free will. We see how people create media, how the media shape the culture in which people live, how the media reflect the culture in which they exist, and how individuals and groups use the media as their own means to shape, redefine, and change culture.

TEXTS

Why do the adventures in a fairy tale often begin once a mother dies? Are blacks more often portrayed as professionals or criminals in television sitcoms? How often are Asians the lead characters in mainstream films? Who generally initiates conversation, men or women, in US (or Mexican) soap operas? These questions are all examples of textual analysis, analysis of the content of media in its various forms, one of the important strands of study to materialize in the wake of Gramsci's work.

During the 1960s and 1970s, academic studies focused almost solely on texts—television talk shows, newspapers, and magazine pages. Finally, scholars recognized the importance of finding out how people read and interpret, and are affected by, these texts: Audience studies were born. The field of psychology has expended a lot of time investigating claims about the effects of television on children, and the debate continues. Sociologists have explored the way women read romance novels or how teenage girls interpret images of super-thin models in magazines. For example, in *Reading the Romance: Women, Patriarchy, and Popular Literature* (1987), Janice Radway argues that women exhibit a great deal of individual agency when reading romance novels, which help them cope with their daily lives in a patriarchal society by providing both escapes from the drudgery of everyday life and alternative scripts. We are not just passive receptors of media; as readers or viewers, we experience texts through the lens of our own critical, interpretive, and analytical processes.

BACK TO THE BEGINNING: CULTURAL PRODUCTION

The media don't just spontaneously spring into being. They aren't organic; they're produced. You may have heard the expression "History is written by the winners." Well before something becomes history, it has to happen in the present. Who decides what's news? How are decisions made about the content of television shows? To write his classic *Deciding What's News* (1979a), Herbert Gans went inside the newsrooms at *CBS Evening News, NBC Nightly*

News, Newsweek, and *Time* in the late 1970s. He paid careful attention to the processes by which these news outlets made their decisions on editorial content, "writing down the unwritten rules of journalism," because rules, sociologists know, contain values. Journalists are supposed to be objective, but Gans illustrated the ways in which the mainstream American values journalists had internalized biased the finished product—the news. The notion of "the facts, just the facts, and nothing but the facts" is a worthy idea but to a large degree a farce. Powerful boards of directors regulate the various media; writers, casting agents, directors, and producers decide what goes into sitcoms, soap operas, and after-school specials. Even the huge cache of secret government documents made available on the internet by WikiLeaks in 2010 (and in other dumps since then) was prescreened by individual hackers and researchers. The media are produced by human beings often working for organizations within which they can be influenced by market pressures to publish "fresh" stories rather than follow situations unfolding slowly over time (Usher, 2014).

Increasingly, however, humans are only indirectly "deciding" what's news by designing the artificial intelligence algorithms that figure out which stories to put before our eyes (based on our past browsing behavior). These algorithms were gamed by the campaigns (and Russian operatives) in the 2016 election, causing Facebook to alter its own software to try to better protect users against "fake news." Computers are even writing the news sometimes: Software is able to write simple stories like the recap of a sporting event, based on data from the game itself, and the bots are getting better and better with each passing day. Online activist and entrepreneur Eli Pariser (CEO of Upworthy) and others have become worried about what he calls the "filter bubble" or online echo chamber (Pariser, 2011). Since what we are shown in our newsfeed depends on what we clicked through before (or what our friends liked), we risk never seeing new information that conflicts with our preexisting views—not a healthy algorithm for a robust, democratic society.

Media Effects

In considering the media, mass culture, and subcultures, we can plot the media's effects in a two-dimensional diagram, as shown in Figure 3.1. The vertical dimension indicates whether the effect is intended (i.e., deliberate) or unintended. The horizontal axis depicts whether it is a short- or long-term effect. A short-term, deliberate media effect (section A in the illustration) would be advertising. As a kid, you may have watched Saturday morning cartoons; watch them now and keep track of the number of advertisements for children's food. A child today might see an ad for Cocoa Puffs

FIGURE 3.1 Media Effects

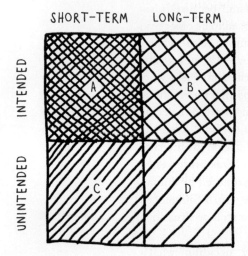

and that same afternoon go grocery shopping with his or her parents. That child is possibly on a sugar crash from an early-morning bowl of Super Frosted Mega Marshmallow Crackle Blasts and is really clamoring for a box of Cocoa Puffs. The advertisers timed it just right, with the pressure on Mom or Dad to buy that cereal.

Section B of Figure 3.1, on the other hand, represents a deliberate, long-term media campaign. Here, a single theme is reinforced through repeated exposure, as in public service announcements or when *Sesame Street* repeats counting and spelling to help educate children. *Sesame Street* was created with the intent of providing educational programming for low-income children who didn't have the same opportunities for day care and preschool as their wealthier peers and it works! This type of campaign is generally used by not-for-profit organizations to educate the public. Other examples include Smokey the Bear (created in 1944, "Only you can prevent forest fires"), Woodsy Owl (created in 1970, "Give a hoot, don't pollute"), and the "This is your brain on drugs" commercials from the Partnership for a Drug-Free America (the original fried egg aired in 1987, and the sequel frying-pan smash aired a decade later). In a new public service announcement from the Ad Council—the nonprofit behind ad campaigns like Smokey the Bear—the advertisement reads "You don't want them responding to your text" next to an ambulance (Ad Council, 2018). Other recent campaigns focus on sexual harassment, suicide prevention, and health problems like type 2 diabetes.

Of course, not all attempts at deliberate, long-term effects are a success. Often, an issue must be framed as a problem before solutions can be advocated, and that can take a long time in a media campaign. For instance, various environmental groups have advocated for taking measurable action to reduce the release of greenhouse gases. However, only in 2013 did the Environmental Protection Agency feel it had enough support in Congress to propose greenhouse gas emission limits for power plants.

Section C of Figure 3.1 represents media with short-term, unintended consequences. An example might be when teenagers play violent video games and then go out and commit crimes almost identical to those portrayed in the game or when a kid listens to heavy metal music with violent lyrics and then commits a school shooting. You hear of such events every so often, and sometimes the media's creator will use the defense that the short-term response was not intended. In an interview, for example, the software producer or musician might be asked, "Did you know that your music is causing teenage boys to commit violent crimes?" And the response will be, "That is not my intention at all. I use violence as a metaphor." Scientific research hasn't yet ruled definitively one way or the other on this controversial subject, but many believe that the media occasionally have short-term, unintended effects.

Finally, section D of the illustration represents the long-term, unintended effects of the media. Many people, not just cultural conservatives, argue that we have been desensitized to violence, sexual imagery, and other content that some people consider inappropriate for mass audiences. In the film industry, for example, the Production Code, also known as the Hays Code, was a set of standards created in 1930 (although it wasn't officially enforced until 1934) to protect the moral fabric of society. The guidelines were fairly strict, and they were a testament to the mainstream ideologies of the time (see the box on pages 106–7). Slowly, however, the power of the code began to erode because of the influence of television and foreign films and the fact that being condemned as immoral didn't prevent a film from becoming a success. In 1967 the code was abandoned for the movie rating system. Over time, we have grown accustomed to seeing sexually explicit material in films, on television, and on the internet. Those who lament this desensitization seek to reinstitute controls over media content.

Mommy, Where Do Stereotypes Come From?

On December 22, 1941, two weeks after the Japanese attack on Pearl Harbor, *Time* magazine ran an article with the headline "How to Tell Your Friends from the Japs." There were annotated photographs to help readers identify characteristics that would distinguish, for instance, friendly Chinese from the Japanese, America's enemies during World War II. The magazine offered the following rules of thumb, although it admitted that they were "not always reliable":

- Some Chinese are tall (average: 5 ft. 5 in.). Virtually all Japanese are short (average: 5 ft. 2-½ in.).

- Japanese are likely to be stockier and broader-hipped than short Chinese.

- Japanese—except for wrestlers—are seldom fat; they often dry up and grow lean as they age. The Chinese often put on weight,

THE RACE AND GENDER POLITICS OF MAKING OUT

"When I'm good, I'm very good, but when I'm bad, I'm better." — ACTRESS MAE WEST

The movie industry's Production Code (1930) enumerated three "general principles":

1. No picture shall be produced that will lower the moral standards of those who see it. Hence the sympathy of the audience should never be thrown to the side of crime, wrongdoing, evil, or sin.
2. Correct standards of life, subject only to the requirements of drama and entertainment, shall be presented.
3. Law, natural or human, shall not be ridiculed, nor shall sympathy be created for its violation.

Specific restrictions were spelled out as "particular applications" of these principles:

- Nudity and suggestive dances were prohibited.

- The ridicule of religion was forbidden, and ministers of religion were not to be represented as comic characters or villains.
- The depiction of illegal drug use was forbidden, as well as the use of liquor, "when not required by the plot or for proper characterization."
- Methods of crime (e.g., safecracking, arson, smuggling) were not to be explicitly presented.
- References to "sex perversion" (such as homosexuality) and venereal disease were forbidden, as were depictions of childbirth.
- The language section banned various words and phrases considered to be offensive.
- Murder scenes had to be filmed in a way that would not inspire imitation in real life, and brutal killings could not be shown in detail. "Revenge in modern times" was not to be justified.
- The sanctity of marriage and the home had to be upheld. "Pictures shall not infer that low forms of sex relationship are the accepted or common thing." Adultery and illicit sex, although recognized as sometimes necessary to the plot, could not be explicit or justified; they were never to be presented as an attractive option.
- Portrayals of interracial relationships were forbidden.

Lucille Ball and Desi Arnaz in the 1950s hit television comedy *I Love Lucy*. Even though their characters were married, they still did not share a bed.

- "Scenes of passion" were not to be introduced when not essential to the plot. "Excessive and lustful kissing" was to be avoided, along with any other physical interaction that might "stimulate the lower and baser element."
- The flag of the United States was to be treated respectfully, as were the people and history of other nations.
- "Vulgarity," defined as "low, disgusting, unpleasant, though not necessarily evil, subjects," must be treated "subject to the dictates of good taste." Capital punishment, "third-degree methods," cruelty to children and animals, prostitution, and surgical operations were to be depicted with similar sensitivity and discretion.

"Rules were made to be broken," the old saying goes, and the film industry did its best to prove the maxim true. Filmmakers like Alfred Hitchcock pushed the boundaries of a ten-second time limit for kisses by filming a lip-lock for the maximum allotted time, panning away, and then returning to the couple still passionately embracing. Rules such as this were slow to change, but not as slow as others. For one thing, you can be sure that the people kissing were a man and a woman, and they were both white. In fact, in the beginning they would have been married, too, but that ideology—the sanctity of marriage—fell away more quickly on the silver screen than did notions about racial and gender hierarchies and stereotypes. The first on-screen interracial kiss, between Sidney Poitier and Katharine Houghton in *Guess Who's Coming to Dinner*, didn't occur until 1967, the same year the US Supreme Court ruled that state laws preventing interracial marriage were unconstitutional. And in a 1968 episode of *Star Trek*, William Shatner and Nichelle Nichols boldly ventured where no one had gone before with the first black–white lip-lock televised in the United States. The first homosexual kiss between two women hit prime time in 1991 (on the TV show *L.A. Law*, which received backlash from advertisers), while the first same-sex kiss between two men did not happen until 2001 on *Will and Grace*. Change in the media takes time; as society's ideologies about what constitutes good and bad love change, the media will reflect those changes by portraying more positive images of people loving whomever they choose.

Tituss Burgess and Mike Carlsen portray a gay couple on Netflix's *Unbreakable Kimmy Schmidt*.

particularly if they are prosperous (in China, with its frequent famines, being fat is esteemed as a sign of being a solid citizen).

- Chinese, not as hairy as Japanese, seldom grow an impressive mustache.

- Most Chinese avoid horn-rimmed spectacles.

- Although both have the typical epicanthic fold of the upper eyelid (which makes them look almond-eyed), Japanese eyes are usually set closer together.

- Those who know them best often rely on facial expression to tell them apart: the Chinese expression is likely to be more placid, kindly, open; the Japanese more positive, dogmatic, arrogant.

How can Gramsci's concept of hegemony help us understand this piece from *Time*? Does this article tell us more about the physical differences between Japanese and Chinese, or about the state of mind of the American public at the time? What values are reflected and projected? Are the descriptors empirical (based on fact) or normative (based on opinion)? This is what racism looks like on paper, and it clearly illustrates America's fear and hatred of the Japanese at that time.

RACISM IN THE MEDIA

The media continue to reflect and perpetuate racist ideologies, even if such examples are not usually as blatant as the 1941 article in *Time*. Sometimes the racism is obvious, and these instances present us with the opportunity to discuss racism in the media. In June 1994 the front cover of both *Time* and *Newsweek* showed the police mug shot of former NFL star O. J. Simpson, who had been arrested for allegedly killing his ex-wife and her friend. *Time* was accused of darkening its image, perhaps implying that those with darker skin are more dangerous, criminal, and evil. In a case whose trial months later would prove to be a racially charged event—Simpson's ex-wife was white, and the police detective overseeing the case had a long history of racial prejudice—the media added fuel to the fire. What purpose was served by darkening Simpson's image? Did the stir surrounding these photos affect the eventual outcome of the trial? Again, we can't state with certainty how one questionable instance in the media affects society, but we can look at the continued negative portrayals of minorities and hypothesize about the cumulative effects of those images.

In early September 2005, just days after Hurricane Katrina had devastated the areas surrounding the Mississippi River basin, two photos quickly began to circulate on the internet amid discussion of racism and the role it played in the reaction (or inadequate government response) to the catastrophe. The first photo, published by the Associated Press, showed a young

The controversial O. J. Simpson arrest photo. *Time* magazine was accused of darkening his features for its cover.

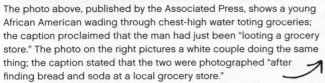

The photo above, published by the Associated Press, shows a young African American wading through chest-high water toting groceries; the caption proclaimed that the man had just been "looting a grocery store." The photo on the right pictures a white couple doing the same thing; the caption stated that the two were photographed "after finding bread and soda at a local grocery store."

African American wading through chest-high water toting groceries; the caption proclaimed that the man had just been "looting a grocery store." The second pictured a white couple doing the same thing; the caption stated that the two were photographed "after finding bread and soda at a local grocery store" (Ralli, 2005). The conclusion these images and their captions conveyed was this: White looters, in their struggle to survive the catastrophe, were not committing a crime, whereas blacks resorting to the same behavior were. Indeed, much of the coverage in the wake of Katrina focused on the looting, vandalism, and other criminal acts that took place. What critics have pointed out, however, is that these people did what most of us would logically do—try to obtain food and water for our suffering families in the absence of competent government assistance and disaster relief. Because the city of New Orleans, which received the majority of the media attention, was 78 percent nonwhite, the victims were frequently portrayed as criminals.

The coverage of Hurricane Katrina, which was exceptional in the amount of criticism that was publicly levied against such racially charged portrayals (unlike, for example, the December 1941 *Time* piece), supports the main thesis of Barry Glassner's *The Culture of Fear: Why Americans Are Afraid of the Wrong Things* (1999). Glassner asserts that, as a culture, we grossly exaggerate the frequency of rarely occurring events, often through amplification of a single instance through media repetition. We tend to divert or redirect our attention from political, economic, and cultural issues that are either taboo or simply too difficult to talk about toward sensational, but rare, events like school shootings and terrorist attacks. The media are the main vehicles through which this process occurs. In the case of Katrina, we blame the victims, the poorest of the poor, for not leaving the city, rather than ask how the government could leave its own citizens stranded like refugees without access to life's basic necessities.

SEXISM IN THE MEDIA

A frequent critique of the media centers on the representation of women. American media in particular, and Western media more generally, are charged with glamorizing and perpetuating unrealistic ideals of feminine beauty. Some argue that repetitive bombardment by these images decreases girls' self-esteem and contributes to eating disorders. Women's magazines have been heavily criticized, although some researchers (such as Angela McRobbie from the United Kingdom) have taken care to show that women who are active, critical readers still enjoy reading women's magazines. However, as the Canadian sociologist Dawn Currie (1999) points out, although girls can choose which magazines, if any, to read and how to critically read them, they can't control the images available to them in those and other texts.

Another focus of feminist media critiques has been images of violence against women. Jean Kilbourne has become one of the most popular lecturers at college and university campuses across America. In 1979 she released a film titled *Killing Us Softly: Advertising's Image of Women*, in which she examines the ways in which women are maimed, sliced, raped, and otherwise deformed in advertising images. One classic example is a photo that shows the image of a woman's body in a garbage can, with only her legs and a fantastic pair of high heels on her feet visible. The message is clear: These shoes are, literally, to die for. Kilbourne's point is clear, too: Such images help sustain a kind of symbolic violence against women. In this critique, advertising does not just reflect the underlying culture that produced it but also creates desires and narratives that enter women's (and men's) lives with causal force.

Of course, there's always room for innovation. Some girls (with the help of their parents) have responded by creating their own magazines that focus on topics other than makeup, clothing, and boys, as mainstream teenage magazines do. For example, *New Moon Girls* is written and edited by girls aged 8 to 13 and contains no advertisements. Likewise, magazines exist for adult women that have more pro-woman messages; *Ms.* magazine was founded in 1971 during the feminist movement to give voice to women and explore women's issues. Because such magazines don't accept advertising from huge makeup companies and designer fashion houses, however, they are often less economically viable than mainstream women's magazines, which carry ads on as many as 50 percent of their pages. *Bitch* magazine ("It's a noun; it's a verb; it's a magazine"), which has been around since 1996, is a self-declared feminist response to pop culture. It is supported by advertisers but is a not-for-profit publication.

Some advertisers have responded to feminist critiques of the media with new approaches. In 2005 Dove, a manufacturer of skin-care products, launched a new series of ads backed by a social awareness program called the Campaign for Real Beauty. Instead of models, the ads featured "real" women complete with freckles, frizzy hair, wrinkles, and cellulite. The images were intentionally meant to offer a contrast to the images we're accustomed to seeing. And, as Dove's advertisers have said, "firming the thighs of a size 2 supermodel is no challenge" (Triester, 2005). The latest version of Dove's Real Beauty campaign shows side-by-side sketches of the same woman as drawn by a sketch artist. The first image is drawn based on the woman's description of herself and is always less attractive than the second image, which is drawn based on someone else's description of the woman, revealing women's inner negative body images. But can calling attention to the way women describe themselves change this inner dialogue? Or will it simply point out another unattractive flaw—low self-esteem?

A 2005 billboard from Dove's Real Beauty advertising campaign, featuring women who are not professional models.

On the one hand, a pessimist might point out that these "real beauty" ads use nontraditional models merely as a way to be novel; typical models are all so uniformly perfect, they've become boring. Also, Dove isn't telling women that they're beautiful whether or not they have dimpled thighs and therefore they don't need firming lotion. Rather, the manufacturer's message is that it's okay to need firming lotion because you're not the only one who does. Furthermore, the women in the campaign are never overweight, yet 70.7 percent of adult Americans are overweight or obese (Centers for Disease Control and Prevention, 2017). An optimist, on the other hand, might view this ad campaign as Dove's effort to be socially responsible and to provide alternative images of beauty that aren't based on the stick-thin, perfect bone structure, wrinkle-free, supermodel ideal. Because you're an active, critical reader, and not a passive receptor of the media, I'll let you decide.

Political Economy of the Media

In the United States, we (politicians especially) spend a lot of time talking about freedom, particularly freedom of the press. The freedom to say whatever you want is often upheld as one of the great markers of the "land of the free." The press, however, is hardly free. Most broadcasting companies are privately owned in the United States, are supported financially by advertising, and are therefore likely to reflect the biases of their owners and backers. (Compare this model with the United Kingdom's, where the British Broadcasting Company [BBC] cannot accept private funding; households must pay fees for owning television sets; and these fees help cover the BBC's operating costs. This system is beginning to change, however, because of increasing economic pressures, such as competition from satellite television.)

In 2017, just two corporations—Alphabet, the parent company of Google, and Facebook— generated 20 percent of global advertising revenue (Zenith USA, 2018). Ownership alone does not equal censorship, but when the majority of the media lie in the hands of a few players, it is easier to ignore or purposely suppress messages that the owners of the media don't agree with or support. For example, Apple's App Store is the dominant player in the sales of apps for smartphones. Apple demands to review every app to ensure that the content is not something that they "believe is over the line," which they go on to explain is something developers will just know when they cross it. This type of vague policy tends to promote a "chilling effect" whereby developers—not knowing where this mystical "line" is—choose to avoid any content they think might be at all objectionable. Receiving Apple's approval is important because once developers have made an app for Apple's iOS platform, they cannot sell it anywhere but the App Store. Is Apple protecting its shoppers or ruling the app developers through a combination of monopoly power and shadowy threats? As corporate control of the media becomes more and more centralized (owned by fewer and fewer groups), the concern is that the range of opinions available will decrease and that corporate censorship (the act of suppressing information that may reflect negatively on certain companies and/or their affiliates) will further compromise the already-tarnished integrity of the mainstream media.

The internet, to some extent, has balanced out communications monopolies. It's much easier to put up a website expressing alternative views than it is to broadcast a television or radio program suggesting the same. The Center for Civic Media at MIT, led by Ethan Zuckerman, works to leverage the internet for the promotion of local activism. One of the MIT projects—VGAZA or Virtual Gaza—allowed Palestinians in the Gaza Strip to document crises and share local stories globally while under embargo. But the internet is not beyond the realm of political economy. Yelp.com, for example, reports that updates to Google's search algorithm place Google+ reviews higher in Google's search results than reviews for the same venues on other sites, shifting web traffic to Google+ at the expense of competing review sites like Yelp and Thrillist (Leswing, 2015).

CONSUMER CULTURE

America is often described as a consumer culture, and rightly so. In my interview with sociologist Allison Pugh, she pointed out that "corporate marketing to children is a 22 billion dollar industry." She then added, "Children 8 to 11 ask for between two and four toys [for Christmas], and they receive eleven on average!" (Conley, 2011a). Sales on major patriotic holidays (Veterans Day, Memorial Day, Presidents' Day) thrive as a result of the notion that it is our duty as American citizens to be good shoppers. As Sharon Zukin points out in her book on shopping culture, *Point of Purchase: How Shopping Changed American Culture* (2003), 24 hours after the terrorist attacks of September 11, 2001, Mayor Rudy Giuliani urged New Yorkers to take the day off and go shopping. It's the tie that binds our society; everyone's got to shop. Malls are our modern-day marketplaces—they are where teenagers hang out, where elderly suburbanites get their exercise, and where Europeans come as tourists to see what American culture is all about. The term consumerism, however, refers to more than just buying merchandise; it refers to the belief that happiness and fulfillment can be achieved through the acquisition of material possessions. Versace, J. Crew, and real estate agents in certain hip neighborhoods are not just peddling shoes, jeans, and apartments. They are also selling a self-image, a lifestyle, and a sense of belonging and self-worth. The media, and advertising in particular, play a large role in the creation and maintenance of consumerism.

DIGITAL.WWNORTON.COM/YOUMAYASK6CORE

To see my interview with Allison Pugh, go to
digital.wwnorton.com/youmayask6core

CONSUMERISM

the steady acquisition of material possessions, often with the belief that happiness and fulfillment can thus be achieved.

ADVERTISING AND CHILDREN

The rise of the consumer-citizen has been met with increasing criticism, but how does our society produce these consumer-citizens? Canadian author and activist Naomi Klein published *No Logo: Taking Aim at the Brand Bullies* in 2000; in this book Klein analyzes the growth of advertising in schools. Pepsi and Coca-Cola now bargain for exclusive rights to sell their products within schools, and brand-name fast foods are often sold in cafeterias. The logos of companies that sponsor athletic fields are displayed prominently. This

These third-grade students at an elementary school in San Clemente, California, seem to be very concentrated on their Google Chromebook laptops. Do you think Google is doing a good thing by providing low-cost technology to school districts, or is it creating millions of consumer-citizens?

has been commonplace in many colleges and universities for some time now, but the increasing presence of advertising in middle and high schools should also be noted.

One striking example is Google Classroom. Google gives out its low-cost Chromebook laptop computers to many schools and school districts, charging only a $30 annual management fee. The computers include a suite of web-based software applications ranging from Google Docs to Google Classroom (a learning management system). Going from basically nothing in 2012, Google had shipped almost 8 million devices to schools by 2016, with the trend line showing no signs of flattening out anytime soon. For example, the city of Chicago, the third-largest school district in the country, saved about $1.6 million per year in technology costs by switching to Google (Singer, 2017). Sounds like a win for everyone from the company whose motto was "Don't be evil," right? Well it is, in many ways, but some parents worry about the data their children are providing the company whose financial model is based almost solely on advertising revenue. Since students will migrate their educational accounts over to their noneducational Google accounts, the company will have "watched" the developmental trajectory of these students throughout most of their childhood and now be able to better target them with ads.

The result of all of this advertising is the creation of a self-sustaining consumer culture among children—albeit one that plays out differently for low-income and high-income families. Let's hear Pugh discuss her research with me again on this point:

> I found, for low-income parents, a practice of what I ended up calling "symbolic indulgence." They couldn't afford everything that a middle-class family might consider part of an adequate resource to childhood. So, they might not have blocks, or a bike. They might not have those basics, but they would have the thing that kind of gave the child something to talk about at school [such as a Gameboy]. And so it would be these highly [socially] resonant items [that parents would purchase]. (Conley, 2011a)

In other words, these highly symbolic purchases of "in" toys or devices gave low-resource families an avenue to feel as though they were able to participate in the broader American consumer culture. Meanwhile, ironically, middle-class parents downplayed their consumerism:

> "I'm not materialistic. I'm not one of those bad parents you read about on TV...never being able to say no to my kids," [they would say]. So what I found for them is the systematic practice of "symbolic deprivation." The kid would have an enormous amount of stuff. There would be mostly yeses in that child's life. But there would be particular things that that child didn't have, so that they [the parents] could really kind of convince me, and convince themselves, that they were honorable people. (Conley, 2011a)

Pugh emphasizes that these dynamics are not just about corporate advertising but about the local social systems in which kids and their families find themselves. So to fix the problem, we have to change that dynamic by (somehow) making people less afraid of being different, or perhaps more realistically by diminishing difference itself through sameness—by students wearing school uniforms, for example.

CULTURE JAMS: HEY CALVIN, HOW 'BOUT GIVING THAT GIRL A SANDWICH?

People can take back the media or use the media for their own ends. In that sense, Rockin' Rollen, whom we met at the beginning of the chapter, was hacking media coverage with pro-Christian messages to override the lack of attention to Christianity he found on TV. Culture jamming (a term that evolved from radio jamming, another form of guerrilla cultural resistance that involves seizing control of the frequency of a radio station) is the act of co-opting media in spite of itself. Part of a larger movement against consumer culture and consumerism, it's based on the notion that advertisements are basically propaganda. Culture jamming differs from appropriating advertisements for the sake of art and sheer vandalism (where the sole goal is the destruction of property), although advertisers probably don't care too much about this latter distinction. Numerous anticonsumerist activist groups have sprung up, such as *Adbusters*, a Canadian magazine that specializes in spoofs of popular advertising campaigns. For example, it parodied a real Calvin Klein campaign (which advanced the career of Kate Moss and ushered in an age of ultra-thin, waiflike models) with a presumably bulimic woman vomiting into a toilet. *Adbusters* also sponsors an annual Buy Nothing Day (held, with great irony, on the day after Thanksgiving, known in retail as "Black Friday," the busiest shopping day of the year), which encourages people to do just that—buy nothing on this specific day of the year—so that they can reclaim their buying power and focus on the noncommercial aspects of the holiday, such as spending time with family and friends.

CULTURE JAMMING
the act of turning media against themselves.

Two satirical ads from *Adbusters* magazine. How do these ads critique or subvert the tobacco and fashion industries?

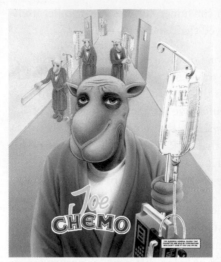

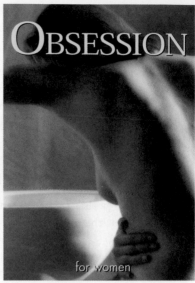

Another *Adbusters* spoof caricatured the legendary Joe Camel, the anthropomorphic advertising icon of Camel cigarettes from 1987 until 1997, when R. J. Reynolds, the tobacco firm that conjured up the character, voluntarily stopped using his image after receiving complaints from Congress and various public-interest groups that its ads primarily targeted children. (In 1991, a *Journal of the American Medical Association* study found that more five- and six-year-old kids recognized Joe Camel than they did Mickey Mouse or Fred Flintstone [Fischer et al., 1991].) In the *Adbusters* spoof, "Joe Chemo" is walking down a hospital hallway with an IV, presumably dying of cancer caused by smoking.

Conclusion

The chapter opened with a description of Rockin' Rollen, the Rainbow Man. Now perhaps you can see how he fits into a discussion of culture and media. He began his 15 minutes of fame by acting the role of an enthusiastic fan at a sporting event. He was ahead of his time in terms of being famous for absolutely nothing. (Note the rise in reality show stars—the ultimate paradox, as the participants are selected because they're "normal" and we then get hooked on watching their "real" lives.) After Rollen's religious conversion, he was almost genius in the way he co-opted the media for his own ends— namely, the promotion of his religious messages. That Rollen was ultimately sentenced to three life sentences (a little harsh for a nine-hour standoff with the police after he took a hostage) perhaps speaks to the fundamental nature of media and culture in our society. Remember that one job of the

POLICY

WHAT'S IN A NAME?

I named my kids E and Yo Xing Heyno Augustus Eisner Alexander Weiser Knuckles, so forgive me if I didn't see what all the fuss was about when Kanye West and Kim Kardashian West named their kids North, Chicago, and Saint. After all, unlike in countries such as France or Japan, there is no US law that constrains what we can name our offspring; only names that would be considered abusive can be stopped. Names present a unique measure of culture: There are few rules, and no institutions attempt to directly influence our choices, unlike almost every other aspect of culture from food to film to fashion. Thus trends in names are as close to a pure, unmediated, reflective mirror of societal culture as we can get.

In that light, one could say it was almost inevitable that West and Kardashian West chose unique names for their children. On the one hand, there's a long tradition of celebrities marking their status by giving atypical names to their children. Remember Moon Unit Zappa? Born in 1967, she was perhaps one of the first notable celebrity offspring given a "weird" moniker. The 1960s was the Age of Aquarius, after all. Besides her own siblings, she was followed by uncapitalized "america," the child of Abbie Hoffman; and Free, the spawn of Barbara Hershey and David Carradine, just to mention a few. Fast-forward to Gwyneth Paltrow's daughter, Apple Martin, and it should come as no surprise that celebrities do things differently.

The more interesting sociological phenomenon that North West embodies, as the daughter of a black man, is the rise of unique black names.

Around the same time celebrities started thinking up names that otherwise served as nouns, verbs, or adjectives, African Americans began to abandon long-standing naming patterns. Until the civil rights movement, a typical black name might have been Franklin or Florence. But then Black Power happened. Blacks wanted to assert their individuality and break ties from the dominant society, so the proportion of unique names—those that appear in birth records only once for that year—shot up.

Kim Kardashian West and Kanye West's decision to name their daughter "North" represents a larger sociological trend of parents giving their children unique names.

Until about 1960, the proportion of unique names for white Americans hovered around 20 percent for girls. For blacks it had always been higher—around 30 percent. During the 1960s, the number of white girls with unique names started to inch up to about 25 percent, but for blacks, it literally skyrocketed, peaking around 1979 at more than 60 percent for girls and reaching almost 40 percent for boys in 1975 (based on Illinois data). Harvard sociologists Stanley Lieberson and Kelly S. Mikelson, who examined this trend in a 1995 paper, followed it only through the 1980s. But I'd be willing to guess that the practice has continued at a similar rate.

You might think that all this name coinage would lead to gender confusion in kindergartens. Though I am not advocating gender rigidity, there is some evidence that gender-ambiguous names can cause problems for boys. Economist David Figlio (2007) found that "boys named Sue" tend to get into more trouble at school around sixth grade, when puberty hits.

But it turns out that even unique names are gendered. When Lieberson and Mikelson gave a list of unique names they found in the Illinois database to respondents, the vast majority identified the gender of the actual child—for example, Cagdas (boy) or Shameki (girl)—correctly. My own experience mirrors this. Nobody mistakes Yo for a girl's name. Meanwhile, three other Es, who heard about my daughter's name from my public musings, wrote to me. (So much for unique...) Two of them were female, bringing the total to 75 percent female. I only wish I had 25 other kids so I could test the gender of every letter in the alphabet. If you think I'm crazy, move to Paris.

sociologist is to see what is usually taken for granted as actually socially constructed. In a way, Rollen committed the crime of disrupting our status quo and violating the unspoken norms of our culture. Religious fanatics are often seen as freaks or terrorists in our culture—in other cultures, they may be viewed quite differently. We rarely question the media's right to control what we see (to be fair, they are for-profit businesses). Did Rollen have a right to position himself strategically behind home plate at major league baseball games? How was his John 3:16 sign different from an "I ♥ MOM" or a "Reverse the Curse" sign?

This chapter has presented some new ways of looking at culture: how we construct it, how it affects us, and what this means for understanding ourselves and the world in which we live. Do you now have a new understanding of culture? Can you now see your own culture through a critical lens? What have you previously taken for granted that you can now view as a product of our culture? Can you now look at the media in a different way? Don your critical thinking cap and put some of the stuff you've just learned into practice.

QUESTIONS FOR REVIEW

1. Thinking about "reality" television shows and people who are well known primarily on social media platforms, define *celebrity* and *cultural icon*. Are people celebrities because they are talented, or can anyone achieve this status? How might this question parallel debates about high culture versus low culture?

2. How does Herbert Gans's *Deciding What's News* (1979a) help us understand the way that cultural production simultaneously reflects and creates our world?

3. Goths are visible as a subculture, in part because of their taste in music and fashion. Using these criteria, identify another subculture. In which ways might this group's values oppose the dominant culture?

4. A student holds the widespread cultural belief in upward social mobility through education and therefore studies thoroughly before an upcoming sociology exam. How might this represent an example of hegemony?

5. The term *culture* is complex, in part because it is used in numerous (sometimes contradictory) ways. Use three of the definitions of culture from this chapter to illustrate how a Shakespearean play might be considered "culture."

6. Let's consider how we are part of a consumer culture. Think about a consumer good you recently acquired and care about, such as an item of clothing. Does this item in any way help establish or demonstrate who you are, what you are about, and how you perceive yourself? How so?

7. How do social media platforms such as Instagram and YouTube potentially change the dynamics of media coverage in the context of a system controlled by a few large companies? Could Instagrammers and YouTubers be considered culture jammers? Explain your answer.

PRACTICE

SUBCULTURE WARS

Some observers think the internet has been a homogenizing force that drives all eyeballs to the latest meme and creates winner-take-all cultural markets. Others suggest that by allowing geographically dispersed individuals to organize around common interests, the internet and social media create a cultural garden where a thousand flowers can bloom. Maybe it's both.

TRY IT!

How do subcultures appropriate and reinterpret mainstream cultural memes? Pick an interest from the list below and two subcultures. Search these combinations and see how these subcultures form unique communities and practices around that hacked theme—for example, Goth Cats versus Emo Cats. You're welcome to go to Google, YouTube, Reddit, Pinterest, or other online sites to research the combinations.

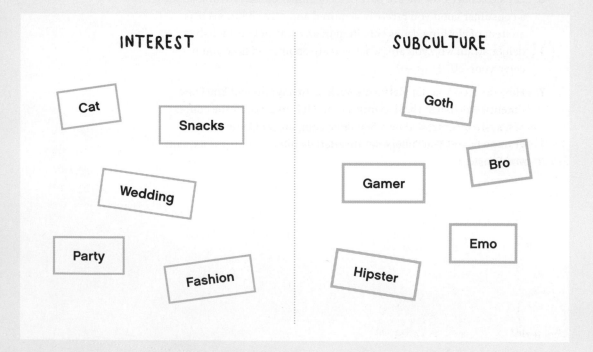

INTEREST

Cat

Snacks

Wedding

Party

Fashion

SUBCULTURE

Goth

Bro

Gamer

Emo

Hipster

THINK ABOUT IT

How do the two subcultures put their unique stamps on something otherwise generic and mainstream? What are the chances, do you think, that the subcultural interpretation will go viral and influence the mainstream?

SOCIOLOGY ON THE STREET

How can we make assumptions about people before we even meet them? What assumptions do you make if you know a person's name? Watch the Sociology on the Street video to find out more: **digital.wwnorton.com /youmayask6core.**

WANT MORE PRACTICE?

Complete the InQuizitive activity for this chapter at digital.wwnorton.com /youmayask6core

PARADOX

4

THE MOST IMPORTANT ASPECTS
OF SOCIAL LIFE ARE THOSE
CONCEPTS WE LEARN WITHOUT
ANYONE TEACHING US.

Socialization and the Construction of Reality

Think back to your first day of college. What did you do upon arriving at the classroom? Presumably, you sat in a chair. You probably opened a notebook and took out either a pen or a pencil, or maybe you fired up a laptop. When the professor walked in and called the class to attention, you stopped talking to the person next to you (or, if you didn't, you at least knew that you should). When handed a stack of syllabi, you took one and passed along the rest. You did not sit on anyone's lap. A million dollars says that you were wearing clothes. Another safe bet is that you did not physically assault anyone. Is all of this an accurate description of what occurred?

Congratulations! You've been properly socialized. So how did I know what you did, and more important, how did you know what to do? Why did you sit in a chair and not on the floor? What if no furniture had been in the room? If a blackboard hung on one wall, you probably sat facing it in the absence of desks. Why? How did you know to bring paper and something with which to write? Did you receive an e-mail earlier in the week with explicit instructions telling you to do so? Why did you put on clothes this morning? When someone hands you a stack of papers, he or she might say, "Take one and pass them along," but even if that person doesn't, you still know what to do. You've internalized many unwritten rules about social

SOCIALIZATION

the process by which
individuals internalize
the values, beliefs, and
norms of a given society
and learn to function as
members of that society.

behavior and public interaction. We call the process by which you learn how to become a functioning member of society socialization.

Imagine Mr. Spock, Tarzan, or an android trying to disguise himself as a college freshman. The droid would have to be programmed in minute detail with an endless list of possible reactions to potential situations. Think for a moment about the differences between your knowledge of how to respond to the following situations versus the responses of our droid: knowing how to answer when someone asks "What's up?"; knowing to shift one's knees when someone else needs to slip in or out of a row of seats in the lecture hall; knowing to wait your turn when asking a question in the lecture or discussion section; knowing how to react when someone yells "Fire!" during class and then screams, versus a student shouting the same word during a final exam and then laughing. Consider the difference between these two possible scenarios at a pizzeria: (1) Someone asks for dough, holding out a $5 bill to the cashier; (2) someone asks for dough, pointing a gun at the same cashier. You know that in the second scenario it is not a wheat-based product to which that person refers. The droid doesn't. He hasn't been socialized. Think how much you have needed to learn, how much knowledge you have internalized and processed, in order to understand that making direct eye contact with someone in a crowded elevator is inappropriate, whereas such behavior at a crowded party might be acceptable. The former might be considered creepy; the latter, flirting.

One famous test in the computer science field of artificial intelligence is the Turing Test, in which a subject is asked to have two parallel conversations. One exchange occurs with an actual human, the other with a computer. Both are conducted by instant messaging or some other text-based platform. If the subject can't reliably distinguish the computer from the living human, then the computer is said to have passed the Turing Test (named for the scientist Alan Turing, who first proposed the test in a 1950 research paper). It was not until 2014 that a computer passed this simple test (fooling 33 percent of judges for five minutes—Turing had seen the threshold at 30 percent) (*Guardian*, 2014). Now imagine adding facial expressions, body language, and other nonverbal communication cues to the test criteria. It quickly becomes clear why to be human is to be socialized and that true artificial intelligence is still a long way off.

Interestingly, while the field of artificial intelligence gets better and better at Turing Tests, psychiatrist Fadi Haddad worries that humans may be getting less socially nuanced. Working in New York City surrounded by people who often avoid interacting with one another, he wonders whether "soon we are all going to be autistic." He remembers riding the subway "ten years ago [when] people used to look at each other and say hi sometimes if you clicked eyes with a stranger.... [T]oday nobody does that. Everybody has a cell phone or does something like this [his face goes blank and he stares off at nobody in particular], and that's very autistic behavior ... that's what we see in New York. I think everybody here is autistic." Sociologist Erving Goffman, though, would call politely ignoring fellow subway and elevator riders "civil inattention" and argue that those who exercise civil inattention on the subway are well socialized. So how do we decide when interaction or inattention is the socially optimal option? If we choose to interact, how do we know which interactions are acceptable? We know how to behave because we are socialized. But what is socialization and how does it work?

Socialization: The Concept

Socialization, then, as defined by Craig Calhoun in the *Dictionary of the Social Sciences* (2002, p. 47), is "the process through which individuals internalize the values, beliefs, and norms of a society and learn to function as its members." Starting from when you were born, the interactions between you and the rest of the world have shaped who you are. Presumably, as a baby, you were wrapped in either a pink or a blue blanket to signify your being a girl or a boy. At some point, you were potty trained—a critical expectation of our society. For babies, the primary unit of socialization is generally the family. As children grow older and enter the educational system, school becomes a key location in their socialization. This is where you probably learned to sit facing the front of the classroom. You learned to raise your hand before asking or answering a question. You learned not to talk when the teacher was speaking and what the punishment would be if you failed to follow this rule. By the time you entered college, you did not have to be told explicitly to do these things because you had *internalized* the rules that govern situations in which the people we designate students and teachers operate. As the examples that open this chapter illustrate, however, we learn more than explicit sets of rules. We learn how to interact on myriad levels in an endless number of situations.

You recognize the limits of your socialization when you find yourself in a new situation and aren't quite sure how to behave. If you've been shy and bookish in high school and you find yourself among a new set of party-going friends at a raucous fraternity party in college, you may not know just what to wear or how to behave. You may take as your model what you've seen on television and in the movies, but the party you are attending may not be quite the same. If, on the other hand, you've been going to similar parties, maybe even at the very same fraternity, with an older brother or sister for the past year or two, this particular party will not present any sort of anxiety. You'll know what to wear, which bathroom to use, whether or not it is cool to post photos to Instagram, what music to like, and which songs deserve an eye-roll. Your previous experience as a tagalong may now make you a leader among your freshman peers.

Early in elementary school you were taught to raise your hand to speak in class. Can you think of other examples of internalized behavior?

Limits of Socialization

Although socialization is necessary for people to function in society, individuals are not simply blank slates onto which society transcribes its norms and values. Twins are often used to support one or the other side of the nature-versus-nurture debate, because they allow us to factor out genetics. Twins living hundreds or even thousands of miles apart may simultaneously experience the same pain in their right arms—score one point for nature. Take another set of twins, however, who were separated at birth in the early 1900s. One of them was raised as a Jew and the other became a Nazi—score one for nurture. So which theory is correct? Both and neither. In sociology, we tend to think less about right and wrong and more about which theories are more or less helpful in explaining and understanding our social world. The concept of socialization is useful for understanding how people become functioning members of society. Like most theories, however, socialization can be limited in its explanatory power. Have you ever heard someone discussing a "problem child"? The conversation might go something like this: "I just don't know what's wrong with him. He comes from such a nice family. And his older brothers are such nice, respectful boys." How did this child go astray? The primary unit of socialization, the family, seems to have been functional. The other children went on to lead happy and productive lives. What else might explain the youngest son's delinquency? For starters, human beings have agency. This means that while we operate within limits that largely are not of our own making (e.g., we cannot choose our parents or siblings, and US law requires that all children receive schooling), we also make choices about how to interact with our environment. We can physically walk out of the school, fall asleep in class, or run away from home.

"HUMAN" NATURE

Is there such a thing as human nature? It's physiology that prompts you to urinate, but it's socialization that tells you where and when to do so. We are largely shaped by interaction, such that without society the human part of human nature would not develop. We can observe this in children raised by animals or denied human contact. Take the case of "Anna," a young girl whose true age was unknown but estimated at about five years and who had been found in torturous conditions, bound to a chair in an attic, where she had been left almost completely alone in the dark since birth. Her nutritional requirements had been minimally met, and it is believed that her only regular human contact was when her mother delivered these small meals. She could not properly speak or use her limbs, or even walk, when she was discovered. She did not respond to light or sounds the way children normally do. With some care and attention, the nurses at the treating hospital were able to provoke some giggles and coos by tickling her. Eventually, she was able to make speech-like sounds and gain more control of her body, but she never developed to the level of most children her age and died a few years later. Some of the doctors examining her wondered if she had been born mentally disabled, but ultimately, they opined that she could not have survived such an environment unless she had been healthy at birth.

The professionals who cared for Anna and followed her case reached five general conclusions:

1. Her inability to develop past an "idiot level of mentality...is largely the result of social isolation."

2. "It seems almost impossible for any child to learn to speak, think, and act like a normal person after a long period of isolation."

3. When she is compared with other cases of isolated children, the similarities "seem to indicate that the stages of socialization are to some extent necessarily related to the stages of organic development."

4. "Anna's history...seems to demonstrate that human nature is determined by the child's communicative social contacts as much as by his organic equipment and that the system of communicative symbols is a highly complex business acquired early in life as the result of long and intimate training."

5. Theories of socialization are neither right nor wrong in this case "but simply inapplicable." (Davis, 1940, pp. 554–65)

What this case illustrates is that "human nature" is a blend of "organic equipment," the raw materials we are physically made of, and social interaction, the environment in which we are raised. In what other ways can we look at the role of socialization and social structures in shaping our behaviors?

How does a small child playing peekaboo demonstrate the social process of creating the self?

Theories of Socialization

Now that we have identified this process called socialization, we can turn to some of the theories about how it works.

ME, MYSELF, AND I: DEVELOPMENT OF THE SELF AND THE OTHER

Have you ever seen a little girl cover her eyes with her hands and declare, "You can't see me now!" She does not realize that just because she cannot see you doesn't mean you cannot see her. She is incapable of distinguishing

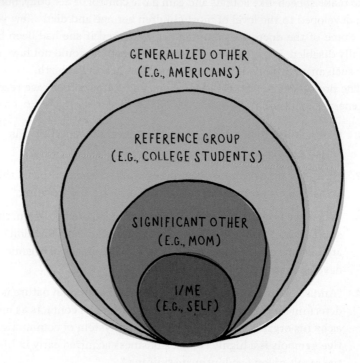

FIGURE 4.1 Mead's Stages of Social Development

GENERALIZED OTHER
(E.G., AMERICANS)

REFERENCE GROUP
(E.G., COLLEGE STUDENTS)

SIGNIFICANT OTHER
(E.G., MOM)

I/ME
(E.G., SELF)

SELF

the individual identity of a person as perceived by that same person.

I

one's sense of agency, action, or power.

ME

the self as perceived as an object by the "I"; the self as one imagines others perceive one.

between *I* and *you* and has not yet formed an idea of her individual self. How does the concept of the self develop? Sociologists would argue that it emerges through a social process. Perhaps the first full theory of the social self was developed by Charles Horton Cooley, who coined the term *the looking-glass self.* According to Cooley in *Human Nature and the Social Order* (1922), the self emerges from our ability to assume the point of view of others and thereby imagine how they see us. We then test this "theory" of how we are perceived by gauging others' reactions and revise our theory by fine-tuning our "self concept."

In the 1930s, George Herbert Mead further elaborated the process by which the social self develops (Figure 4.1). Infants only know the I—that is, one's sense of agency, action, or power. Through social interaction, however, they learn the me—that is, the self as a distinct object to be perceived by others (and by the I). Imagine, for example, that you are taking care of your two-year-old cousin Joey. He wants a cookie. You explain to him that you will happily give him a cookie just as soon as you go to the bathroom. He starts shouting, "I want cookie! I want cookie!" And you are ready to scream, "But I have to pee!" In a more rational moment, you might argue with your adorable cousin: "How would you feel if I demanded a cookie immediately when you entered the house after a six-hour car ride with no rest stops along the way and your bladder was about to explode?" Although such a line of logic might work with your cousin's seven-year-old sister, it will not work with little Joey because he has yet to develop a sense of the other—that is, someone or something outside of oneself. In Joey's mind, there is no *other*, the *you* who hasn't seen a toilet in six hours; there is only the *self* who wants a cookie. This is why we allow children a certain amount of leeway. If your best friend behaved like this, however, you would accuse him or her of being childish, and you would be correct.

How do we move beyond the self? To function as fully adult members of society, we need to be able to recognize that other people have wants, needs, and desires that are sometimes similar to and sometimes different from our own. Thus, imitation, play, and games are important components of childhood development. When a child imitates, he or she is just starting to learn to recognize an other. That's what peekaboo is all about. Eventually, kids understand that you are still there when they cover their eyes. They can then advance to play, according to Mead. During play, children are able to make a distinction between the self and the other. Suppose that you are playing with your cousin Joey, who is now five years old, and he says, "I'll be the cop, and you be the bad guy." He's recognizing you as the other, the bad guy, who has a different set of motives, responses, and actions from the good guy, the cop. Eventually, children move beyond play to formal games.

Games involve a more complex understanding of multiple roles; indeed, you must be able to recognize and anticipate what many other players are going to do in a given situation. Have you ever watched a swarm of toddlers try to play soccer? It's a mess. They're content to kick the ball; that's enough for them. They can't pass, let alone coordinate an offensive attack on an empty goal. It takes an understanding of the other to coordinate passing: You have to consider where to place the ball and calculate whether your teammate can make it to that spot in time for the pass to occur. Games such as soccer involve more than just hand–eye (or foot–eye) coordination. They involve a sophis-

ticated understanding of the various positions others can occupy—that is, they require a theory of social behavior, knowing how others are likely to react to different situations. The goalie is not likely to leave the box in front of the goal, even if I tempt him or her with the ball. But a defender or midfielder is likely to pursue me if I am not controlling the ball very well. We have learned to anticipate these behaviors from repeated experience in the context of a particular sort of constrained social interaction— namely, a soccer game.

OTHER

someone or something outside of oneself.

Why are games an important part of child development? What do team sports like soccer teach us about multiple roles?

As our parents, siblings, friends, teachers, and soccer coaches socialize us, we learn to think beyond the self to the other. According to Mead, however, that brings us only halfway to being socialized. The final step is developing a concept of the generalized other, which represents an internalized sense of the total expectations of others in a variety of settings—regardless of whether we've encountered those people or places before. In this way, we should be able to function with complete strangers in a wide range of social settings. For example, it is the perception of the generalized other that keeps you from taking off your pants to lounge more comfortably in the park on hot summer days, that (it's hoped) keeps you from singing on the bus no matter how much you are enjoying the song on Spotify, and that keeps you from picking a wedgie in public. We can, and do, continually update our internal sense of the generalized other as we gather new information about norms and expectations in different contexts.

For instance, as a child, you may have been chided by your parents not to pick your nose. You have to be taught that this is unacceptable behavior, because otherwise it might seem perfectly reasonable. You have something in your nose, it's bothering you, you want to remove whatever is stuck there, and, lo and behold, your finger is just the right size. It's remarkably convenient. However, you learn not to pick your nose because it's a socially unacceptable action. But wait! Sometimes you might catch your dad picking his nose in the bathroom at home. Now you revise the original lesson and realize what is most important: not to pick your nose in public. You may even have been taught explicitly that certain activities that are acceptable in private are unacceptable in public. The concept of the generalized other shapes our actions by our internalization of what is, and is not, acceptable in different social situations. Some of us may have internalized the notion that nose picking is inherently disgusting and should never be done anywhere. Some of us may have internalized the notion that nose picking is a little gross but believe it's okay when done in private; the action should just not be performed in public. If you walked outside right now and saw a child picking his nose, you might think, "Eww," but laugh because he is, after all, a child. On the other hand, if his mom was picking her nose too, you might give her a look of disapproval for violating an established social norm.

People may also intentionally violate established norms. It is not uncommon for children to touch their genitalia in public (watch any group of four- or five-year-olds and you will quickly spot who needs to go to the bathroom). As a society, however, not touching or exposing your genitals in public is a well-established norm (a law, in fact). The exhibitionist who exposes himself or masturbates in public does so not because he was poorly socialized. He has a keen understanding of the generalized other and violates the norm with the explicit purpose of soliciting a reaction from a generalized other (or a very specific other, in some cases).

Agents of Socialization

FAMILIES

For most individuals, the family is the original source of significant others and the primary unit of socialization. If you have siblings, you may develop the sense that older and younger children are treated differently. The general impression is that the younger siblings get away with more. Parents, having already gone through the experience of child rearing, may relax their attitudes and behaviors toward later children (even if unconsciously). Note, too, that socialization can be a two-way street. Information doesn't always flow from the older to the younger family members. For example, the children of immigrants—who are immersed in the US school system while their parents may maintain less contact with mainstream American communities—are likely to take on the role of an agent of socialization instead of the other way around, teaching their parents the language and other tools of cultural assimilation. In a similar vein, a study by economist Ebonya Washington (2008) showed that members of Congress who had daughters were more likely to vote for feminist measures. Perhaps daughters socialize their parents into being more sensitive to women's concerns. Alternatively, it could be that when legislators have their own daughters' future to worry about, they vote for legislative changes that will help them (and, by extension, women in general). So it's hard to say whether this is really socialization at work or merely a change in rational, selfish calculations. Meanwhile, sociologist Emily Rauscher and I found that for the average American, daughters made parents more politically conservative, specifically with respect to views about sexuality (Conley & Rauscher, 2013).

The socialization that occurs within the family can be affected by various demographics. Parents of different social classes socialize their children differently. For example, when asked what values they want their children to have, middle-class parents are more likely to stress independence and self-direction, whereas working-class parents prioritize obedience to external authority (Kohn & Schooler, 1983). Indeed, sociologists have long recognized that parents' social class matters, but how exactly this privilege is transmitted to children (beyond strictly monetary benefits) has been less clear. To better understand this process, ethnographer Annette Lareau spent time in both black and white households with children approximately 10 years of age. She found that middle-class parents, both black and white, are more likely to engage in what she calls "concerted cultivation." They structure their children's leisure time with formal activities (such as soccer leagues and piano lessons) and reason with them over decisions in an effort to foster their kids' talents.

According to Annette Lareau, how do working-class and middle-class families structure their children's free time differently? What are the results of these different socializing behaviors?

Working-class and poor parents, in contrast, focus on the "accomplishment of natural growth." They give their children the room and resources to develop but leave it up to the kids to decide how they want to structure their free time. A greater division between the social life of children and that of the adults exists in such households (Lareau, 2002). Whereas middle-class parents send their kids off to soccer practice, music lessons, and myriad other after-school activities, kids in poor families spend a disproportionate amount of time "hanging out," as has been observed by Jason DeParle in *American Dream* (2004), his chronicle of three families on public assistance struggling through the era of welfare reform in Milwaukee. Likewise, a 2006 study by Annette Lareau, Eliot Weingarter, and this author shows the same statistical results: Outside of school, disadvantaged children spend 40 percent more time in unstructured activities than their middle-class counterparts.

Middle-class kids, on the other hand, spend their days learning how to interact with adult authority figures, how to talk to strangers, and how to follow rules and manage schedules. From a very young age, they are taught to use logic and reason to support their choices by mirroring their parents' explanations of why they can or cannot get what they want. Low-income parents, Lareau found, were more likely to answer their children with "Because I said so," instilling respect for authority but missing an opportunity to help their children develop logical reasoning skills commonly used in adult interactions. Middle-class kids discover the confidence that comes with achievements such as learning to play the piano or mastering a foreign language. Whether they actually have fun is unknown, but they are certainly socialized into the same kind of lifestyles that their parents hope them to have as adult professionals.

In fact, it should be no surprise that the rise of the "overscheduled" child comes during a period when, for the first time in history, higher-income Americans work more hours than lower-income Americans. If we flip the equation and look at leisure time, it seems that getting lots of education might limit your fun time. Highly educated women spend the fewest hours at leisure (30.3 hours every week), and men with a high-school education or less spend the most leisure time (39.1 hours), though half of the leisure gap is due to the difficulty less-educated men and women have finding full-time jobs (Attanasio et al., 2013). Professional parents familiarize their children with the kind of lives they expect them to lead as adults.

An important question, however, is how these different parenting strategies may or may not affect the long-term outcomes of kids. Luckily, I was able to speak to Annette Lareau upon the publication of her updated edition of *Unequal Childhoods* (2011), for which she went back and followed up with many of the families she had originally studied. Here's what she told me:

> There really were no surprises. I would say it was sad. Many of the working-class and poor children had wanted to do well, their parents had wanted them to do well, but things had not worked out. Not one was in the professional sector. [For] the middle-class families, it wasn't always easy sailing [either]. Garrett Talinger got his heart broken. Melanie Hamlin had a friend killed in a car accident. It's not that bad things don't happen to middle-class kids; they do. But in terms of being launched, in terms of their life chances, they were much more likely to go to college, and their parents helped them and supervised them in college. So their parents helped them choose their college classes, they gave them advice on their majors, and so the parents continued to provide guidance and help as they went into adulthood. Alexander Williams, the middle-class African American boy, is now going to become a doctor. Garrett Talinger became a high-level manager. He has a suit and tie, his face is shining. (Conley, 2011b)

DIGITAL.WWNORTON.COM/YOUMAYASK6CORE

To see my interview with Annette Lareau about *Unequal Childhoods*, go to digital.wwnorton.com/youmayask6core

This was so interesting, I probed further: What exactly happened in the poor and working-class families' lives that caused the kids, despite their high aspirations, to drop out of high school or go no further than a high-school diploma?

> To apply to college involves many, many, many steps. If they wanted to go to college, did you take, say, advanced algebra? Did you take the [other] classes you needed to apply to college? Then did they take the SAT and the ACT? Did they apply to college? And if they got in, did they go? And so by the end you have these diversions in pathways between the parents who are middle class and the parents who are working class, despite the aspirations.

Lareau went on to provide a concrete example of these hurdles in the case of a working-class girl who was not included in the book due to space reasons:

> She applied to colleges, but she applied to colleges that were two and three hundred points above her SAT. Now, if she had been my daughter, could I have gotten her into a school? Probably. I would have found a school that would have taken a child who had a learning disability, or had low scores, and I could have placed her in college. It would have taken a lot of work. But in her system, her mother depended on the [high] school to help her daughter go to college. So the mother didn't see the applications, and the mother had a car, but she didn't go on college tours. It was up to her daughter and the school. And that is a reasonable decision. Her daughter was rejected everywhere, and she ended up going to community college for a semester and dropping out. And [her experience follows] a very typical pattern for working-class families.

Though Lareau can't definitively discern cause and effect here—What if Tara's SAT scores had been higher? Or if she had lucked out with a fabulous guidance counselor?—things might have turned out differently for Tara, despite the "natural growth" strategy her mother followed. But by showing how social stratification actually worked on the ground, Lareau tells a pretty convincing—if depressing—story.

SCHOOL

When children enter school, the primary locus of socialization shifts to include reference groups such as peers and teachers. In addition to helping you learn the three *R*s, one of the teacher's main goals is to properly socialize you—teaching you to share, take turns, resolve conflict with words, be quiet when necessary, and speak when appropriate. Walk into any kindergarten classroom and compare it with a third-grade class. What are the main differences? For starters, there is probably a lot more order, and less noise, among the third-graders. When you were young and needed to leave the classroom as a group (whether for recess, gym, or music class), you probably had to line up and follow your teacher. Did you have to do this in high school? Highly unlikely, because teenagers are able to get themselves from one classroom to the next, whereas five-year-olds are not.

When students resist classroom behavior norms, many parents and teachers turn to medication. Psychiatrist Fadi Haddad, whom we heard from in

DIGITAL.WWNORTON.COM/YOUMAYASK6CORE

To see my interview with Fadi Haddad, go to
digital.wwnorton.com/youmayask6core

the chapter opener, treats kids who have been referred to him because their parents or teachers think they have attention-deficit hyperactivity disorder (ADHD). He has identified four factors leading to the uptick in ADHD cases in the United States. Only one of them is the mental health of the kid; the rest are rooted in the broader social context. He sees that "the demands of the schools are built on certain curricul[a]" that are "not very flexible... they want the kid to behave in a certain way." If a kid does not behave, "it's easy to say this kid has ADHD and that's why he is not successful in school rather than, oh, well, maybe this kid, his brain is functioning a little bit different, and if we change the curriculum, he will be a brilliant kid and he will succeed."

A compounding parallel issue is that American culture is "work, work, work, work. So parents start work at 8:00 in the morning. They don't come home before 5:00 or 6:00 in the evening... and they want kids to do their homework. If the kid is not able because of many different reasons.... This kid might have symptoms of depression, and that's why he's not able to concentrate on his work. It's much easier for the parents to say, 'Well, my son has ADHD and he needs medication' rather than dealing with other issues that [are] affecting them." The school structure and the long work hours parents face run into a third concern. Haddad is a psychiatrist specializing in children's mental health, but, as he says:

> [A]ny doctor can prescribe medications for ADHD, and many people, because [of] the difficulty of accessing child psychiatrists and the expense—because child psychiatrists are very expensive—say,

"well, I cannot afford going to a child psychiatrist. I will go to my pediatrician." And they'll sit with the pediatrician for five minutes, and the pediatrician gets convinced, "all right, you have ADHD, so let me give you medications for ADHD." So we see half of the kids taking ADHD medications, [although] many of them would benefit from something else. (Conley, 2014c)

Ultimately, Haddad says, it is easier to medicate the children to fit with the socialization norms than to change society. As he sums up the standardizing impact of socialization: "[A]nybody who's different, anybody who's functioning on [a] different level, we do not accommodate them." For active young kids, "we just claim that, 'okay, *you* have a problem. You need help.' And the easiest problem to have is ADHD because the medications are easy to prescribe." Socialization may not come easily for anyone, but it is especially difficult for those who find conformity a challenge.

Schools, however, teach us more than how to show up prepared for class, and all schools are not created equal. In *Preparing for Power* (1985), Peter W. Cookson Jr. and Caroline Hodges Persell explore how private prep schools indoctrinate the students who attend them into a world of social status and privilege. The researchers "document how the philosophies, programs, and lifestyles of boarding schools help transmit power and privilege and how elite families use these schools to maintain their social class." Admission to these schools is highly competitive and far from democratic. "Legacies" (children who come from families where at least one other member has attended the same school) make up more than half (54 percent) of the attendees on average and as many as 75 percent at some schools. The nonacademic part of the curriculum ranges from upper-class sports such as crew and sailing to cultural education in the form of study abroad and smaller field trips overseas. Probably the most important aspect of prep-school education is that it links students into social networks—helping them get into the top colleges and hobnob with students from wealthy and powerful families— that they will have access to and benefit from for the rest of their lives.

PEERS

Once we reach school age, peers become an important part of our lives and function as agents of socialization. Adolescents, in particular, spend a great deal of their free time in the company of peers. Peers can reinforce messages taught in the home (even the most liberal of friends will probably expect you to wear clothing when you hang out with them after school) or contradict them. Either way, conformity is generally expected; hence the term *peer pressure*. Walk into any high-school cafeteria and you'll observe the power of conformity among peer groups. If you have ever tried drugs or alcohol, did you do so by yourself? Probably not. Even when we are being deviant, we often do so in the company (and often at the suggestion) of others. This is why parents express concern about their children hanging out with the "wrong crowd." They're not just being paranoid.

Adolescents do tend to be more open to their friends' advice than to their parents'. A 2002 study of young adolescents (by Wood et al.) found, on the one hand, that friends were a major source of information on dating and sex. This makes sense. No matter how cool your parents are (or try to be), when you're 15, you probably don't want to talk to them about your love life. On the other hand, the same study found that even though adolescents obtained much of their information on sex from friends, they didn't necessarily believe it. In fact, they were less likely to believe what they learned from friends and the media and more likely to name parents and sex educators as sources of reliable information. It seems as if most adolescents have already learned that they shouldn't believe everything they hear.

ADULT SOCIALIZATION

The socialization we receive as children can never fully prepare us for the demands that we will face as adults. *Adult socialization* simply refers to the ways in which you are socialized as an adult. When you work at a restaurant, you learn, for instance, what your job responsibilities are, how to take orders and place them in the kitchen, and how to carry a tray full of drinks. As a result of your prior socialization, you probably already knew that you should not talk back to customers. Similarly, your plans upon graduation from college might include moving out of your parents' house, finding a job, and maybe starting a family. Right now, you may not know everything you need to in order to function in each of those situations, but you will learn. Some roles, like some jobs, take more preparation than others—you have to go to law school to become an attorney and medical school to become a doctor. Did you ever wonder why parenting schools don't exist?

Resocialization is a more drastic form of adult socialization. When you change your environment, you may need some resocialization. If you plan to live in another country, you may have to learn a new language and new ways of eating, speaking, talking, listening, or dressing. If you went to a single-sex high school and are attending a co-ed college or university, this change will probably require some resocialization, depending on the extent to which you interacted with the opposite sex during your high-school years. The most drastic case of resocialization would be necessary if you had suffered a terrible accident and lost all of your memory. You would need to relearn everything—how to hold a fork and knife, how to tie your shoes, how to engage in conversation. You would be completely childlike once again—and any inappropriate nose picking would be forgiven.

RESOCIALIZATION

the process by which one's sense of social values, beliefs, and norms are reengineered, often deliberately, through an intense social process that may take place in a total institution.

TOTAL INSTITUTIONS

↑

Marines training at Parris Island. How is Marine boot camp an example of a total institution?

The term total institution refers to an institution that controls all the basics of day-to-day life. Members of the institution eat, sleep, study, play, perhaps even bathe and pray together. Boarding schools, colleges, monasteries, the army, and prisons are all total institutions to varying degrees—prisons being the most extreme (see Chapter 6). Sociologist Gwynne Dyer's *War* (1985) provides some insight into the total institution of the US Marine Corps. Marine boot camp strips down much of the prior socialization of recruits and resocializes them to become Marines. Think about how boot camp treats new enlistees: Certain parts of their identity are erased. Uniforms are handed to them and males' heads are shaved. Enlistees live, eat, sleep, bathe, exercise, and study together. Their every move is watched and critiqued, from how they hold a weapon to how they tuck in the corners of their bedsheets. New Marines— and recruits in all branches of the armed forces—also must be resocialized to accept the notion that it is okay, even necessary, to kill. Most of us have learned quite the opposite, of course—that under no circumstances should we kill another person. Basic training is not only about teaching recruits new skills; it's about changing them so that they can perform tasks they wouldn't have dreamed of doing otherwise. It works by applying enormous physical and mental pressure on men and women who have been isolated from their normal civilian environment and constantly placed in situations where the only right way to think and behave is that espoused by the Marine Corps. For people to be effective soldiers, they must learn a new set of rules about when it is okay and/or necessary to kill (that is, when they are on active duty, not on leave and at the pizza parlor with their family).

TOTAL INSTITUTION

an institution in which one is totally immersed and that controls all the basics of day-to-day life; no barriers exist between the usual spheres of daily life, and all activity occurs in the same place and under the same single authority.

Social Interaction

STATUS

a recognizable social position that an individual occupies.

To talk about how institutions and society as a whole socialize us, we need a language to describe social interaction. Robert Merton's role theory provides just such a vocabulary. The first key concept for an understanding of role theory is status, which refers to a recognizable social position that an individual occupies. The person who runs your class and is responsible for grading each student has the status of professor. Roles, then, refer to the duties and behaviors associated with a particular status. You can reasonably expect your professor to show up on time, clothed, and prepared for class. You can also expect that he or she has a fair amount of education, often a PhD, and is knowledgeable about the material. This may not always be true, but it is a fairly standard set of role expectations for someone with the status of professor. Roles are complicated, however, and relatively few of them materialize with handbooks and clear sets of expectations. Sometimes we experience role strain, the incompatibility among roles corresponding to a single status. An example of this is the old dictum for university faculty, particularly young professors, to "publish or perish." On the one hand, they need to stay on top of their research, write articles, give lectures, and attend conferences to stay abreast of current topics and remain active in the academic world. On the other hand, they have to teach and perform all the teaching roles that come with the status of professor—preparing lectures, running class sections, meeting with students, grading papers, and writing letters of recommendation.

When your professor arrives at class tomorrow morning with bags under her eyes and a look on her face that seems to say, "No amount of coffee will help me at this point," you may be tempted to conclude, "Oh, she's experiencing role strain." But what if her exhaustion results not from the time demands of simultaneously grading her midterms and preparing for an upcoming research conference, but rather from the turmoil that ensued last night when the family dog destroyed her daughter's biology project and she had to stay up all night helping her daughter redo it? In this case, role conflict is the culprit. It is not the roles within her status as professor that are the root of her problems, but rather the tensions between her role as professor and her role as mother. Whereas *role strain* refers to conflicting demands within the same status, *role conflict* describes the tension caused by competing demands between two or more roles within different statuses. Each one of us, at any given time, enjoys numerous statuses. These statuses (and their corresponding roles) can and do change over time and between places. When you started college, how did your status change? The obvious answer is that you went from being either a high-school student or perhaps an unskilled worker to becoming a college student. If you graduated from high school and went straight to college, you traded in one status for another. If you still maintain a full- or part-time job, you have added another status. The term status set refers to all the statuses you have at any given time.

To obtain a better sense of how roles and statuses function, try the following experiment: Write down as many answers to the question "Who am I?" as you can. Compare your answers with your classmates'. With remarkable similarity, the lists will include statuses such as brother, sister, daughter, boyfriend, student, lifeguard, babysitter, roommate, and so forth. This occurs because we know ourselves in our social roles, in the ways in which

ROLE

the duties and behaviors expected of someone who holds a particular status.

ROLE STRAIN

the incompatibility among roles corresponding to a single status.

ROLE CONFLICT

the tension caused by competing demands between two or more roles pertaining to different statuses.

STATUS SET

all the statuses one holds simultaneously.

we relate to others. You most likely will have listed the key components of your status set.

There are some other key terms that you must learn to understand role theory. Sociologists often make a distinction between an ascribed status and an achieved status, which basically amounts to what you are born with versus what you become. Another way to think about it is in terms of involuntary versus voluntary status. Your age, race, and sex are all largely ascribed statuses, whereas your status as a juggler, drug dealer, peace activist, or reality television aficionado is an achieved status. Sometimes, one status within our status set stands out or overwhelms all the others. This is called a master status. Examples might include being unemployed, being lesbian, being disabled, or any status that overshadows other statuses. Master status roles can be ascribed, like disability, or achieved, like being a celebrity. The key characteristic of a master status is that people tend to interact with you on the basis of that one status alone.

GENDER ROLES

One of the most popular lines of thought to evolve from role theory has been the concept of gender roles, sets of behavioral norms assumed to accompany one's identity as masculine, feminine, or other. In their critiques of role theory, gender theorists such as Candace West and Don Zimmerman (1987) have argued that the statuses of male/female have distinct power and significance that role theory doesn't adequately capture (you can read more about this in Chapter 8 on sex and gender). We can see how gender fits into larger theories of socialization. From the moment they leave the womb, babies usually wear either pink or blue to designate their sex. There are signs, balloons, greeting cards, and e-cards announcing "It's a boy!" or "It's a girl!" These seemingly silly differences—babies don't care if they wear blue or pink—create a context for older kids and adults to treat babies not as an undifferentiated group of babbling incoherents but as boys and girls. Studies (Lewis et al., 1992) have shown that people interact with babies

ASCRIBED STATUS

a status into which one is born; involuntary status.

ACHIEVED STATUS

a status into which one enters; voluntary status.

MASTER STATUS

one status within a set that stands out or overrides all others.

GENDER ROLES

sets of behavioral norms assumed to accompany one's status as male or female.

differently based on whether they are boys or girls, commenting on how "big" and "strong" baby boys are and how "pretty" baby girls may be, or closely snuggling with female babies rather than holding them facing outward, as with male babies, so that they might see the world.

Not just family members but the larger social world interact with boys and girls very differently and, consequently, socialize them into different roles and into the schema of a gender binary itself (the idea that one has to pick between two genders). Take a stroll through your

How do the displays at this New York City toy store serve as an example of the ways that we learn gender roles through socialization?

local toy store, and you will glimpse the function of toys and play—both very important to the development of children—in creating and maintaining gender roles. The toys for playing house will likely be displayed in boxes showing images of little girls pushing and pulling pink irons and purple vacuum cleaners. The boxes for toy stoves will depict girls cooking. Baby dolls will be wrapped in pastel colors, and their packaging will show little girls cradling them. Now meander through the section of the store displaying tool sets, workbenches, and toys related to outdoor activities. On the packages, you will note little boys dressed in bright primary colors, wielding hammers and fishing poles. The action-figure aisle will be similarly gendered (think Barbie versus G.I. Joe). (Admittedly, these are generalizations. Some images will show boys playing house and girls constructing a bridge, but the vast majority will adhere to what we perceive as traditional gender roles.)

In high school, gender-role socialization continues as peers police each other. Sociologist C. J. Pascoe spent a year in a working-class high school in California finding out just how teens enforce gender norms on one

Who are you? What are the different roles in your status set? For example, singer Beyoncé's statuses include mother, daughter, and partner.

another. She discovered what she calls "fag discourse," a term that describes the near-continuous use of the term *fag* or *faggot* as an insult teenage boys use against one another to curtail improper behavior. She explained in a recent interview how it worked. Someone could get called a fag if, as she said, "you danced, if you cared about your clothing, if you were too emotional, or if you were incompetent." What's more, actually being gay turned out to be beside the point. As Pascoe explained: "What I came to realize was that they used *fag* as an insult to police the boundaries of masculinity. It wasn't about same-sex desire. In fact, when I asked them about same-sex desire, one boy said, 'Well being gay is just a lifestyle; you can still throw a football around and be gay.'" This "masculinity policing" is no joking matter. A boy she called Ricky was "targeted so relentlessly that he dropped out of school" (Conley, 2009c).

If boys are constantly policing the boundaries of masculinity with homophobic insults, how are girls being socialized into their gender roles? At the school in Pascoe's study, girls did not insult each other. Instead they had to put up with constant sexual harassment from the boys. As the boys looked for behaviors that would prove they were not fags, they used girls as unwitting resources with which to perform aggressive sexuality by describing to one another either how they could "get" girls or the outlandish and often violent sexual escapades that would then follow. Further, they engaged in rituals of forceful touching, often physically constraining girls' movements. Pascoe describes one hallway scene in which she "watched one boy walk down the hallway jabbing a girl in the crotch with his drumstick yelling, 'Get raped, get raped'" (Conley, 2009c). In this case, gender socialization looks a lot like gender-based bullying. Do these scenes remind you of high school? Was your gender performance policed by your peers?

For some, constant gender-boundary policing has catastrophic effects. Ricky was subject to the steadiest, most virulent fag discourse in Pascoe's study and he dropped out. An 11-year-old in Massachusetts committed suicide after being relentlessly teased; many of the insults he received were part of fag discourse. And Pascoe points out that 90 percent of school shooters who go on rampage school shootings have been subject to homophobic harassment and teasing (Conley, 2009c). Even when it does not end so tragically, such gender socialization during childhood may form the roots of adult gender performances—ranging from the fact that men are more likely to sexually harass women in the workplace than the reverse to the simple fact that men are 33 percent more likely to talk over or interrupt a woman than they are a man (Shore, 2017).

The Social Construction of Reality

In a February 19, 2006, *New York Times* op-ed essay titled "Mind over Splatter," Vassar professor Don Foster commented on the debate surrounding a Jackson Pollock painting that was recently alleged to be a fake. Is the painting any less artistic for not being a real Pollock when it was, after all, good enough to fool so many people for so long? In a similar vein, Shakespeare's Juliet once inquired, "What's in a name? That which we call a rose by any other name would smell as sweet." The implication is that something is what it is, apart from what we call it or who made it. But do we agree? This question speaks less to the essential nature of things and more to the social construction of reality. Something is real, meaningful, or valuable when society tells us it is.

So what does it mean to say that something is socially constructed? This question is less a debate about what is real versus fake and more an explanation of how we give meanings to things or ideas through social interaction. Two good ways of understanding how we socially construct our reality are to compare one society over different time periods and to compare two contemporary societies. Let's look at some examples.

In US society we take for granted that childhood is a critical and unique stage of life. It wasn't always considered so, however; this view came about through a series of cultural, political, and economic changes. In preindustrial times children were expected to care for younger siblings and contribute to the household in other ways as soon as they were physically capable. Toys specifically for children, one of the cultural markers of childhood, did not exist. The early years of one's life were not regarded as a time for play and education. They represented a time of work and responsibility, like most of life's course. However, the development of the industrial factory led to the need for schools, a place where children might spend their days, because parents now left home to go to work. This change in parents' lifestyles meant that a separate sphere had to be created for children. People's lives became more segregated by *age,* and because of this separation, childhood began to be taken seriously and protected. For example, child labor laws sprang up to protect children from hazardous working conditions.

Which one of these paintings is a "real" Jackson Pollock? Does it matter? (For the record, the real one is on the right.) How is the controversy over the Pollock paintings an example of the social construction of reality?

SYMBOLIC INTERACTIONISM

a micro-level theory in which shared meanings, orientations, and assumptions form the basic motivations behind people's actions.

Adolescence has had a similar history. The notion that a distinct phase exists between childhood and adulthood is relatively new. This important change evolved during the 1950s. After decades of financial hardship and war, America's booming economy led to a new freedom in the popular culture. Access to higher education broadened as well. For many Americans, this extension of educational opportunity delayed the onset of adult responsibilities—such as employment and raising families. With the possibility of an extended adolescence, teenagers (biologically able to reproduce but delayed in their assumption of adult sexual roles) emerged as a discrete social category. Concurrently, cultural changes were marked by the advent of rock and roll, doo-wop, and other popular forms of music. With more access to radios and, to a lesser extent, televisions, and with more free hours after school, teens found the time and means to consume this emerging music culture. Today we can see the life course becoming even further subdivided, with the advent of terms such as *tween*, which refers to the time between childhood and one's teenage years (roughly, ages 10 to 12) and *emerging adulthood*, which is between age 18 and the late twenties.

As an example of cross-cultural differences in the social construction of reality, consider the following question: What constitutes food? We tend to think of insects as interesting, necessary to the ecosystem, annoying, or maybe simply gross. However, we generally do not regard them as food. In other cultures, dishes such as fried grasshoppers and chocolate-covered ants are considered a delicacy. Which perspective is "right"? In fact, Americans may be an exception in not eating insects, which, I am told, are delicious. Insects are nutritious (high in protein, low in fat and calories, and a good source of many vitamins and minerals) and economical, but they are inherently neither gross nor delicious. The point is that foods, which we take for granted as natural or self-evident, are assigned meanings and values in different cultural contexts.

We might call the process by which things—ideas, concepts, values—are socially constructed symbolic interactionism (see Chapter 1), which

Lunch? Cicadas, grasshoppers, and other insects on skewers for sale in Donghuamen Night Market in Beijing, China.

suggests that we interact with others using words and behaviors that have symbolic meanings. This theory has three basic tenets:

1. Human beings act toward ideas, concepts, and values on the basis of the meaning that those things have for them.

2. These meanings are the products of social interaction in human society.

3. These meanings are modified and filtered through an interpretive process that each individual uses in dealing with outward signs.

Symbolic interactionism can be very useful in understanding cultural differences in styles of social interaction. A classic example of this would be the distance two people stand from one another when conversing. Just because these boundaries are symbolic does not mean that they aren't real. In an interaction between a tourist and a local, standing too close or too far away may make one of the parties feel uncomfortable, but in an international business meeting or in peace talks between nations, symbolic interactions take on a greater level of importance. Similarly, in our culture, we generally believe that looking someone in the eye while talking to them indicates respect and sincerity. In some other cultures, however, it is considered extremely rude to look someone directly in the eye. When you raise your glass in a toast with others and say, "Cheers," you can generally focus your eyes wherever you want. In many European countries, however, it is highly impolite and may be regarded as a sign of dishonesty not to establish eye contact while touching glasses.

Comprehending the three basic tenets of symbolic interactionism listed

How does symbolic interactionism help us understand the differences in greetings among various cultures? Pictured here are Bedouins touching noses, Malian men with their arms around each other, and the Belgian royal family celebrating the prince's eighteenth birthday.

earlier is key to seeing how the process of social construction is both ongoing and embedded in our everyday interactions. Symbolic interactionism as a theory is a useful tool for understanding the meanings of symbols and signs and the way shared meanings—or a lack thereof—facilitate or impede routine interactions. Let's look at another theory of human interaction, Erving Goffman's dramaturgical theory of society, which laid the groundwork for symbolic interactionism by using the language of theater as a paradigm to formally describe the ways in which we interact to maintain social order.

DRAMATURGICAL THEORY

DRAMATURGICAL THEORY

the view (advanced by Erving Goffman) of social life as essentially a theatrical performance, in which we are all actors on metaphorical stages, with roles, scripts, costumes, and sets.

All the world's a stage,
And all the men and women merely players.

—William Shakespeare, *As You Like It*

We might say that the dramaturgical theory of society has its roots in William Shakespeare, but we generally credit Goffman with expounding this theory in *The Presentation of Self in Everyday Life* (1959). He argued that life is essentially a play—a play with a moral, of sorts. And this moral is what

Goffman and social psychologists call "impression management." That is, all of us actors on the metaphorical social stage are struggling to make a good impression on our audience (who also happen to be actors). What's more, the goal is not just to make the best impression on others; we often actively work to ensure that others will believe they are making a good impression as well. This helps keep society and social relations rolling along smoothly (without the need for too many retakes).

Because we are all actors with roles, according to Goffman's dramaturgical theory, we also need scripts, costumes, and sets. Think again about your first day in this college classroom. You knew your role was student, and presumably the professor understood his or her role as well. The professor handed out the syllabus (a prop), talked about the general outline of the class for the semester (a script), and maybe gave a short lecture (a performance). What if he had instead started talking about his love for Batman and all things related to Batman? You might have indulged him briefly, thinking that he seemed a bit loony but would eventually relate this tangent to sociology. If your professor, however, had spent the entire lecture talking about Batman, how would you feel? Shocked? Confused? The professor would have deviated from the script generally followed in a college classroom. Was anyone wearing a tuxedo or dressed like an action hero? Probably not, because those are not the costumes we wear to a sociology lecture. Think back to your status set for a minute and ask yourself what the stages, costumes, scripts, and other props associated with each of your roles are. For some, the answer is fairly clear.

We act out some of our roles much more intentionally than others, such as when we are a member of a sports team. If you are on the basketball team, you play on a basketball court (the stage or set), wear a uniform (the costume), follow the rules (the script), and play with a ball (a prop). Often, we play several roles simultaneously, and if we have been properly socialized and don't suffer from too much role conflict, we can transition in and out of roles with a certain amount of ease. Presumably, you understand that the classroom is not the appropriate stage on which to act out an intimate scene with your partner. If you did, one of your classmates might heckle you to "get a room." In dramaturgical theory, the translation would be, "This is the wrong stage. Find the right one."

Another important part of Goffman's theory is the distinction between front-stage and backstage arenas. If you've ever participated in a school play or some other type of performance, you know all about front stage and backstage. Here the meanings are quite literal—a curtain separates these areas, so the border between the front stage and the backstage is clearly delineated. If you've ever worked in a restaurant, you see very clearly a similar line of demarcation. As a waiter you are all smiles out in the dining room. Whenever the customer asks for something, you reply politely with a "Yes, ma'am" or "Right away, sir." Back in the kitchen, however, you might complain loudly to the rest of the staff, criticizing your customer's atrocious taste for ordering escargot with his cheeseburger.

For people in certain professions, the distinction between front stage and backstage is more literal. In the news we sometimes hear about celebrities who have been caught saying something inappropriate because they thought the camera or microphone was off—they believed they were comfortably backstage. Whoops! The higher your profile, particularly if you have a master status such as a celebrity or politician, the greater the portion of your daily world that takes on the designation of front stage. You are always under close scrutiny, so you are always expected to perform your role. In other situations, the lines between front stage and backstage become more blurred. Has a professor ever caught you off guard while you were talking with friends about her class? You might have thought that you were backstage when joking with your friends in the student union, but the unfortunate, unexpected appearance of your professor turned the situation into a front-stage experience.

FACE

the esteem in which an individual is held by others.

Face, according to Goffman, is the esteem in which an individual is held by others. We take this notion very seriously; hence the idea of saving face, which is essentially the most important goal of impression management. If your best friend walked into the classroom right now and sat down beside you with a big smear of chocolate on his face, you would tell him. If the professor walked in sporting the same chocolate smear, however, you might not say anything. Because of the difference in status, and therefore power, between you and the professor, you may decide it's not your place, your role, to say something. If few students are in the class and the professor is your adviser with whom you have a fairly amicable relationship, you might indicate to him that he's got a little something on his cheek. Have you ever encountered a stranger with toilet paper stuck to the bottom of her shoe or with the fly of his pants unzipped? Did you say something? If you did, you were probably motivated by the thought that, if it were you, you would want someone to say something. If you didn't, perhaps you thought calling attention to the toilet paper or exposed underwear would embarrass the person. Or maybe you just simply felt it wasn't any of your business to interfere with a stranger. So you can see how no absolute, fixed scripts exist. (That said, one common rule of thumb is to speak up if the person can do something about it. For example, you would tell a man that he has chocolate on his face because he can wipe his face and remove the chocolate. On the other hand, you would not tell him that his shirt is ugly, because he probably cannot change his shirt at that moment.)

Similarly, if your professor walked in tomorrow with a black eye, you might ask him what happened. Not wanting to admit he fell asleep grading papers and smacked his face on the desk, he might simply say, "You should see the other guy." This is a common joke people make to save face in light of injuries about which they do not wish to speak. There are many subtleties to learn in life, and nobody explicitly teaches us all of these things. Like driving a car, we generally learn by doing and by sometimes making mistakes.

Indeed, mistakes, called breaches, are themselves an important part of the game. When there's a breach in an established script, we work hard to repair it and move forward. Humor is a useful tool in getting the script back in place. So if your professor lets out an enormous belch in the middle of a lecture, how would you react? You might be horrified; you might laugh nervously. You might delight a little because she has been so arrogant all semester and you are pleased to see her off script, not the perfectly polished professor. If the professor manages to recover and says, "Note to self: Skip the Mountain Dew before lecture next time!" it may smooth over the situation a little (but just a little, because let's face it, what happened was funny) and allow class to continue without a hitch. If the same thing happens during your next class, however, you are just going to assume that your professor is either poorly socialized and extremely uncouth, has severe gastrointestinal problems, or needs to scale back on carbonation.

After Sony Pictures Entertainment's internal computer system was hacked, executive Amy Pascal suffered from scandal when her e-mails and other personal information were released and her backstage life became public.

Sometimes, the stakes are so high or the situation is so new that we seek explicit guidance. People can and do make careers out of negotiating social scripts and methods for saving face—think about Miss Manners, Emily Post, or Dear Abby. Not wanting to act inappropriately but unsure of how to behave in situations where we lack a script, we may appeal to professional etiquette experts to coach us through a scene and give stage directions. The truth is, however, that human social interaction is too complex for a single script to work universally.

The art of tact even involves, on occasion, breaking the rules to make others feel comfortable. One particular fable tells the story of a peasant who is invited to the palace for dinner. Unknowingly, she picks up the bowl of water meant for handwashing and begins to drink from it. Many of the nobles at the table chuckle and mutter disparaging comments under their breath. But the queen, who invited the poor woman to dine with her, instead picks up her own silver finger bowl and drinks the water from it. She breaks a formal rule to preserve the face of the invited guest. Moreover, the social situation dictates that only the queen herself, as the most powerful figure at the table, can do that to repair the situation and save face for everyone.

There are some ways to generalize, however. For example, we almost always need to begin our scripts in specific ways; we generally can't just plunge into our lines. (Imagine a colleague getting on the elevator, not saying hello or even looking directly at you, but launching straight into the story of what happened to her on the way to work that day while staring up at the floor number indicators.) Goffman uses the term *opening* to signal the start of an encounter—that is, the first bracket. The closing bracket marks the end of an encounter. Sometimes, we need a nonverbal bracket to commence an encounter, a signal to cease our civil inattention. *Civil inattention* means refraining from directly interacting with someone, even someone you know, until an opening bracket has been issued. For example, you might clear your throat before speaking to somebody. You might catch his or her eye. You might stand up. You might purposely go to the bathroom first, so you can pass that person on the way back and start a conversation. You hope that you will bump into the person and catch his or her eye. We have to signal to people; we have to warn them that we are going to break civil inattention and initiate an encounter.

Openings can be awkward, but it may be even more difficult to end situations. How many times have you gotten off the phone saying, "I should get going." Where? To do what? Or you may say, "Well, I should let you go." Maybe you do hear sirens and screams in the background and you are genuinely concerned that your conversation is keeping the person on the phone from more pressing matters. Under less extreme circumstances, you may be phrasing your desire to end the conversation as a kind gesture to your interlocutor. Sometimes, we resort to formal closings. A ringing bell often indicates the end of a class period. In the absence of a bell, or even prior to its ringing, however, you may close your notebook, start to pack your books, and pick up your coat to signal to the professor that, although the lecture on socialization is fascinating, you have another class in 10 minutes that is a 15-minute walk away. There are *given gestures* that signal a closing, such as putting on your coat. Or, at the end of a meal, you may rub your stomach and say, "Boy, that was delicious. I'm stuffed!" to indicate that you have finished eating but did very much enjoy the food. However, *given-off gestures* also exist, unconscious signals of our true feelings. If you grimace every time your fork reaches your mouth, you communicate something else entirely regarding your thoughts about the meal. Many of our gestures and brackets are nonverbal. This is why it is often more difficult to end a conversation on the phone than in person: Much of our toolbox is not available to us.

ETHNOMETHODOLOGY

At this point, you should be thinking more critically about the everyday interactions that we often (necessarily) take for granted. In the 1950s and 1960s Harold Garfinkel (1967) developed a method for studying social interactions, ethnomethodology, that involves *acting* critically about them.

ETHNO-METHODOLOGY

literally "the methods of the people"; this approach to studying human interaction focuses on the ways in which we make sense of our world, convey this understanding to others, and produce a shared social order.

Ethnomethodology literally means "the methods of the people" (from *ethnos*, the Greek word for "people"). Garfinkel and his followers became famous for their "breaching experiments." They would send their students into the social world to see what happened when they breached social norms. In one example, Garfinkel sent his students home for the weekend and told them to behave as if their parents' home were a rooming house where they paid rent. Imagine what would happen if you sat down at the kitchen table and demanded to know when dinner was typically served and what days of the week the bed linens were changed. Similarly, a New York City professor instructed his students to ask people on the subway for their seats without offering any reason. Some students simply could not take this action. Others did it, but lied and said that they were not feeling well when they asked for the seat. Try your own breaching experiments. What do you normally do when you get in an elevator? You face forward and watch the numbers above the door. Next time you get into an elevator, face backward. See how the other people in the elevator react. A few weeks into the semester, students tend to sit in the same seats in the same class, particularly if it's a small group, even if they do not have assigned seats. The next time you are in such a situation, take someone else's usual seat and see how he or she reacts. (Come on, that's an easy one!)

Let's think through some of the reactions to breaches. Take the example of the student on the subway asking a stranger to give up his seat. What would you do if this happened to you? Would you give up your seat to the student? You might just get up and offer it to him, because you assume he would not ask for it unless he had a good reason (even if you don't see a cast on his leg). If the stranger were elderly, on crutches, or pregnant or had a small child, you might be more willing to get up—in fact, on most forms of public transportation, seats are marked as reserved for exactly these kinds of people. When this class assignment was originally handed out, note that some of the students simply could not bring themselves to do it. Others lied, saying they were sick, because being ill provides a valid excuse to sit down. Why is that? What's the big deal in asking someone to give you a seat? Well, you might decide, you just don't typically do such a thing. It's not normal behavior. But note that as a society, we construct rules and meanings for what constitutes normal.

Now contemplate the previously mentioned elevator scenario. How would you react if you stepped into an elevator and the only other person inside was facing away from the door (assuming the elevator only opens on one side)? You might look at the wall to see if something is there at which the person is looking and that you can't see. You might try to stand farther away from that person. If someone else gets on the elevator, maybe you would gauge her reaction. If she gave you a look that said, "What's this guy doing?" you might give a sympathetic look, even the hint of a smile, to indicate your agreement: "Yeah, crazy, huh?" Then you could breathe a little

A scene from the film *Borat*. What established scripts did Sacha Baron Cohen's character Borat violate by going on an elevator naked? How did the unsuspecting woman on the elevator try to cope with the breach?

easier, knowing that the guy facing backward in the elevator is abnormal, and you, the other passenger, and anyone else who steps inside and faces forward are normal, although there's no social imperative to face forward rather than backward in an elevator.

If you walked into the classroom today and someone was sitting in the seat you consider "yours," how would you feel? Maybe you wouldn't think twice about it, but chances are you would notice, have an emotional reaction—be it annoyance, confusion, even anger—and get to class earlier the next time to claim what's rightfully yours.

NEW TECHNOLOGIES: WHAT HAS THE INTERNET DONE TO INTERACTION?

What happens when we are faced with entirely new situations? We have no rules, no scripts, no established social norms. How do people know what to do or what constitutes appropriate behavior? Usually, some continuity exists between situations, so we can draw on our previous knowledge (just as you could anticipate the norms of a college classroom even though you had never been in one before). But what about something like the internet, something that humans created, which in turn creates social situations never before possible? Let's think about what happens.

Take social media, for example. Tinder, Reddit, Second Life, and other online forums provide an interesting test of the dramaturgical model. The potential anonymity of the internet allows us to portray ourselves however we choose. Online I can play the role of a 57-year-old stay-at-home mom who was an international college badminton champ and who enters semi-annual pesto-making competitions. Is that who I am? Some online dating and networking websites allow us to craft our own presentation of self, alter our identities, and thereby create new "realities." We've largely removed the stage, costumes, props, vocal inflections, and other nonverbal cues, so we must depend entirely on the scripts with which we are presented. We even develop new ways of communicating—for example, by using emojis to substitute for nonverbal cues to indicate tone (e.g., that we're joking in a text). 😊 😜

The Internet has changed society in other ways, such as forcing us to develop new technologies to prevent identity theft and create secure online transactions for shopping and banking. Certain aspects of the internet have also altered the nature and details of crime. Software that allows for music sharing necessitates the development of a new set of ethics. Is it okay to download music free from the internet? It's illegal in most cases, for sure, but is it unethical? Are there certain circumstances in which it might be okay? For example, when Radiohead made its album *In Rainbows* available for download and left the amount of payment up to customers, should folks have paid or was it just fine to download without donating? The new technologies are changing what we as a society mean by the term *stealing*. Other websites, such as eBay, allow for the sale of stolen goods in ways previously unforeseen. Intentionally or not, people may foster underground economies by purchasing stolen goods in a way not possible before the advent of online auctions. Clicking on an item and entering your credit card number is a completely different social interaction than walking into a dark alley and buying something that literally fell off the back of a truck.

POLICY

ROOMMATES WITH BENEFITS

Eager to throw off my nerdy past and reinvent myself at college, I wrote "party animal" on my roommate application form where it asked incoming freshmen whether they wanted to bunk with a smoker or a nonsmoker. When I told my mother about this later, she laughed and bought me a T-shirt that sported the image of Spuds MacKenzie, the 1980s Budweiser beer mascot, under the words "The original party animal."

I ended up with Tony from Sacramento, a very quiet, Republican son of a judge. (I suppose it's good policy to separate the party animals from those who request them.) I learned to appreciate his taste in music (U2 and the Smiths, as opposed to my predilection for reggae and jazz), and we agreed to disagree about politics during the reelection campaign of Alan Cranston, then one of the most liberal members of the US Senate. I had never met anyone like Tony. And I'm pretty sure he hadn't come across many half-Jewish, Democratic children of New York artists. We learned to get along that first year at Berkeley, and every now and then even tried on each other's values and beliefs, just to see how they fit.

Today I am sad that most of my students will not experience what I did back when Facebook CEO Mark Zuckerberg was in diapers. While the internet has made it easy to reconnect with the lost Tonys of our lives, it has made it a lot more difficult to meet them in the first place, by taking a lot of randomness out of life. We tend to value order and control over randomness, but when we lose randomness, we also lose serendipity.

As soon as today's students receive their proverbial fat envelope from their top-choice college, they are on Facebook meeting other potential freshmen. By the time the roommate application forms arrive, many like-minded students with similar backgrounds have already connected and agreed to request one another as roommates.

It's just one of many ways in which digital technologies now spill over into non-screen-based aspects of social experience. I know certain people who can't bear to eat in a restaurant they haven't researched on Yelp. And Google and Facebook, of course, tailor content to exactly what they think you want to find.

But this loss of randomness is particularly unfortunate for college-age students, who should be trying on new hats and getting exposed to new and different ideas. Which students end up bunking with whom may seem trivial at first glance. But research on the phenomenon of peer influence—and the influences of roommates in particular—has found that there are, in fact, long-lasting effects based on whom you end up living with your first year.

David R. Harris, a sociologist at Cornell, studied roommates and found that white students who were assigned a roommate of a different race ended up more open-minded about race (Harris & Sim, 2002). In an earlier study, the economist Bruce Sacerdote (2001) found that randomly assigned roommates at Dartmouth affected each other's GPAs.

(Of course, influences can sometimes be

Patti Kilroy, right, and Bliss Baek in their dorm room at NYU after first contacting each other through Facebook. Does this process of connecting online reduce the diversity and randomness of the college experience?

negative. Roommates can drive each other's grades up or down. In 2003, researchers at four colleges discovered that male students who reported binge drinking in high school drank much more throughout college if their first-year roommate also reported binge drinking in high school [Duncan et al., 2005; Eisenberg et al., 2013; Kremer & Levy, 2008].)

These studies are important because we know that much education takes place outside the formal classroom curriculum and in the peer-to-peer learning that occurs in places like dorm rooms.

Other than prison and the military, there are not many other institutions outside of college that shove two people into a 100-square-foot space and expect them to get along for nine months. Can you think of any better training for marriage?

In fact, in my research with Jennifer A. Heerwig, we found that Vietnam-era military service actually lowers the risk of subsequent divorce (Conley & Heerwig, 2011). It's possible that the military teaches you how to subsume your individual desires for the good of the collective—in other words, how to get along well with others.

The drive to tame randomness into controllable order is a noble impulse, but letting a little serendipity flourish isn't such a bad thing. Nor is getting to know someone different from yourself. All colleges should follow the lead of Hamilton College in New York, where roommate choice is not allowed. And if you end up with the roommate from hell? You'll survive, and someday have great stories to tell your future spouse, with whom you'll probably get along better.

Conclusion

In the nature-versus-nurture debate, sociologists have generally fallen firmly on the side of nurture. At least that's what most of us study. Socialization helps us understand and explain how it is that babies—those wrinkled, sometimes alien-looking creatures newly emerged from their mothers' wombs—become people who attend college. How much stake do we put in preserving normality, the status quo? We already talked about tact. Little children are tactless all the time. I once handed my grandfather a wrapped present and said, "Happy Father's Day! It's a shirt!" It was cute because I was 5, but what if I had done the same thing at 15? Would it have been cute? Somewhere along the way, we learn the unwritten rules.

On May 23, 1999, the World Wrestling Federation (WWF) aired a live pay-per-view event called *Over the Edge*. When a harness malfunctioned, wrestler Owen Hart, who was being lowered into the ring, fell 78 feet to his death. Although television viewers did not see the live footage of his death, once Hart was removed and sent to the hospital (he was pronounced dead on arrival), the program continued. The decision to go forward with the event sparked conversation and criticism far beyond the wrestling world. Some were shocked, even outraged, that the producers did not cancel the program, whereas others stood by the old adage, "The show must go on."

Paramedics try to resuscitate professional wrestler Owen Hart after a deadly fall in Kansas City, Missouri.

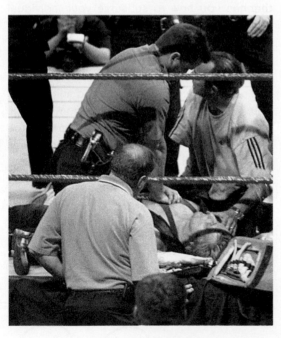

Why the radically different reactions? Well, for starters, we do not have a social script for this type of incident. Death is a serious matter, to be sure, but how should death be handled when it occurs on live television? We tend to seek a return to normalcy when things do not go as planned. Trying to quickly transition the situation back to normal (and probably to keep their paying customers happy), the WWF continued the live event. Others were disgusted by this seeming insensitivity. The following night the association aired a two-hour tribute to the fallen wrestler.

This is an extreme example and a tragic one, but we witness more mundane, even comical, breaches on a fairly regular basis. When someone important commits them, they draw yet more attention. The point is that we all have a stake in things going a certain way. When things don't go according to plan, depending on the severity, nature, and context of the deviation, we must find some way to recover, and recover we usually do. Whether it is a small breach in our shared understanding of turn-taking etiquette or a major rupture to the fabric of society, people work hard to broker consensus with respect to shared meaning. The alternative is nothing less than chaos and social insanity.

QUESTIONS FOR REVIEW

1. How does George Herbert Mead's concept of "generalized other" explain why, at the beginning of class, you became silent when your professor started speaking?

2. How does the case of "Anna" affect your assessment of early-socialization programs like Head Start?

3. School plays an important role in our socialization. Think about the way socialization works: What are some of the things we learn from schooling (e.g., the first years in elementary school), and how does this learning differ from what we are taught by our teacher? How are things like gender performance shaped in school?

4. Parents of different social classes socialize children differently. For example, middle-class parents are more likely to stress independence and self-direction, whereas working-class parents prioritize obedience to external authority. Using this example, how does socialization through families potentially reproduce social inequality?

5. You are a university student, but you also wait tables at a restaurant. One evening, one of your professors happens to come in for a meal (awkwardly, on a first date!). Use role theory to describe the interaction (and possible role conflict) that ensues.

6. What do sociologists mean by "social construction of reality"? How does the idea of social construction bring into question certain elements of everyday life, like gender roles?

7. Let's imagine you use file-sharing networks for music downloading and to discuss your favorite music subgenres with people throughout the world. How does this differ from, for example, speaking only with employees at your local music store? Think about the way technology affects how you interact, the characteristics of the people with whom you're interacting, and how different ways of interacting might affect socialization.

PRACTICE

ROLE CONFLICT AND ROLE STRAIN

Part of modern social life involves seamlessly managing the inevitable conflicts and strains that arise due to the many hats we wear. Sociologist Robert Merton identified two forms of this phenomenon: (1) role strain, where the tension is between two roles (i.e. duties or scripts) associated with a particular status; and (2) role conflict, which stems from competing demands arising from different statuses.

TRY IT!

Make a list of a few of your statuses, e.g. student, roommate, waitress. For each of the statuses, list three roles that each entails. Here's one example:

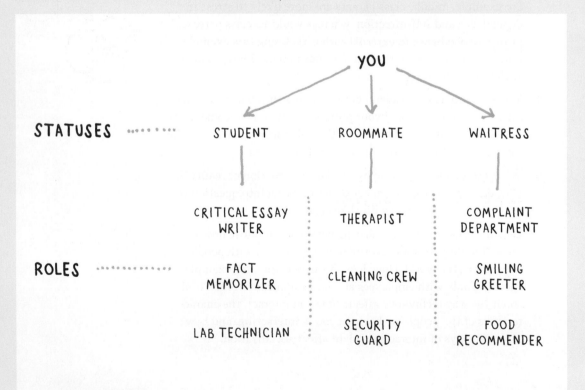

YOU

STATUSES	STUDENT	ROOMMATE	WAITRESS
	CRITICAL ESSAY WRITER	THERAPIST	COMPLAINT DEPARTMENT
ROLES	FACT MEMORIZER	CLEANING CREW	SMILING GREETER
	LAB TECHNICIAN	SECURITY GUARD	FOOD RECOMMENDER

THINK ABOUT IT

When you lay out these various roles, you can see that the role of "complaint department" is in conflict with the role of "food recommender." You have to upsell the food and make it sound appealing—but then you have to take seriously complaints about the undercooked burger you just recommended.

ROLE STRAIN: COMPLAINT DEPARTMENT ←——→ FOOD RECOMMENDER

We can also see evidence of role strain. Let's say your essay is due by midnight. You made some progress, but the door to your dorm room springs open and your roommate is weeping because her boyfriend just dumped her. What do you do?

ROLE CONFLICT: CRITICAL ESSAY WRITER ←——→ THERAPIST

How do you manage the role strains and conflicts that you've identified? Merton talked about compartmentalization, for example, in which you create a firewall between two conflicting statuses.

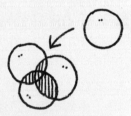

SOCIOLOGY ON THE STREET

Why does breaking social norms make others uncomfortable or even hostile? Watch the Sociology on the Street video to find out more: **digital.wwnorton.com/youmayask6core.**

WANT MORE PRACTICE?

Complete the InQuizitive activity for this chapter at digital.wwnorton.com /youmayask6core

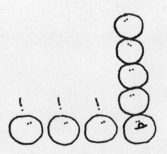

5

PARADOX

THE STRENGTH OF WEAK TIES:
IT IS THE PEOPLE WITH
WHOM WE ARE THE LEAST
CONNECTED WHO OFFER US
THE MOST OPPORTUNITIES.

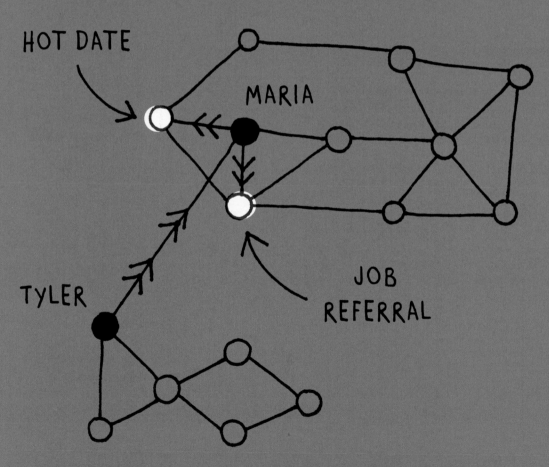

HOT DATE

MARIA

JOB
REFERRAL

TYLER

Groups and Networks

Who is Satoshi Nakamoto? Nobody really knows for sure. Candidates have included a deceased extropian cryptographer on the West Coast of the United States named Hal Finney (extropians are a group of people dedicated to finding a way to live forever), a Hungarian American recluse, an Australian academic, and others. What is for sure is that whoever uses the pseudonym Satoshi Nakamoto is one of the 50 richest people in the world, due to the value of the bitcoin that person possesses (worth about $20 billion at the time of this writing). In late 2008, someone using the name Satoshi Nakamoto published a white paper online titled "Bitcoin: A Peer-to-Peer Electronic Cash System." The idea was bold: Instead of keeping financial records in a central system, like a bank, all financial transactions would appear on a public shared ledger, the "blockchain." It wasn't until January 2009, at the depths of the financial crisis and the Great Recession, that Nakamoto released the software to make operational the concept of a peer-to-peer anonymous form of currency. In the first "block" issued in the blockchain, Nakamoto embedded the following message: "The Times 3 January 2009 Chancellor on brink of second bailout for banks." And, voila, the first major cryptocurrency was born.

Since the launch of bitcoin, the number of cryptocurrencies has skyrocketed. New ones include ethereum, litecoin, and zcash. Once mainly used for the "dark web," or online trade in illegal goods, bitcoin (or other electronic currency) are now accepted as payment by a rapidly growing number of mainstream businesses. Many residents of countries with unstable currencies who may have once tried to shift their assets to dollars or euros now use bitcoin to avoid the possibility that runaway inflation will decimate their savings.

↑

In Seoul, South Korea, monitors display exchange rates of cryptocurrencies including bitcoin (top left) and others such as ethereum and litecoin.

The CEO of America's largest bank, JPMorgan Chase, once scoffed at bitcoin, claiming that he would fire anyone who traded in it, but he now admits the power of the blockchain concept. Cryptocurrencies represent an amazing feat of technical savvy—specifically the use of digital signatures and hash functions—to make sure all transactions are on the up-and-up. (For a great overview on how they work from a technical perspective, see **youtube.com /watch?v=bBC-nXj3Ng4**). But they also represent the triumph of social networks.

All currencies rely on trust within a social network to some extent. After all, if everyone suddenly decided not to accept British pounds sterling or Swiss francs, the value of these fancy pieces of paper would evaporate. But traditional currencies are all backed by a government—usually a central bank—that regulates the amount in circulation and "guarantees" their authenticity. Even pseudo-currencies like airline points (which exceeded the total value of dollars or euros or pounds back in 2005 [Clark, 2005]) require a central authority to lend them value—namely the airline. What's unique about bitcoin and other cryptocurrencies is that their authority and value is not centralized at all but rests in the social network itself by virtue of the fact that a coin is merely an entry into a public ledger (or blockchain) that is recorded multiple times across the entire network of users' computers. Yet the owners (and miners) of bitcoin can remain completely anonymous, like our protagonist, Satoshi Nakamoto. Welcome to the power of social networks.

This chapter explores some of the basic theories about group interaction and how it shapes our social world. We'll look at the connections between groups: how size and shape matter, what roles group members play, and how the power of groups works compared with individuals and other institutions. We'll also discuss organizations and how they both react to and create social structure.

Social Groups

Unless you live alone in the woods, and perhaps even then, you are a member of many social groups. Social groups form the building blocks of society and most social interaction. In fact, even the self evolves from groups. Let's start by talking about the various types and sizes of groups. In his classic work "Quantitative Aspects of the Group," sociologist Georg Simmel (1950) argues that without knowing anything about the group members' individual psychology or the cultural or social context in which they are embedded, we can make predictions about the ways people behave based solely on the number of members, or "social actors," in that group. This theory applies not just to groups of people but also to states, countries, firms, corporations, bureaucracies, and any number of other social forms.

JUST THE TWO OF US

Simmel advances the notion that the most important distinction is that between a relationship of two, which he calls a dyad, and a group of three, which he calls a triad. This is the fundamental distinction among most social relations, he argues, and it holds regardless of the individual characteristics of the group's members. Of course, personality differences do influence social relations, but there are numerous social dynamics about which we can make predictions that have nothing to do with the content of the social relations themselves.

The dyad has several unique characteristics. For starters, it is the most intimate form of social life, partly because the two members of the dyad are mutually dependent on each other. That is, the continued existence of the group is entirely contingent on the willingness of both parties to participate in the group; if either person leaves, the dyad ceases to be. This intimacy is enhanced by the fact that no third person exists to buffer the situation or mediate between the two. Meanwhile, the members of a dyad don't need to be concerned about how their relationship will be perceived by a third party.

For example, we might consider a couple the most intimate social arrangement in our society. Both people must remain committed to being in the dyad for it to exist, and if one partner leaves, the couple no longer exists. There can be no secrets—if the last piece of chocolate cake disappears and you didn't eat it, you know who did. You could withhold a secret from your dyadic partner, but in terms of the actions of the group itself, no mystery lingers about who performs which role or who did what. Either you did it, or the other person did.

In a dyad, symmetry must be maintained. There might be unequal power relations within a group of two to a certain extent, but Simmel would argue that in a group of two an inherent symmetry exists because of the earlier stipulation of mutual dependence: The group survives

DYAD

a group of two.

TRIAD

a group of three.

Dyads are the foundation of all social relationships. Why are they the most intimate relationship, according to Georg Simmel?

only if both members remain. Even in relationships where the power seems so clearly unequal—think of a master and a servant or a prisoner and his captor—Simmel argues that there's an inherent symmetry. Yes, the servant may be completely dependent on the master for his or her wages, sustenance, food, and shelter, but what happens to the master who becomes dependent on the labor that the servant performs? Of course, forcible relationships might develop in which one of two parties is forced to stay in the dyad, but to be considered a pure dyad, the relationship has to be voluntary. Because a dyad could fall apart at any moment, the underlying social relation is heightened.

The dyad is also unique in other ways. Because the group exists only as long as the individuals choose to maintain it in a voluntary fashion, the group itself exerts no supra-individual control over the individuals involved. For example, whereas a child might claim, "She made me do it" and shamelessly tell on her older sister, a member of a dyad is less likely to say, "I was just following orders" or "The whole group decided to go see *Mission: Impossible 7*, and I really didn't want to, but I went anyway." The force of a group is much stronger when three or more individuals are part of the group.

Let's take a real-life example of how the characteristics of a dyad play out and see why they matter. Think about divorce. One point in a marriage at which the divorce rate is especially high is when a first child is born (and not just because of the parental sleep deprivation that arrives with a newborn). The nature of the relationship between the two adults changes. They have gone from being a dyad to becoming a triad. Perhaps the parents feel a sudden lack of intimacy, even though the baby is not yet a fully developed social actor. On the flip side, a husband or wife might begin to feel trapped in a marriage specifically because of a child. All of a sudden, group power exists—a couple has evolved into a family, and with that comes the power of numbers in a group.

AND THEN THERE WERE THREE

This brings us to the triad, distinguished by characteristics you can probably infer by now. In a triad, the group holds supra-individual power. In other words, in a group of three or four, I can say, "I'm really unhappy, I hate this place, I hate you, and I'm leaving," but the group will go on. The husband may walk out on his wife and children, but the family he's abandoning still exists. He's ending his participation in the group, but the group will outlast his decision to leave it. Therefore, the group is not dependent on any one particular member.

What's more, in a triad, secrets can exist. Who left the cap off the toothpaste? If more than two people live under the same roof, you can't be sure. Politics is another aspect inherent in a group of three or more. Instead of generating consensus between two individuals, now you have multiple points of view and preferences that need to be balanced. This allows for power politics among the group's members. Simmel refers to three basic forms of political relations that can evolve within a triad depending on what role the entering third party assumes (Figure 5.1). The first role is that of mediator, the person who tries to resolve conflict between the other two and is sometimes brought in for that explicit purpose. A good example would be a marriage counselor. Rather than go to therapy, couples having marital problems often start a family because they believe a baby will bring them back together. Unfortu-

In what ways are triads more complex than dyads? What are the possible roles of triad members?

nately, as most couples come to realize sooner rather than later, a baby cannot play the role of a mediator. Rather, the dynamics of the unhappy family may turn into a game of chicken: Which parent is more devoted to the child? Which dyad forms the core of the group, and which person will be left out or can walk away more easily?

A second possible role for the incoming third member of a triad is that of *tertius gaudens* (Latin for "the third that rejoices"). This individual profits from the disagreement of the other two, essentially playing the opposite role from the mediator. Someone in this position might have multiple roles. In the previous example, the marriage counselor plays the part of the mediator, but she is also earning her wages from the conflict between the couple. Maybe she encourages continued therapy even after the couple appear to have resolved all their issues, or perhaps she promotes their staying together even though they've already decided to get a divorce.

The third possible role that Simmel identifies for a third party is *divide et impera* (Latin for "divide and conquer"). This person intentionally drives a wedge between the other two parties. This third role is similar to *tertius gaudens*, the difference between the two being a question of intent and whether the rift preexisted. (If you've ever seen or read Shakespeare's *Othello*, there is no better example of *divide et impera* than the way Iago, counselor to Othello, uses the Moor's insecurities to foster a rift between him and his wife, Desdemona, in order to strengthen Iago's own hand in court politics. The play ends tragically, of course.)

To take an example with which millions of Americans are familiar, let's return to the case of the triad formed when a romantic couple has a child but experiences strife and separates. What happens when the couple divorces? What role does the child play? A child could play any of the roles mentioned above. In the original dyad of the biological parents, the child

MEDIATOR

the member of a triad who attempts to resolve conflict between the two other actors in the group.

TERTIUS GAUDENS

the member of a triad who benefits from conflict between the other two members of the group.

DIVIDE ET IMPERA

the role of a member of a triad who intentionally drives a wedge between the other two actors in the group.

FIGURE 5.1 Political Relations within a Triad

MEDIATOR

The mediator attempts to resolve conflict between the other two members of the triad and is sometimes brought in for that explicit purpose.

TERTIUS GAUDENS

Latin for "the third that rejoices." This individual profits from the disagreement of the other two actors, essentially playing the opposite role from the mediator.

DIVIDE ET IMPERA

Latin for "divide and conquer." This person intentionally drives a wedge between the other two parties.

can be a mediator, forcing his or her parents to work together on certain issues pertaining to his or her care. A child can be "the third who rejoices" from the disagreement of the two, profiting from the fact that he or she might receive two allowances or extra birthday presents because each parent is trying to prove that he or she loves the child more. A complicated *divide et impera* situation could develop if one of the child's parents enters into a second marriage, in which the kid remains the biological child of one parent and becomes the stepchild of the other. In this case, all sorts of politics arise because of the biological connection between the one parent and the child-versus-the-marital-love relationship between the two adults. The relationship between the nonbiological parent and the stepchild, who have the weakest bond, may be difficult. The situation could unfold in any number of ways. Many domestic comedies (think *The Parent Trap*) are based on the premise of a young, angst-ridden prankster playing the role of *divide et impera* between his or her parent and the new stepmother or stepfather.

When contemplating how these theoretical concepts work within actual social interaction, keep in mind that these groups—dyads and triads—don't exist in a vacuum in real life. In discussing the politics of a stepfamily, we're talking about a household where there's a stepparent, a biological parent, and a child. Beyond our textbook example, in real life, this triad probably doesn't function so independently. There is likely another biological parent living elsewhere, maybe another stepparent, maybe siblings. Most social groups are very complex, and we need to take that into consideration when we attempt to determine how they operate. This is why we start with Simmel's purest forms. The interactions that take place in groups of two and three become the building blocks for those in much larger groups.

SIZE MATTERS: WHY SOCIAL LIFE IS COMPLICATED

One key insight is that as the number of people (nodes) in the group increases geometrically (2 + 1 = 3; 3 + 1 = 4; 4 + 1 = 5), the complexity of analyzing that group's ties (edges) increases exponentially (2 × 2 = 4; 4 × 4 = 16; 5 × 5 = 25). A two-person group has only one possible and necessary relationship; a tie must exist between the two people for there to be a group. In a triad, a sum of three relationships exists, with each person within the group having two ties. And again, for it to be a group of three, we're not talking about possible connections but actual ones. Each person in the triad has to have two ties—breaking a single tie turns the triad into two dyads. Even if one of the ties between two members is weaker, it is so well reinforced by the remaining two ties that it is unlikely to fade away, a feature known as the "iron law" of the triad, or more technically, "triadic closure." When you move beyond triads to groups of four or more, something different happens. To create a group of four, there must be at least four relationships, but you can have as many as six. Figure 5.2 shows this exponential rise in possible relationships graphically.

In a diagram with four people (A, B, C, and D), we can cross out the diagonals and the group will still exist. Everyone may have only two relationships as opposed to three (the number of possible relationships). A and D might never have spoken to each other, but the group will continue to function. The tendency, however, is for these possible relationships to become actual relationships. You and your roommate are in separate chemistry classes, but your lab partners happen to be roommates. If you both become friendly with your lab partners outside of class, chances are that you will meet your roommate's lab partner and vice versa. Such social ties between friends of the same friend tend to form. This can be good or bad. You may form a study group and, if your roommate's lab partner's boyfriend is a chemist, you (and everyone else) may benefit from his help. On the other hand, if the two break up and you start dating the chemist, it might make future labs a little uncomfortable for your roommate.

FIGURE 5.2 Relationship between Group Size and Complexity

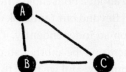

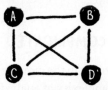

ONE POSSIBLE RELATIONSHIP

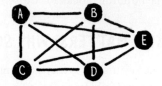

THREE POSSIBLE RELATIONSHIPS

SIX POSSIBLE RELATIONSHIPS

TEN POSSIBLE RELATIONSHIPS

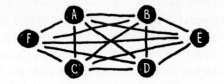

FIFTEEN POSSIBLE RELATIONSHIPS

LET'S GET THIS PARTY STARTED: SMALL GROUPS, PARTIES, AND LARGE GROUPS

SMALL GROUP

a group characterized by face-to-face interaction, a unifocal perspective, lack of formal arrangements or roles, and a certain level of equality.

Groups larger than a dyad or triad, according to Simmel, can be classified into one of three types: small groups, parties, and large groups. A small group is characterized by four factors. The first is *face-to-face interaction*; all the members of the group at any given time are present and interact with one another. They are not spread out geographically. Second, a small group is *unifocal*, meaning that there's one center of attention at any given time. Turn-taking among speakers occurs. A classroom, unless it's divided and engaged in group work, should be unifocal.

Classes usually don't qualify as small groups because a third characteristic of small groups is a *lack of formal arrangements* or roles. A study group, though, might qualify if you decide shortly before an exam to meet with some of your classmates. You need to agree on a place and time to meet, but otherwise there is no formal arrangement. In the classroom, however, a professor is in attendance, as are teaching assistants and students—all of whom play official roles in the group. The roles generally encountered in classes also contradict the fourth defining characteristic of a small group,

equality. After all, you won't be giving your professor a grade, nor will she get into trouble, as you might, for arriving late at a planned lecture. Yours isn't a reciprocal and equal relationship. Within a small group, as in a dyad, there is a certain level of equality. Only in a dyad can pure equality exist, because both members hold veto power over the group. However, in a small group, even if the group will continue to exist beyond the membership of any particular member, no particular member has greater sway than the others. No one member can dissolve the group. If someone in your study group gets tired and falls asleep on his book, you and your classmates can continue to study without him.

When does a small group become a party? If you have ever hosted a party, you know that the worst phase is the beginning. You're worried that people might not show up: The party has started, only three people are there (everyone, after all, tends to arrive fashionably late), and you start to wonder, "Is this going to be it?" You have to keep a conversation going among three people; you refill their glasses as soon as they take one sip. If you're drinking alcohol (only if you're of legal drinking age, of course), you may consume more at the beginning of the party because you are nervous about it not going well. When does your small gathering officially evolve into a party, so that you can relax and enjoy yourself? Simmel would say that a party, like a small group, is characterized by face-to-face interaction but differs in that it is *multifocal.* Going back to the example of your sociology study group, if two people begin to talk about Margaret Mead's theory of the self while the rest of you discuss the differences between role conflict and role strain, then it is bifocal; if another subgroup splits off to deliberate on reference groups, then it has become multifocal. According to Simmel, you've got yourself a party! So when you're hosting your next party, you'll recognize when it has officially started.

What makes the study group on the left a small group, and how is it different from the cocktail party on the right?

PARTY

a group that is similar to a small group but is multifocal.

↑

How is this classroom an example of a large group?

LARGE GROUP

a group characterized by the presence of a formal structure that mediates interaction and, consequently, status differentiation.

PRIMARY GROUPS

social groups, such as family or friends, composed of enduring, intimate face-to-face relationships that strongly influence the attitudes and ideals of those involved.

The last type of group to which Simmel makes reference is the large group. The primary characteristic of a large group is the presence of a *formal structure* that mediates interaction and, consequently, *status differentiation*. When you enter a classroom, it should be clear who the teacher is, and you comprehend that she has a higher status than you have in that specific social context.

The professor is an employee of the university, knows more than you know about the subject being taught, and is responsible for assigning you a grade based on your performance in the course. You might be asked to complete a teacher evaluation form at the end of the semester, but it's not the same thing as grading a student. You and your professor aren't equals. The point is that the inherent characteristics of a group are determined not just by its size but also by other aspects of its form, including its formal, bureaucratic structures (if it has any). Whether a group stays small, becomes a party, or evolves into a large group may depend on numbers, but it also may depend on the size and configuration of the physical space or technological platform that mediates interactions between and among group members, preexisting social relationships, expectations, and the larger social context in which the group is embedded.

PRIMARY AND SECONDARY GROUPS

Simmel wasn't the only one who tried to describe the basic types of groups. Sociologist Charles Horton Cooley (1909) emphasized a distinction between what he called primary and secondary groups. Primary groups are *limited in the number of members*, allowing for face-to-face interaction. The group is an end unto itself, rather than a means to an end. This is what makes your family different from a sports team or small business: Sure, you want the family to function well, but you're not trying to compete with other families or manufacture a product. Meanwhile, primary groups are *key agents of socialization*. Most people's first social group is their family, which is a primary group. Your immediate family (parents and siblings) is probably small enough to sit down at the same dinner table or at least gather in the same room at the same time. Loyalty is the primary ethic here. Members of a primary group are *noninterchangeable*—you can't replace your mother or father. And while you have strong allegiances to your friends, your primary loyalty is likely to be to your family. Finally, the relationships within a primary group are *enduring*. Your sister will always be your sister. Another example of a primary group might be the group of your closest friends, especially if you've known each other since your sandbox years.

The characteristics of secondary groups, such as a labor union, stand in contrast to those of primary groups. The group is *impersonal*; you may or may not know all of the members of your union. It's also *instrumental*; the group exists as a means to an end, in this case for organizing workers and lobbying for their interests. In a secondary group, affiliation is *contingent*. You are only a member of your union so long as you hold a certain job and pay your dues. If you change jobs or join another union, your membership in that earlier group ends. Because the members of a secondary group change, the roles are more important than the individuals who fill them. The shop steward, the person chosen to interact with the company's management, may be a different person every year, but that position carries the same responsibilities within the group regardless of who fills it. A sports team is another example of a secondary group, although if you're also close friends with your teammates and socialize with them when you're not playing sports, the line between a primary and secondary group can become blurred.

SECONDARY GROUPS

groups marked by impersonal, instrumental relationships (those existing as a means to an end).

GROUP CONFORMITY

Although we tend to put a high value on individuality in American culture, our lives are marked by high levels of conformity. That is, groups have strong influences over individual behavior. In the late 1940s the social psychologist Solomon Asch carried out a now-famous series of experiments to demonstrate the power of norms of group conformity. He gathered subjects in a room under the pretense that they were participating in a vision test, showed them two images of lines, and asked which ones were longer than the others and which were the same length (Figure 5.3).

The trick was that only one person in each room was really a research subject; the rest of the people had been told ahead of time to give the same wrong answer. While a majority of subjects answered correctly even after they listened to others give the wrong answer, about one-third expressed serious discomfort—they clearly struggled with what they thought was right in light of what everyone else was saying. Subjects were the most confused when the entire group offered an incorrect answer. When the group members gave a range of responses, the research subjects had no trouble answering correctly. This experiment

FIGURE 5.3 The Asch Test

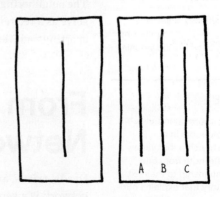

SOURCE: Asch, 1956.

demonstrates the power of conformity within a group. More troubling instances of group conformity may be seen in cases of collective violence such as gang rape, which tends to occur among tightly knit groups like sports teams or fraternities (Sanday, 1990).

IN-GROUPS AND OUT-GROUPS

IN-GROUP

another term for the powerful group, most often the majority.

In-groups and out-groups are another broad way of categorizing people. The in-group is the powerful group, most often the majority, whereas the out-group is the stigmatized or less powerful group, usually the minority (though the numbers don't have to break down this way). For example, in the United States, heterosexuals are the in-group in terms of sexuality (both more powerful and numerically greater), whereas homosexuals, bisexuals, and those who have other nonnormative sexual identities fall into the out-group. However, in South Africa, despite being a minority group, whites are the in-group because of their enormous political and economic power (the legacy of colonialism and apartheid), whereas blacks are the out-group despite their greater numbers. The significance of in-groups and out-groups lies in their relative power to define what constitutes normal versus abnormal thoughts and behavior.

OUT-GROUP

another term for the stigmatized or less powerful group, the minority.

REFERENCE GROUPS

REFERENCE GROUP

a group that helps us understand or make sense of our position in society relative to other groups.

We often compare ourselves to other groups of people we do not know directly in order to make comparisons. For example, your class might compare itself to another introduction to sociology class, which has a take-home midterm and an optional final. If your class has a 20-page term paper and a two-hour comprehensive final exam, you might feel as if you face an unfair amount of work. If your class has no external assignments, however, and your final grades are based on self-assessment, you might feel comparatively lucky. In either case, the other class serves as a reference group. Reference groups help us understand or make sense of our position in society relative to other groups. The neighboring town's high school or even another socioeconomic class can serve as a reference group. In the first instance, you might compare access to sporting facilities; in the second, you might compare voting patterns.

SOCIAL NETWORK

a set of relations— essentially, a set of dyads—held together by ties between individuals.

TIE

the connection between two people in a relationship that varies in strength from one relationship to the next; a story that explains our relationship with another member of our network.

From Groups to Networks

Dyads, triads, and groups are the components of social networks. A social network is a set of relations—a set of dyads, essentially—held together by *ties* between individuals. A tie is the content of a particular relationship. One way to think about "the ties that bind" is as a set of stories we tell each other that explain a particular relationship. If I ask you how you know a specific person, and you explain that she was your brother's girlfriend in the eighth grade, and the two of you remained close even after her relationship with your brother ended, that story is your tie to that person. For every person in your life, you have a story. To explain some ties, the story is very simple: "That's the guy I buy my coffee from each morning." This is a uniplex tie. Other ties have many layers. They are multiplex: "She's my girlfriend. We have a romantic relationship. We also are tennis and bridge partners. And now that you mention it, we are classmates at school and also fiercely competitive opponents in Trivial Pursuit."

A narrative is the sum of stories contained in a set of ties. Your university or college is a narrative, for example. Every person with whom you have a relationship at your university forms part of that network. For all your college-based relationships—those shared with a professor, your teaching assistant, or classmates—your school is a large part of the story, of the tie. Without the school, in fact, you probably wouldn't share a tie at all. When you add up the stories of all the actors involved in the social network of your school—between you and your classmates, between the professors and their colleagues, between the school and the vendors with whom it contracts—the result is a narrative of what your college is. Of course, you may have other friends, from high school or elsewhere, who have no relationship to your school, so your college is a more minor aspect of those relationships.

In Chapter 4 we talked about the power of symbols. What would you need to do if you wanted to change the name of your school? You and your friends could just start calling it something else. But you'd probably have to hire a legal team to alter the contracts (a form of tie that is spelled out explicitly in a written "story") that the school maintains with all its vendors. You would need to alert alumni and advise the departments and staff at school to change their stationery, websites, and any marketing materials. You would have to advise faculty to use the new name when citing their professional affiliation. You would need to contact legacy families (let's say a family whose last seven generations have attended this college) and inform them of the name change. Do you now have a sense of how complicated it would be to change this narrative? If you try to make a change that involves a large network, the social structure becomes very powerful. Ironically, something abstract like a name can be more robust than most of the physical infrastructure around us.

EMBEDDEDNESS: THE STRENGTH OF WEAK TIES

One important dimension of social networks is the extent to which they are embedded. Embeddedness refers to the degree to which a social relationship is reinforced through indirect paths (i.e., friends of friends) within a network.

NARRATIVE

the sum of stories contained in a set of ties.

EMBEDDEDNESS

the degree to which social relationships are reinforced through indirect ties (i.e., friends of friends).

The more embedded a tie is, the stronger it is. That is, compared to a relationship with someone whom only you yourself know, a tie to someone who also knows your mother, your best friend, and your teacher's daughter is more likely to last. It may feel less dramatic and intimate, but it's robust and more likely to endure simply by virtue of the fact that it's difficult to escape. You will always be connected to that person—if not directly, then through your "mutual" friends. However, the counterpoint to this dynamic lies in what sociologist Mark Granovetter (1973) calls the strength of weak ties, referring to the fact that relatively weak ties—those not reinforced through indirect paths—often turn out to be quite valuable because they bring novel information. Let's say you're on the track team and that's your primary social group. Occasionally you now see an old classmate from high school whom you didn't really know (or run in the same circles with) back then, because he also attends your college, but he's on the volleyball team. If the track team isn't doing anything on Friday night but the volleyball team is having a party, you've got an "in" through this relationship that's much weaker than the ones you've got with the other members of the track team. That in is the strength of the weak tie you maintain with the classmate from home. If you end up taking your friends from the track team to the volleyball party and they become friends with your old friend, the tie between you and your old classmate is no longer weak because it is now reinforced by the ties between your old friend and your new track friends—regardless of whether you are actually more intimate with him or not. This example helps explain why the structure of the network is more complicated than the tie from one individual to another. Even though your friendships did not change at the volleyball party, the fact that people in your social network befriended others in your social network strengthens the ties between you and both of the individuals.

The strength of weak ties has been found especially useful in job searches (Granovetter, 1974). In a highly embedded network, all the individuals probably know the same people, hear of the same job openings, maintain the same contacts, and so on. However, your grandparents' neighbor, whom you see every so often, probably has a completely different set of connections. The paradox is that this weak tie provides the most opportunities. When Granovetter (1973) interviewed professionals in Boston, he determined that among the 54 respondents who found their employment through

STRENGTH OF WEAK TIES

the notion that relatively weak ties often turn out to be quite valuable because they yield new information.

Small business owners exchanging business cards at a speed networking event. In this limited amount of time, participants hope that making as many weak ties as they can will eventually help them expand and improve their businesses.

personal network ties, more than half saw their contact person "occasionally" (less than once a week but more than once a year). Perhaps even more surprising was the fact that the runners-up in this category were not people whom the respondents saw "often" (once a week or more); it was those they saw "rarely" (once a year or less), by a factor of almost two to one. Additional research finds that weak ties offer the greatest benefits to job seekers who already have high-status jobs, suggesting that social networks combine with credentials to sort job applicants and that strong ties may be more useful in low-status, low-credential job markets (Wegener, 1991).

As seen in Figure 5.4, by linking two otherwise separate social networks, the weak tie between Natalie and Emily provides new opportunities for dating—not only for them but for their friends as well. Their tie bridges what we call a structural hole between the two cliques, a gap between network clusters, where a possible tie could become an actual tie or where an intermediary could control the communication between the two groups on either side of the hole. In the figure, Jenny is a social entrepreneur bridging a structural hole because the people on the left side of the network diagram (Emily and Jason) have no direct ties with the people on the right (Michael, Doug, and Jeff). Their ties are only indirect, through Jenny. Assuming that the two sides have resources (romantic or otherwise) that would complement each other's, Jenny is in a position to mediate by acting as a go-between for

STRUCTURAL HOLE

a gap between network clusters, or even two individuals, If those individuals (or clusters) have complementary resources.

FIGURE 5.4 The Strength of Weak Ties

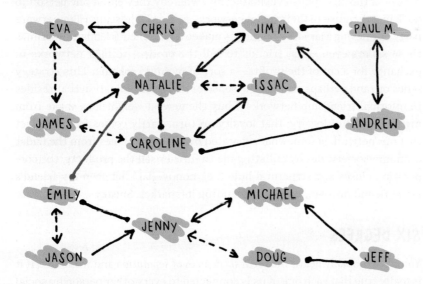

Natalie and Emily are social entrepreneurs; they command information to which the rest of their respective networks do not have access.

the groups. When a third party connects two groups or individuals who would be better off in contact with each other, that third party is an "entrepreneur," and he or she can profit from the gap. (Sounds like the *tertius gaudens* in the triad? It should.)

When sociologist Ronald Burt (1992) studied managers in a large corporation, he found that those with the most structural holes in their social networks were the ones who rose through the company ranks the fastest and farthest. This notion can be expanded to explain how a great deal of profit making occurs in today's economy. At one extreme is the totally free market, in which there are no structural holes; no restriction on information exists, and all buyers and sellers can reach one another—think eBay. At the other extreme is the monopoly, in which one firm provides necessary information or resources to a multitude of people (i.e., maintains and profits from a gaping structural hole). And then there is everything in between these extremes: everyone from shipping magnates to spice traders to mortgage brokers to multilevel marketers (see later in the chapter). Take real estate agents as an example. They earn their money by contractually maintaining (or creating) a structural hole. By signing up sellers, the real estate agents prevent the sellers from directly engaging in a transaction with potential buyers. Recently, the social network possibilities facilitated by the internet (discussed later in the chapter) have done much to erode the power of brokers—for example, by driving once-powerful travel agents into near extinction (ditto for stockbrokers).

At the same time, corporations have leveraged people's existing social networks through multilevel marketing, whereby they enlist one person to hock merchandise to their social network, such as by hosting a Tupperware party or offering Mary Kay cosmetics makeovers to their friends. Sometimes these sellers even enlist friends to sell the product to *their* networks in exchange for a cut of the profits—a sort of pyramid scheme. This strategy relies on (and perhaps strains) the sense of trust and obligation that resides in informal friendship networks. I buy the newest weight-loss shake from my friend out of loyalty; that loyalty, in turn, partly comes from the fact that our network is somewhat separated by structural holes from the wider economy/society. But by enlisting me to further sell the products, the corporation closes any structural hole (i.e., nonoverlap) between my friend's network and my own, thereby expanding its market. Sneaky.

SIX DEGREES

You have probably heard the term *six degrees of separation* and wondered if it is really true that each one of us is connected to every other person by social chains of no more than six people. The evidence supporting the six degree theory came out of research undertaken in the 1960s by Stanley Milgram, whose colleagues were pestering him about why it always seemed like the strangers they met at cocktail parties turned out to be friends of a friend. Milgram decided to test the reach of social networks by asking a stockbroker in Boston to receive chain letters from a bunch of folks living in Lincoln, Nebraska. The Lincolnites could send letters only to friends or relatives who they believed would be likely to know someone who might know someone who might know the guy in Boston. About 20 percent of the letters eventually made it to Boston, and the average trip length was just over five people, hence the idea that in the United States there are no more than five people between any set of strangers, or six degrees of separation.

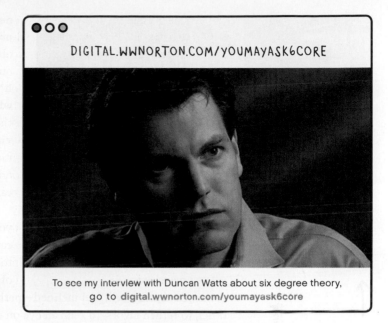

DIGITAL.WWNORTON.COM/YOUMAYASK6CORE

To see my interview with Duncan Watts about six degree theory, go to digital.wwnorton.com/youmayask6core

Duncan Watts (2003) noticed that Milgram's findings applied only to the letters that made it to their final destination. What about the letters that did not complete the journey? Were their chains quite a bit longer than six steps, thus making our six degree theory more like a twelve degree theory? Watts set up a similar—this time worldwide—experiment using e-mail and statistical models to estimate global connectedness and found that Milgram was not quite right. In Watts's words, "It's not true that everyone is connected to everyone else, but at least half the people in the world are connected to each other through six steps, which is actually kind of surprising" (Conley, 2009b). Furthermore, Watts was able to test the commonsense notion that there are some people out there who just seem to know everyone and that it must be through these superconnected people that the rest of us are able to say we're only six degrees from Kevin Bacon (or whomever). But instead he found that when it comes to whom we know, "the world's remarkably egalitarian," and that superconnectors played almost no role in getting the e-mail forwarded all the way to its destination (Conley, 2009b).

SOCIAL CAPITAL

Having many weak ties is one form of what sociologists call *social capital*. Like human capital, the training and skills that make individuals more productive and valuable to employers, social capital is the information, knowledge of people or things, and connections that help individuals enter preexisting networks or gain power in them. Consider the importance of

SOCIAL CAPITAL

the information, knowledge of people or things, and connections that help individuals enter, gain power in, or otherwise leverage social networks.

Neighbors in the Central City section of New Orleans gather for their weekly domino game. Communities with thick webs of connection tend to thrive, with lower crime rates and more volunteer involvement.

networking in endeavors such as preventing neighborhood crime or obtaining a good job. As it turns out, the cliché holds a lot of truth: It's not just what you know but whom you know. Weak ties may be the most advantageous for an individual; however, for a community, many dense, embedded ties are generally a sign of high levels of social capital.

This concept makes sense when you think about it. Dense social capital means that people are linked to one another through a thick web of connections. As a result of these connections, they will feel inclined—perhaps even impelled—to help each other, to return favors, to keep an eye on one another's property. The more connections there are, the more norms of reciprocity, values, and trust are shared. After all, there is no such thing as total anonymity: Even if you don't know someone directly, chances are that you are only one or two degrees removed from him or her.

In this way, strong social capital binds people together; it weaves them into a tight social fabric that can help a community thrive. "You tell me how many choral societies there are in an Italian region," notes social capital scholar Robert Putnam, "and I will tell you plus or minus three days how long it will take you to get your health bills reimbursed by its regional government" (Edgerton, 1995). After years of research in Italy, Putnam determined that different regions of the country varied widely in their levels of participation in voluntary associations. As it turns out, the strength of participation in a region was a fairly good predictor of the quality and efficiency of its regional government (and, in turn, its economic growth).

The United States and Social Capital If it's true that social capital is correlated with economic and political health, some critics would say that the United States may be in trouble, especially when we compare our recent past with what some see as America's golden age of joining.

In the 1830s, Alexis de Tocqueville wrote *Democracy in America* based on his visit to the United States from France. Tocqueville was surprised to find that America was, in his words, a "land of joiners." By this, Tocqueville meant that Americans frequently came together to join voluntary associations. "Americans of all ages, all conditions, all minds constantly unite," Tocqueville wrote. "Not only do they have commercial and industrial associations in which all take part, but they also have a thousand other kinds:

religious, moral, grave, futile, very general and very particular, immense and very small." In democratic societies such as the United States, Tocqueville observed that citizens enjoyed greater equality than citizens in aristocratic societies. Although Tocqueville praised this equality, he also believed that it made democratic citizens independent and weak, so organizations made citizens politically stronger. After all, what good is one vote? He wrote that democratic citizens "can do almost nothing by themselves, and none of them can oblige those like themselves to lend them their cooperation. They therefore all fall into impotence if they do not learn to aid each other freely" (Tocqueville, 1835). Voluntary associations in Tocqueville's land of joiners were a way for independent citizens to assist one another.

The propensity of Americans to join voluntary groups—the Parent Teacher Association at your high school, the local softball league, or a knitting club—has puzzled sociologists, historians, and political scientists since Tocqueville first wrote about America as a land of joiners. Why are Americans so likely to join groups? Our participation in elections and formal political processes is one of the lowest in the world, which makes the question all the more intriguing. Some, like Tocqueville, suggest that the uniquely egalitarian nature of American democracy has made Americans more likely than Europeans to enlist in voluntary organizations (Doyle, 1977). Other scholars suggest that America's unique pattern of settlement is responsible for high levels of voluntary organizations. In particular, the town square culture of early New England, in which people came together in town squares to discuss and debate current civic issues, created a long-lasting culture of voluntary association (Baker, 1997). Still others point to America's identity as a land of immigrants who formed voluntary organizations to unite with other immigrants who shared similar cultural or political values (Gamm & Putnam, 1999).

Despite some high-profile exceptions like the record-setting Women's March on Washington the day after Trump was inaugurated, overall, voluntary participation in civic life has taken a turn for the worse, and as a result, the nation's stock of social capital is at risk. In his best seller *Bowling Alone: The Collapse and Revival of American Community* (2000), Putnam traces the decline of civic engagement in the last third of the twentieth century. We are more loosely connected today than ever before, he says, experiencing less family togetherness, taking fewer group vacations, and demonstrating little civic engagement.

In *Loose Connections: Joining Together in America's Fragmented Communities* (1998), Robert Wuthnow finds that people are worried about what they see as a breakdown of families and neighborliness and a concurrent rise in selfishness. When Wuthnow surveyed a random sample of Americans, fewer than half believed that their fellow citizens genuinely cared about others. Even worse, some studies suggest that more and more people think their neighbors are inherently untrustworthy (Lasch, 1991). A considerable

Bowling alone? Although the overall number of bowlers has increased, the number of people bowling in groups has dropped. Are we seeing a decline in social engagement?

majority of Americans believe that their communities are weaker than ever before (Wuthnow, 1998). Without communal ties of civic and religious participation, Americans have lost trust in their neighbors. Interpersonal trust matters; it is essential for building the complex social structures necessary for a functioning democracy and economy.

Indeed, more and more people are bowling alone. Actual bowling activity was on the rise at the end of the twentieth century, when Putnam and Wuthnow were writing—the total number of bowlers in America increased by 10 percent from 1988 to 1993—but *league* bowling dropped a whopping 40 percent in the same time frame. This is no good for bowling-lane owners, but it's also bad news for democracy in America. Bowling alone is part of a more general trend of civic disengagement and a decline in social capital. It's happening in PTAs, the Red Cross, local elections, community cleanups, and labor unions. Even membership in the Boy Scouts has decreased by 26 percent since the 1970s. More people live alone. Some go so far as to say that friendships have become shallower, and the phenomenon of deep, enduring friendships is an increasing rarity (Flora, 2013; Wuthnow, 1998). As civic participation withers, activities once performed by communities have moved toward private markets.

Who's to blame for America's fading civic life? The social entrepreneur who has too many structural holes to maintain? Parents? Our school systems, television, the internet? A combination of all these and other factors works in conjunction with broad social trends, creating an increasingly differentiated, specialized, urbanized, and modern world. Social institutions must adjust to the flexibility, sometimes called the liquidity, of modernity by becoming more fragmented, less rigid, and more "porous" (Wuthnow, 1998). It becomes easier to come and go, to pass through multiple social groups such as churches, friends, jobs, and even families. Gone is the rigid fixity of finding and holding onto a lifelong "calling." Similarly, a majority of graduating college students (60 percent) attend more than one school before receiving a degree (Zernike, 2006).

This mobility and flexibility take a toll on people's lives. More than in previous eras, people report feeling rushed, disconnected, and harried, so it's no surprise that civic responsibilities are pushed aside. It's not that Americans don't care. In fact, they join more organizations and donate more money than ever before. They just don't give their time or engage in face-to-face activities (Skocpol, 2004). One possible explanation for this trend is the rise of online associations. We may be showing up less because the internet makes it easy to form new groups whenever we feel the need. It also allows for social connection (of some form) without requiring face-to-face contact. The explosion of websites like Facebook, Twitter, and Instagram is an important example of this phenomenon. Political activism has moved online as well, with the Barack Obama, Howard Dean, and Bernie Sanders campaigns leading successful

Web-based fund-raising efforts that brought new donors into the political process. What's more, activist fund-raisers intent on rapidly providing aid and relief to Haitians following the January 2010 earthquake invited donations via text messaging. The American Red Cross was able to raise more than $22 million through text messaging alone (Strom, 2010); meanwhile, the website gofundme.com has allowed folks to raise a total of $5 billion.

This single mother and teacher in Tampa, Florida, created a GoFundMe page requesting $17,000 to pursue a master's degree in mental health counseling. Why might crowdfunding—that is, requesting small amounts of money from a large number of people—be more effective than asking face-to-face?

Lastly, before we blame the internet for a decline in social capital and civic life, we should note that the trend toward giving more but showing up less predates the web. It is probably more a result of increased work hours and other pressures to keep up in an age of rising inequality.

The social pendulum swinging between loneliness and unity may be starting to turn back toward togetherness, with slightly different characteristics than we are used to seeing. Sociologist Eric Klinenberg (2013) set out to study people who live alone, trying to understand the costs and benefits of leaving some of the ties that bind dangling free. He found that being old and living alone is indeed a lonely reality and a growing problem for women, who have always tended to outlive their husbands but are now aging alone in neighborhoods that their children have left for jobs and lives elsewhere. But for young and middle-aged people, living alone is a lifestyle frequently filled with friends, dates, co-workers, volunteer work, and plenty of socializing. People who live alone are more likely to volunteer than people the same age who are living with partners or families. Maybe civic participation is not dying, after all.

Understandably, Putnam's claims have ignited much controversy. Some researchers have noted that even if an increase in participation occurred after 9/11, it was short-lived among adults (Sander & Putnam, 2010). Others claim the opposite, insisting that social capital never declined as Putnam declared. Rather, it has simply become more informal. Wuthnow, for

CASE STUDY:
SURVIVAL OF THE AMISH

Pennsylvania's Lancaster County attracts more than 5 million visitors a year with its Pennsylvania Dutch country charm. Horse-drawn wagons carry visitors over covered bridges toward historic museums, colonial homes, and restaurants. Lancaster thrives on tourists' curiosity about the Amish, who first settled the land in 1693. Today about 59,350 Amish live in nearby homogenous farm communities. As far as appearances go, they certainly meet tourists' expectations.

Wearing straw hats and black bonnets, riding in horse-drawn buggies, the Amish provide great photo opportunities. Children are taught in private schoolhouses, typically one room, only until the eighth grade, at which point they work full-time on the family farm. The lives of the Amish

An Amish barn raising in Tollesboro, Kentucky.

revolve around going to church, tilling the earth, and working for the collective good. They value simplicity and solidarity. How, visitors wonder, have Amish communities in Pennsylvania and other states survived in our fast-paced society?

The Amish certainly appear to be relics of the past, but looks can be deceiving. Although the Amish place primary importance on agriculture, they do so in anything but a premodern fashion. The Amish, especially those living close to urban areas, have adopted farming innovations such as the use of insecticides and chemically enhanced fertilizers. Their homes are stylishly modern, with sleek kitchens and natural gas-powered appliances.

Because Lancaster County has experienced a certain degree of urban growth and sprawl, the Amish are not immune to the hustle and bustle of commerce. In fact, many Amish are savvy business owners. The number of Amish-owned micro-enterprises more than quadrupled between 1970 and 1990. By 1993 more than one in four Amish homes had at least one nonfarm business owner (Kraybill and Nolt, 1995). The result of this economic growth is a curious mix of profit-seeking entrepreneurship with a traditional lifestyle. Imagine finding out that your local rabbi, priest, or imam doubled as a high-rolling stock trader during the week.

How can these seemingly incompatible spheres of religious tradition and commerce coexist? This is the question Donald Kraybill set out to answer when he studied 150 Amish business entrepreneurs in Lancaster County (1993; in

1995 with Nolt). Not only do the Amish trade with outsiders, Kraybill documented, but they do so quite successfully: About 15 percent of the businesses he studied had annual sales exceeding half a million dollars. Their success rate is phenomenal: Just 4 percent of Amish start-ups fail within a decade, compared with the 75 percent of all new American firms that fail within three years of opening. That is, in a time when the majority of new American business ventures flop, virtually all Amish businesses succeed. Are the Amish just naturally better at conducting commerce? Is it something in their faith, or self-discipline? Perhaps the answer lies in the community's hand-pumped water?

The answer is, none of the above. The secret to Amish success turns out to be the way they strategically combine their traditions with the rest of the modern world. As the Amish have become increasingly entangled in the economic web of contemporary capitalism, they have held onto their cultural traditions by maintaining an ideologically integrated and homogenous community. They are distinctly premodern and un-American in that they believe in the subordination of the individual to the community. They have rejected the prevalent American culture of rugged individualism, the notion of "every man for himself," often said to be the basis for successful entrepreneurship, in favor of "every man for the greater good." Individuals are expected to submit not only to God but also to teachers, elders, and community leaders. Their social fabric firmly binds people together. Whereas fashion for many people is a means of self-expression, the dark, simple clothing of the Amish signals membership in and subordination to the community. They have no bureaucratic forms of government or business; rather, they operate in a decentralized, loose federation of church districts. They reject mass media and automobiles and limit their exposure to diverse ideas and lands. The outcome is homogeneity of belief, unified values, and not surpris-

Kimberly Hamme works on billing for her online business, Plainly Dressed, in her Paradise Town, Pennsylvania, home office. Hamme sells what most people would call Amish clothing. She does most of her business over the internet.

ingly, dense social capital. Amish businesses, like the people themselves, are tightly enmeshed in social networks, such as church and kinship systems, that provide economic support. The typical Amish person has more than 75 first cousins, most of whom shop in the same neighborhood. Add to that the taboo on bankruptcy within the community, and Amish businesses would be a dream come true for investors—that is, if the Amish were to accept outside capital (they don't).

Rather than marvel at how this culture maintains its centuries-old traditions while functioning in the business world, we should look at how they succeed in business regardless of being Amish. Kraybill and Nolt found that the cultural restraints of being Amish do indeed thwart business opportunity to some extent. Amish business owners aren't allowed to accept financial capital from outsiders or to prosecute shoplifters (to do so would single out lawbreakers and go against community solidarity). The success of Amish entrepreneurship is a telling example of the power of social capital.

example, argues that modernization brings about new forms of "loose connection" but hardly a disappearance of all connection. Nor does it necessarily follow that modern Americans are any worse off than when connections were tight. Things change, but that does not always mean they change for the worse. Though Putnam laments the loss of face-to-face communal ties, in the past three decades we've witnessed an explosion of non-place-based connections: Think of the rich social life occurring on social media, including Twitter, blogs, and Facebook.

With the exit of the old (such as the Elks, Rotary, and other fraternal and civic organizations), in have come new kinds of clubs—large national groups like netroots political groups such as MoveOn.org, which one can join by mail or online, and informal support groups like Weight Watchers and hybrids like Meetup.com, where interest groups form online but then meet face to face. Furthermore, the trend of declining social capital is falsely linear, as if from the 1970s to today civic society has moved in one simple direction (downhill). More probable is that civic engagement moves like a pendulum, swinging back and forth between privatism (as in the 1920s) and heightened public consciousness (as in the 1930s). Right now may feel like the end of social capital, but perhaps we're just at a low point on a constantly shifting trend line. The calls to save social capital may be a form of projected nostalgia, a misplaced romanticizing of the past.

Sociologist Michael Gaddis has been using network structure to look at another challenging aspect of social capital: Even for people with a healthy number of ties to friends, family, and community, not all social capital is equal. As Gaddis points out, "Everyone knows friends, co-workers, family members, but the important part of social capital...is the resources that are linked to you through these networks. Do I know someone who knows

Volunteers build an elementary school playground in New Orleans. After Hurricane Katrina devastated the Gulf Coast in 2005, thousands of college volunteers participated in the rebuilding efforts.

someone who has a job opening and could refer me? Can I access those resources?" He looked at kids growing up in "low-income families, one-parent families" who applied to the Big Brothers Big Sisters program where they hoped to be assigned an adult mentor with whom to spend time. Because of high demand, only some students got to participate, making it possible for Gaddis to compare students who got mentors with those who wanted them but were not able to get them. It turned out that "mentors with higher education levels, higher income" were "able to make greater changes" among the little brothers and sisters (Conley, 2013b). It is not how many people you know that gives you greater social capital but the resources associated with the people you know and their willingness to share those resources with you that increases your social capital.

DIGITAL.WWNORTON.COM/YOUMAYASK6CORE

To see my interview with Michael Gaddis about the impact of social networks on low-income children, go to digital.wwnorton.com/youmayask6core

The picture we've arrived at is that of a complex social world where the decay of some forms of civic life is accompanied by the eventual emergence of new ways of building communities. Americans living in modern, urban, anonymous, and loosely connected communities carve out new social spaces, in turn creating a different kind of social fabric that holds together our republic. People adapt to what's new, retain what they can of the old, and negotiate within global forces and local communities.

Network Analysis in Practice

Researchers take the concepts we've discussed so far—embeddedness, the iron law of the triad, and network position—and apply them to real-world contexts in order to understand how group life shapes individual behavior. Network analysts also map out social relationships to better understand transmission phenomena such as the spread of disease, the rise and fall of particular fads, the genesis of social movements, and even the evolution of language itself.

THE SOCIAL STRUCTURE OF TEENAGE SEX

According to sociologists Lisa Wade (2017) and Kathleen Bogle (2008), "hooking up" has replaced going steady on campus; "friends with benefits" are preferred over girlfriends and boyfriends, with all their attendant demands and the corresponding commitment (see Chapter 8). On the other hand, Wade also reports that about one-third of college students do not (or cannot) participate in this romantic culture on campus. In fact, the proportion of teens and young adults who are not having sex at all has risen across the globe in recent years (*Economist*, 2018b).

Here are some nationally representative numbers for the United States: About 50 percent of American teenagers over the age of 15, when interviewed by researchers, have admitted to engaging in sexual intercourse. (Boys probably tend to exaggerate their sexual experience, and girls probably downplay it.) A good number of those who have not yet had intercourse are still sexually active in other ways: Approximately one-third have "had genital contact with a partner resulting in an orgasm in the past year." Bluntly put, what this means is that a good two-thirds of American teens are having sex or participating in some form of sexual activity. Teenagers' romantic relationships tend to be short term compared with adults', averaging about 15 months, so there is a fair amount of partner trading. Survey research among students in college, where hook-up culture is prevalent, found that 70 percent use condoms when they engage in vaginal/penile intercourse. That 70 percent is "a lot less than a hundred, but a lot more than zero," notes principal investigator Paula England (Conley, 2009d). To top that off, most adolescents with a sexually transmitted infection "have no idea that they are infected." All of these factors combine to make American teenagers a breeding ground for sexually transmitted infections, which have increased dramatically in this age group in the last decade.

So what's a public health officer to do? During the administration of George W. Bush, the religious right and conservative policy makers suggested the "virginity pledge" and other abstinence policies as a solution. As it turns out, the pledge does delay the onset of sexual activity on average, but when the teenagers who take it eventually do have sex, they are much more likely to practice unsafe sex (Bearman & Brückner, 2001; Brückner & Bearman, 2005). Among the many problems in designing safe-sex or other programs to reduce the rate of STIs among teenagers is the fact that we knew very little about the

Despite sensational media reports about teenage hook-ups, monogamous couples such as J. D. and Elysia, both 14, of Yorktown High School in Arlington, Virginia, are more typical. What else does research reveal about high-school sexual relationships?

FIGURE 5.5 Analysis of High-School Sexual Relationships

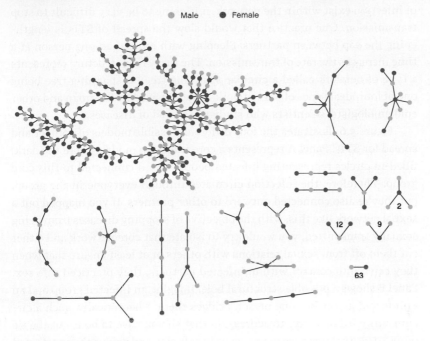

● Male ● Female

2

12 ● 9

63

SOURCE: Bearman et al., 2004

sexual networks of American adolescents until quite recently. It would not be far off to say that we knew more about the sexual networks of aboriginal tribes on Groote Eylandt than we did about those of American teenagers.

One component of the National Longitudinal Survey of Adolescent Health, conducted by J. Richard Udry, Peter Bearman, and others from 1994 to 1996, investigated the complete sexual network at 12 high schools across the nation, including the pseudonymous Jefferson High School, whose 1,000-person student body is depicted in Figure 5.5 (Bearman et al., 2004). They focused their analysis on Jefferson, because its demographic makeup, although almost all white, is fairly representative of most American public high schools and, more important, it is in a fairly isolated town, so less chance exists of the sexual networks spilling over to other schools.

The pink dots represent girls and the green ones boys. The dyad in the lower right-hand corner tells us that there are 63 couples in which the partners have only had sex with each other. There are small, comparatively isolated networks consisting of 10 or fewer people, and then one large ring that encompasses hundreds of students. Preventing the transmission of infection in the small networks is much simpler than preventing the spread in the large ring. One young man in the ring has had nine partners, but even if you persuaded him to use condoms or practice abstinence, you still wouldn't address most of the network. Your action might positively impact the people immediately around him. But to the extent other origin points of infection exist within the network, it is going to be very difficult to stop transmission. One practice that would slow the spread of STIs is lengthening the gap between partners: Sleeping with more than one person at a time increases the rate of transmission. The fuzzy ring structure represents a type of network called a circular spanning tree—a spanning tree being one of four ideal types of sex networks hypothesized by Bearman and other epidemiologists (scientists who study the spread of diseases).

Figure 5.6 illustrates the four different possible models of contact and spread for STIs. Panel A represents a core infection model, where the dark, filled-in circles representing infected people are all connected to this core group. Therefore, the infection circulates through everyone in the group, but they're also connected outward to other partners. If you mapped out a sexual network like this, with the objective of stopping diseases from being sexually transmitted, you would try to isolate that core network and either cut them off from sexual relations with others or at least ensure that when they came into contact with uninfected partners, they practiced safe sex. Panel B shows a possible structural hole. Imagine an infected group and an uninfected group, but one person bridges them. Theoretically, such a circumstance is fairly easy to address, in that all you have to do is cut the tie or persuade that one person to engage in safe sex, and you've thus protected the uninfected population. Panel C depicts an inverse core model, representing the way much of AIDS transmission occurs in populations where

FIGURE 5.6 Models for Spread of Sexually Transmitted Infections

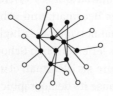

Panel A: core infection model

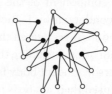

Panel C: inverse core model

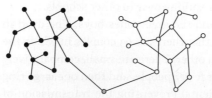

Panel B: bridge between disjointed populations

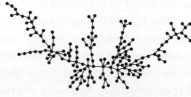

Panel D: spanning tree

SOURCE: Bearman et al., 2004.

men routinely spend long periods of time away from their families, such as long-haul truckers and some men from African villages. The traveling men visit prostitutes in the city or at truck stops, acquire the virus from one, transmit it to another, and then bring the virus back home. The infected members are not connected directly with one another (i.e., the prostitutes are not having sex with each other); rather, the individuals at the periphery of the core (the men who solicit the prostitutes) connect the core members to each other and possibly beyond the group to other populations. The last network model, illustrated in panel D, is the spanning tree model, in linear rather than circular form. This is essentially how power grids are laid out: There's a main line, and branches develop off of that line. It is difficult to completely stop transmission along a spanning tree model. That's why we design electrical grids to be a series of spanning trees so that if one circuit fails the power can continue to flow around it, although as anyone who has experienced a blackout knows, specific sections can be left without power if they are severed from the rest of the tree. In terms of STI transmission, you could initiate some breaks that would split the tree into two groups, but that's not going to completely isolate the infection. If you attack something on a branch, you're not doing anything to the rest of the network. There's no key focal point that allows you to stop the spread.

ROMANTIC LEFTOVERS

When Bearman and his colleagues analyzed the sexual habits of teenagers, they uncovered another rule that governed this social network. They found lots of examples of triads, where partners are traded within groups. But the main rule that seemed to govern these relationships was "no cycles of four," which means you do not date the ex of your ex's current boyfriend or girlfriend. The most interesting aspect of the rule, sociologically speaking, is that no one was consciously aware of this pattern. The researchers interviewed many students and not one of them directly stated, "Of course not, you don't date the ex of your ex's new flame." Yet it's the single taboo that governs everyone. Figure 5.7 (next page) illustrates this rule of thumb graphically.

At time 1, Matt and Jennifer are dating, as are Jareem and Maya. At time 2, Jennifer and Jareem date. The rule suggests that Matt and Maya will never date. Why is that? Once Jennifer and Jareem start dating, if Matt and Maya decide to date each other, they are relegating themselves to secondary social status, as if they were "leftovers." The practical, take-home lesson in all of this is that if you want to date the ex of your ex's new crush, act before your ex does. If you're Jennifer and you wish to prevent your old boyfriend Matt from going out with Maya, quickly start dating Jareem, because then Maya and Matt will never date. But extend the "no seconds" rule to thirds or more, and the taboo erodes. The fact that no students in the high school studied were aware of the rule is what makes this kind of

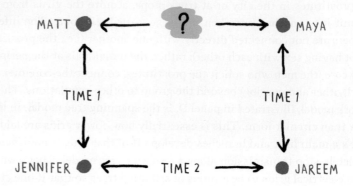

FIGURE 5.7 Romantic "Leftovers"

MATT ← ? → MAYA

TIME 1   TIME 1

JENNIFER ← TIME 2 → JAREEM

SOURCE: Bearman et al., 2004.

social norm possible: It's not conscious. This is a good example of how social structures govern individual-level behavior, and it speaks to some of the limitations of interpretive sociology. If the researchers had taken a more Weberian approach and asked students how they choose partners and, more important, why they don't date certain people, they probably wouldn't have discovered this rule. The researchers could see it only by taking a bird's-eye view and analyzing this structure with mathematical tools.

Organizations

I've mentioned several times that sociology—here network analysis—can be applied not just to individuals but to all social actors, which may include school systems, teams, states, and countries. In the contemporary United States, companies and organizations are important social actors. In fact, thanks to the Fourteenth Amendment, they have identities as legal persons: They sponsor charitable causes; they can sue and be sued; they even have birthdays.

Organization is an all-purpose term that can describe any social network—from a club to a Little League baseball team to a secret society to your local church to General Motors to the US government—that is defined by a common purpose and that has a boundary between its membership and the rest of the social world. *Formal* organizations have a set of governing structures and rules for their internal arrangements (the US Army, with its ranks and rules), whereas *informal* organizations do not (the local Meetup.com group for ambidextrous tennis players [an actual group!]). Of course, a continuum exists, because no organization has absolutely no rules, and no organization has a rule for absolutely everything. Therefore, the study of

ORGANIZATION

any social network that is defined by a common purpose and has a boundary between its membership and the rest of the social world.

organizations focuses mainly on the social factors that affect organizational structure and the people in those organizations.

ORGANIZATIONAL STRUCTURE AND CULTURE

Have you ever heard the phrase *the old boys' club*? The term is used to refer to exclusive social groups and derives from fraternities, businesses, and country clubs that allow only men—specifically, certain groups of elite men—to join. These groups have their own customs, traditions, and histories that make it difficult for others to join and feel as if they belong, even when the "boys" aren't being deliberately hostile. The term organizational culture refers to the shared beliefs and behaviors within a social group and is often used interchangeably with *corporate culture*. The organizational culture at a slaughterhouse—where pay is low, employees must wear protective gear, the environment is dangerous, and animals are continuously being killed—is probably very different from the organizational culture at a small, not-for-profit community law center. The term organizational structure refers to how power and authority are distributed within an organization. The slaughterhouse probably has a hierarchical structure, with a clear ranking of managers and supervisors who oversee the people working the lines. The law center, however, might be more decentralized and cooperative, with five partners equally co-owning the business and collaborating on decisions. How an organization is structured often affects the type of culture that results. If a business grants both parents leave when a new child enters the home, allows for flextime or telecommuting, or has an on-site child-care center, those structural arrangements will be much more conducive to creating a family-friendly organizational culture than those of a company that doesn't offer such benefits.

The growth of large multinational corporations over the course of the last 100 years has affected organizational structure. One example of this impact can be seen in *interlocking directorates*, the phenomenon whereby the members of corporate boards often sit on the boards of directors for multiple companies. In 2013, for example, the boards of insurance companies such as AIG, Humana, MetLife, and Travelers frequently had board members who also served on the boards of drug and medical device makers such as Johnson & Johnson, DuPont, Abbott Laboratories, and Dow Chemical. (The website **theyrule.net** allows you to create an interactive map of companies' and institutions' boards of directors.) Does it matter that these people sit together on the same boards? The problem, critics argue, is that we then allow a select group of people—predominantly rich, white men—to control the decisions made in thousands of companies. Such people also have ties to research institutions and elected officials that may compromise their objectivity and create conflicts of interest. Capitalism, after all, is based on competition, but if board members on interlocking directorates favor the other companies to which they are connected, suppliers may not be

ORGANIZATIONAL CULTURE

the shared beliefs and behaviors within a social group; often used interchangeably with *corporate culture*.

ORGANIZATIONAL STRUCTURE

the ways in which power and authority are distributed within an organization.

competing on a level playing field when bidding for contracts. Or worse, take the situation that might develop when a board member of a drug maker asks his friend and fellow board member at a health insurance company to give preferential coverage to his company's drugs over a competitor's drugs. This can lead to higher prices for consumers who need to purchase the competitor's drugs. Another situation might arise in which one of these two board members knows a former member of Congress through board service together who they can use as a lobbyist to see that federal health programs like Medicare and Medicaid also give preferential treatment to a particular drug, costing taxpayers more than they would have paid without the pressure arising from these relationships. This type of situation can lead to what sociologist C. Wright Mills called a "power elite" or aristocracy. (Concern over the consolidation of control in the media industry, for example, is discussed in Chapter 3.)

INSTITUTIONAL ISOMORPHISM: EVERYBODY'S DOING IT

Networks can be very useful. They provide information, a sense of security and community, resources, and opportunities, as we saw illustrated by Granovetter's concept of weak ties. Networks can also be constraining, however. Paul DiMaggio and Walter Powell (1983), focusing on businesses, coined the phrase *institutional isomorphism* to explain why so many businesses that evolve in very different ways still end up with such similar organizational structures. Isomorphism, then, is a "constraining process that forces one unit in a population to resemble other units that face the same set of environmental conditions" (Hawley, 1968). In regard to organizations, this means that those facing the same conditions (say, in industry, the law, or politics) tend to end up like one another.

> ISOMORPHISM
>
> a constraining process that forces one unit in a population to resemble other units that face the same set of environmental conditions.

Let's consider a hypothetical case: A new organization enters into a fairly established industry but wants to approach it differently, perhaps a bank that wants to distinguish itself from other banks by being more casual or more community oriented. The theory of isomorphism suggests that such a bank, when all is said and done, will wind up operating as most other banks do. It's locked into a network of other organizations and therefore will be heavily influenced by the environment of that network. The same is true for new networks of organizations. A group of not-for-profits might spring up in a specific area. Because all will face the same environmental conditions, they will likely be, in the final analysis, more similar than different, no matter how diverse their origins. DiMaggio and Powell are part of a school of social theory referred to as the new institutionalism, which essentially tries to develop a sociological view of institutions (as opposed to, say, an economic view). In this vein, networks of connections among institutions are key to understanding how the institutions look and behave. These theorists would argue that all airlines raise and lower their fares at the same time, for example, not because they are independently reacting to pure market forces but because symmetry, peer pressure, social signaling, and network laws all govern the organizational behavior of these Fortune 500 companies to the same extent that these forces affect the sex lives of seniors at Jefferson High School. Pretty scary, huh?

POLICY

RIGHT TO BE FORGOTTEN

When I was in high school, I thought I would try to be cool and up my social status by swiping a bottle of my father's booze and bringing it along on a school trip. I had little experience with this sort of activity, so I just grabbed whatever I saw. My plan backfired: It turned out that I had stolen dry vermouth, typically used only in minute amounts for mixing martinis or Manhattans. I had no idea that vermouth was not something generally drunk straight. (Though, in an ironic twist, it turns out that it is actually quite sophisticated to have it alone with an orange peel, I have since learned.) The moment after I whispered conspiratorially to one of the cool kids that I had booze and showed him the bottle in my possession, laughter spread across the chartered bus. For the rest of my high-school career, I was labeled "vermouth man."

Thankfully for me, I went to college 3,000 miles away where almost no others from my high school attended, so I was able to reinvent myself. Today, however, with social media, almost everything teenagers do is recorded for posterity, leaving digital traces. (This is one of the main changes

to adolescent social life documented by danah boyd; see my interview with her in Chapter 2.) College counselors tell students to set their privacy settings on the most strict level when it's application season. Students themselves often use "senior names" on social media to disguise their real identities. These solutions to the privacy problem don't always work. Just ask the 10 students who had their admissions to Harvard revoked when it was revealed that they had shared sexually explicit and racially offensive memes within their "Harvard memes for horny bourgeois teens" Facebook group (Natanson, 2017).

But even if these short-term fixes work, what about the rest of their lives? A dumb or offensive comment that makes it onto a website or an embarrassing party photo that tags you might trail you like a digital ball and chain each time you try to apply for a new job or date a new suitor. Is there a right to be forgotten just like the right to free speech or the right to a trial by one's peers?

The European Commission seems to think so. The EU has drafted regulations guaranteeing the right to "obtain from the controller [i.e., the data collector or manager, such as a search engine or online database], the erasure of personal data relating to them and the abstention from further dissemination of such data, especially in relation to personal data which are made available by the data subject while he or she was a child or where the data is no longer necessary for the purpose it was collected for, the subject withdraws consent, the storage period has expired, the data subject objects to the processing of personal data or the processing of data does not comply with other regulation." What this—and the results of a lawsuit against Google in Spanish court—essentially means is that search engine firms now provide Europeans with a form they can fill out to request removal of links to information they'd rather not have out there, so to speak. The "right to be forgotten" is now considered a fundamental human right in some jurisdictions.

Sounds reasonable, no? Why should this generation enjoy any less privacy than prior ones just because technology has changed? Well, not so fast. As with almost any "right," there are competing interests. Does your right to be forgotten trump my right to know about you as a potential employer, neighbor, or boyfriend? Where does it end if we allow whitewashing of historical records? While we might all agree that revenge porn should be banned, what about links to bad reviews? Or even tags by former flames in photos that are searchable but might upset our new partners? The line is hard to draw. Perhaps, like sealed criminal records, the age of majority should be used to wipe the slate clean? What about material that is moved to jurisdictions that have looser laws (like the United States, which privileges free speech to a greater extent than Europe does)? The global reach of the internet's many-headed hydra makes it hard to slice off one head of information without another popping up elsewhere.

While this battle between privacy advocates and free-speech diehards is likely to be fought out for many years on multiple fronts, one thing is certain: The changing nature of social media and online networks means it's probably better to think more carefully about swiping a bottle of booze than I did back in the dark ages of 1985.

Should there be a universal "right to be forgotten"? What are some arguments for, and some arguments against, data erasure on social networking websites?

Conclusion

What do we learn from the formal analysis of group characteristics and social networks? Simply knowing the formal characteristics of a group helps us understand much of the social dynamics within it. Is it a dyad or a triad? What is the proper reference group for a particular social process?

Is this group a primary or secondary group—and what does that mean for my obligations to it? Likewise, we can use network analysis in micro- and macro-level studies. You could carefully weigh the potential consequences of dating your best friend's ex by mapping out your social network and anticipating shifts in ties that might transpire. Or you could analyze President Richard Nixon's strategy of "triangulation" of the Soviet Union and Communist China during the early 1970s using the iron law of the triad. Sociologists use network analysis to study everything from migration to social movements to cultural fads to global politics.

QUESTIONS FOR REVIEW

1. Cryptocurrencies such as bitcoin rely on a technology called "blockchain": every transaction between any two individuals is recorded in a public ledger. How does this illustrate the concept of embeddedness?

2. What are some of the benefits, problems, and obstacles to implementing a right to be forgotten on the web?

3. If getting a job is all about connections, how does the work on the strength of weak ties round out our understanding of this phenomenon? How does nepotism (hiring family members) fit into this discussion?

4. An undecided voter who knows little about the political candidates reads the result of the latest poll on voting day and sees that other voters seem inclined to choose one of the candidates. According to Asch's work on conformity, how might the poll affect the voter's behavior? Do you think media sources should release polls shortly before an election?

5. If you could choose your position in a social network, would you want to bridge a structural gap? Why might the manager of a company try to prevent the development of structural gaps between the company's various departments? Would a high-school nurse be more likely to encourage or discourage structural gap formation?

6. What is an organizational structure? Describe the organizational structure at your school or workplace and determine how this structure might affect the organizational culture.

7. A new coffee shop opens in your neighborhood, which already has two other coffee shops. The new coffee shop offers free internet access to customers. Within a few weeks, the two other shops offer free internet access as well. Explain how this example might illustrate Paul DiMaggio and Walter Powell's concept of institutional isomorphism.

PRACTICE

HOW TO DISAPPEAR

In 1995, on a red eye to Poland, I fell asleep and had my passport stolen. For all I know, there is another Dalton Conley who is still off drinking vodka and eating kielbasa with my papers. That happened before RFID chips, computerized global databases, and biometrics. Today, it would be a lot harder for me to mimic what my thief did—that is, either disappear or assume a new identity. Back then, records of documents were more easily falsified and one didn't leave digital traces with every trip to the supermarket. Before 9/11, you could even check into a hotel with no ID. But in today's hyper-connected world, it's much harder to fall off the grid.

TRY IT!

Let's say you wanted to disappear. What are the steps you'd take to drop off the face of the earth— at least according to the social networks in which you're embedded? Here are the steps I would take:

Sell all my worldly possessions for cash on Craig's List

↓

Buy a burner phone

↓

Hitchhike to Mexico (certain border crossings
don't check ID in that direction)

↓

Learn Spanish

↓

Find a job picking fruit

↓

Never contact anyone from my prior life ever again

THINK ABOUT IT

Disappearing is not just shutting down your social media and saying goodbye to your friends. How would you actually function in society if you had to be completely on the DL—invisible to banks, the government, and every other institution?

SOCIOLOGY ON THE STREET

How has social media significantly changed the number and strength of our weak ties? Watch the Sociology on the Street video to find out more: **digital.wwnorton.com/youmayask6core**.

WANT MORE PRACTICE?

Complete the InQuizitive activity for this chapter at digital.wwnorton.com/youmayask6core

PARADOX

6

IT IS THE DEVIANTS AMONG US WHO HOLD SOCIETY TOGETHER.

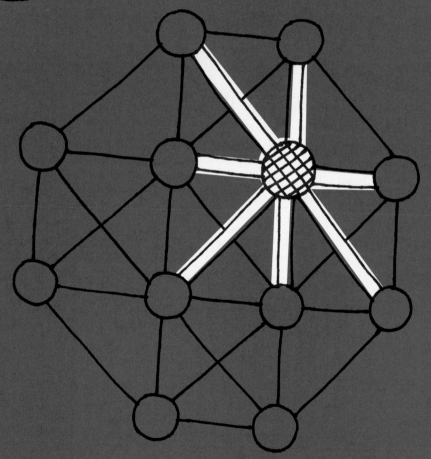

Animation at digital.wwnorton.com/youmayask6core

Social Control and Deviance

197

Smiley got his nickname because he always smiled at the most improbable times. When teachers yelled at him because he had trouble paying attention, he smiled. When his smile upset the teachers and incited more yelling, he kept smiling. When he and a friend broke into a car for a place to sleep after his abusive parents kicked him out, he smiled. He was surely smiling at the pretty girls with lipstick-reddened lips and pompadours whom he and a couple of homies decided to visit in the neighborhood of a rival gang. He was likely still smiling when eight rival gang members showed up, bringing guns to what might have been a fistfight. Smiley was shot in the head. His homie Victor scooped him off the pavement, bits of fresh brain matter clinging to his new sneakers. Victor cradled the dying Smiley in the backseat on the way to the emergency room; there was no time to wait for an ambulance that might never show up at their ghetto address.

When the police arrived at the hospital, they assigned partial blame to Victor, threatening to charge him as an accessory to murder for being present at the scene. Victor, in anger and disbelief, instead demanded that the police find the real shooter. "What for?" one of them replied, "We want you to kill each other off."

Homeboy Victor is now Associate Professor Victor Rios at the University of California, Santa Barbara. The other 67 homies from the original gang have not had Victor's success: 4 were murdered; 9 were permanently injured, most with handguns; 12 became addicted to drugs, sometimes living on the streets and panhandling to survive. Just 2 graduated from high school; only Victor went on to college. If it is surprising that Victor, a gang-member son of a single mother on welfare in an impoverished neighborhood, made it into one of the most heavily credentialed occupations, what does that tell us about the relationship between deviance and social mobility?

Because Victor is now a sociologist, I asked him what he has to say about getting out of the ghetto and into the ivory tower. "A lot of times people say to me, 'Oh, Professor Rios, you're so unique. You have all these qualities, you made it out. You pulled yourself up by the bootstrap. You made it out of the ghetto and now you are here.' And my response is, 'Well, part of it is hard work, but everyone works hard.' For example, my mom, she's washed dishes 30 years of her life, working 10 hours a day, and she still makes 10, 11 dollars an hour. That's hard work. But she never progressed." He argues that hard work is not enough; there must be an opportunity trajectory leading out of poverty.

Victor clearly recalls when he started to recognize systemwide short-comings: "I was fortunate to find a teacher that cared, and she heard what happened [to Smiley], and she reached out to me and got me mentors from the university—students, college students, that wanted to go help the ghetto kids." These students provided enlightenment through the sociological imagination; they helped Victor see that there was a system at work larger than himself, larger even than his community: "I was actually living in a world of poverty that wasn't necessarily just produced by the way that people in my community acted, but it was also produced by a larger system of racism, classism, and segregation" (Conley, 2009e).

In his book *Punished: Policing the Lives of Black and Latino Boys* (2011), Rios examines the way the current aggressive policing strategies have effectively criminalized young boys in poor neighborhoods. Police and parole officers are stationed in schools and community centers, the spaces in which education and mentoring traditionally occur, creating a self-fulfilling prophecy where teens are assumed to be criminals, treated with suspicion bordering on aggression, and watched closely until caught in some criminal act. Is flooding crime-ridden neighborhoods with aggressive policing the right thing to do? Or does sending more police to a neighborhood where criminal activity is concentrated simply increase the number of people who get caught, closing routes out of that neighborhood by saddling its residents with criminal records?

The goal of this chapter is to examine how society coheres and why some people transgress the boundaries of normality. For starters, how do people who presumably started life as innocent little kids end up in prison? How did Rios, an impoverished gang member growing up amid

DIGITAL.WWNORTON.COM/YOUMAYASK6CORE

To see my interview with Victor Rios, author of *Punished*, go to digital.wwnorton.com/youmayask6core

violence and theft, end up getting a PhD and becoming a professor? We can then turn to broader questions: Why doesn't society at large look like the violence-ridden Oakland, California, neighborhood of Victor's youth? Why do most of us choose to sacrifice some of our personal interests for the sake of the social whole? How can we explain how society achieves predictable order, and what role does the criminal justice system play? Theorists of social deviance have produced various answers to these questions.

What Is Social Deviance?

Social deviance, loosely understood, can be taken to mean any transgression of socially established norms. It can be as minor as farting in church or as serious as murder, so long as it consists of breaking the rules by which most people abide. Minor violations are acts of informal deviance, such as picking your nose. Even if no one will punish you, you sense it is somehow wrong. At the other end of the continuum, we have formal deviance, or crime, which is the violation of laws enacted by society. When deviant persons are caught in deviant acts, depending on the seriousness of the offense, they are subject to punishment. For instance, farting in church might result in glares from your fellow churchgoers, whereas a crime like theft may result in formal state-sanctioned punishments such as fines or community service, or for more severe violations, imprisonment, death, or even torture.

Because social norms and rules are fluid and subject to change, the definitions of what counts as deviance are likely to vary across contexts. Even seemingly obvious cases of deviance, such as killing another person, may not be so clear upon further investigation. When a soldier kills an enemy combatant, that act is considered heroic. But if the same soldier kills his or her spouse, the act is considered heinous and punishable by a long prison sentence. Persons and behaviors have been variously called deviant, depending on what culture and historical period they happen to fall in. When women engaged in premarital sex in the Puritan colonies, they were subject to social exile; when women do so today in some Islamic countries, they risk public execution by stoning. In contemporary American society, however, sex outside of marriage is common and largely accepted. Similarly, 50 years ago, it was a crime for African Americans to share water fountains and swimming pools with white citizens; most people would consider such legislation unthinkable bigotry today. Just four decades ago, homosexuality was illegal nationwide. Gay bars were frequent targets of police raids, and sexual orientation was grounds for excluding immigrants (a rule that held until the Immigration Act of 1990; Foss, 1994). As recently as 2003, the US Supreme Court struck down Texas's criminalization of homosexual sex in *Lawrence v. Texas*; before the ruling, it was punishable by arrest and a $500

SOCIAL DEVIANCE

any transgression of socially established norms.

CRIME

the violation of laws enacted by society.

fine. Changes in laws signal shifting social values and changes in social norms, such as increased tolerance of sexual and racial diversity.

In the popular conception, deviance typically takes the form of blatant rule-breaking or lawlessness. But deviance does not necessarily require outlandish activity on the part of social misfits. Deviance is a broad concept covering everything from answering your cell phone during a lecture to murder. Both formal deviance and informal deviance are collectively defined and often subject to a range of punishments. For example, transgender persons—those who do not have a categorical masculine or feminine identity—are often subject to informal and formal punishments for their gender deviance, although most are perfectly law-abiding citizens. The same goes for ethnic minorities, people who bike down crowded city streets, and those who sing along with their smartphones on the subway or bus. Even reserved, soft-spoken people, if they find themselves in a noisy crowd, may find that because their behavior goes against the grain, they become, for the moment, social deviants. Yet, deviance is sticky; it cannot be turned off and on so easily. Deviance can be a powerful label, capable of reproducing the entrenched social inequalities that punishment is often supposed to correct.

Functionalist Approaches to Deviance and Social Control

Imagine that society is a single, complex organism with many internal organs that perform specific tasks. The state is society's brain, its decision-making center where legislators contemplate the morality of laws and communicate legislative decisions to other social organs charged with implementation.

All of society's organs are necessary to keep the social organism alive and healthy, and these organs are composed of groups of individuals, or cells. A functionalist approach explains the existence of social phenomena by the functions they perform. In this framework, the state develops because society needs a decision-making center to help organize and direct social life. All of society's various parts, or organs, are defined by their functions and are arranged according to the needs of the social organism.

Émile Durkheim, author of *The Division of Labor in Society* (1893/1997), used such a functionalist approach to explain social cohesion, the way people form social bonds, relate to each other, and get along on a day-to-day basis. Durkheim's thesis is that there are two basic ways society can hold together, or cohere, which he called mechanical and organic solidarity. Mechanical or segmental solidarity—which characterized premodern society—was based on the sameness of the individual parts. In this model, the functional units are like cargo containers; each container is like all the others and can perform the same function as the next. Cohesion stems from the reliable similarity of the parts. In a state of organic solidarity—which characterizes modern society—social cohesion is based on interdependence, because the members in this type of social body perform different, specialized functions, and this increased mutual dependence among the parts is what allows for the smooth functioning of the whole. The physical analogy here is a machine in which each part is different and none would be so meaningful outside the context of the machine.

In premodern society, people were held together by sameness. A peasant farmer in feudal times might have eked out a living by tilling a small plot of land, planting seeds, and then harvesting crops. The peasant next door also eked out a living by tilling, planting, and harvesting. Slight variations existed from one farm to the next, but the farmers' life conditions, and particularly their day-to-day experiences in the field, were roughly the same. The farmers' conversations probably would not have been strained or awkward; the two neighbors would have had plenty of advice and commiseration to share. Such a situation would be an example of mechanical solidarity.

SOCIAL COHESION

social bonds; how well people relate to each other and get along on a day-to-day basis.

MECHANICAL OR SEGMENTAL SOLIDARITY

social cohesion based on sameness.

ORGANIC SOLIDARITY

social cohesion based on difference and interdependence of the parts.

Social norms and the punishments for violating them change over time and from place to place. Whether it was executing women as witches in the seventeenth century, enforcing Jim Crow laws in the segregated South, or prosecuting John Geddes Lawrence and Tyron Garner for engaging in a same-sex relationship, our definitions of what constitutes deviance change.

Farmers in premodern society would not have struggled to relate to one another; their sense of sameness bred mechanical solidarity. In contrast, most workers in the industrial and postindustrial economy perform such specialized tasks that they relate to one another through organic solidarity.

As Western society became more industrialized in the late eighteenth and nineteenth centuries, however, workers developed specialized skills, allowing them to perform particular tasks better and more quickly. This division of labor resulted in a dramatic increase in productivity, but with the efficiency and high productivity of specialization come some negative consequences. Specialized workers have less and less in common with one another, and they may be less able to understand each other. For example, if you held a highly specialized position within the economy as, say, a techno-artist punk rocker, you might find yourself thinking that no one understands what it's like to be you. As a techno-artist punk rocker, you are isolated and alienated because your highly specialized position—your roles within the economy and society more generally—makes it difficult to find common ground with others. If you sit down on a commuter train next to an investment banker and try to talk with her, an awkward conversation might ensue, in which each of you tries to comprehend the daily activities of the other. You may share your morning commute and nothing more. Labor specialization divides us by making it more difficult to form social bonds. However, specialization of tasks enables everyone to do something best. Maybe you are *the best* Japanese humane-certified chicken farm sex determiner (actually called a chick sexer, and yes, they make six figures), because, among other reasons, you don't really have much competition. This high degree of labor specialization makes us all interdependent, creating organic cohesion.

Distinguishing between these two types of social solidarity leads us to our first insight into social deviance. When individuals commit acts of deviance, they offend what Durkheim calls the collective conscience, meaning the common faith or set of social norms by which a society and its members abide. Without a collective conscience—a set of common assumptions about how the world works—there would be no sense of moral unity, and society would quickly dissolve into chaos. So when individuals break from the collectively produced moral fabric, societies are faced with the task of repairing the gash in the fabric by realigning the deviant individual through either punishment or rehabilitation.

The way in which social realignment is achieved, Durkheim concludes, depends on the type of solidarity holding that particular society together. Premodern societies, where people are united by sameness, tend to be characterized by punitive justice: making the offender suffer and thus defining the boundaries of acceptable behavior. Such punishment might involve collective vengeance. If someone in a medieval village committed adultery, stole vegetables, or murdered someone, the villagers would gather together, perhaps in a rowdy, pitchfork-wielding mob, to punish the criminal who had offended the collective conscience. The act of collective group punishment might have culminated in storming the offender's house or hanging the poor sap in public. In either case, the group punishes the criminal in an act of collective vengeance. The collective probably wasn't interested in hearing the criminal's side of the story: the facts of his personal life, his possible motivations for committing the crime, or the details of his regrettable childhood. Mechanical social sanctions both reinforce the boundaries of acceptable behavior and unite collectivity through actions such as hanging, stoning, or publicly chastising a former member of the community. It is important to note that it is this collective action of vengeance that, by uniting the group through its perpetration and associated emotions of revenge, guilt, and so on, is key to the production of cohesion and unity. (Of course, there are always exceptions, such as the premodern Amish [see Chapter 5] who don't even bother to prosecute shoplifters.)

Two women accused of collaborating with the Nazis are marched through the streets of Paris in the summer of 1944. The mob ripped the women's clothes, then painted swastikas on their shorn heads in order to punish them. How is this an example of a mechanical social sanction?

A similar form of justice can be administered by the state through a more formal process. When the infamous Oklahoma City bomber, Timothy McVeigh, was put to death, he was, in effect, murdered by the collectivity, by the citizenry of the United States. To opponents of the death penalty, McVeigh's execution amounted to state-sponsored murder. To others, however, his death signaled that killing innocent people is not acceptable behavior. Although only a small team actually carried out McVeigh's execution, his death was an act in which—theoretically at least—we all participated. When we collectively and publicly put McVeigh to death, we were united in our reinforcement of social norms: Paradoxically, his deviance helped keep our society together.

Organic solidarity, in contrast, by differentiating individuals, produces social sanctions that focus on the individual—that is, they are tailored to the specific conditions and circumstances of the perpetrator. This response to deviance is rehabilitative, meaning that the response is designed to transform the offender into a productive member of society. In the modern mind-set, we care about the rapist's or murderer's motivations and regrettable childhood. We treat the criminal as an individual who can be "fixed" if we can root out the causes of, and triggers for, his or her criminality. For example, what happens if a drug addict steals car radios to support her cocaine habit? The court may order the addict into rehab in hopes of reintegrating her into the productive mainstream.

Punishments may also be restitutive—that is, they attempt to restore the status quo that existed before an offense or event. Tort law is an example of restitutive social sanctions. For instance, let's say a negligent contractor doesn't bother to install grating at the bottom of a swimming pool. A little

Eric Todaro (left) and Richard Grooms, inmates at a state penitentiary in Oregon, work on a General Education Diploma (GED) test. According to Durkheim, why would prisons provide educational programs and other rehabilitative tools?

girl playing at the edge of the pool is sucked into the drain, almost drowns, and ends up with brain damage. Restitutive social sanctions, or tort law, force the contractor (or his insurance company) to pay millions of dollars to the parents of the little girl. The money attempts to reestablish social equilibrium by repaying the parents for what they have lost. (Of course, in this case at least, money is a wholly inadequate salve.)

In the United States, we consider ourselves a modern society, yet both forms of social sanctions, mechanical and organic, still lurk within the US justice system. We don't only try to rehabilitate our criminals and reimburse victims. We also still employ the death penalty in many states. Does Texas, where approximately one-third of the executions in the United States have taken place, have a more premodern division of labor than Wisconsin, which does not punish crimes with the death penalty? Probably not. Did the division of labor revert to primitive, subsistence levels all of a sudden in 1976, when the Supreme Court reinstated the legality of the death penalty? Of course not. Furthermore, some traditionally liberal states such as California and Oregon still have the death penalty on their books. At the same time, we also have a signature form of modern, organic social sanctions as embodied by our elaborate court system. Durkheim doesn't argue that these forms of social sanctions are mutually exclusive—that they can't exist together. In fact, he would expect to find much more of one form of social sanction, but not only one form. Both forms, however, help hold us together by reinforcing the boundaries of normal, socially acceptable behavior.

To make his case, Durkheim analyzed different historical penal and moral codes to discern the ratio of premodern sanctions to modern sanctions throughout history. He examined the Code of Hammurabi in Babylon dating back to 1750 B.C.E. as well as the Pentateuch, the first five books of the Old Testament. He studied more recent sanctions in the Magna Carta, the Napoleonic Code in France, and the South American Drago Doctrine. He argued that, as history progressed and the division of labor developed, the ratio of premodern to modern sanctions changed to favor modern, less punitive sanctions. So although the United States may still permit its states to apply the death penalty, Durkheim would hypothesize that as our labor market changes to favor even more specialization, we might expect the death penalty to fizzle out. In fact, approximately a dozen states have repealed the death penalty in recent decades, but we should be cautious in drawing overall conclusions about the relationship between division of labor and forms of punishment by focusing on one particular form of sanction.

SOCIAL CONTROL

We have explored the paradox of deviance in Durkheim's *The Division of Labor in Society*—the idea that deviance, and specifically the act of collective punishment, holds us together. But what makes us good law-abiding citizens in the first place? Social control is what sociologists refer to as the set of mechanisms that create normative compliance, the act of abiding by society's norms or simply following the rules of group life. Other sociologists have tried to explain how social control works on individuals to induce compliance to social norms.

Sociologists classify mechanisms of social control into two categories. The first are called formal social sanctions. In most modern societies, these formal sanctions would be rules or laws prohibiting deviant criminal behavior such as murder, rape, and theft. These sanctions are formal, overt "expressions of official group sentiment" (Meier, 1982). Informal social sanctions are based on the usually unexpressed but widely known rules of group membership. Have you ever heard someone use the expression "an unwritten rule"? Informal social sanctions are the unwritten rules of social life. So, hypothetically, if you loudly belch in public, you will probably be the object of scowls of disgust. These gestures of contempt at your socially illicit behavior are examples of informal social sanctions, the ways we keep each other in check by watching and judging those around us. As discussed in Chapter 4, the process of socialization is largely responsible for our acquisition and understanding of these unspoken rules of group social life. Through years of trial and error, we have internalized the rules of the social game.

The idea behind informal social sanctions is that we are all simultaneously enforcing the rules of society and having them enforced on us. How does this work? At the same time we are watching or observing others, others are watching us, too. All of us are both spectators and objects of spectacle. We are all the agents of a diffuse, watchful gaze, as in the 1881 painting *Luncheon of the Boating Party* by Pierre-Auguste Renoir, where everyone is gazing at someone else, but no two people make direct eye contact. And beyond watching, all of us can grant rewards for others' good behavior in the form of smiles and encouragement, but we can also sanction with dirty looks, snide comments, and worse. In this way, we are all chipping in our small contributions to the construction of the social whole.

Think about the neighborhood watch groups that preceded the widespread adoption of electronic home security systems. People in a community banded together and agreed to keep an eye out for trespassers, burglars, and suspicious strangers. The idea behind these groups was that neighborhoods could maintain social order by overtly declaring their intention to visually monitor their turf. Sometimes such groups issued signs or decals for doors and windows to signal to unsavory characters that the neighborhood was being "protected" by watchful eyes. But this kind of activity doesn't have to be so formalized. The urban theorist Jane Jacobs (1961) coined the term *the eyes and ears of the street* to describe the fact that, ideally, in mixed-use (i.e., commercial and residential) neighborhoods, the thread of social control is implicitly woven into daily life. Through their windows, grandmothers watch children playing baseball on the street below, on the alert for any trouble. During the course of a busy weekday, a shopkeeper might notice a group of teenagers, who should be at school, loitering outside his store and report this to their parents or the school.

In any society, many agents of both formal and informal social control exist. Our local neighbors act as the primary agents of informal social control, whereas the state, or the government, often has a hand in the construction of formal social sanctions by making laws.

Pierre-Auguste Renoir's *Luncheon of the Boating Party* (1881).

The police are an obvious example of an agent of state social control formed for the protection of the public. The police patrol public parks and neighborhoods after nightfall and contain public protests to ensure social order.

Informal social control is the bedrock on which formal social control must rest. If the police go on strike, for example, whether chaos will ensue depends on the degree of social cohesion in a given community. Likewise, without strong informal social norms, the police are relatively helpless. Consider the difficulty police officers have tracking down and arresting those who loot and commit acts of vandalism during an urban riot. Or the difficulty in solving or prosecuting a crime if witnesses are not willing to step forward and testify. If an entire community relaxes its informal social control, formal social control inevitably fails. Lately, the boundary between informal and formal social control is perhaps blurring a bit with the proliferation of surveillance cameras installed by private and public entities.

A NORMATIVE THEORY OF SUICIDE

After making his observations on social solidarity and social control in *The Division of Labor in Society*, Durkheim next applied his ideas to the sociological study of suicide, perhaps the most individual act of deviance. Or is it? If you were asked to explain what causes a person to die by suicide, you might say mental illness, depression, drug addiction, or perhaps a catastrophic event. These accounts reflect our perception of suicide as something intensely personal, mediated by individual life circumstances or caused by chemical imbalances and emotional disorders. And these explanations may contain a piece of the answer. However, Durkheim believed these individualistic accounts of suicide were inadequate.

FIGURE 6.1 A Normative Theory of Suicide

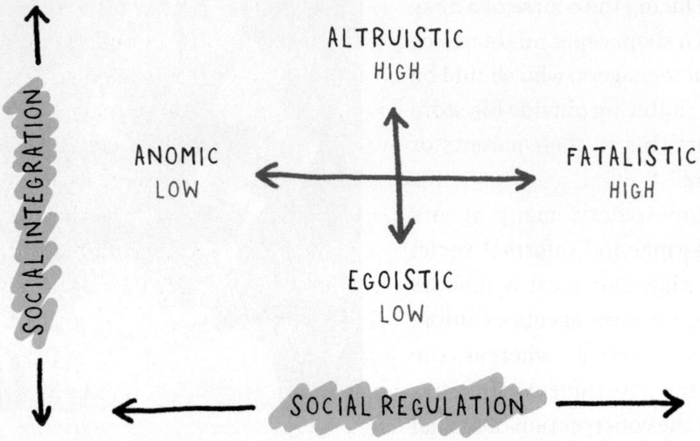

In his 1897 book *Suicide*, Durkheim sought to explain how social forces beyond the individual shaped suicide rates. According to Durkheim, suicide is, at its root, an instance of social deviance. By observing patterns in suicide rates across Europe (just as he had previously discerned patterns in penal codes), Durkheim developed a normative theory of suicide.

Durkheim proposed that by plotting "social integration" on the *y*-axis and "social regulation" on the *x*-axis of a Cartesian coordinate system, we can better see how social forces influence suicide rates (Figure 6.1). Social integration refers to the degree to which you are one with your social group or community. A tightly knit community in which members interact with each other in a number of different capacities—say the coach of your child's Little League team is also your dentist and you are the dentist's mechanic—is more socially integrated than one in which people do not interact at all or interact in only one role. Social regulation refers to how many rules guide your daily life and what you can reasonably expect from the world on a day-to-day basis—the degree to which tomorrow will look like today, which looks like yesterday. To be at low risk for suicide (and other deviant behavior), you need to be somewhere in the middle: integrated into your community with a reasonable (not oppressive) set of guidelines to structure your life. If you go too far in either direction along either axis, you have either too much or too little of some important facet of "normal" life.

Let's say you drop down the *y*-axis significantly in the direction of egoism. You are not very well integrated into your group. And Durkheim argues that, because others give your life meaning, you would feel hopeless. You wouldn't be part of some larger long-term project, would feel insignificant, and would be at risk of committing egoistic suicide. We all need to feel as if

SOCIAL INTEGRATION

how well you are integrated into your social group or community.

SOCIAL REGULATION

the number of rules guiding your daily life and, more specifically, what you can reasonably expect from the world on a day-to-day basis.

EGOISTIC SUICIDE

suicide that occurs when one is not well integrated into a social group.

we have made a difference in other people's lives or produced something for their good that will endure after we have died.

Durkheim demonstrated the prevalence of this phenomenon by using statistics about suicide rates across different religious groups. Although many Western religions formally prohibit suicide, rates varied substantially across religious affiliations. Durkheim found that throughout Europe, Protestants killed themselves most often, followed by Catholics and then Jews. Why? Protestantism is premised on individualism. In most Protestant denominations, there isn't an elaborate church hierarchy as there is in the Catholic church, and Protestants are encouraged to maintain a direct personal relationship with God. By changing the individual's relationship to God (and therefore of the individual's relationship to the Church), Protestantism also stripped away many of the integrative structures of Catholicism, putting its members at greater risk for egoistic suicide. (More recent research shows that today the greater distinction is between people who are religiously affiliated and those who are not, the latter experiencing significantly higher levels of suicide.)

But why did Jews have the lowest suicide rate of any major European religious group if all these religions prohibited suicide? Judaism may be less structured than Catholicism, but historically speaking, Jews have remained a persecuted minority group. When a social group is persecuted or rejected by the so-called mainstream, its members often band together for protection from persecution. For example, in the United States today, African Americans have one of the lowest suicide rates, probably as a result of their bonding as a minority group, united in a common struggle against a history of oppression.

Too little social integration increases the risk of suicide, but too much social integration can also be dangerous. A person who strays too far up the y-axis might commit altruistic suicide, because a group dominates the life of that individual to such a degree that he or she feels meaningless aside from this social recognition. Think about Japanese ritual suicide, sometimes called seppuku (sometimes colloquially known as hara-kiri). In this scenario, samurai warriors who had failed their group in battle would disembowel themselves with a sword rather than continue to live with disgrace in the community. Durkheim uses the example of Hindu widows in some castes and regions of India, who were expected to throw themselves on their husbands' funeral pyre to prove their devotion. This practice, called suttee or sati, symbolized that a woman properly recognized that her life was meaningless outside her social role as a wife. Official efforts to ban suttee commenced as early as the sixteenth century, and it is now illegal and extremely rare in India.

Altruistic suicide can be a more personal, less ritualistic, choice, as well. For example, do you think that suicide rates in the military are higher

ALTRUISTIC SUICIDE

suicide that occurs when one experiences too much social integration.

among enlisted soldiers or officers? Enlisted soldiers might be the obvious choice because they experience a lower standard of living and less social prestige, but statistics show that the suicide rate is, in fact, higher among officers. Why? Too much social integration. The identity of officers—their sense of honor and self-worth—is more completely linked to his role in the military. Enlisted soldiers, in contrast, are not as responsible for group performance. More likely, these soldiers perceive their military service as a job and still identify strongly with their other, civilian roles, so they have a lower risk of altruistic suicide.

As we move to the *x*-axis, social regulation, why or how would social regulation influence the suicide rate? Imagine you commute to school every day. You rise at approximately 7:00 A.M., leave the house by 8:00, and take the 8:10 bus to school. But now imagine that some days the bus is late, some days the bus is early, and sometimes it just doesn't come at all. You have no way of knowing when or if the bus will come. You get tired of standing on a cold, lonely corner waiting for a bus that may never come, and after a while you might just stop trying to get to class at all. What's the point of waking up if you won't make it to class on time despite your effort? You have developed a sense of learned helplessness, a depressed outlook in which sufferers lack the will to take action to improve their lives, even when obvious avenues are present.

At the heart of learned helplessness is the sense that we are unable to stave off the sources of our pain, that we have no control of our own well-being. Durkheim studied a similar condition, which he termed *anomie*. Literally meaning "without norms," anomie is a sense of aimlessness or despair that arises when we can no longer reasonably expect life to be more or less meaningful. Our sense of connection between our actions and values is eroded, because too little social regulation exists. Durkheim labeled suicide that resulted from insufficient social regulation anomic suicide. For example, after the stock market crashed in 1929, many businessmen jumped out of skyscraper windows to their deaths. These stockbrokers and investors may have felt that they did everything right and still ended up destitute. For them, the connection between what they thought was the right thing to do—work hard on Wall Street—and just rewards was severed. They had no idea how to cope with the changes.

ANOMIE

a sense of aimlessness or despair that arises when we can no longer reasonably expect life to be predictable; too little social regulation; normlessness.

ANOMIC SUICIDE

suicide that occurs as a result of insufficient social regulation.

Durkheim's argument about anomic suicide seems intuitive when it refers to negative events rupturing our everyday lives, but the same principle also holds for positive life events. For example, many lottery winners report spells of severe depression after winning millions of dollars, displaying another case of anomie (Nissle & Bshor, 2002). If a very poor, frugal man wins the lottery, all of his money-saving habits instantaneously become irrelevant, unnecessary, or even a bit silly. Maybe he previously structured his Sundays this way: Walk down to the corner store just as it's about to close to pick up a castoff of the Sunday paper, painstakingly cut coupons from the circulars for hours, and then plan a visit to each of three local grocery stores to find the best deals throughout the week. Maybe he always took lunch to work in reused brown paper bags to save money. Now he has $5 million in his checking account and no behavioral template (something that social processes yield) for life as a wealthy man. His difficulties do not revolve around the quantity of material resources or standard of living suddenly available to him. They rest in how meaningless the rules he previously used to give meaning to his life now seem.

The final coordinate is fatalistic suicide, which occurs when a person experiences too much social regulation. Instead of floundering in a state of anomie with no guiding rules, you find yourself doing the same thing day after day, with no variation and no surprises. In 10 years, what will you be doing? The same thing. In 20 years? The same thing. You have nothing to look forward to because you reasonably expect that nothing better than this will ever happen for you. This type of suicide usually occurs among slaves and prisoners. You might imagine that slaves and prisoners would commit suicide because of their physical hardships, but Durkheim's research suggested that it is more accurate to understand their suicidal deaths as resulting from the suffocating tyranny of monotony.

Early feminists wrote of the "problem that has no name": that life as a 1950s suburban stay-at-home mother was a stifling routine, the same every day, for as long as the imagination could conjure. Sylvia Plath, a feminist poet and writer, committed suicide in 1963 shortly after the publication of her novel *The Bell Jar*, in which the semiautobiographical character Esther's fatalism is palpable: "I saw the days of the year stretching ahead like a series of bright white boxes, and separating one box from another was sleep, like a black shade. Only for me, the long perspective of shades that set off one box from the next had suddenly snapped up, and I could see day after day after day glaring ahead of me like a white, broad, infinitely desolate avenue" (1971, p. 143). Women's roles in society were tightly controlled, and thus they were at higher risk for fatalistic suicide.

According to many theories of deviance, what happens at the group level affects what happens at the individual level. For example, Durkheim hypothesized that members of minority groups were more socially integrated within their group, and therefore Jews in Europe had lower suicide

FATALISTIC SUICIDE
suicide that occurs as a result of too much social regulation.

rates. Perhaps minority solidarity inspires feelings of belonging and love between family and nonfamily alike within the group. Because feeling loved and wanted is generally regarded as a good thing, we could say that this solidarity makes group members happy or at least staves off depression. Less depressed, happier people commit suicide less often than those who feel unimportant, worthless, and hopeless. Group dissimilarities start at the macro level (group solidarity), filter down to the individual level (feelings of depression or happiness), and can be detected again in the aggregate (differential suicide rates).

SOCIAL FORCES AND DEVIANCE

In keeping with Durkheim's attempt to discover the social roots of suicide and other forms of deviance, sociologist Robert Merton pioneered a complementary theory of social deviance. Instead of stressing the way sudden social changes lead to feelings of helplessness, Merton argued that the real problem behind anomie occurs when a society holds out the same goals to all its members but does not give them equal ability to achieve these goals. Merton's strain theory, advanced in 1938, explains how society gives us certain templates for acting correctly or appropriately. More specifically, we learn what society considers appropriate goals and appropriate means of achieving them. The *strain* in strain theory arises when the means don't match up to those ends; hence, Merton's theory is also called the "means-ends theory of deviance." When someone fails to recognize and accept either socially appropriate goals or socially appropriate means (or both), he or she becomes a social deviant.

If you have decided to pursue a college education, presumably to land a decent job, you are probably what Merton terms a conformist. A conformist accepts both the goals and strategies to achieve those goals that are considered socially acceptable. Your goal is to earn a good living, maybe start a family, and take long, exotic vacations where you will take photos with a fancy camera and then post them online to impress your friends. You've decided to pursue this lifestyle through a better education, deliberate cultivation of the right social network, and hard work.

Now let's say that you go to class every day, take minimal notes, and read just enough to earn a passing grade. All you want is to get by and be left alone. You don't care how much money you will earn, as long as it's enough to pay for your studio apartment and other small monthly bills. You've accepted society's acceptable means (you're still going to college, after all), but you've rejected society's goals (the big house, the 2.3 kids, and the new car). You've rejected the idea of getting ahead through hard work, the American dream. You are a ritualist, a person who rejects socially defined goals but not the means.

If, however, you yearn to be rich and famous but don't have the scruples,

STRAIN THEORY

Robert Merton's theory that deviance occurs when a society does not give all of its members equal ability to achieve socially acceptable goals.

CONFORMIST

individual who accepts both the goals and the strategies that are considered socially acceptable to achieve those goals.

RITUALIST

individual who rejects socially defined goals but not the means.

INNOVATOR

social deviant who accepts socially acceptable goals but rejects socially acceptable means to achieve them.

patience, or economic resources to get there by using socially acceptable means, you may be an innovator. Let's say you are particularly interested in buying a mansion and expensive jewelry, and marrying a gorgeous husband or wife. Instead of slaving away on Wall Street for years, you sell drugs, fence stolen goods, and make a few friends in the Mafia.

Among those who reject *both* means and goals are retreatists and rebels, although the boundaries between the two are not always so cut and dried. Retreatists completely stop participating in society. This type is perhaps illustrated by adventurer Christopher Johnson McCandless, the subject of Jon Krakauer's best-selling 1996 book *Into the Wild*, who simply decided not to play the game and moved to the Alaskan woods, where he lived without running water or electricity. (I won't tell you what happens to him in case you want to read the book or watch the 2007 film adaptation.) A rebel also rejects both traditional goals and traditional means but wants to change (or destroy) the social institutions from which he or she is alienated. One example is Ernesto "Che" Guevara, the Argentine Marxist who famously fought

RETREATIST

one who rejects both socially acceptable means and goals by completely retreating from, or not participating in, society.

REBEL

individual who rejects both traditional goals and traditional means and wants to alter or destroy the social institutions from which he or she is alienated.

Conformist

Ritualist

Innovator

Retreatist

Rebel

Which type are you? Do you follow socially accepted means and goals? According to Robert Merton, you're a conformist. Doing the bare minimum? You're probably a ritualist. If you're like WorldCom CEO Bernard Ebbers and want to earn big rewards but have few scruples about how you reach them, you're an innovator. You're a retreatist if, like members of a self-supporting commune, you reject all means and goals of society. You're a rebel, like Che Guevara, if you not only reject social means and goals but also want to change society itself.

for communism in Cuba. His disgust at the impoverished conditions he encountered as a doctor traveling through Latin America led to the formation of a guerrilla group. Che chose to fight the government rather than, say, propose new legislation or raise money for a new hospital.

Symbolic Interactionist Theories of Deviance

Whereas Durkheim and Merton focused on the ways different parts of the organic social body function together, another school of sociologists in the 1960s and 1970s took a different approach to the study of deviance. Working in the tradition of symbolic interactionism (see Chapters 1 and 4), a term coined by Herbert Blumer (1969), these sociologists stressed the particular meanings individuals bring to their actions, rather than the broader social structures of which they are unwittingly a part. To determine why people commit crimes, or to seek the root causes of deviance in a given society, a symbolic interactionist would look at the small and subtle particulars of a social context, and the beliefs and assumptions people carry into their everyday interactions. Functionalist theories, such as Durkheim's and Merton's, are sometimes called macro theories because they seek to paint the social world in wide brushstrokes: generalizable trends, global or national forces, and broad social structures. At the opposite end of the spectrum, micro theories such as symbolic interactionism zoom in on the individual. Symbolic interactionism takes seriously our inner thoughts and everyday interactions with one another, including how others see us and how we respond to our surroundings.

LABELING THEORY

As a child, did you ever shoplift, trespass, or forge a document? If you were slow and tactless enough to get caught, your parents probably gave you a stern lecture and maybe grounded you. Chances are, however, that over time the incident slowly faded into the background and you stopped feeling guilty. You probably never came to think of yourself as a shoplifter or trespasser, as a criminal or social deviant. You simply made a stupid mistake, never to be repeated. Let me give you an example from my own past. One afternoon when I was in junior high school, a friend and I decided to play "fireman, waterman," a game that required one person (the fireman) to flick lit matches into the air, while the waterman tried to extinguish them with a plant mister. To make a long story short, I was the fireman and I won. As you might expect, the incident ended badly—specifically, one of the stray, airborne matches set my friend's apartment on fire. After extinguishing the blaze, fire department officials questioned me about my role in starting the fire, but ultimately, I was absolved from blame and the fire was declared an accident. My parents, although obviously shaken and disappointed, never formally punished me. The trauma of the incident, they said, was lesson enough. Now imagine that, instead of being pardoned, I was held accountable for my role in starting the fire and sent to a juvenile corrections facility. Do you think I still would have followed the same life trajectory, eventually going to college and graduate school, and becoming a sociology professor and textbook author?

The labeling theory of social deviance offers insight into how people become deviants. According to this theory, individuals subconsciously notice how others see or label them, and their reactions to those labels over time form the basis of their self-identity. It is only through the social process of labeling that we create deviance by assigning shared meanings to acts. We all know that stealing, trespassing, and vandalizing are wrong, abnormal, and criminal behavior, but these acts are considered deviant only because of our shared meanings about the sanctity of private property and our rules about respecting that property. Although the fire caused by my unfortunate stint in pyrotechnics was labeled an accident, it might have been termed a crime just as easily. There was and is nothing inherently deviant about setting an apartment on fire, accidentally or otherwise. In 1963, Howard S. Becker, a proponent of labeling theory, made precisely this point, arguing that individuals don't commit crimes in a vacuum. Rather, social groups create deviance, first by setting the rules for what's right and wrong, and second by labeling wrongdoers as outsiders. Offenders are not born; they are made. They are a consequence of how other people apply rules and sanctions to them.

Social groups create rules about the correct or standard mode of conduct for social actors. When these rules are broken, society's reaction to the act determines if the offense counts as deviance. Take for example the case of opioid use. When opioid use was dominated by heroin and disproportionately afflicted black Americans, it was criminalized. However, when such drug addiction soared among whites, it was seen more as a health issue than one of crime or deviancy and was referred to in medical terms such as "epidemic." Becker also argues that not only are deviant acts created by a process of labeling but *deviants* are also created by a process of labeling. If you break a rule—say, accidentally burn down an apartment—but your violation is not labeled a crime or otherwise recognized as deviant, then *you* are not recognized as a criminal or deviant. We become (or don't become) deviant only in interaction with other social actors. It is this social reaction to an act, and the subsequent labeling of that act and offender, that create social deviance.

To illustrate this process of becoming deviant, Becker interviewed

LABELING THEORY

the belief that individuals subconsciously notice how others see or label them, and their reactions to those labels over time form the basis of their self-identity.

How did Howard Becker apply labeling theory to the use of marijuana?

50 marijuana users in "Becoming a Marihuana User" (1953)—in a period when not only was marijuana illegal in all states, but it was quite socially stigmatized. He began his study with a simple inquiry: Are marijuana users different from non-users? Is there something about the individual psychology of marijuana users that causes them to smoke marijuana? The answer given by most scientists, politicians, and parents in the 1950s was a hearty *yes*. But Becker argues that chronic marijuana use results from a process of social learning. Before people light up, they must first learn how to smoke marijuana. More important, they must then redefine the sensations of marijuana use as "fun" and desirable. Just as setting an apartment on fire is not inherently a crime, getting high is not "automatically or necessarily pleasurable" (Becker, 1963). In fact, one user he spoke with never could learn to enjoy pot:

> It [marijuana] was offered to me, and I tried it. I'll tell you one thing. I never did enjoy it at all. I mean it was just nothing that I could enjoy. [Becker: Well, did you get high when you turned on?] Oh, yeah, I got definite feelings from it. But I didn't enjoy them. I mean I got plenty of reactions, but they were mostly reactions of fear. [You were frightened?] Yes. I didn't enjoy it. I couldn't seem to relax with it, you know. (p. 240)

A person trying marijuana is usually able to redefine the experience when smoking with long-term users who assure the novice that the physical effects are normal, even enjoyable. One of Becker's interviewees, a veteran smoker, recalled how a beginner "became frightened and hysterical" after smoking for the first time. An older user told the newbie, "I'd give anything to get that high myself" (Becker, 1963). This comment and others like it apply new meaning to the sensations of getting high. For some, marijuana smoking becomes pleasant only through a social process. How objects and sensations become meaningful (or pleasant) through social processes is the focus of Becker's study, in contrast to explanations that focus on what objects *are*. Similarly, the taste of alcohol probably didn't appeal to your palate when you first tried it; however, because alcohol had a certain social allure, being associated with either adulthood or rebellion, you may have learned to appreciate its taste (perhaps before you reached the legal drinking age).

Another example of a social process may help illustrate the power of labeling theory. As part of a psychology experiment, a group of eight adults with steady employment and no history of mental illness presented themselves at different psychiatric in-patient hospitals and complained of hearing voices (Rosenhan, 1973). Each of the pseudo-patients was admitted after describing the alleged voices, which spoke words like "thud," "empty," and "hollow." In all cases, the pseudo-patients were admitted with a diagnosis of schizophrenia. The researchers instructed the pseudo-patients, once they were hospitalized, to stop "simulating any symptoms of abnormality." Each pseudo-patient "behaved on the ward as he 'normally' behaved." Still, the doctors and staff did not suspect that the pseudo-patients were imposters. Instead, the treating psychiatrists changed their diagnosis of schizophrenia to "schizophrenia in remission." Something even more troubling than the misdiagnosis occurred. Once the pseudo-patients had been labeled insane, all their subsequent behavior was interpreted accordingly. One male pseudo-patient, for example, gave the hospital staff a truthful account of his personal history. During his childhood he had been close to his mother but not his father. Later in life, he became close with his father and more distant from his mother. His marriage was "characteristically close and warm," with no persistent problems. He reported rarely spanking his children. Sounds like a fairly typical guy, right? This is how the hospital record described the pseudo-patient's personal history:

> This white 39-year-old male...manifests a long history of considerable ambivalence in close relationships, which begins in early childhood. A warm relationship with his mother cools during adolescence. A distant relationship to his father is described as becoming very intense. Affective stability is absent. His attempts to control emotionality with his wife and children are punctuated

In experiments such as David L. Rosenhan's or films like *Shock Corridor* (left), people with no history of mental illness are admitted into psychiatric hospitals. These examples raise questions about the stickiness of labels. What are some of the consequences of being labeled a deviant?

THE STANFORD PRISON EXPERIMENT AND ABU GHRAIB

The force of labels and roles can affect us very quickly. The Stanford Prison Experiment, conducted by Philip Zimbardo in 1971, provides insight into the power of such social labels and how they might explain the incredibly inhumane acts of torture, most involving violence and humiliation, committed at Abu Ghraib, the American-run prison in Iraq. Zimbardo, a psychology professor at Stanford, rounded up some college undergraduate men to participate in an experiment about "the psychology of prison life." Half the undergraduates were assigned the role of prisoner, and half were assigned the role of prison guard. These roles were randomly assigned, so there was nothing about the inherent personalities of either group that predisposed them to prefer one role over the other.

To simulate the arrest and incarceration process, the soon-to-be prisoners were taken from their homes, handcuffed, and searched by actual city police. Then all the prisoners were taken to "prison"—the basement of the Stanford psychology department set up with cells and a special solitary confinement closet. The guards awaited their prisoners in makeshift uniforms and dark sunglasses to render their eyes invisible to inmates. Upon arrival at the prison, all the criminals were stripped, searched, and issued inmate uniforms, which were like short hospital gowns. The first day passed without incident, as prisoners and guards settled into their new roles. But on the morning of the second day, prisoners revolted, barricading themselves in their mock cells and sparking a violent confrontation between the fictitious guards and prisoners that would ensue for the next four days. From physical abuse, such as hour-long counts of push-ups, to psychological violence, such as degradation and humiliation, the guards' behavior verged on sadism, although just days before these same young men were normal Stanford undergrads.

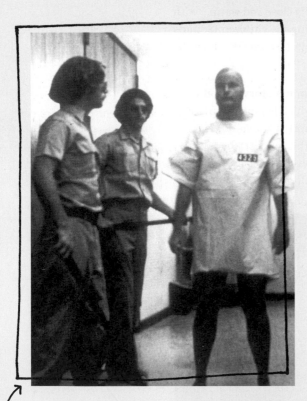

Philip Zimbardo's Stanford Prison Experiment.

This Iraqi detainee in Abu Ghraib prison was hooked up to wires after American soldiers made him stand on a box. How can Zimbardo's experiments help us understand the torture at Abu Ghraib?

The prisoners quickly began "withdrawing and behaving in pathological ways," while some of the guards seemed to relish their abuse. The original plan was for the experiment to last 14 days, but after only 6 days it spiraled out of control and had to be aborted.

The lesson, claims Zimbardo, is that good people can do terrible things, depending on their social surroundings and expectations. When thrown into a social context of unchecked authority, anonymity, and high stress, average people can become exceptional monsters. It's a phenomenon Zimbardo calls the "Lucifer effect," and it offers insights into how the atrocities at Abu Ghraib prison in Iraq became possible (Zimbardo, 2007). In 2005, when the media made public the horrifying images of Iraqi prisoners being degraded and abused—some naked and on their knees inches from barking dogs; some wearing hoods and restrained in painful, grotesque positions; some forced to lie atop a pile of other naked prisoners—with grinning American soldiers looking on, many commentators (and military officials) sought to explain the abuses as a case of "bad apples" among otherwise good soldiers. Bad apples don't just arise out of nowhere, however, nor are people inherently malicious or brutal by nature. Zimbardo's experiment, it seemed, offered a viable explanation.

In 2018, however, the Stanford Prison Experiment was again in the news when journalist Ben Blum published an article calling the experiment a "sham," questioning its ethical and methodological basis. Blum claimed the study was engineered to encourage the "guards" to humiliate and abuse the "prisoners." In response, Zimbardo forcefully defended his experiment, insisting that its goal was not to faithfully recreate the conditions of a prison—but rather, that the experiment was meant to be a "cautionary tale of what *might* happen to any of us if we underestimate the extent to which the power of social roles and external pressures can influence our actions" (2018). Sounds pretty sociological to me.

Wherever we stand on this debate, Zimbardo's quote confirms what we already know about socialization: when given limitless power under high stakes and in an environment of extreme uncertainty—as happened to the soldiers at Abu Ghraib—abuse can become the norm, and people who are otherwise good can do evil things.

by angry outbursts and, in the case of his children, spankings. And while he says that he has several good friends, one senses considerable ambivalence embedded in those relationships also. (Rosenhan, 1973, p. 253)

Even though the pseudo-patient was acting "normally" shortly after his admission, the "abnormal" label stuck and continued to color the staff's perceptions and diagnoses. Just in case you are curious, the pseudo-patients were kept by the hospitals for 7 to 52 days, with an average stay of 19 days. Sticky label, indeed.

Labeling theorists believe that the fact that deviant labels stick no matter what the circumstances has important consequences for behavior. If, after committing a crime or hearing voices, you were labeled deviant and people treated and thought about you differently, how would you feel? Like the same person but just someone who, say, was arrested for possessing drugs? Or would the criminal label become part of the way you thought about yourself? Labeling theorists call the first act of rule breaking (which can include experiences or actions such as hearing voices, breaking windows, or dyeing one's hair neon orange) primary deviance. After you are labeled a deviant (a criminal, a drug addict, a shoplifter, even a prostitute), you might begin behaving differently as a result of the way people think about and act toward you. Others' expectations about how you *will* act affect how you *do* act. Secondary deviance refers to deviant acts that occur after primary deviance and as a result of your new deviant label. For example, a deviance researcher interviewed a woman in her sixties who had been under psychiatric care for some 20 years (Glassner, 1999). This is how she recalls the process of becoming mentally ill:

> The first time I was taken to a psychiatrist for help was when I was getting depressed over a miscarriage. I had tried for many years to have that baby, and finally I was pregnant and planning to be a mother and all, and then I lost it. Anyhow, they told me I was "deeply depressed," but it didn't really mean much to me. I figured I'd get over it once I got pregnant again or something. But when I went home, everybody treated me differently. My husband and my mother had met with the social worker, who explained that I had this problem, with depression and all. From then on I was a depressive. I mean, that's the way everyone treated me, and I thought of myself in the same way after a while. Maybe I am that way, maybe I was born that way or grew up like that, but anyhow, that's what I am now. (p. 73)

First, the woman sees her experience with depression as a passing phase, something she will recover from in time. Then a psychiatrist labels her "deeply depressed." When she arrives home, family members treat her differently, and "from then on" she is "a depressive." You can see how the woman's initial experience is redefined by her interaction with the psychiatrist and how the psychiatrist's label eventually becomes an integral part of the woman's identity. Of course, these labels themselves are socially structured by group stereotypes that can often serve as default cognitive categories.

STIGMA

We've seen how primary deviance can snowball into secondary deviance, whereby a few initial wrong steps or unlucky breaks can define a person's future actions and reactions. It doesn't end there. Secondary deviance can quickly become social stigma. A stigma is a negative social label that not only changes others' behavior toward a person but also alters that person's self-concept and social identity. We frequently say that certain behaviors, groups, identities, and even objects are stigmatized. Mental illness still carries a stigma in our society. Pedophilia carries a bigger one.

Having a criminal record can also carry a stigma, as can a person's race. In 2001 Devah Pager, then a graduate student in sociology, conducted a study to determine the effects of race and a criminal record on employment opportunities. She dispatched potential job candidates (African American and white male college students who had volunteered for the experiment) with similar résumés to apply for entry-level service positions. Half the participants of each race were told to indicate a prior felony conviction on their job applications. Pager (2003) found that of those with no criminal record, the white applicants were more likely to get a response (34 percent compared with 14 percent) but also that the white men with a supposed felony conviction were more likely to be called back than the black men who did not report a criminal record (17 percent compared with 14 percent). The employers in Pager's study were slightly more willing to consider a white applicant with a felony record than a black applicant with a clean history. So not only is a criminal record a stigma that deters potential employers but the race of a job applicant continues to matter. One possible explanation, reasoned Pager, is that

> employers are attempting to select the best candidate for the job, but are affected by all kinds of pervasive and largely unconscious stereotypes that result in privileging or preferring a white candidate to a black candidate. . . . Employers have these negative stereotypes based on really pervasive media imagery, for example, of young black men involved with the criminal justice system or acting out in some negative way. There's lots of information we get about all of the negative characteristics that we might attribute to the African American population. . . . They talked about black men as being lazy

DIGITAL.WWNORTON.COM/YOUMAYASK6CORE

To see my interview with Devah Pager, author of *Marked*, go to
digital.wwnorton.com/youmayask6core

and dangerous and criminal and dressing poorly; they were very candid about all of these negative characteristics that they attributed to black men. But then, right after that, when I asked them about their experiences over the past year with black applicants or black employees, they had a much harder time coming up with concrete examples of those general attitudes, and for the most part, employers reported having very similar kinds of experiences with their black and white employees. (Conley, 2009f)

Pager's research reveals how stigmas have real consequences, shaping the map of opportunities for men with criminal records, and illuminates the challenges faced by black men, whose skin carries a stigma of deviance into the labor market whether they've been in prison or not. Of course, in this case, because the testers were trained with a script, the issue of internalized self-stigma doesn't come up to influence their interactions. But such internalized stigma is itself an important source of disadvantage and stress for marginalized groups.

BROKEN WINDOWS THEORY OF DEVIANCE

As labeling theory predicts, the ways other people see you affect your behavior and overall life chances. So, too, does the way you see your social surroundings, according to the broken windows theory. In 1969 Zimbardo conducted another experiment in which he and his graduate students abandoned two cars, leaving them without license plates and with their hoods propped up, in two different neighborhoods (Wilson & Kelling, 1982). The first car was abandoned in a seemingly safe neighborhood in Palo Alto, California, where Stanford University is located. The second car was left in the South Bronx in New York City, then one of the most dangerous urban ghettos in the country. Unsurprisingly, the abandoned car in Palo Alto remained untouched, whereas the car deposited in the South Bronx lost its hubcaps, battery, and any other usable parts almost immediately. However, it was the next stage in the experiment that offers valuable insight into how social context affects social deviance. Zimbardo went back to the untouched car in Palo Alto and smashed it with a sledgehammer. He and his graduate students shattered the windshield, put some dents in the car's sides, and again fled the scene. What do you think happened to the smashed and dented car in the "safe," rich neighborhood? Passersby began stopping their cars and getting out to further smash the wreck or tag it with graffiti. The social cues, or social context, influenced the way people, even in a rich neighborhood, treated the car.

In the South Bronx, the overall social context—involving many broken windows, graffiti, and dilapidated buildings—encouraged deviant acts at the outset because the neighborhood setting of decay and disorder signaled to residents that an abandoned car was fair game for abuse. In Palo Alto, where neighborhood conditions were clean and orderly, people were unlikely to vandalize an abandoned car. However, when the vehicle was already mangled, that cue of turmoil signaled to people that it was okay to engage in the otherwise deviant act of vandalism. The broken windows theory of deviance explains how social context and social cues impact the way individuals act—specifically, whether local, informal social norms allow such acts. When signals seem to tell us that it's okay to do the otherwise unthinkable, sometimes we do. The broken windows theory of deviance has inspired some politicians to institute policies that target the catalysts for inappropriate behavior (vandalism, burglary, and so forth). In fact, one of Zimbardo's graduate students who participated in the car experiment, George Kelling, later worked as a consultant for the New York City Transit Authority in 1984, devising a plan to crack down on graffiti. The underlying assumption of the plan was that the continued presence of graffiti-covered cars served as a green light for more graffiti and perhaps even violent crimes. The Transit Authority then launched a massive campaign to clean up graffiti, car by car, to erase the signs of urban disorder, in the hope of reducing subway crime. In the mid-1990s New York City mayor Rudy Giuliani also initiated a campaign of "zero tolerance" for petty crimes such as turnstile jumping, public urination, the drinking of alcohol in public, and graffiti. Today, the city's newer subway cars are "graffiti-proof," meaning that spray paint doesn't

BROKEN WINDOWS THEORY OF DEVIANCE

theory explaining how social context and social cues impact whether individuals act deviantly—specifically, whether local, informal social norms allow deviant acts.

People inspect an abandoned car in the South Bronx. Philip Zimbardo placed this car in New York City and left another near Stanford University in Palo Alto, California. The car near Stanford went untouched for days, but the car pictured here was relieved of its hubcaps and other parts almost immediately.

adhere to the metal exterior of the cars. Indeed, both petty and serious crimes have dropped dramatically in New York City since the 1990s (Kelling & Sousa, 2001), although some criminologists dispute the causal impact of the Giuliani strategy.

Crime

As noted earlier, crime is a more formal sort of deviance, not only subject to social sanction but also punishable by law. Sometimes an act falls clearly at one end of the spectrum or the other. Situations of self-defense aside, killing a person is generally agreed to be criminal; nose picking, however unpalatable, isn't. In other cases, though, the distinction is not so clear. Whereas you might get strange looks for wearing aluminum-foil hot pants to class, if you show up with no clothes on at all, you will likely be sent to a mental health care provider for assessment or arrested and thrown in jail for indecent exposure. But what if your hot pants weren't foil at all but just silver paint? Would this be indecent exposure?

Crime runs the gamut in type and degree. Most of us think of violence or drugs when we contemplate crime, but there are many other ways of breaking the law.

STREET CRIME

You don't make up for your sins in church. You do it in the streets.
—"Johnny Boy" Civello in *Mean Streets*

STREET CRIME

crime committed in public and often associated with violence, gangs, and poverty.

Street crime generally refers to crime committed in public. The term invokes specific images, however, of violent crime, typically perpetrated in an urban landscape. Both historically and today, street crime is often associated with gangs, and currently it is also associated with both disadvantaged minority groups and poverty. Just why people are drawn to a career of street crime has long been a favorite question of social scientists and is still a matter of heated debate. Explanations for the fluctuations and trends in violent crime rates also vary widely. One theory, for example, posits that street crime rises and falls in relation to the availability of opportunity within the legitimate economy. When this is lacking, people turn elsewhere. A wrinkle on this theory is provided by "differential opportunity theory" (Cloward & Ohlin, 1960). Differential opportunity theory states that in addition to the legitimate economic structure, an illegitimate opportunity structure also exists that is unequally distributed across social classes. That is, some groups have more opportunities than others in the illicit economy and it is really the ratio of risks and rewards in the formal and black market economies that influences participation in crime (rather than just the opportunities in the mainstream economy alone).

One way, then, to reduce crime would be to raise the costs of working in the illegitimate economy, thereby lowering the net returns. Such a strategy lies behind tougher sentencing policies, such as "three strikes" laws (if you are convicted of three felony crimes, you are imprisoned for life), which aim simultaneously to deter criminals and to incarcerate habitual offenders. The focus of such policies tends to be violent crime; lately, however, such laws have been extended to nonviolent crimes, such as drug dealing. Another strategy would be to increase the returns to entry-level opportunities in the legitimate economy; raising the minimum wage is one way to do this. Either strategy—decreasing the returns to the illegitimate economy or increasing the returns to the legitimate economy—shortens the distance, or differential, between the two economies.

Many people attribute the decrease in crime rates to the adoption of a community policing ideology, in which police officers are viewed (and view themselves) as members of the community they serve, and not to the punitive nature of prison life. Rather than enforce laws from the outside in, as community members walking the streets, police can work to develop reliable relationships and bonds of trust with community residents. On a practical level, this means that greater police presence is felt in areas where community policing is in effect. (However, convincing evidence on this causal claim is still lacking; that said, community policing probably doesn't make crime rates any worse.)

WHITE-COLLAR CRIME

Bernard L. Madoff rose through the social ranks of wealth on Long Island, then Manhattan, then London, and eventually became the chairman of the NASDAQ exchange, which was based on technology his privately held firm had developed. In 2009, he was sentenced to 150 years in prison—a life sentence for the then-71-year-old. His crime was taking money from investors and fabricating great rates of return to pump up his reputation and keep money coming in when, in fact, his funds were losing money. Returns to investors were paid not from any actual gains their initial investments were accruing, but from the new money coming in the door. Madoff lived an opulent lifestyle, employing well-connected friends and family to keep a steady stream of new investments coming in. During the economic downturn in 2008, Madoff could not keep up with investors' demands to cash out. An investigation by the Federal Bureau of Investigation (FBI) led to his arrest. Madoff estimated that he had lost $50 billion, making his the largest fraud in American history.

Bernie Madoff.

Infractions such as fraud are called white-collar crime, a term coined by sociologist Edwin Sutherland in 1939. It is typically committed by a professional (or professionals) in his or her (or their) capacity in the professional world against a corporation, agency, or other professional entity. According to the FBI (2003),

> [w]hite-collar crimes are categorized by deceit, concealment, or violation of trust and are not dependent on the application or threat of physical force or violence. Such acts are committed by individuals and organizations to obtain money, property, or services, to avoid the payment or loss of money or services, or to secure a personal or business advantage.

Although street crime is the most prevalent type of crime, looking strictly at the numbers, white-collar crime has greater financial impact. The FBI's prosecution of white-collar criminals resulted in billions of dollars in restitution (repayment) and hundreds of millions in fines in 2011—and those totals reflect only successful prosecutions. In contrast, the FBI has estimated that in 2013 the total loss from robbery, burglary, larceny-theft, and motor vehicle theft combined was $16.6 billion (FBI, 2014a, 2014b).

A particular type of white-collar crime is corporate crime, offenses committed by the officers (CEOs and other executives) of a corporation. In the late spring of 2012, JPMorgan Chase trader Bruno Iksil (also known as the London Whale) placed a giant risky trade and ended up losing $6.2 billion for the bank. Losing billions on a trade is deviant but not illegal. Iksil was fired, but avoided arrest. US federal prosecutors have indicted his boss and his assistant for conspiring to cover up the losses by filing false reports. Penalties for their crimes include a $920 million fine for JPMorgan, but no individuals did time (Stewart, 2015). Despite being the worst financial crisis since the Great Depression, nobody went to jail for the shady practices that led to the 2007–8 meltdown of the banking system.

INTERPRETING THE CRIME RATE

Although it may seem, after a glance at the gruesome headlines in today's newspapers, that crime is getting worse, the truth is that deviance has always been present. Kai Erikson, in *Wayward Puritans* (2005), demonstrates how even America's seemingly most upstanding and God-fearing community, the Puritans, produced some bad apples—drunkards, adulterers, and thieves. Erikson's point, however, was that while what is defined as deviant may change over time, deviance is forever with us. For example, Erikson found "crimes against the Church," particularly by Quakers in the late 1650s, to be of central concern to the Puritan community. Religious or moral offenses,

FIGURE 6.2 Total US Violent Crime Rate, 1960–2016

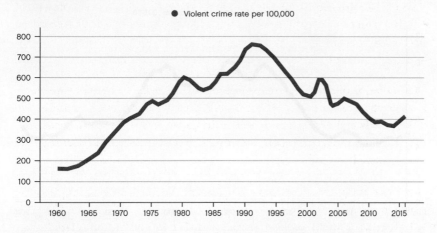

Violent crime rate per 100,000

SOURCES: FBI, 2015, 2017.

such as drunkenness and adultery, were of primary concern over economic or political ones—at least those were the crimes taken most seriously and probably reported most frequently during the Puritan era. Erikson's central thesis is that a relatively stable amount of deviance may be expected in a given community, but what counts as deviance evolves depending on the type of society or historical period we examine. Think back to our early definition of social deviance; it is, after all, a social construct and subject to change over time and across cultural values.

If the definitions of deviance and crime are always changing, how can we discern if the crime rate is going up or down? In 1960, the total crime rate was around 160.9 per 100,000 people. By 1992, that number had soared to a little over 757, but it has now fallen back to around 401. Figure 6.2 shows the violent crime rate in the United States from 1960 to 2016.

How can so much change occur in the crime rate in so little time? Is the situation really rapidly degenerating into complete chaos? Probably not. To make a statement about the crime rate going up or down, we have to know what is factored into the crime rate. For instance, the definition of assault may change to include more minor offenses. Recently, one local government proposed including brawls at sporting events in the assault category. And if the classifications of violent crimes are changing, it's difficult to compare the crime rate over time. It is also possible that levels of crime reporting by victims fluctuate. In times of economic recession, people might be less likely to report crime because they feel helpless, depressed, or apathetic. Alternatively, a presidential address on fighting terrorism or injustice might make us more diligent in defending the social order.

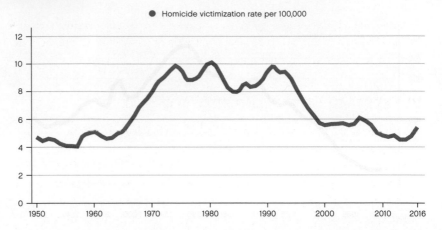

FIGURE 6.3 Homicide Victimization Rate, 1950–2016

● Homicide victimization rate per 100,000

SOURCES: FBI, 2015, 2017.

The point is that the crime rate changes in response to fluctuations in how society classifies and reacts to deviance. In fact, reporting bias may work in the opposite direction from the actual crime rate. Imagine a neighborhood where crime is common. People might be reluctant to report being pickpocketed on the subway when they have just read about three murders the same day. Conversely, in a situation where crime rates are perceived to be low, such as at the opera house, you might be more likely to report a missing wallet. In this way, the reporting of petty crimes may vary inversely to the rate of serious crimes. For these reasons, criminologists (experts trained in studying crime) usually reject the overall crime rate as a reliable indicator of trends in crime. Instead, criminologists use the murder rate to make statements about the overall health of society. Why do they use these statistics instead of violent crime rates? For one thing, it's difficult to fake a murder—either there is a body or there isn't. The murder rate is the best indicator we have of crime in general (Figure 6.3), but even it is not immune to broader changes in society.

For example, if you were shot today in the United States, you would have a much better chance of surviving the gunshot injury than you would have had in 1960. It could be that many more people are arriving at the hospital with bullet wounds, but far fewer of those victims are dying on the operating table thanks to advances in medical technology. The greater survival rate complicates our interpretation of the murder rate (Harris et al., 2002). Has there been a decline in the number of murder attempts, or are doctors just better able to treat bullet wounds? It's difficult to say definitively. We can use the murder rate as a gauge to gain a general sense of crime fluctuations, but precise statistics are practically impossible.

Crime Reduction

Thus far, we have studied some of the functions and social origins of criminal behavior, as well as the prevalent methods of measuring crime. So how do crime fighters fight crime?

DETERRENCE THEORY OF CRIME CONTROL

In 1971 President Richard M. Nixon ushered in America's first War on Drugs. Linking drug use to violent crimes and the decay of social order, the federal and state governments instituted an array of harsher penalties for the possession or sale of illegal substances. The philosophy of deterrence, or deterrence theory, suggests that "crime results from a rational calculation of the costs and benefits of criminal activity" (Spohn & Holleran, 2002). If you know you can make a quick buck by selling cocaine (maybe enough to pay your rent, say, or put food on the table), you might be tempted to engage in such illicit activities. But what if, instead of a slap on the wrist, a person caught with small quantities of cocaine is sentenced to a minimum of 15 years in prison? The temptation to sell cocaine might be reduced because you know that the cost of getting caught is higher than the potential benefits of selling drugs. Let's say you are arrested and wind up in prison for 15 years. Would you choose to sell cocaine again? How might the criminal justice system prevent you from doing so?

Specific deterrence is what the system attempts when it monitors and tries to prevent known criminals from committing more crimes. An example would be the prison parole system, because the criminal remains under supervision and is being specifically deterred from committing another crime. When he or she is released and freed from direct control after parole has ended, however, the effect is one of general deterrence, whereby criminals who have been punished for a previous offense may opt not to commit more crimes for fear of going back to jail. Another example of general deterrence would be if, from word on the street, you learn that dealing cocaine carries a prison sentence of 15 years and decide not to risk it. Deterrence theory suggests that crime in general, and recidivism in particular, can be reduced through both specific and general deterrence. Recidivism is the "reversion of an individual to criminal behavior" after involvement with the criminal justice system (Maltz, 2001).

However, deterrence theory—in practice—has other, unintended consequences that may lead to more crime. For one thing, increased supervision and stricter parole codes have made it more likely that offenders will commit technical violations against their parole terms. Because the system has increased its surveillance of former prisoners, requiring time-consuming meetings with parole officers, the system creates better odds for technical

DETERRENCE THEORY

philosophy of criminal justice arising from the notion that crime results from a rational calculation of its costs and benefits.

RECIDIVISM

when an individual who has been involved with the criminal justice system reverts to criminal behavior.

slipups and for catching and punishing them. This is suggested by studies showing that rates of new criminal offenses (as opposed to technical violations) for parolees tend to resemble those of people with comparable sociodemographic characteristics who have never been incarcerated. In addition, the prison experience might not have the intended rehabilitative effect. Could prisons make it more difficult for offenders to return to the straight and narrow?

Let's first think back to Durkheim and his theory of anomie, which affirms that when our normal lives are disrupted and we can no longer rely on things being relatively stable, we are more likely to commit suicide or engage in other deviant acts. Going to prison for 10 or 15 years might have just this effect; it would be very difficult to "find stable employment, secure suitable housing, or reconcile with...family" afterward. Also, reintegration into a community after release from prison is extremely difficult because of "the absence of...informal social controls and strong social bonds" (Spohn & Holleran, 2002). Furthermore, while in prison, drug offenders rarely receive the kind of substance abuse treatment that addicts need. They are thus more likely to revert to drug use upon release. Therefore, imprisonment may be particularly counterproductive for drug offenders, and getting tough on crime may inadvertently breed criminals. In addition, as low-level offenders interact daily with serious criminals in prison, they may become socialized by these new peers, adopting their attitudes and behaviors.

Ex-convict turned prison reentry social worker (and later lawyer) Marc Ramirez has an insider's perspective on the system, arguing that the violent, punitive culture of incarceration is counterproductive to rehabilitating, productively socializing, and preparing prisoners to give back to society.

> We're paying top dollar to incarcerate people when there are cheaper alternatives. There's been study after study on the [rehabilitative] effects of education, but yet we take Pell Grants and education programs out of prison. Family, having a family base [is critically important for successful reentry], but then these people in prisons are so far away from their families, that they lose ties. We make visiting so difficult for families, you know, it's hard to maintain the family. The cost of making a phone call from a lot of prisons is prohibitive. (Conley, 2014d)

Ramirez is angry that the options facing newly released prisoners are suffocatingly limited. He remembers seeing prisoners "terrified to leave prison because they have not prepared, they have no support system. They don't know what they're going to do...there were so many people who would come back. And my thing was always like wow, so many of you guys have had breaks your way and I can't get a break. You gotta do better. You gotta want better. But you're not really given the tools to do better...the system is kind of designed to fail." Ramirez's story suggests that providing more housing, employment, and counseling to former convicts as they reenter society may help reduce the number of former prisoners who commit more crime upon release.

Sociologists and geographers have also examined the impact of incarceration on the communities prison inmates call home (Fagan et al., 2003; Williams, 2005). In New York City the incarcerated population is disproportionately drawn from a small handful of neighborhoods, a pattern that is sustained even as crime rates have dropped dramatically. While they are in prison, inmates cannot make positive contributions to the community in terms of providing steady incomes, starting families, or building up social networks that lead to employment in legitimate industries. At the community level, high rates of incarceration among community members create conditions for continued poverty, as those left behind must support families on fewer salaries both during and after incarceration. From Pager's research earlier in this chapter, we know that former felons have a difficult time finding employment. Blocked access to legitimate employment creates conditions for sustained poverty. At the same time, historical crime rates are used to determine current police involvement, which means that once-crime-ridden neighborhoods will continue to receive disproportionate formal surveillance and have a higher likelihood that their residents will be arrested.

DIGITAL.WWNORTON.COM/YOUMAYASK6CORE

To see my interview with Marc Ramirez, go to digital.wwnorton.com/youmayask6core

GOFFMAN'S TOTAL INSTITUTION

The high rates of recidivism in the United States bring us back to labeling theory, which suggests that the process of becoming a deviant often involves contact with, even absorption into, a special institution such as a prison or mental health institution. And as labeling theorists have shown, our interaction with others very significantly impacts the formation of our personal identity. Erving Goffman, writing in the symbolic interactionist tradition, theorized about how institutions such as prisons and mental health hospitals often become breeding grounds for secondary deviance, providing an important link in the reproduction of deviance through their effects on inmates and patients (Goffman, 1961).

TOTAL
INSTITUTION

an institution in which
one is totally immersed
and that controls all the
basics of day-to-day life;
no barriers exist between
the usual spheres of daily
life, and all activity occurs
in the same place and
under the same single
authority.

Presumably, most of us sleep, play, and work in different places, with different people and rules structuring our interactions at each location. For example, you probably spend the evening in your home or dorm, relatively undisturbed, and leave in the morning for class, where you are expected to dress appropriately and respect your professors. Total institutions are distinguished by "a breakdown of the barriers separating" these "three spheres of life" (sleep, work, play). In total institutions, "all aspects of life are conducted in the same place and under the same single authority" (Goffman, 1961). In total institutions such as prisons and mental hospitals, life is highly regimented, and the inmates take part in all scheduled activities together. The inmates have no control over the form or flow of activities, which are chosen by the institutional authorities to "fulfill the official aims of the institution." Thus, if the official aim of prisons is rehabilitation, prisoners might be required to attend at least one self-improvement class a day or engage in some sort of productive labor. All of these activities occur in the same place with the same group of people every day.

In Chapter 4 we examined the various theories of socialization and the development of the self through our interactions with other social actors. All your life you have been slowly accumulating knowledge about who you are in the world. Once you enter a total institution such as a prison, a process that strips away your sense of self begins. As Goffman puts it, "a series of abasements, degradations, humiliations, and profanations" quickly commences. Once people enter a prison or a mental institution, they are closed off from their normal routines and cease to fulfill their usual social roles. Because the roles people play are important to the way they perceive themselves, this separation from the world erodes that sense of identity. In

Inmates in an Arizona jail. The local sheriff requires all of the county's inmates to work seven days a week. They are fed only twice a day, are denied recreation, and receive no coffee, cigarettes, salt, pepper, or ketchup.

the Stanford Prison Experiment, what first occurred when the "inmates" arrived at the "prison" in the basement of the psychology department? They were issued uniforms and given numbers in place of names. The mandated homogeneity of prisoners (and patients) results in an erasure of self. The total institution simultaneously strips away clothes, personal belongings, nicknames, hairstyles, cosmetics, and toiletries—all the tools that people use to identify themselves. The inmates no longer have control over their environment, peer group, daily activities, or personal possessions. It's not hard to see how this process leads to a sense of helplessness. The total institution rapidly destroys a prisoner's sense of self-determination, self-control, and freedom.

The authorities in these total institutions are given the unenviable duty of "showing the prisoners who's boss" and expediting the process of prisoner degradation to ensure "co-operativeness" (Goffman, 1961). Wardens must "socialize" inmates into compliance through informal "obedience test[s]" or "will-breaking contest[s]" until the inmate "who shows defiance receives immediate visible punishment, which increases until he openly 'cries uncle' and humbles himself." When Zimbardo locked a group of seemingly normal college kids into the basement of a Stanford building (discussed in the box on pages 220–21), he found that not only were the inmates affected but the guards also felt impelled to use increasingly violent and inhumane means to solidify their authority and keep the prisoners in check. We can see these patterns in everyday life, too. When given the responsibility of watching younger siblings, have you ever secretly (or not so secretly) delighted in bossing them around? Because the roles we play are important to how we think about ourselves, it is hardly possible for the self-images of both guards and prisoners to remain unaffected by the prison environment.

FOUCAULT ON PUNISHMENT

On March 2, 1757, Damiens the regicide was condemned "to make the amende honorable [a kind of ritual abasement] before the main door of the Church of Paris," where he was to be "taken and conveyed in a cart, wearing nothing but a shirt, holding a torch of burning wax weighing two pounds"; then, "in the said cart, to the Place de Greve, where, on a scaffold that will be erected there, the flesh will be torn from his breasts, arms, thighs, and calves with red-hot pincers, his right hand, holding the knife with which he committed the said parricide, burnt with sulfur, and, on those places where the flesh will be torn away, poured molten lead, burning oil, burning resin, wax and sulfur melted together and then his body drawn and quartered by four horses and his limbs and body consumed by fire, reduced to ashes and his ashes thrown to the winds." (Foucault, 1977, p. 1)

The execution of Robert-François Damiens, a French servant who attempted to assassinate King Louis XV at Versailles in 1757.

Besides possibly turning your stomach, the above passage vividly illustrates the dramatic shift in penal practices from the eighteenth century to the present day. How do we conceive of punishment nowadays? We usually think of prisons, juvenile detention centers, and probation. In *Discipline and Punish* (1977), the French theorist Michel Foucault examines the emergence of the modern penal system and how this system represents a transformation in social control. How did the modern prison system emerge? And what functions does it serve in the disciplining of modern life?

When Robert-François Damiens was publicly tortured and then eventually drawn and quartered for trying to kill King Louis XV, the target of punishment was Damiens's body. The entire public spectacle revolved around Damiens suffering for his wrongdoing, culminating in his death. Damiens was even put to death holding the same knife he used to attack the king. According to Foucault (1977), this gruesome "violence against the body" exemplifies a premodern form of punishment, which is concentrated on the body and associated with the crime committed. So, for example, if you kill someone, you are publicly executed. If you steal something, perhaps your fingers or hand will be cut off. This "eye for an eye" mentality is similar to that represented by Durkheim's mechanical social sanctions.

We might like to believe that modern punishment came about because prison reformers lobbied for more humane penal tactics that aimed to reform the criminal through rehabilitation. We no longer (with the exception of the death penalty) publicly violate the criminal's flesh by, for example, pouring molten lead on his excoriated body. Punishment takes place in private, away from the public eye, and it leaves the criminal's body intact. (Although much violence, including rape, does occur inside prison walls, the state does not formally sanction it and is supposed to protect prisoners from such attacks.) Foucault claims that modern punishment has as its target what he calls "the soul" of the prisoner. The soul, for Foucault, is the sum of an individual's unique habits and peculiarities: what makes me *me* and you *you*. Such a penal system tries to understand the individual and his or her abnormalities to correct or reform bad habits. (Again, this is a highly stylized view of the history of criminal justice, given that in the United States, some jurisdictions still impose the death penalty and our government has even tortured political detainees. At the very least, we have witnessed an incomplete Foucaultian transformation.)

By "reforming the soul," Foucault means the use of experts such as social workers, psychologists, and criminologists to analyze and correct individual behavior. How does a prisoner become eligible for parole? The *New York State Parole Handbook* (New York State Division of Parole, 2007) indicates that "parole 'readiness'" depends on the inmate's "good prison behavior," involvement in "prison programming" for education and skills acquisition, and substance abuse counseling, all to "make important strides in self-improvement." After a criminal leaves prison on parole, the newly released prisoner is assigned a field parole officer who is in charge of monitoring the whereabouts of the parolee and guiding his or her reentry into community life. The handbook also indicates that a parole officer must help parolees "develop positive attitudes and behavior" and "encourage participation in programs for self-improvement." Parolees are, in principle, scrupulously supervised by parole officers and sometimes subject to unscheduled visits at work or home. This is all part of what Foucault would consider the modern face of penal practices.

How did the transformation in penal practice, from punishment targeted at the body to reform of the soul, take place? Foucault believes that this transformation is linked to changes in how social control operates more generally, which in turn lead to innovations in penal practices. Foucault uses the following example to illustrate the way modern punishment is organized and its implications for modern social control:

> The prisoners' day will begin at six in the morning in winter and at five in the summer. They will work for nine hours a day throughout the year. Two hours a day will be devoted to instruction. Work and the day will end at nine o'clock in winter and at eight in summer.... At the first drum-roll, the prisoners must rise and dress in silence, as the supervisor opens the cell doors. At the second drum-roll, they must be dressed and make their beds. (Foucault, 1977, p. 6)

Foucault's example, extracted from a contemporary prisoners' timetable in France, contrasts strikingly with the way poor Damiens was punished several centuries earlier. Foucault's point, however, is that this sort of

The Stateville Penitentiary in Illinois was built along the principles of Bentham's panopticon, a model for a prison in which inmates would always be visible.

regimentation happens in not only prisons but also society at large. Penal practices are indicative of how social control is exercised outside prison walls. Disciplinary techniques are modes of monitoring, examining, and regimenting individuals that are diffused throughout society. Foucault gives many examples of where and how this discipline takes place both in and out of prisons—in the military, in schools, in medical institutions, and so forth. For example, when you first entered school, perhaps even when you entered college, you were probably required to take a series of standardized tests. You were also required to visit your pediatrician for shots and a checkup in which he or she meticulously documented your growth and health. Once in school, you were made to sit in straight, orderly rows (so the teacher could see all students at all times) and then were issued a report card every term. If you were chronically disruptive or inattentive in class, you might have been sent for special testing by experts to determine if you had a learning or behavioral disorder. These are all examples of how you have been subjected to various modes of discipline that monitor, examine, and regiment individuals.

Foucault used the imagined architectural design of the panopticon as a metaphor for this march toward total surveillance and control. Jeremy Bentham, an English philosopher, devised the panopticon as a prison design.

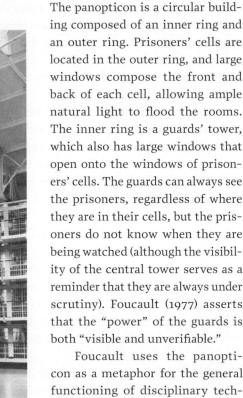

The panopticon is a circular building composed of an inner ring and an outer ring. Prisoners' cells are located in the outer ring, and large windows compose the front and back of each cell, allowing ample natural light to flood the rooms. The inner ring is a guards' tower, which also has large windows that open onto the windows of prisoners' cells. The guards can always see the prisoners, regardless of where they are in their cells, but the prisoners do not know when they are being watched (although the visibility of the central tower serves as a reminder that they are always under scrutiny). Foucault (1977) asserts that the "power" of the guards is both "visible and unverifiable."

Foucault uses the panopticon as a metaphor for the general functioning of disciplinary techniques in society. Therefore, when the modern prison system emerged, based on monitoring, examining, and regimenting individual prisoners, there was a "gradual extension of the mechanisms of discipline," and they "spread throughout the whole social body," which led to "the formation of what might be called in general the disciplinary society." Remember all the steps you had to go through to gain entrance to kindergarten? The tests and checkups? And then the report cards, the parent–teacher conferences, the tidy rows of desks? These are the sorts of panoptic (literally, "all-seeing") disciplinary techniques diffused throughout the social body. In Foucault's words, "our society is one not of spectacle, but of surveillance."

THE US CRIMINAL JUSTICE SYSTEM

At various points in history, the US criminal justice system has fluctuated between two approaches to handling criminals: rehabilitation and punishment. Since the 1890s, argues Frank Allen (1981), the rehabilitative ideal has lost most of its significance and appeal because of shifting cultural values among Americans. Despite what Durkheim predicted, the concept of punishment ("lock 'em up and throw away the key") has largely replaced rehabilitation, winning political and popular favor and influencing criminal justice policy. The result? In 2015, about 6.8 million people were on probation, parole, or in state/federal prison, about 2.8 percent of US adults or 1 in every 37 (Kaeble & Glaze, 2016; Figure 6.4). We are in an era of what sociologists call "mass incarceration."

Not all Americans have been affected equally by the policy of mass

FIGURE 6.4 Size of Prison Population, 1980–2015

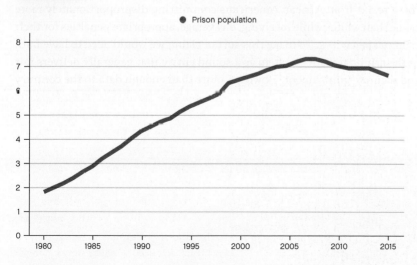

SOURCES: Glaze & Kaeble, 2014; Kaeble & Glaze, 2016.

incarceration, however. Among black men, 16.6 percent were current or former prisoners in the early 2000s, compared with 7.7 percent of Hispanic men and 2.6 percent of white men. Among women, a similar pattern holds, although far fewer women are incarcerated (1.7 percent of black women, 0.7 percent of Hispanic women, and 0.3 percent of white women). The lifetime chances of imprisonment for men and women combined are 3.4 percent for whites, 10 percent for Hispanics, and 11.3 percent for blacks. If current incarceration rates don't change, a whopping 32 percent of black males are estimated to serve time in a state or federal prison during their lifetime, compared with 17 percent of Hispanic men and 5.9 percent of white men. These numbers of prisoners are strikingly high for an industrialized democracy, and the United States is the only industrialized nation in the world to use capital punishment (US Department of Justice, 2007).

Similar to imprisonment rates, justice on death row is not color-blind (see Figure 6.5). In 1998 a study of Philadelphia death penalty cases revealed ample evidence of some form of racial discrimination: Either the race of the defendant or the race of the victim came to bear on the outcome of the case (Baldus et al., 1998). First, the race of the murder victim matters. For example, in a study of the North Carolina justice system, researchers found that the odds of a murderer receiving a death sentence rose 3.5 times if the victim was white, holding constant other relevant factors (Unah & Boger, 2001). Second and more obviously, the race of the accused matters. Contrary to popular belief about black-on-white violence, interracial murders are fairly rare. As of January 2018, out of 1,466 executions carried out in the United States since 1976 (when the Supreme Court reinstated the death penalty), 287 involved cases of black-on-white homicide, but only 20 involved executions arising from cases of white-on-black murder (Death Penalty Information Center, 2018b). One hypothesis would assert that these statistics could result from African Americans committing disproportionately more crime than whites while receiving, on average, appropriate penalties for their criminal behaviors. To make this sort of claim, we would need to have faith in the criminal justice system as a sound entity that generally delivers fair and accurate punishment. However, more than enough data to the contrary

FIGURE 6.5 Number of Executions by Race, 1976–2017

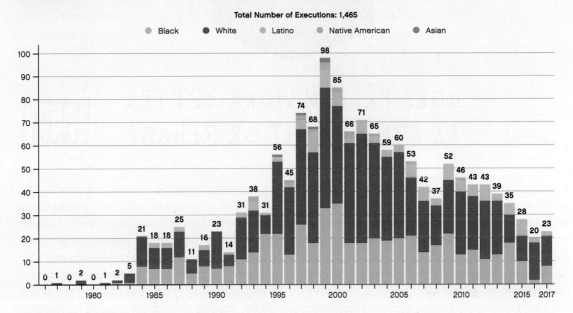

Total Number of Executions: 1,465

● Black ● White ● Latino ● Native American ● Asian

SOURCES: Death Penalty Information Center, 2017, 2018a.

exist. Between 1973 and December 31, 2017, 161 people were exonerated from death row: Of these, 84 were black, 62 white, and 13 Latino (Death Penalty Information Center, 2018c). In 2014 alone, seven persons were released from death sentences, after an average of 30 years behind bars, their careers and families having been permanently altered (Death Penalty Information Center, 2015). Some of these exonerations resulted from new DNA evidence. In a sense, the new technology has shed light on the flaws in the system, raising the question "How many cases were wrongly judged before the deployment of DNA testing?" This evidence of injustice has led to the abolition of the death penalty in Illinois, Connecticut, New Jersey, New York, New Mexico, Maryland, Delaware, and Nebraska since 2007.

DOES PRISON WORK BETTER AS PUNISHMENT OR REHAB?

With prison overcrowding resulting from our extremely high incarceration rate (see page 239), one of the questions running through criminal justice research is the proper role of time behind bars. Does going to prison give inmates access to services and treatment that makes them better equipped to be productive members of society, thereby reducing the number of crimes they might commit in the future? Is it plausible to think that individuals who are incarcerated may spend their time kicking drug habits, gaining religion, avoiding negative influences in their community, and obtaining an education and emerge less likely to commit crime when they are released?

Or perhaps the very thought of serving hard time works to deter crime? If so, we might expect the strongest deterrent effect to be among those who have already dealt with the harsh realities of prison life. After all, who would want to go back to jail after being released and tasting the freedom that most of us take for granted? This is the principle behind "scared straight" programs for youth who are teetering on the path of righteousness. By showing teenagers in trouble what prison is "really" like, such interventions hope to take any glamorous shine off a life of crime.

But it could also be the case that prison ends up being like "college for criminals"—a chance to network with other convicts, learn new illicit skills, and harden oneself to mainstream social norms, all while missing out on work experience and educational opportunities while gaining the stigma of having done time. In other words, perhaps prison actually creates crime by increasing recidivism.

At first blush it would seem that prison does indeed breed crime. After all, one of the best predictors of future arrest is past incarceration. However, it could obviously be the case that those criminals sent up the river, so to speak, are the ones most beyond hope of reform. That is, they would have committed more crimes regardless of whether they went to prison, received probation, or were let off completely. We could, of course, look at offenders who commit the same crime, comparing those who go to prison and those who don't, to see how they fare afterward. But the problem is that there may be subtle differences in the two groups that judges knew about but that we cannot hope to adequately measure from administrative records.

A solution to this problem comes from the fact that some states and localities randomly

Inmates at San Quentin State Prison in an adult education class. Why do prisons spend taxpayer money to fund educational opportunities for inmates?

assign judges to cases. Like anyone else, judges are human beings with certain tendencies. Specifically, some judges are harsher in their sentencing than others. So it's like assigning convicts to a randomized medical treatment: Some get the "treatment" (the harsh judge who sends them to prison) and some get the "placebo" (the more lenient judge who gives them probation). By examining variability in outcomes among convicted criminals based on the type of judge they were assigned (not on their own characteristics), we can identify the effects of incarceration—and, by extension, the effects of the massive investment we've made in the prison system.

As it turns out, for adults, it doesn't make much difference to their probability of committing a future crime (and getting caught) whether or not they go to prison. (This lack of a difference suggests that taxpayers could get roughly the same crime rate but save a bunch of money on

prison overhead by sentencing most criminals to house arrest. Of course, for this to work we have to assume that the deterrent effect against first-time offenses is minimal as well.)

For youth, the story is even starker: Across the United States, more than 130,000 juveniles are detained each year, and on any given day 70,000 minors are in formal detention. But as it turns out, locking these kids up makes them less likely to complete high school and more likely to commit crimes as adults (as evidenced from the random assignment to judges). In other words, far from being "scared straight," kids sent to detention are simply prepped for criminal life. Of course, this means that not only is locking kids up expensive, but it generates future costs in the form of more crime to deal with when they grow up.

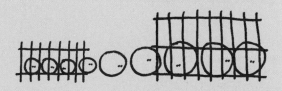

Conclusion

Training a sociological lens on deviance requires a careful and slow review of not only the immediate causes and effects of deviance but also the broader social forces that undergird and define it. Sometimes, doing so takes us to paradoxical places, such as Émile Durkheim's finding that some degree of crime is healthy for a society, insofar as crime unites the social body by allowing us all to rally against a common enemy. Sometimes, doing so reveals the unintended consequences of labels such as "deviant." And quite often, sociologists study tragedies in order to seek the social answers to violence and death. Sociology does not hope to provide comprehensive accounts of criminal acts or injustice, but it does seek to dig deeper for concrete answers in our social world.

QUESTIONS FOR REVIEW

1. How does Philip Zimbardo's Stanford Prison Experiment help explain the behavior documented at Abu Ghraib prison?

2. Describe two potential roles—as punishment or rehab—that prisons could play in society. What evidence from this chapter suggests that prisons "work" to rehabilitate people? To scare or punish people?

3. Explain the difference between mechanical and organic solidarity. How do deviants hold us together in both types of society?

4. We usually think of suicide as solely the result of an intensely personal decision. What is Émile Durkheim's explanation for suicide? Define egoistic suicide and describe why, according to Durkheim, more Protestants commit suicide than Catholics.

5. A student wants to achieve good grades but is not interested in studying for exams; instead, the student finds various ways to cheat. How does Robert Merton's strain theory explain this behavior, and which "type" does the student exemplify?

6. How does the broken windows theory of deviance support the claim that the definition of *deviant* depends on social context? What happened to social control mechanisms like formal and informal social sanctions in Philip Zimbardo's study involving cars?

7. Why don't criminologists use the crime rate as an indicator of crime trends? What do they use instead, and why?

8. Describe Émile Durkheim's theory of the collective conscience and explain how it is related to punishment. How does Michel Foucault's focus in *Discipline and Punish* (1977) differ from Durkheim's work regarding punishment in premodern and modern societies?

9. Explain how a fraternity house could be considered a total institution.

EVERYDAY DEVIANCE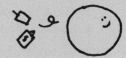

One legal scholar has claimed that the average American citizen commits three felonies a day (Silverglate, 2009). This may sound like a gross overestimate, but underneath lies a provocative argument: that our society has become overcriminalized by federal prosecutors who take vague laws and misconstrue seemingly ordinary activity as criminal. Regardless of what you think of this argument, I hope this chapter has helped convinced you that crime and deviance are part of the fabric of our society and that breaking the rules is more common than some care to admit.

TRY IT!

Write a list of what petty crimes you commit on a regular basis. I'll start us off.

ON A TYPICAL DAY, I MAY (OR MAY NOT) HAVE:

JAYWALKED

SPED WHILE DRIVING (AND BROKEN OTHER MOTOR VEHICLE RULES)

STREAMED A TV SHOW ON A SKETCHY WEBSITE

LOITERED

ROBBED A BANK (JUST KIDDING)

THINK ABOUT IT

Have you ever been cited or arrested for any of your activities? If not, why not?
If so, did it affect your future behavior?

Do you think everyone (i.e., people in different demographics) could get
away with the same things you do? Why or why not? How do informal social norms
and formal laws mesh or clash?

SOCIOLOGY ON THE STREET

Formal and informal social sanctions allow people to
coexist with strangers in crowded public spaces. When do
people break these sanctions, and how do others enforce
sanctions? Watch the Sociology on the Street video to find
out more: **digital.wwnorton.com/youmayask6core.**

WANT MORE PRACTICE?

**Complete the InQuizitive
activity for this chapter
at digital.wwnorton.com
/youmayask6core**

7

INEQUALITY IS THE RESULT
OF ABUNDANCE.

Stratification

Sarah Katz, meet Jeff Rutgers. While it turns out that first names probably don't predict success all that well, last names tell us not only about our chances for success today, they link us to the successes (or lack thereof) of our distant ancestors. By analyzing the distribution of last names in the professional directories of lawyers and doctors, economic historian Gregory Clark (2014) is able to assess the degree to which societies demonstrate social mobility or its opposite, social reproduction.

As it turns out, by being an Ashkenazi Jew (those who populated Europe), Sarah Katz enjoys about an eightfold greater chance of being a medical doctor today than does Jessica Smith (Smith being the most common surname in the United States). Jeff Rutgers, meanwhile, by virtue of being descended from a male ancestor who attended an Ivy League or similarly old university sometime between 1650 and 1850, enjoys an approximate threefold advantage. Also of advantaged pedigree are Hiro Suzuki, by virtue of having a Japanese surname, and Patricia Winthrop, who shares a name with a rich family of the 1920s.

Perhaps unsurprisingly, the actor Denzel Washington does not hail from a family of doctors and lawyers. Washington is among the most common African American surnames and displays the mark of racial inequality such that black Washingtons are two-thirds less likely to be represented among the directory of physicians than they should be based on overall numbers in the population. Ditto for Annie Begay and John Yazzie, who have the two most common Native American names and are 94 percent less likely to be doctors compared with the US average! While racial inequality for these groups is well known, I'd like to introduce you to Ralph Gagnon, who I bet you didn't know was also disadvantaged even though he is white.

According to Clark, the descendants of New France settlers—that is, the mostly French people who came through Québec, Canada, and other northern areas—have 40 percent lower odds of becoming a doctor. Though this is, of course, less than the disadvantage of being black, Latino, or Native American, it is also perhaps more surprising because we think of the French as not only white but a relatively high-status group. Indeed, some of our most snooty words come from French (Hors d'oeuvres, anyone?). What's more, people of French origin have been associated with an elite status since the time of the Norman conquest of England in 1066.

More than figuring out what your name says about your postgraduation chances, Clark's analysis is interesting because it challenges some of our most cherished assumptions about social mobility. Through his analysis of names in Sweden, India, medieval Europe, the modern United Kingdom, the United States, Chile, and other societies, Clark finds that the intergenerational correlation in status is much higher than we had previously thought. Typically, sociologists and economists examine the extent to which parent-and-child measures of social class, such as income, are correlated. Zero would mean that it makes no difference whatsoever to your income what your parent earned. One would mean that there was an absolutely rigid caste system where each generation perfectly reflects the success of the prior one. Most societies obviously fall in the middle somewhere. The latest research on income, for example, has busted the myth that the United States, despite being more unequal than most European societies, has higher rates of economic mobility. Indeed, the United States consistently ranks as the least mobile rich country in the world when it comes to income.

But Clark's analysis tells us that the US parent–child correlation in "social class" overall is about .84, compared to .3 to .6 in income depending on the study. To explain the discrepancy, he suggests that earlier work focused on one measure of social class, making it look like there's a lot more mobility than there actually is. Take Microsoft founder Bill Gates Jr., for example. He dropped out of college even though his father graduated from law school, so by education measures Gates Jr. would seem to be downwardly mobile. But measure income or wealth, of course, and he is upwardly mobile in a huge way. However, Clark's contention is that in some underlying, summary measure of class or status, most of us reflect our parents' social position a lot more than we may care to believe. This, he argues, is the only reason he would see the effects of what your ancestors did 15 generations ago on what you might be doing for a living now.

Perhaps even more surprising than the fact that the rates of social mobility are lower than we thought they were is that, according to Clark, they don't vary all that much across time and place. Medieval Europe with its feudal system had more or less the same degree of social exchange between classes as does modern Britain today. The unequal, deregulated United States enjoys more or less the same fluidity as social democratic Sweden on the one hand, and on the other hand is not significantly more mobile than India, whose caste system is thought to be the most rigid social structure of all. (Though discrimination based on castes has been outlawed in India since 1950, there are still enormous inequalities by caste membership.) Even conscious efforts to totally remodel society, including the slaughtering of elites, can only temper slightly the degree of social inheritance. Despite the purges of the Cultural Revolution, massive land redistribution, and complete political transformation, Communist China under Mao Zedong reduced its intergenerational persistence to only about .7, down from .8.

So if we value social mobility, what's a poor old sociologist to do in light of Clark's analysis? Maybe we should all give up and just change our last names to Potros, a Coptic Christian name from Egypt. As it turns out, in the United States at least, Copts are the most successful group of all. One version of Coptic lore claims that they are the true descendants of the pharaohs. If that's right, then maybe the long arm of social history is even stronger than Clark imagined.

Views of Inequality

To answer questions about stratification, we need a conceptual framework through which to view inequality. In the following section, I present three possible frameworks and leave it to you to decide which you think is most appropriate to explain the intersection of various dimensions of stratification. We'll start with the Enlightenment in the eighteenth century and move forward from there.

JEAN-JACQUES ROUSSEAU

Writing prolifically from 1750 to 1782, Jean-Jacques Rousseau greatly influenced the political ideas of the French Revolution and the development of socialist thought. Seeing humankind as naturally pure and good, Rousseau appealed to biology and human instincts to explain social outcomes. For Rousseau, it is only through the process of building society and repressing this pure natural character that social problems develop. Specifically, Rousseau sees the emergence of private property, the idea that a person has the right to own something, as the primary source of social ills. If we were to strip away the elements of society that result from the institution of private property—the competition, isolation, aggression, and hierarchical organization—only social equality, a condition in which no differences in wealth, power, prestige, or status based on nonnatural bases exist, would remain. According to Rousseau (2004), there are two forms of inequality: physical (or natural), which "consists in a difference of age, health, bodily strength, and the qualities of the mind or of the soul," and social (or political), which

> depends on a kind of conventional inequality and is established or at least authorized by the consent of men. This latter consists of

STRATIFICATION

the hierarchical organization of a society into groups with differing levels of power, social prestige, or status and economic resources.

SOCIAL EQUALITY

a condition in which no differences in wealth, power, prestige, or status based on nonnatural conventions exist.

Jean-Jacques Rousseau.

the different privileges which some men enjoy to the prejudice of others, such as that of being more rich, more honored, more powerful, or even in a position to exact obedience. (p. 15)

Rousseau acknowledges that a certain amount of natural inequality will always exist between people: Someone is always going to be better than someone else at kicking a ball, hunting large game, doing math, or seeing long distances. Some of these inequalities are a simple result of aging, as a basketball player in her twenties will almost certainly have greater physical abilities than a basketball player in her eighties. In contrast, social inequality is a result of privileges and uneven access to resources (i.e., private property) and will eventually lead to social ills. There are only so many resources available in any society and, more broadly, in the world as a whole.

Imagine that all the resources available in a given society are represented in a single pizza, cut into 10 slices, and 10 hungry people are sitting around the pie. In Rousseau's ideal natural world, where people are inherently good and lack any notion of personal ownership, each individual would get one slice. But in a world where private property is the norm, the division is likely to be significantly less equitable. For whatever reason—maybe some of the 10 have more guile, maybe some paid more than others, maybe one person's family owns the pizza parlor—the pie is distributed unevenly among the group. Rousseau sees such inequality as ultimately detrimental and as a catalyst for social conflict.

Imagine yourself as one of the people who get no pizza, even though there are 10 of you and 10 slices. Imagine further that you have a hungry family who would certainly have appreciated that slice of pizza. Now imagine watching the person sitting across from you slowly enjoying 5 steaming slices, one by one. Would this make you jealous? Would it make you spiteful? Would it make you subvert your essentially good nature in trying to steal a slice? Might you even be willing to hurt the pizza hoarder to feed yourself and your family? Might the pizza hoarder then defend his personal property and maybe even hurt you to keep your hands off it? According to Rousseau, he might also buy off the biggest and strongest of those who have no pizza with two slices in order to protect his rights.

THE SCOTTISH ENLIGHTENMENT AND THOMAS MALTHUS

Rousseau saw the move away from the pure state of nature as an extremely negative historical development, complaining that "man was born free, and he is everywhere in chains." But thinkers of the later Enlightenment, including Adam Ferguson and John Millar of Scotland and Thomas Malthus of England, saw inequality as good, or at least necessary. These three agreed that inequality arises when private property emerges and that private property emerges when resources can be preserved, because it is only through surpluses that some people are able to conserve and increase their bounty. This leads to the paradox of this chapter: Inequality is a result of surplus.

When individuals or groups in society become more efficient and productive, they can gather, hunt, or grow more than they themselves can consume at a given time. They can then conserve such a surplus, whether that means turning milk into cheese, making berries into preserves, or putting their money into a hedge fund. In whatever way they choose, they can preserve current resources and transform them into assets, a form of wealth that can be stored for the future. (In fact, the word *asset* comes from the French legal expression *aver assetz*, meaning "to have enough.") Whereas previously the incentive might have been to share the wealth when you couldn't store it—to generate reciprocal goodwill for the rainy day when you don't have enough to eat and your neighbor does—now when property can be preserved, the incentive is for individuals to hoard all of it, so that when a rainy day comes along, personal savings are available. Alternatively, if a resource-rich individual decides to distribute surplus wealth, he or she might decide to extract power, promises, and rewards in return for what he or she provides.

For Ferguson and Millar, such social developments resulting from the establishment of private property represent a huge improvement in society, because private property leads to higher degrees of social organization and efficiency: If an individual can preserve and accumulate resources and become more powerful by storing up assets (private property), he or she will have much more incentive to work. Thanks to personal incentives, people won't just slack off after they have accumulated what they need for the day's survival, but will instead work to build up society, and as a by-product they will improve civilization. For the Scottish Enlightenment thinkers, inequality was a prerequisite for social progress, and social progress was a prerequisite for the development of civilization—the greatest goal toward which humankind could strive.

Malthus also had a positive view of inequality, but for a different reason. In 1798, Malthus published an anonymous treatise titled "An Essay on the Principle of Population as It Affects the Future Improvement of Society," arguing that human populations grow geometrically (multiplicatively, like rabbits), while our ability to produce food increases only arithmetically (much more slowly). Simply put, his theory suggests that a rising number of people on the planet will eventually use up all the available resources and bring about mass starvation and conflict. Take the pizza example from above: With 10 people and 10 slices, the pie could be divided evenly. But what if there were 20 people or 50 people? Or 100 people? As the number of people increases, each person gets less and less, even if the pie is divided evenly. Humankind, Malthus believed, was similarly destined to live in a state of constant near-death misery, as population growth always pushed society to the limits of food availability.

Because of these dire trends, Malthus believed inequality was good, or at least necessary, for avoiding the problem of massive overpopulation and hence starvation. Inequality, from his perspective, kept the population in check. After all, for Malthus, the main problem with the world, especially with England, was the number of people in it, and anything that tended to restrain population growth was supremely good. In this vein, he denounced soup kitchens and early marriages while defending the effects of smallpox, slavery, and child murder (Heilbroner, 1999). Malthus believed that over-population would create more and more human misery, and therefore the most logical solution would be to allow the population to thin itself out naturally rather than to exacerbate the problem by reducing the levels of inequality, a measure that would "temporarily" ease the condition of the "masses," thereby causing their numbers to swell even more. Such a condition is today called a Malthusian population trap—a situation in which population growth leads not to abundance but to increased poverty.

In an interview for this book, I asked Jeffrey Sachs of the Earth Institute at Columbia University if he thought that sub-Saharan Africa might be facing a Malthusian trap as the per capita land available for farming dwindles, and if so, what can be done to get out of the trap. Sachs captured the complexity of the big picture, so I'll let him speak for himself:

> In Africa there's been a partial transition from high mortality to low mortality, from maybe 600 of every 1,000 children dying before their fifth birthday to a situation where it's perhaps around 150 per 1,000 dying. Still remarkably high by global standards, but way down from what it was. But the fertility rates especially in the rural areas remain high—as many as five or six, sometimes even seven per woman. That means rapid population growth. The woman that has six children, [of whom] two die, has on average four children who grow up to adulthood. Two of those will be daughters. Each woman is replacing herself with two daughters in the next generation. That's a doubling of the population from one generation to the next, called a gross population number. That's extraordinary. You can't keep ahead of that in terms of economic development, and certainly not in ecology. So what do you do about that? What do you do to accelerate the reduction of fertility? Save the children. It seems paradoxical. But when the children are staying alive, the families say, "Ah! It's safe to have fewer children. I don't need to have six for insurance. My children will survive."
>
> Make sure the girls stay in school, especially secondary school. A girl who has a chance to go to secondary school will get married several years later. By then she will be more empowered, have a market value in the economy, say, "I'm gonna go out and get a job. I don't want to get married, and also I'm not gonna let my father choose to marry me at age twelve," which is how it might

Farmworker Stella Machara and her sons Kudakwachi and Simbaracha in front of their home in Zimbabwe. According to Jeffrey Sachs, why is sub-Saharan Africa stuck in a Malthusian population trap? To see more of the conversation with Sachs, visit **digital.wwnorton .com/youmayask6core**.

be normally. Make sure there are contraceptives available. Family planning is not a freebie. It absolutely requires training, skilled community work, health workers, the physical availability of contraception; and that's more than an impoverished woman, in a patriarchal setting, with lots of children dying in a rural area, is going to somehow find her way to on her own. She may have zero cash. History has shown free availability of family planning and contraceptive services is part of the set of things that are necessary to lower fertility that also includes child survival, girls' education, women's awareness, and community health workers. That can lead to a remarkably accelerated demographic transition, and that's the kind of holistic approach that successful societies have done. (Conley, 2009g)

GEORG WILHELM FRIEDRICH HEGEL

A third view of inequality comes from the German philosopher Georg Wilhelm Friedrich Hegel, who viewed history in terms of a master–slave (sometimes called a master–servant) dialectic. The word dialectic means a two-directional relationship—one that goes both ways, like a conversation between two people. One person talks, putting out an idea or thesis. Then the other responds, pointing out some problems with the thesis or posing a counterposition, an antithesis. Then the original speaker responds, and it is hoped, the two arrive at a synthetic arrangement constructed from elements of the original position and the strongest counterpoints.

In Hegel's master–slave dialectic, the slave is dependent on the master because the master provides food, shelter, and protection. In this way, the

DIALECTIC

a two-directional relationship, following a pattern in which an original statement or thesis is countered with an antithesis, leading to a conclusion that unites the strengths of the original position and the counterarguments.

slave is akin to a child raised by the master. However, as Hegel observed, the master is also dependent on the slave, who performs the basic duties of survival until the master can no longer function on his own. He doesn't remember how to grow his own food, prepare meals, or even get dressed without help. Basically, the master would not be able to function if left to fend for himself. Thus the master–slave nexus becomes a relationship of mutual dependency.

But Hegel was writing in the early nineteenth century. There really aren't masters and slaves today, are there? In fact, while we think of slavery as a dreadful practice left on the ash heap of the past, researchers of modern human trafficking believe that there are actually more enslaved people now—estimates suggest 45.8 million people (Global Slavery Index, 2016)—than there were at the height of the transatlantic slave trade (Bales, 2016). These slaves often toil in ecologically devastating industries like brick making, charcoal production, and strip-mining according to Kevin Bales, author of *Blood and Earth* (2016).

What's more, according to the CIA, the United States is no exception to this phenomenon, with forced laborers constituting tens of thousands of entrants to the country each year. Modern human trafficking often involves females who are kept as domestic workers and/or sex workers in cities around the world, but some tomato pickers in Florida and construction workers in Dubai have also been considered part of the modern trade in humans (Ehrenreich & Hochschild, 2004; Estabrook, 2009; Human Rights Watch, 2006).

Florencia Molina, a victim of human trafficking, became virtually enslaved at a dressmaking shop on the outskirts of Los Angeles, where she worked up to 17 hours a day, seven days a week, and lived there too, without the option of showering or washing her clothes.

The path to becoming a slave is often paved with a mixture of deceit and desperation. Poor families in rural areas are perhaps too eager to believe the promises of men traveling through their villages, who offer to find good, safe work for their daughters in the city. Once the girls arrive, they are often raped by these men to "teach" them how to serve clients, given a meager room, watched over to prevent escape, and told they must work to repay the cost of their travel and lodging. Some end up staying this way for life, because the stigma of prostitution is something they cannot bear to take back to their families and they have no other way to make money.

In the case of the tomato pickers, the story has a more optimistic ending. After pressure from the Coalition of Immokalee Workers, 30,000 tomato pickers in Florida now receive an extra penny per pound and are better protected from verbal and sexual abuse following a three-year campaign. The campaign started with a boycott of Taco Bell and has now won support from Taco Bell's parent company, Yum! Brands, as well as Burger King, Trader Joe's, Chipotle, Whole Foods, Subway, and Walmart (Greenhouse, 2014). In this sense, it took a coalition of determined outsiders to break the dependencies between the exploited people who needed their abusers for lodging and the abusers who thought they could not stay in business if they charged their corporate purchasers that extra penny per pound.

Hegel views history as marching steadily forward from a situation of few masters and many servants or slaves—such as monarchies and empires, not to mention the entire feudal system—to a society with more and more free men and women, a situation of democracy and equality. His was an optimistic view of history, to say the least. According to Hegel, notions of inequality are constantly evolving in a larger historical arc. He saw this as a trajectory that would eventually lead to equality for all (or very nearly all). We will see later that many sociologists have had a bone or two to pick with this position, as well as with those laid out by Rousseau, Ferguson, Millar, and Malthus.

Standards of Equality

Assuming for the moment that we do value equality in society, what kind of equality do we want? There are different ideologies or belief systems regarding equality, so it's important to recognize that when various groups use the rhetoric of equality, particularly in the political sphere, they may be talking past each other or jostling for advantage. What did Thomas Jefferson mean when he proclaimed in the Declaration of Independence that "all men are created equal"? And what exactly do we mean by equality in the twenty-first century?

EQUALITY OF OPPORTUNITY

One standard of equality is termed **equality of opportunity**. Let's think of society as a game of Monopoly, in which all of the players try to maximize their holdings of wealth (the game's houses and hotels) and their income (the rent collected from other players who land on built-up properties after an unfortunate roll of the dice). In this game, the person with the most money in the end is the winner. Monopoly follows the rules of equality of opportunity. Sure, one player may wind up flat broke and another player may control 95 percent of the wealth, but the rules were fair, right? Everyone had an equal chance at the start. Assuming nobody cheated, any differences were a result of luck (the dice) and skill (the players' choices). The same goes for society. The mere existence of inequality is not at issue here. Some people

EQUALITY OF OPPORTUNITY

the idea that everyone has an equal chance to achieve wealth, social prestige, and power because the rules of the game, so to speak, are the same for everyone.

Civil rights activists in San Francisco demonstrate against Jim Crow laws in 1964. Why did the demand for equal opportunity resonate with most Americans?

BOURGEOIS SOCIETY

a society of commerce (modern capitalist society, for example) in which the maximization of profit is the primary business incentive.

EQUALITY OF CONDITION

the idea that everyone should have an equal starting point.

have more wealth and income than others, and some people enjoy greater social prestige or power than others, but the rules of the game are still fair. We all go into the game of society, as we do into a game of Monopoly, knowing the rules, and therefore any existing inequality is fair as long as everyone plays by the rules.

This is the standard model for what equality means in a bourgeois society, such as modern capitalist society. Although the word *bourgeois* is often used pejoratively, here I mean it as a neutral description—a society of commerce in which the maximization of profit is the primary business incentive. Almost all modern capitalist societies purport to follow an ideology of equal opportunity. The equal opportunity ideology was adopted by civil rights activists in the 1960s, who argued that the rules were unfair. For instance, the segregation of public spaces and other Jim Crow laws (a rigid set of antiblack statutes that relegated African Americans to the status of second-class citizens through educational, economic, and political exclusion) were seen as incompatible with the bourgeois notion of fairness or equality of opportunity.

The rallying cry for equality of opportunity resonates with our overall capitalist ideology. Unequal opportunity clearly stifles meritocracy, a system in which advancement is based on individual achievement or ability. For example, if you were to have a heart transplant, would you not want the most knowledgeable heart surgeon with the most dexterous hands? Or would you settle for the one born to rich parents who finagled their child's way into medical school? Under a system of equal opportunity, everyone who is willing can compete to become a heart surgeon, and eventually the most talented will rise to fill the most demanding, important positions.

EQUALITY OF CONDITION

A second standard of equality is a bit more progressive and is termed equality of condition. Going back to our Monopoly analogy, imagine if at the beginning of the game, two players started out with an extra $5,000 and already owned hotels on Boardwalk and Park Place, and Pennsylvania and North Carolina Avenues, respectively, while the other two players started out with a $5,000 deficit and owned no property. Some would argue that the rules of the game need to be altered in order to compensate for inequalities in the

Middlebury College student Laura Blackman received help from the Posse Foundation, which recruits inner-city kids for elite colleges. How are affirmative action policies an example of equality of condition?

relative starting positions. This is the ideology behind equality of condition. An example of this ideology is affirmative action, which involves preferential selection to increase the representation of women, racial and ethnic minorities, and/or people from less privileged social classes in areas of employment, education, and business from which they have historically been excluded (see the policy discussion at the end of this chapter). Affirmative action is when a person or group in a gatekeeper position, such as a college admissions office or the human resources department in charge of hiring new workers, actively selects some applicants who have faced the steep slope of an uneven playing field, often due to racism or sexism in society. Sometimes these policies can enhance efficiency if they result in fairer competition; other times they may result in a trade-off between efficiency and equality in favor of the latter. We may accept that certain groups with slightly lower SAT scores are admitted to colleges because they faced much greater obstacles to attain those SAT scores than other groups of test takers who enjoyed more advantages. Such "social engineering" may be based on race, class, gender, or any other social sorting category, as long as the proponents can make a justifiable claim that the existing playing field is tilted against them.

EQUALITY OF OUTCOME

The final and most radical (or rather most antibourgeois and anticapitalist) form of equality is equality of outcome, the position that each Monopoly player should end up with the same amount regardless of the fairness of the "game." This admittedly would make for a dull round of Monopoly: In this scenario, the players on Baltic Avenue would make just as much as those on Boardwalk, and some mechanism in the game—say, a central banker or all

EQUALITY OF OUTCOME

the idea that each player must end up with the same amount regardless of the fairness of the "game."

the players pooling their resources—would ensure game play that lived up to Karl Marx's maxim in the *Critique of the Gotha Programme* (1999), "From each according to his ability, to each according to his needs." This standard is concerned less with the rules of the game than with the distribution of resources; it is essentially a Marxist (or Communist) ideology. Like Rousseau's idea that there will always be physical inequalities—some will be better at math, others will be better at physical work, and so on—equality of outcome is the idea is that everyone contributes to society and to the economy according to what they do best. For instance, someone who is naturally gifted at math might become an engineer or a mathematician, whereas someone better at physical work might build infrastructure such as roads and bridges. A centrally organized society calls on its citizens to contribute to the best of their ability, yet everyone receives the same rewards. The median income is everyone's income regardless of occupation or position.

Under equality of outcome, or equality of result, the individual incentives touted by the Scottish Enlightenment thinkers are eliminated. Nobody earns more power, prestige, and wealth by working harder. In this system, the only incentive is altruistic; you are giving to society for the sake of its progress and not merely for your own betterment (although arguably, it may be in the self-interest of the individual to see society progress as a whole). Facebook founder Mark Zuckerberg and his wife, Priscilla Chan, would probably have to give up 99.99 percent of their wealth rather than the "mere" 99 percent they have pledged to charity to help equalize outcomes (Brandom, 2015), whereas a single mother working two jobs in order to provide food, shelter, and day care for her children is likely to take a bit of a step up. Critics worry that without the selfish incentives of capitalism, all progress would halt, and sloth might take over. Unless an oppressive rule emanated from some central agency, collective endeavors would be doomed to failure on account of the free rider problem, the notion that when more than one person is responsible for getting something done, the incentive is for each individual to shirk responsibility and hope others will pull the extra weight. (You might have encountered this sort of situation while collaborating on group projects for school.) Critics of this system argue that even when there is a relatively strong central planning agency to assign each worker a task and enforce productivity, such an administrative mechanism cannot make decisions and allocate resources as efficiently as a more free-market approach does.

FREE RIDER PROBLEM

the notion that when more than one person is responsible for getting something done, the incentive is for each individual to shirk responsibility and hope others will pull the extra weight.

Forms of Stratification

Thus far we have discussed standards of equality and inequality only in general terms, but there are many dimensions and forms by which this equality or inequality can emerge in a given society. For example, a society can be stratified based on age. In many tribes, for instance, age determines the social prestige and honor that the group accords an individual. (By contrast, some argue that in the youth-obsessed culture of the United States, age is a source of dishonor.) Those younger than you would place great weight on your words and in some cases treat them as commands. Furthermore, birth order can be the basis of inequality. Such a system was prevalent as late as the nineteenth century in European society, where the firstborn male typically inherited the family estate; however, in certain parts of the world, birth order still plays a role in resource distribution. Gender or race and ethnicity also can be the predominant forms social stratification takes. Ultimately, many dimensions exist within which inequality can emerge, and often these dimensions overlap significantly.

In Western Australia, Aboriginal elder Brandy Tjungurrayi wears a pencil through his nose—a sign of his senior status.

Stratified societies are those where human groups within them are ranked hierarchically into strata, along one or more social dimensions. Many sociologists and philosophers believe that there are four ideal types of social stratification—estate system, caste system, class system, and status hierarchy system—although all societies have some combinations of these forms and no type ever occurs in its pure form. In addition to these, some sociologists propose a fifth ideal type, an elite–mass dichotomy. Each system has its own ideology that attempts to legitimate the inequality within it.

ESTATE SYSTEM

The first ideal type of social stratification is the estate system. Primarily found in feudal Europe from the medieval era through the eighteenth century, and in the American South before the Civil War, social stratification in estate systems has a political basis. That is, laws are written in a language in which rights and duties separate individuals and distribute power unequally. For example, in the antebellum American South, many states required land-ownership for voting privileges. Europe also historically restricted voting rights to landowners. Before reforms, political participation depended on the social group (the estate) to which you belonged. There was limited mobility among the three general estates—the clergy, the nobility, and the commoners (the commoners were typically further divided into peasants and city dwellers)—and each group enjoyed certain rights, privileges, and duties. In certain

ESTATE SYSTEM

a politically based system of stratification characterized by limited social mobility.

eras, it was possible for a rich commoner to buy a title and become a member of the nobility. And often, one son or daughter of a noble family would join the clergy and become part of that estate. Therefore, there was some mobility, but social reproduction—you are what your parents were and what your children will be—generally prevailed. (We'll examine the concept of social mobility more closely below.)

CASTE SYSTEM

Another type of stratification is a caste system. As opposed to having a political basis, the caste system is religious in nature. That is, caste societies are stratified on the basis of hereditary notions of religious purity. Today the caste system is primarily found in South Asia, particularly in India. The historical legacy of the caste system still dominates rural India despite it having been constitutionally abolished for almost 70 years and despite affirmative action policies to help those of lower caste origins. Although contradictory opinions exist, origins of the caste system are rooted in Hinduism and a division of labor predetermined by birth. Social hierarchy was further entrenched by preferential treatment of upper caste members during British colonialism. The result was rigid social stratification of four main castes or *varnas*: Brahman (priests and scholars), Kshatriya (soldiers), Vaishya (merchants and farmers), and Shudra (servants class). Excluded completely from the caste system are both Adivasi (India's indigenous people) and Dalit or "Untouchables" who are considered the least pure of all class distinctions. Over time, the caste system in India has evolved into a complex matrix of thousands of subcastes. Each of the major castes is allowed to engage in certain ritual practices from which the others are excluded. For instance, the Shudra caste is not permitted to study ancient Hindu texts while Dalits are prohibited from entering temples or the performance of *any* rituals that confer purity. This places them at the bottom of the social hierarchy and typically leaves them with occupations seen as impure, such as the cremation of corpses or disposal of sewage, reinforcing their "untouchable" social status.

There has historically been little mobility between caste ranks in India. This is in part because castes are largely endogamous—communities in which members generally marry within the group. Because of the resulting problems with classifying children born from intercaste marriages, the caste system would simply fall apart if high levels of exogamy (marriage between castes) occurred. The strict divisions would begin to look more like a spectrum and would eventually fade away. This *social closure* is a powerful method of maintaining the caste hierarchy within Indian society. For example, if you are born into the Dalit or "Untouchable" caste, you will likely marry a Dalit and your children will be Dalits. An increase in exogamy is creeping in, but social mobility remains low, especially for Dalits.

Although there is little-to-no individual mobility in caste systems, one unique fact of the system is that an entire caste could obtain a higher position in the hierarchy by adopting practices and behaviors of the upper castes. This process is called sanskritization. Typically practices that are adopted are those that pertain to a higher degree of religious purity (vegetarianism or fasting, for example). At the same time a caste will eschew aspects of its own traditions that are considered impure (animal sacrifice or the consumption of alcohol). In this

way, an entire group of people can rise in social status in one to two generations. Sometimes such an attempt works, and sometimes it doesn't. During the colonial period, when the British governed South Asia, the Vaishya, the second-lowest caste in Pakistan, adopted Christianity, the religion of the British, in an attempt to jump ahead in the hierarchy. Their attempt was not successful. However, by becoming Christian, the Vaishya caste did enjoy a unique fate after the partition of Pakistan from India. Pakistan became a Muslim state in which Islamic law (the shar'ia) was enforced. One of the rules of the shar'ia is that Muslims are forbidden from imbibing or serving alcohol (or any mind-altering substances). So who got the jobs serving foreigners in the hotel bars? The Vaishya, who were now Christian. Their efforts to jump ahead in the caste system may have failed, but they ended up with fairly decent jobs in a relatively impoverished country by capitalizing on westernization.

↑

Islamic law prohibits Muslims from serving or imbibing alcohol, so the Christian Vaishya handle beer distribution in Pakistan. This employee inspects a bottle of vodka at the Murree Brewery Company, Pakistan's oldest public company and only brewery.

CLASS SYSTEM

A third type of stratification is the class system, an economically based hierarchical system characterized by cohesive, oppositional groups and somewhat loose social mobility. Class means different things to different people, and there is no consensus among sociologists as to the term's precise definition. For instance, some might define class primarily in terms of money, whereas others see it as a function of culture or taste. Some people barely even notice it (consciously at least), whereas others feel its powerful effects in their daily lives. So what is class? Is it related to lifestyle? Consumption patterns? Interests? Attitudes? Or is it just another pecking order similar to the caste system? Some controversy has existed about whether class is a real category or a category that exists in name only. As in the caste

CLASS SYSTEM

an economically based hierarchical system characterized by cohesive, oppositional groups and somewhat loose social mobility.

Why is the concept of class problematic? For example, Oseola McCarty, of Hattiesburg, Mississippi, worked as a washerwoman and had only a sixth-grade education. However, she donated $150,000 to the University of Southern Mississippi from her savings.

PROLETARIAT

the working class.

and estate systems, the lines that separate class categories in theory should be clearly demarcated, but there have been problems in drawing boundaries around class categories—for instance, upper class, middle class, and so on. Some scholars have even argued that class should be abandoned as a sociological concept altogether. So let's try to clear up some of the misconceptions.

Unlike other systems, a class system implies an economic basis for the fundamental cleavages in society. That is, class is related to position in the economic market. Notions of class in sociological analysis are heavily influenced by two theorists, Karl Marx and Max Weber. In Marxist sociological analyses, every mode of production—from subsistence farming to small-scale cottage industries to modern factory production to the open-source information economy—has its own unique social relations of production, basically the rules of the game for various players in the process. Who controls the use of capital and natural resources? How are the tasks of making and distributing products divided up and allocated? And how are participants compensated for their roles? In tips? Hourly wages? In-kind goods? Profit or rent? Thus, to talk about class in this Marxist language is to place an individual into a particular group that has a particular set of interests that often stand opposite to those of another group. For example, workers want higher wages; employers (specifically, capitalists) wish to depress wages, which come largely out of profits.

In this sense, class is a relational concept. That is, one can't gain information about a person's class by simply looking at his or her income (as in, "That person made only $13,000 last year; therefore, she is working class"). Class identity, in fact, does not correspond to an individual at all but rather corresponds to a role. A person may pull in a six-figure salary, but as long as he owns no capital (i.e., stock or other forms of firm ownership) and earns his salary by selling his labor to someone else, he finds himself in the same category as the lowest-paid wage laborer and antagonistic to "owners," who may net a lot less income than he does. And an individual may, over the course of her career, change class positions as she changes jobs. Class positions of jobs themselves—that is, the roles with respect to the production process—do not change, however.

Indeed for Marx, it all boils down to two antagonistic classes in a fully mature capitalist society: the employing class (the bourgeoisie or capitalist class) and the working class (the proletariat). The proletariat sells its labor to the bourgeoisie in order to receive wages and thereby survive. But, according to Marx, the bourgeoisie extract surplus value from the proletariat, even when a few of the proletarians make high incomes. As this is a fixed-pie or zero-sum view of economic production, an exploitative and hence inherently conflict-ridden relationship exists between the two classes. The central aspect of Marxist class analysis is this exploitation—capitalists taking more of the value of the work of laborers than they repay in wages.

Because the two-class model does not appear to adequately describe the social world as we find it in most modern capitalist economies, more recent Marxist theorists such as Erik Olin Wright have elaborated this basic model with the concept of contradictory class locations. Wright suggests that people can occupy locations in the class structure that fall between the two "pure" classes. For instance, managers might be perceived as both working class and capitalist class: They are part of the working class insofar as they sell their labor to capitalists in order to live (and don't own the means of production), yet they are in the capitalist class insofar as they control (or dominate) workers within the production process. Conversely, the petit bourgeoisie, a group including professionals, craftsmen, and other self-employed individuals or small-business owners, according to Wright, occupy a capitalist position in that they own capital in the form of businesses, but they aren't fully capitalist because they don't control other people's labor. The issue of class definition could also be further complicated by multiple class locations (multiple jobs), mediated class locations (the impact of relationships with family members, such as spouses, who are in different class locations), and temporally distinct class locations (for instance, many corporations require that all their managers spend a couple of years as a shop floor worker before becoming a manager). Marxists use these distinctions to analyze how various classes rise and fall under the capitalist system.

Max Weber takes an alternative view of class. He argues that a class is a group that has as its basis the common life chances or opportunities available to it in the marketplace. In other words, what distinguishes members of a class is that they have similar value in the commercial marketplace in terms of selling their own property and labor. Thus, for Weber, property and lack of property are the basic categories for all class situations. If you have just graduated from law school, you may have no current income or wealth, but you enjoy a great deal of "human capital" (skills and certification) to sell in the labor market, so you would clearly be in a different class from someone with little education but a similar current economic profile. But if you owned a company that you inherited, for instance, you would belong to yet another (higher) class. Weber's paradigm is distinct from the Marxist class framework, where the basic framework is antagonistic and exploitative; rather, for Weber, class is gradated, not relational. Put another way, your

BOURGEOISIE

the capitalist class.

CONTRADICTORY CLASS LOCATIONS

the idea that people can occupy locations in the class structure that fall between the two "pure" classes.

class as a newly minted attorney does not affect or determine my class as an accountant, computer programmer, or day laborer.

STATUS HIERARCHY SYSTEM

STATUS HIERARCHY SYSTEM

a system of stratification based on social prestige.

The fourth type of stratification system, a status hierarchy system, has its basis in social prestige, not in political, religious, or economic factors. In classical sociology, Weber contributed most heavily to the modern-day sociologist's understanding of status. For Weber, status groups are communities united by either a positive or negative social estimation of their honor. Put more simply, status is determined by what society as a whole thinks of the particular lifestyle of the community to which you belong. In this sense, those with and without property can belong to the same status group if they are seen to live the same lifestyle. Let's use the example of professors. There are certain things that professors typically do in common: They attend conferences to discuss scholarly issues, they teach college courses, and they tend to read a lot. This leads society as a whole to confer a certain status on professors, without placing the sole focus on income, which can vary widely among professors.

But even this lumping of professors (or other occupational categories) into one group can obscure huge status distinctions within that particular group. Some faculty are lucky enough to be on what's called the tenure track, a system designed to protect academic freedom. Once tenure-track faculty achieve tenure, they obtain a level of job security that is similar to civil servants and others who cannot easily be fired. Unless an entire school goes bankrupt, they can lose their jobs only if they grossly violate the terms of their employment by, say, not showing up to teach or sexually harassing students or falsifying their research. Compare that level of security to the grow-

Can you tell which professors are tenured and which aren't? (Spoiler alert: you probably can't based simply on their appearance.)

ing ranks of adjunct professors, who these days are the workhorses of academia. They get paid by course taught, typically without benefits; usually have to teach many more courses than full-time or tenured professors; and often do not know whether they will have enough work until right before a semester starts. Though they are also professors— usually with the same credentials as their tenured peers—their lifestyle shares some similarities with itinerant laborers, salespeople who work on commission, or others who enjoy only a minimal level of predictability in their income flows. Of course, the converse can also hold true. Namely, various individuals in a society who earn similar incomes may not have much in common in terms of lifestyles (and therefore status). Think again of a professor and, say, a plumber, who may earn the same annual incomes but have very different day-to-day experiences of work and orientations to the world.

Although a status group can be defined by something other than occupation, such as a claim to a specific lifestyle of leisure (skate punks) or membership in an exclusive organization that defines one's identity (the Daughters of the American Revolution), much work by sociologists has been related to occupational status. After all, work is one of the most centrally defining aspects of our lifestyle. In the 1960s, for example, Peter M. Blau and Otis Dudley Duncan (1967) created the Index of Occupational Status by polling the general public about the prestige of certain occupations (an abbreviated version based on more recent data appears in Table 7.1, next page). Many folks have since refined this rank ordering, but the story is much the same. What is particularly interesting is that the hierarchy is largely stable over time and across place. Occasionally, new jobs need to be slotted in (there were no web designers in the 1960s), but they generally are slotted in fairly predictable ways based on the status of similar occupations and do not create much upheaval within the overall rank ordering. The scores on the Duncan scale (as it is known) range from 0 to 96, with 0 being the least prestigious and 96 the most prestigious.

Table 7.1 shows that occupations with very different characteristics may have similar prestige scores. For instance, actors and firefighters enjoy roughly the same amount of prestige despite being very different in both their daily labor and their job security (firefighters are usually civil servants with job protection, while actors piece together a living through gig work, often supplementing). Blau and Duncan observed that five-sixths (just over 83 percent) of the explanation for people's status ratings of occupations was attributed to the education necessary for the position and not the income corresponding to that position.

Although we have emphasized that status may have its basis in occupation, it can also be formed through consumption and lifestyle, though these factors are often closely linked. This means that there should be a tendency for status differences between groups to be finely gradated and not relational: Fundamentally antagonistic status groups such as capitalists and laborers or owners and renters do not exist. Rather, people exist along a status ladder, so to speak, on which there is a lot of social mobility. Often, individuals seek to assert their status or increase their status not just through their occupation but also through their consumption, memberships, and other aspects of how they live. They might try to generate social prestige by driving a fancy car, living in a gated community, wearing stylish clothes, or using a certain kind of language.

TABLE 7.1 The Relative Social Prestige of Selected US Occupations

Occupation	Prestige Score	White-collar occupation	Blue-collar occupation	Occupation	Prestige Score	White-collar occupation	Blue-collar occupation
Physician/surgeon	86	X		Welder	41		X
Lawyer	74	X		Data entry operator	41	X	
Architect	72	X		Farmer/rancher	40		X
Dentist	72	X		Carpenter	39		X
Member of the clergy	68	X		File clerk	36	X	
Registered nurse	66	X		Child-care worker	35		X
Secondary-school teacher	66	X		Auto body repairperson	32		X
Veterinarian	62	X		Retail salesperson	31	X	
Sociologist	61	X		Truck driver	31		X
Police officer	58		X	Aircraft mechanic	31		X
Actor	55	X		Cashier	30	X	
Firefighter	53		X	Taxi driver/chauffeur	29		X
Realtor	48	X		Waiter/waitress	28		X
Machinist	47		X	Bartender	25		X
Musician/singer	47	X		Door-to-door salesperson	23		X
Construction equipment operator	46		X	Janitor	22		X

SOURCE: Frederick, 2010.

ELITE–MASS
DICHOTOMY
SYSTEM

a system of stratification
that has a governing elite,
a few leaders who broadly
hold power in society.

ELITE-MASS DICHOTOMY SYSTEM

The final stratification system is the elite–mass dichotomy system, with a governing elite, a few leaders who broadly hold the power in society. Vilfredo Pareto, in *The Mind and Society* (1983), took a positive view of elite–mass dichotomies, whereas C. Wright Mills saw much to dislike in such systems. For Pareto, when a select few elite leaders hold power—as long as the elites are the most able individuals and know what they are doing—the masses are all the better for it. This imbalance, where a small number of people (say 20 percent) cause a disproportionately large effect (more like 80 percent), has come to be known as the Pareto Principle or the 80/20 rule. The basis for Pareto's argument is that individuals are unequal physically, intellectually, and morally. He suggests that those who are the most capable in particular groups and societies should lead. In this way, Pareto believes in a meritocracy, a society where status and mobility are based on individual attributes, ability, and achievement. Pareto opposed caste systems of stratification that create systematic inequality on the basis of birth into a specific group. He criticized societies based on strict military, religious, and aristocratic stratification, arguing that these systems naturally tend to collapse. Concerning such systems, he argues that aristocracies do not last, and that, in fact, history is a "graveyard of aristocracies" (Pareto & Finer, 1966).

Along these lines, the ideal governing elite for Pareto is a combination of foxes and lions—that is, individuals who are cunning, unscrupulous, and innovative along with individuals who are purposeful and decisive, using action and force. This applies not only to the political realm but also to the economic realm, and the masses will be better off for it. The whole system works over time, in Pareto's view, when there is enough opportunity and social mobility to ensure that the most talented individuals end up in the elite and that it does not become a rusty aristocracy.

Mills takes a much more negative view of the elite—mass dichotomy, arguing that it is neither natural nor beneficial for society. In Mills's view, the elite do not govern the way Pareto claims they do. Mills argues in *The Power Elite* (2000) that there are three major institutional forces in modern American society where the power of decision making has become centralized: *economic institutions* (with a few hundred giant corporations holding the keys to economic decisions), the *political order* (the increasing concentration of power in the federal government and away from the states and localities, leading to a centralized executive establishment that affects every cranny of society), and the *military order* (the largest and most expensive feature of government). According to Mills, "families and churches and schools adapt to modern life; governments and armies and corporations shape it; and, as they do so, they turn these lesser institutions into means for their ends." The elite for Mills are simply those who have most of what there is to possess: money, power, and prestige. But they would not have the most were it not for their positions within society's great institutions. Whereas Pareto views elite status as the reward for the talent that helped certain individuals rise through the ranks of society, Mills sees the unequal power and rewards as determining the positions. And whereas Pareto sees a benefit in having power centralized in a large, otherwise ungovernable society, Mills warns of the dangers. For Mills, such a system hurts democracy by consolidating the power to make major decisions into the hands (and interests) of the few.

MERITOCRACY

a society where status and mobility are based on individual attributes, ability, and achievement.

The power elite is further stratified for Mills; at the inner core of the power elite are those individuals who interchange commanding roles at the top of one dominant institutional order with those in another. In Mills's view, as the interactions among the big three power elites increase, so does the interchange of personnel among them. For example, after serving as President Barack Obama's budget director, Peter Orszag took a top management position at Citigroup, a firm that benefited substantially from the government bailout after the 2008 financial crisis. Who replaced him in the White House? Jacob Lew, who was budget director under President Bill Clinton and later worked for Citigroup himself. Likewise, back in 1961, Robert McNamara left the presidency of Ford Motor Company to serve as secretary of defense under President John F. Kennedy. Today, of course, business elites like Trump and his son-in-law, Jared Kushner, are running our political affairs. Likewise, we have at least three military generals serving in elite political positions: at the time of this writing, John Kelly is White House chief of staff, John Bolton is national security adviser, and James Mattis is secretary of defense. Such exchange across domains is of central importance to Mills, because as the elite increasingly assume positions in one another's domains, the coordination among the elite becomes more and more entrenched.

2016 presidential candidate Hillary Clinton, former president Bill Clinton, and their daughter, Chelsea Clinton, who has worked as an NBC news correspondent and management consultant. How do the Clintons represent modern power elites?

The inner core also includes the professional go-betweens of economic, political, and military affairs. These include individuals at powerful legal firms and financial institutions, such as corporate lawyers and investment bankers. The outer fringes of the power elite, which change more readily than the core, are individuals who count in the decisions that affect all of us but who don't actually make those decisions.

Let's take, for example, the business sector in America. More than 80 percent of the 1,000 largest corporations shared at least one director with another large company, and on average any two of the corporations were connected by fewer than four degrees of separation (Davis, 2003). Wall Street is particularly densely connected. Most troubling is the fact that many US Treasury secretaries, including the current one, Steven Mnuchin, have worked at a single firm: the investment bank Goldman Sachs. As a result of these networks, which Mills both saw and predicted, decisions by

INCOME VERSUS WEALTH

Most people, when they consider economic status, think of *income*: money received by a person for work, from transfers (gifts, inheritances, or government assistance), or from returns on investments. Recent trends in earnings in the United States suggest that there is an increasing divergence or inequality between the bulk of the people and the rich, but especially between the super-rich and the merely rich. For instance, from 1950 to 1970, for every dollar earned by the bottom 90 percent of the American population, those in the top 0.01 percent earned an additional $162. If that sounds like a big distinction, that's nothing compared with more recent data. Between 1990 and 2002, for every dollar earned by those in the bottom 90 percent, each taxpayer at the top (and this would include Bill Gates) took home $18,000 (Johnston, 2005).

A recent trend in sociological analysis, however, is to analyze stratification in terms of wealth ownership. What is *wealth* in relation to income? Wealth is a family's or individual's net worth (total assets minus total debts). For the majority of American families, assets include homes, cars, other real estate, and business assets, along with financial forms of wealth such as stocks, bonds, and mutual funds. Put simply, wealth is everything you own minus debts such as mortgages on homes, credit card debt, and, as most of you will probably have, debt from student loans. One way to think about the difference between income and wealth is to imagine income as a stream or river of money flowing through a family's hands. Wealth, by contrast, is a pool of collected resources that can be drawn on at specific times, a financial reservoir. In 2015, half of the world's wealth was held by just 1 percent of its residents (Kersley and Stierli, 2015).

various companies become increasingly similar. Many studies have suggested that these elite networks share practices, principles, and information that account for some of the surprising conformity in approaches to corporate governance and ethics. For example, the response to the 2008 financial crisis was evidently mapped out in a room with a dozen or so top bankers and the president of the New York Federal Reserve, Tim Geithner, who later became Treasury secretary in the Obama administration.

How Is America Stratified Today?

SOCIOECONOMIC STATUS

an individual's position in a stratified social order.

INCOME

money received by a person for work, from transfers (gifts, inheritances, or government assistance), or from returns on investments.

WEALTH

a family's or individual's net worth (i.e., total assets minus total debts).

UPPER CLASS

a term for the economic elite.

Sociologists often use the phrase socioeconomic status to describe an individual's position in a stratified social order. When sociologists talk about socioeconomic status, they are referring to any measure that attempts to classify groups, individuals, families, or households in terms of indicators such as occupation, income, wealth, and education. Although the boundaries between socioeconomic categories are not sharply defined, the lay public generally divides society into the upper class, the middle class, the working class, and the poor. Because these are common terms, we need to take them seriously, even if they are not of scientific origin and lack sufficient rigor.

THE UPPER CLASS

Generally, the upper class refers to the group of individuals at the top of the socioeconomic food chain. In practice, however, the term is used to describe diverse and complex concepts. Historically, the upper class was often distinguished by not having to work. (The economist and social critic Thorstein Veblen dubbed this group "the leisure class.") Its members were able to maintain their lifestyle by collecting rent and/or other investment returns. They were the aristocracy, the wealthy, the elite, the landowners. The only way to join this sphere was by birth or (occasionally) marriage. The upper class was the basis for Marx's capitalist class.

In the United States, "upper class" is associated with income, wealth, power, and prestige. According to some sources, the primary distinguishing characteristic of upper-class individuals is their source of income—generally more from returns on investments rather than wages. Although estimates vary, approximately 1 percent of the US population is considered to fall in this stratum (Figure 7.1). From 1980 to 2014, the average income in the bottom half of earnings remained $16,000 (adjusted for inflation). Over the same period, the average earnings of the 1 percent tripled—from $420,000 to $1.3 million (Piketty et al., 2018). By 2012, the top 0.1 percent owned 22 percent of the total wealth in the United States, up from 7 percent in the late 1970s (Saez & Zucman, 2016). In 2016 the CEOs of America's largest companies made 271 times that of the typical worker, compared to 59 times the average in 1989 (Mishel & Schieder, 2017). In the period between 2009 and 2012, the top 1 percent of Americans saw their income grow by 31.4 percent, but the income of all other Americans barely grew at all, increasing just 0.4 percent (Saez, 2013). Over and above income levels, the upper class is also distinguished by prestige and power, which can be used to promote personal agendas and influence everything from political decisions to consumer

FIGURE 7.1 Distribution of Net Worth in the United States, 2016

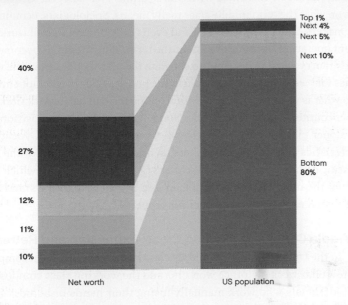

SOURCE: Wolff, 2017.

trends. This is of particular importance because, as noted earlier, members of the upper class often wear many hats. As Dennis Gilbert states in *The American Class Structure in an Age of Growing Inequality* (1998),

> The members of the tiny capitalist class at the top of the hierarchy have an influence on economy and society far beyond their numbers. They make investment decisions that open or close employment opportunities for millions of others. They contribute money to political parties, and they often own media enterprises that allow them influence over the thinking of other classes. (p. 286)

THE MIDDLE CLASS

Although those in the upper class have very real influence and control, the effects of this are sometimes limited in the public perception. In the 2007 article "The American People: Social Classes" published in the online magazine *Life in the USA* (a "complete web guide to American life for immigrants and Americans"), Elliot Essman speaks for much of the mainstream media when he asserts that the implications of such power are not far-reaching in our daily lives. "The very rich control corporations and have some political power, but the lifestyle and values of the very wealthy do not have much impact on the country in general," he states. "America is a middle class nation."

The United States is often thought of as a middle-class nation—so much so that depending on how the question is phrased, almost 90 percent of Americans have self-identified with this stratum. That said, there is no consensus on what the term middle class really means. Sociologists, economists, policy makers, think tank analysts, and even the public at large all work with different operating assumptions about the term. The categories become particularly blurry when we attempt to separate the middle class from the working class (or "working families," to use the political campaign euphemism).

So what is middle class? If you look up the term in various dictionaries, you'll encounter lots of definitions. Some refer explicitly to position (i.e., below upper class, above lower class); others speak to shared vocational characteristics and values. Still others mention principles. One of the most interesting comes from *Merriam-Webster's Collegiate Dictionary,* which indicates that the middle class is "characterized by a high material standard of living, sexual morality, and respect for property."

The Middle Class and Working Class: Expansion and Retrenchment
In the United States, the middle class has historically been composed of white-collar workers (office workers), and the working class composed of those individuals who work manually (using their hands or bodies). However, this distinction eroded with two trends. First, the post–World War II boom led to the enrichment of many manual workers. In those days, most working-class whites, ranging from factory workers to firemen to plumbers, were able to buy their own homes, afford college for their children, and retire comfortably. The working class became the newly expanded middle class.

From the post–World War II era of the late 1940s through the oil crisis of 1973, this middle class was a large and fairly stable group, maintained by corporate social norms emphasizing equality in pay and salary increases (Krugman, 2005). From 1947 through 1979, the average salary increase was fairly stable across all household income levels—in fact, the lowest-earning 20 percent of households showed the highest average earnings increase, 116 percent.

A second, countervailing trend has also eroded the traditional manual–nonmanual distinction between the working and middle classes: the rise of the low-wage service sector. Since 1973, manufacturing has steadily declined in the United States, and the service jobs replacing factory work have generally been either very high skilled (and rewarded) or relatively low skilled (and therefore not paid very well). This new and expanding group of low-wage service workers challenges the notion that working-class status arises from physical labor. Data-entry clerks, cashiers, paralegals, and similar occupations are technically white-collar jobs yet pull in working-class wages.

Over the past three decades, the income gap between a corporate CEO and a single-mother waitress has grown exponentially, and the relatively stable middle class of previous decades has become increasingly stratified (Figure 7.2). For example, a CEO who makes at least $200,000 is in the top

FIGURE 7.2 Ratio of US CEO Pay to Average Worker Pay, 1965–2016

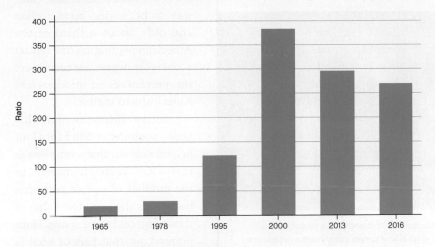

SOURCES: Davis & Mishel., 2014; Mishel & Schieder, 2017.

5 percent of households, and his or her average expected salary increase since 1979 (adjusted for inflation) is about 50 percent. On the other end of the scale, a waitress who makes only $22,000 is in the bottom 20 percent of households, and her income will have increased by 27 percent over the same period. And if we consider an average family—say, a two-parent household with two kids and a middle range annual income of $52,000—then this family would probably have experienced an income increase of only 14 percent since 1979, a sign of what some call the shrinking middle class (Gould, 2014). (Go to **wealthinequality.org** for more information.)

How and why has this differential income growth happened? One factor may be the changing nature of available employment. As the technological sector expands, the majority of jobs are being created either at very high skill levels with correspondingly high salaries (engineers and hedge fund managers) or at very low levels with little room for upward mobility (baristas at Starbucks who serve lattes to those hedge fund managers). And the fastest growing job category for those with no more than a high-school education is food preparation and service.

Furthermore, this bifurcated job growth is reflected in the change in educational expectations of potential employers. Whereas for much of the post–World War II period, a high-school diploma could earn a person (a white man, specifically) enough to support a family, a college degree is now often a minimum requirement for decent employment. But although the number of high-school graduates going to college has increased (to roughly 70 percent in 2016; Bureau of Labor Statistics, 2017b), bachelor's degree completion rates still stand at approximately 59 percent nationally (i.e., lots of students are

DIGITAL.WWNORTON.COM/YOUMAYASK6CORE

To see sociologist Michael Hout discuss the benefits of a
college education, see **digital.wwnorton.com/youmayask6core**

not finishing) (National Center for Education Statistics, 2017a). If a college education is the gateway to becoming middle class and only about a third of the American population has at least a four-year degree, we can expect the percentage of middle-class Americans to shrink.

In an interview for this book, sociologist Michael Hout helped explain that even though there has been stagnation in the number of people receiving bachelor's degrees, the value of those degrees is increasing. Hout pointed out that part of what is holding students back from getting their degree is that the cost of college is increasing and a larger proportion of the cost is borne by individuals and families rather than states. For example, due to state budget shortfalls, the University of California system—one of the biggest and best—announced a tuition hike of 32 percent for the academic year starting in 2010 that was swiftly followed by another 9.6 percent rise in 2011 and is slated to increase another 28 percent by 2019 (McMillan, 2011; Pickert, 2014). Hout goes on to describe the way that sacrificing to pay for college plays out differently across social classes.

> There is a class-specific preference for these kinds of risks. The sons and daughters of college graduates see the benefit of a college education more clearly. They have had it drummed into them since they could talk. . . . Sons and daughters of people who haven't been to college can exercise a certain skepticism about it, and they see in their neighborhoods evidence that it might not pay off. Who leaves the neighborhood? Somebody who has reaped the full benefit of the college education. They're out of sight and out of mind. Who's back in the neighborhood? That kid who maybe dropped out after three years, with a three-figure loan debt, six-figure loan debt. And the presence of those people makes it look like college doesn't pay off. (Conley, 2009h)

Some argue that along with income inequality, income insecurity (or volatility) has also increased for the middle classes. In *The Great Risk Shift*

(2006), Jacob Hacker asserts that the chances of an American family experiencing a 50 percent drop in their annual income from one year to the next were 17 percent in 2002, up from around 7 percent in 1970. A 50 percent salary cut would still leave the CEO mentioned earlier in the top 40 percent of household incomes, but how about our waitress and our middle-income family? Cutting their salary by half would send either into a major financial crisis, maybe even bankruptcy. However, other researchers argue that income instability is not the result of fluctuating earnings or job insecurity but rather stems from two major changes in family dynamics. First, there are more changes to household structure today because of divorce and remarriage; second, women now play a greater role in breadwinning (even exceeding men as a proportion of the workforce at one point during the recent recession), but they also enter and exit the workforce with greater frequency than men. (See Chapter 8 for more about women in the workforce.)

THE POOR

Unlike the fuzzy definitions of other classes, poverty (which gets its own, entire chapter in this book, Chapter 10) has an official, government definition. In 2018, the poverty line for a family of four was $25,100 (US Department of Health and Human Services, 2018). The poor are, ironically, often said to resemble the rich in being more oriented toward the present and therefore less worried about the future than their middle-class and working-class counterparts (although this is highly debated). Day-to-day survival keeps the poor clearly planted in the present. Of course, like any class, the poor are not a unified group. In fact, one distinction often made in political speeches is that between the "working poor" (those who deserve our assistance) and the "nonworking poor" (those who can work but don't and therefore have a weaker moral claim on assistance). This latter group is sometimes called the "underclass." Of course, even these two categories obscure huge distinctions within either group; in fact, poverty is a state that families usually shift in and out of throughout their history, and often a clear distinction does not exist between the working class and the poor.

Global Inequality

One of the main reasons cited for rising income and wealth inequality in the United States is globalization—the rise in the trade of goods and services across national boundaries, as well as the increased mobility of multinational businesses and migrant labor. If the effect of globalization in the United States has been to bifurcate labor into high-skilled and low-skilled jobs, what has been the effect on worldwide inequality? The answer to that question depends on how you frame the analysis. In the long view, there is no question that global income inequality has been steadily rising over the last few centuries. At the start of the agricultural and industrial revolutions, almost the entire population of the world lived in poverty and misery (see the discussion of Malthus earlier in the chapter). Birthrates were high, but so was mortality. Most people barely survived, no matter where they lived.

But then, thanks to technological innovations, food production started to increase dramatically in some areas of Europe. Soon after—during the Industrial Revolution—large-scale factory production took the place of small cottage industries, resulting in the creation of vast, unequally distributed wealth. Simultaneously, European powers began to explore, conquer, and extract resources from other areas of the globe. Fast-forward a few hundred years to the mid-twentieth century, when the world and most of its wealth were carved up and ruled by major Western powers. Enormous global inequalities had emerged through the combination of colonialism and unequal development. Since around 1950, many of these former colonies have gained political independence, but they have lagged well behind the West in terms of income per capita.

Scholars have long been trying to figure out why Europe developed first. Early explanations, dating back to the French essayist Montesquieu (1748/1750), focused on geographic differences between the peoples of Europe and the global south (as the less developed regions of the world are sometimes called). Perhaps because of the south's heat or humidity, Montesquieu associated virtue and rationality with the north; the south he associated with vice and passion. Montesquieu's views are now discarded as a racist account of uneven development, but more recent versions of the geographical explanation focus on differences in the length of growing seasons, the higher variability of water supply (because of more frequent droughts), the types of cereal crops that can grow in temperate zones compared with tropical and subtropical regions, the lack of coal deposits (the first fuel that drove industrialization, long before oil), and perhaps most important, the disease burden in warmer climates (Sachs, 2001).

Economist Jeffrey Sachs explains that Africa's geography made it much more difficult for an agricultural revolution to occur. He prods us to understand the complexity of the constraints on Africa's development before pointing fingers:

> Africa is a continent largely of that rain-fed agriculture, whereas the Green Revolution was based first and foremost on irrigation agriculture. Now is that bad governance that Africa does not have vast river systems? Or is that a matter of the Himalayan Tibetan plateau, which creates the Indus, the Punjab (meaning five rivers, after all), the Ganges, the Yangtze, and so forth? These are functions of physical

geography. Africa has a savannah region, which means a long dry season together with a single wet season typically. (Conley, 2009g)

So Africa was not fortunate when it came to physical geography, and it missed out on the Green Revolution because it could not easily implement irrigation agriculture. But that's not all. If you recall from our discussion of Malthus, Africa also carried a higher disease burden because of the way malaria is transmitted through animal hosts—more deadly in Africa than in Asia. Even with that, Sachs is not through explaining just how Africa came to be so disadvantaged compared with other regions of the world. Important impacts of colonialism would be easy to overlook if we focused only on geography and infectious disease transmission, he says.

> Let's talk more about Africa. Racism. Absolutely pervasive. The slave era, the treatment of white people in the rich north Atlantic vis-à-vis the black people. Let's face it: We want to pretend that's not part of our society, part of our history, part of our view of the hopelessness of regions, part of the way that they were treated unfairly. If you look through the colonial-era memos in Africa about the repression of education levels and so forth by colonial powers, that's also part of the history. Then you come to infrastructure, you come to physical geography, you come to history, and you come to an odd historical context. (Conley, 2009g)

Was the atrocious treatment by colonial powers somehow the fault of African leaders? Does this ever-so-brief primer on the history of African development help you understand that the sociological imagination cannot function properly without a wide array of information about everything from epidemiology to history to geography and beyond?

Others argue that geography doesn't matter as much as social institutions do. In one version of this line of reasoning, it is a strong foundation of property rights, incorruptible judiciaries, and the rule of law in general that predict economic development (Easterly & Levine, 2002). These, in turn, were institutions "native" to Europe and were transferred to the areas where European colonists settled and

DIGITAL.WWNORTON.COM/YOUMAYASK6CORE

Watch an interview with economist Jeffrey Sachs about overcoming poverty in Africa at **digital.wwnorton.com/youmayask6core**

Tremendous global inequalities have emerged through the combination of colonialism and unequal development. What are some of the ways that social scientists explain the gap between rich and poor regions of the world?

lived in significant numbers (Acemoglu et al., 2001), spurring later development in areas such as India and Latin America (compared with much of Africa, where Europeans did not settle, which has lagged in development). Other scholars argue that the types of relationships different colonized regions had with the European powers largely determined their fate today: For example, under the rule of the British Raj, India was endowed with a huge network of railways to move cotton, tea, and other products to market. This network, in turn, helped spur economic growth once India got over the rocky transition to self-government after gaining its independence. By contrast, because of the disease burden in most sub-Saharan African countries, Europeans didn't stay and invest but rather focused on extractive industries, building railways that ran just to and from mines instead of extensive networks of roads and train tracks.

Today that legacy of unequal starting places and economic potentials means that the latest spurt of globalization has engendered an even greater level of income inequality across the north–south divide while creating huge differences within developing countries. Indeed, some regions, such as sub-Saharan Africa and eastern Europe, have become poorer over the last 25 years, whereas in some previously poor regions, such as South and East Asia, income levels have risen. Considering the net change, economist Xavier Salai-i-Martin estimates that there were around half a billion fewer poor people in the world by the turn of the twenty-first century than there were in the 1970s (Salai-i-Martin, 2002). Likewise, he claims, "all indexes show a reduction in global income inequality between 1980 and 1998." There have indeed been global reductions in population-weighted between-country differences, which can largely be attributed to the rapid growth of the Chinese and Indian economies over this period. Within individual countries, however, inequality has generally risen over the same time frame. The story of globalization and inequality is thus a complex and constantly changing one.

Social Reproduction versus Social Mobility

Once we begin to understand the basics of stratification—how members of a society are hierarchically organized along different lines—the next issue is the possibility of those members changing their social position in the hierarchy. This is generally referred to as social mobility, the movement between different positions within a system of social stratification in any given society. Pitirim A. Sorokin (1959) emphasized the importance of not just looking at the mobility of individuals but also examining group mobility. For example, rather than just asking why Ugo, a cashier at Sears, was promoted to regional manager, thereby increasing his income and status, Sorokin suggests we also look at why the group Ugo belongs to—black males of Nigerian descent in their mid-twenties—does or does not appear to be generally mobile.

> SOCIAL MOBILITY
>
> the movement between different positions within a system of social stratification in any given society.

Sorokin noted that social mobility can be either horizontal or vertical. *Horizontal social mobility* means a group or individual transitioning from one social status to another situated more or less on the same rung of the ladder. Examples of this might be a secretary who changes firms but retains her occupational status, a Methodist person who converts to Lutheranism, a family that migrates to one city from another, or an ethnic group that shifts its typical job category from one form of unskilled labor to another. *Vertical social mobility*, in contrast, refers to the rise or fall of an individual (or group) from one social stratum to another. We can further distinguish two types of vertical mobility: *ascending* and *descending* (more commonly termed *upward* and *downward*). An individual who experiences ascending vertical mobility either rises from a lower stratum into a higher one or creates an entirely new group that exists at a higher stratum. Ugo's promotion to regional manager at Sears is an example of rising from a lower stratum to a higher one. By becoming a manager, he has changed his class position, a change that confers both a higher salary and more prestige. Conversely, imagine that for some reason there was an immediate need for translators of Igbo (or Ibo, the language spoken by the Igbo people based in southeast

Nigeria) in the United States. Ugo, who speaks Igbo, along with many other Nigerians who speak the language, would then find better jobs and thereby enjoy higher social status. This would be an example of a new group, Igbo translators, existing in a higher stratum.

Descending vertical mobility can similarly take either of two forms: individual or group. One way to think of these forms is as the distinction between a particular person falling overboard from a ship and the whole ship sinking. Sorokin (1959) asserted that "channels of vertical circulation necessarily exist in any stratified society and are as necessary as channels for blood circulation in the body."

Most sociologists concerned with studying mobility focus on the process of individual mobility. These studies generally fall into two types: mobility tables (or matrices) and status attainment models. Constructing a mobility table is easy, although making sense of it is much more difficult (see Table 7.2 for an example). Along the left-most column of a grid, list a number of occupations for people's fathers (it was traditionally done for males only). Across the top, list the same occupational categories for the sons (the respondents). There can be as many or as few categories as you see fit, so long as

TABLE 7.2 Mobility Table: Father's Occupation by Son's First Occupation

FATHER'S OCCUPATION	SON'S OCCUPATION					
	UPPER NONMANUAL	LOWER NONMANUAL	UPPER MANUAL	LOWER MANUAL	FARM	TOTAL
Upper nonmanual	1,414	521	302	643	40	2,920
Lower nonmanual	724	524	254	703	48	2,253
Upper manual	798	648	856	1,676	108	4,086
Lower manual	756	914	771	3,325	237	6,003
Farm	409	357	441	1,611	1,832	4,650
TOTAL	4,101	2,964	2,624	7,958	2,265	19,912

SOURCE: Hout, 1983.

they are consistent for the parental and child generations. For example, a five-category analysis might include upper nonmanual occupations (managers and professionals), lower nonmanual occupations (administrative and clerical workers, low-level entrepreneurs, and retail salespeople), upper manual occupations (skilled workers who primarily use physical labor, such as plumbers and electricians), lower manual (unskilled physical laborers), and farmworkers. The number of categories can be expanded, and it is not uncommon to see seven-, nine-, or even fourteen-category tables. The key is filling in the boxes. Probably not many people fall into the cell where the father is upper nonmanual and the son is a farmer. However, because of the decline in the agricultural workforce over the course of the twentieth century, you are likely to observe (at least in the United States) a fair bit of movement from farmwork in the parental generation to other occupations in the children's generation. Changes in the distribution of jobs lead to what sociologists call structural mobility, mobility that is inevitable from changes in the economy. With the decline of farmwork because of technology, the sons and daughters of farmers are by definition going to have to find other kinds of work. This type of mobility stands in contrast to exchange mobility, in which, if we hold fixed the changing distribution of jobs, individuals trade jobs such that if one person is upwardly mobile it necessarily entails someone else being downwardly mobile.

As most measures of economic inequality have risen each year since the 1960s, Americans have comforted themselves with the thought that they still live in the land of opportunity. Rates of occupational (and income) mobility in the United States have long been thought to have dwarfed those of European societies with their royalty, aristocracies, and long histories. However, recent research suggests that this may be cold comfort: Some economists have argued that US mobility rates have declined significantly since the 1960s. Others go so far as to say that Americans now enjoy less mobility than their European counterparts. A fierce debate has ensued, because many of these studies compare apples and oranges—different measures, different data, different years. However, a consensus seems to be slowly emerging that mobility rates should be broken down into the two components discussed above, structural and exchange mobility. When we do this, it turns out that rates of "trading places" are fairly fixed across developed societies. By contrast, historically the United States has enjoyed an advantage in growth-induced upward mobility; as the farm and blue-collar sectors withered and white-collar jobs expanded, sons and daughters of manual workers have, by necessity, experienced a degree of upward occupational mobility. However, sociologists and economists debate whether economic growth still drives upward mobility, or whether bifurcated job growth means intergenerational stagnation.

Another common methodology for studying social mobility is the status-attainment model. This approach ranks individuals by socioeconomic

STRUCTURAL MOBILITY

mobility that is inevitable from changes in the economy.

EXCHANGE MOBILITY

mobility resulting from the swapping of jobs.

STATUS-ATTAINMENT MODEL

approach that ranks individuals by socioeconomic status, including income and educational attainment, and seeks to specify the attributes characteristic of people who end up in more desirable occupations.

status, including income and educational attainment, and seeks to specify the attributes characteristic of people who end up in more desirable occupations. The occupational status research of Peter M. Blau and Otis Dudley Duncan (1967), who ranked occupations into a status hierarchy to study social attainment processes, is generally seen as the paradigmatic work in this tradition. Unlike mobility tables, the status-attainment model allows sociologists to study some of the intervening processes. For example, how important is education in facilitating upward occupational shifts in status? How critical is the prestige of a person's first job out of school? How does IQ relate to the chances for upward or downward mobility? The status-attainment model is an elastic one that allows researchers to throw in new factors as they arise and see how they affect the relationships between existing ones, generally ordered chronologically over a typical life course. Generally, it is thought that education is the primary mediating variable between parents' and children's occupational prestige. That said, research shows that parental education and net worth, not occupation or income, best predict children's educational and other outcomes (Conley, 1999). Blau and Duncan didn't measure net worth, but by now it has become a fairly standard factor in many socioeconomic surveys.

Although there is increasing consensus on what aspects of class background matter (and how much they do), there is relatively little understanding of the multiple mechanisms by which class is reproduced (or how mobility happens). For instance, families with a higher socioeconomic status are likely to have more success in preparing their children for school, entry exams, and ultimately the job market because they have greater access to resources that promote and support their children's development. These might include educational toys when the children are very young, tutors in grade school, and expensive test-preparation courses for exams such as the SAT, GRE, LSAT, and MCAT. Once again, let us compare the CEO of a large, profitable US corporation with a single mother working as a waitress at the local diner. The status of the CEO's occupation and his disproportionate income and wealth in relation to the waitress's easily put his socioeconomic status leaps and bounds ahead of hers. Because of this disparity, he can send his children to the top private schools, where they will be funneled toward highly paid positions. In contrast, assuming that the single mother has no other source of income, her children are likely to get a public education without all the extras. This by no means mandates that they are destined for lower-level occupations, but it certainly suggests that if this were a race, the CEO's children started several miles ahead.

CLASS-BASED
AFFIRMATIVE ACTION

The push for class-based affirmative action (for lack of a better term) is partly a response to the slow decline of race-based affirmative action, as evidenced by the Supreme Court's 2014 ruling to uphold the decision of Michigan's voters to ban race-based affirmative action across the University of Michigan system. In addition to the (legally) declining significance of race in admissions, the call for economic considerations in the college admissions process arises from mounting evidence that class has become an increasingly salient driver of academic opportunity (and success).

The statistics about increasing class stratification on American campuses are alarming: "The college-completion rate among children from high-income families has grown sharply in the last few decades, whereas the completion rate for students from low-income families has barely moved" (Bailey & Dynarski, 2011). Moreover, high-income students make up an increasing share of the enrollment at the most selective colleges and universities (Reardon et al., 2012), even when compared with low-income students with similar test scores and academic records (Bailey & Dynarski, 2011; Belley & Lochner, 2007; Karen, 2002).

Class-based admissions policies, then, offer a way to redress unequal access to selective institutions of higher education while also indirectly tackling racial disparities. In addition, class-based policies, if well designed, can help address some of the criticisms of traditional, race-based affirmative action.

One of the most common criticisms of race-based affirmative action is that as currently designed, such admissions policies typically help those minorities who least need it. Whereas before the 1970s race was seen to trump class in determining the life chances for success for the vast majority of African Americans, today it is the reverse pattern that predominates (Reardon et al., 2012). Back in 1967, Peter M. Blau and Otis Dudley Duncan described the process of stratification in the United States in their landmark book *The American Occupational Structure*. In this study, they found that class background mattered little for African Americans compared with whites. Instead, they described a dynamic called "perverse equality": No matter what the occupation of the father of a black man (this was a period of low labor-force participation for women overall, even if black women did work at significant rates), he himself was most likely to end up in the lower, manual sector of the labor market. Meanwhile, in each generation a small, new cadre of professional blacks would emerge seemingly randomly through a dynamic they described as "tokenism"—that is, family background mattered little in predicting who emerged into the small, black professional class.

By the mid-1970s, however, this dynamic had changed. In 1978, sociologist William Julius Wilson described a black community where class stratification was increasingly rearing its head.

What challenges do colleges face in promoting diversity through admissions policies?

Later work confirmed intergenerationally what Wilson observed cross-sectionally: There were increasing class divisions within the black (and Latino) communities, and class background was an increasingly salient predictor of economic success, not just for whites but for minorities as well (Conley, 1999; Killewald, 2013).

Sean Reardon (2013) goes so far as to argue that class disparities have eclipsed racial ones, at least in terms of achievement: "The black-white achievement gap was considerably larger than the income achievement gap among cohorts born in the 1950s and 1960s, but now it is considerably smaller than the income achievement gap. This change is the result of both the substantial progress made in reducing racial inequality in the 1960s and 1970s and the sharp increase in economic inequality in education outcomes in more recent decades." Economist Roland Fryer Jr. (2010) sums this up nicely: "Relative to the 20th century, the significance of discrimination as an explanation for racial inequality across economic and social indicators has declined."

In short, although class divisions within historically underrepresented minority groups are increasing, 1960s identity-group policies treat disadvantaged groups uniformly. The result of such a homogenizing admissions policy is that the most disadvantaged minorities are not helped and intra-racial stratification is enhanced. Thus, either in lieu of, or in combination with, race-based policies, class-based affirmative action could address these inequalities within minority (and majority) communities.

If one were to decide to design a class-based admissions policy, how would one implement such a scheme? Can we just ask students' parents to self-report their "class" or socioeconomic status by disclosing their income, education, and wealth? Here's where the devil lies in the details. Parental income is the most easily verifiable, because colleges can ask for tax returns. But analysis shows that it matters the least in predicting college enrollment and completion. In other words, by designing admissions policies around income, we would be doing less well in helping those who need it most. Parental education is actually the strongest predictor but the least verifiable. (How can you check that someone has *more* education than they report on a form?) Parental wealth, meanwhile, is the most unequal by race—getting us the most racial bang for the buck, so to speak. The typical African American family has about 10 percent of the net worth of the typical white family. Wealth also predicts college outcomes (though not as strongly as parental education). So it would seem to be an ideal measure for balancing concerns about race and class diversity on campus.

However, parental wealth faces an even greater problem of verification, potential gaming, and perverse incentives against savings. For example, because Medicaid has strict asset limits, many families shift assets from one individual to another (or even get divorced) in anticipation of needing Medicaid's long-term care insurance component. Similar shell games may emerge in response to offspring approaching their senior year of high school. Unlike the case of education, however, where self-reporting is the only metric to go by for parental wealth, we can infer a lot based on a few factors that are less apt to be gamed.

First, we can measure the median housing value of a community in which a student was raised. This has been shown to be a very good proxy for individual wealth level. If it were measured for all years from birth, the incentive to move to a poor-value neighborhood in the period just preceding college applications would be minimized. Second, other forms of wealth can be ascertained or imputed through property tax records, estate tax records, and schedules A through D of the federal income tax return. While these individual-level measures could theoretically be gamed, the fact that they are measured over multiple years (as with the address of the applicant) minimizes such potential threats, and when combined with the neighborhood-level measures, such risk is further minimized.

Or, when all is said and done, if we want to maintain or increase diversity on campus, we can do what we are already doing: just asking applicants to check a box on the race question on their application (never asking for verification), while hoping that people are honest and the courts are sympathetic.

Conclusion

Horatio Alger Jr. (1832–1899) was an American author of dime novels that told rags-to-riches stories. The narratives typically depicted a plucky young downtrodden boy who eventually achieved the American dream of success and fortune through tenacious hard work while maintaining a genuine concern for the well-being of others. Alger wrote more than 130 novels with titles such as *Sink or Swim* (1870), *Up the Ladder* (1873), *From Boy to President* (1881), *Making His Way* (1901), *A New Path to Fortune* (1903), and *Finding a Fortune* (1904). Works like these contributed to the national ideology that all Americans could make it if they only pulled themselves up by their own bootstraps.

Although more than 100 years old, Alger's novels still resonate with a faith in mobility that is woven into the national fabric and self-image of Americans. Today the majority of Americans are no longer positive about the possibility of upward mobility. A 2013 poll by the Pew Charitable Trusts found that only 37 percent of Americans thought that their children would

This Horatio Alger Jr. novel features a newsboy who rises to newspaper editor. Are most Americans today likely to achieve upward mobility?

RISEN *from the* RANKS

HORATIO ALGER JR.

be better off than they were (Stokes, 2017), and in 2017 only 36 percent of Americans believed that they themselves have achieved the American dream (Pew Research Center, 2017b). What is so striking is that more and more Americans are buying into the Alger myth—more Americans than 20 years ago believe it possible to start at the bottom and work your way to the top. People generally believe that hard work and education are more important than social connections or a wealthy background. Are Americans overly optimistic or are sociologists just being naysayers? Only more research will tell us for sure.

QUESTIONS FOR REVIEW

1. How does the number of doctors with last names from particular ethnic groups demonstrate the lasting influence of history on intergenerational mobility?

2. How does class-based affirmative action continue to help some racial minority students? What does this tell you about the status of various categories of stratification in America?

3. Whereas inequality is the result of abundance, how does the relationship between the bourgeois and proletariat suggest that abundance is the result of inequality?

4. To talk about the rich, the poor, and the way society is economically stratified sounds like the job of economists. Why should sociologists be interested in stratification? How does a better understanding of stratification potentially contribute to the well-being of society?

5. Why, according to Adam Ferguson and John Millar, is inequality necessary? In what way does their argument anticipate the "free rider problem"?

6. What is "equality of condition," and why did Thomas Malthus argue against striving for this form of equality?

7. What are the ideal types of social stratification and how do they differ? Which one, in your opinion, best describes stratification in the United States?

8. What is the difference between income and wealth? Why might certain sociologists prefer to measure inequality based on wealth instead of income?

9. What is structural mobility, and how does this concept describe the decline of manufacturing jobs in the United States since the early 1970s?

10. According to Max Weber, what do being a teacher, having a cool car, and being a member of a prestigious association have in common?

THE $5,000 TOOTHBRUSH

Scholars from Weber to Bourdieu have talked about lifestyle and consumption choices as an important part of the stratification system. It tends to be the most extreme examples that remind us of so-called conspicuous consumption, a term first coined in 1899 by sociologist Thorstein Veblen in his book *The Theory of the Leisure Class*. For example, in a video for *GQ*'s series "Most Expensivest Shit," rapper 2 Chainz tries out a $5,000 luxury toothbrush, one of the most costly toothbrushes in the world. (His reaction? "Tastes like titanium." [*GQ*, 2014])

TRY IT!

Pick an object you frequently use—an article of clothing, a toothbrush, your phone. Then choose a service you take advantage of, such as getting your hair cut, getting around town, or eating out. Research the most and least expensive of these goods and services, and provide specific examples.

Where do your consumption practices fall in relation to these extremes? Where do your practices fall compared to those of your friends?

FREQUENTLY USED OBJECT

_____

MOST EXPENSIVE	LEAST EXPENSIVE
_____	_____
$	$

FREQUENTLY USED SERVICE

_____

MOST EXPENSIVE	LEAST EXPENSIVE
$ _____	$ _____

THINK ABOUT IT

What geographic areas tend to have more narrow ranges of inequality, with most products priced similarly, and what areas tend to have wider ranges, with extremes on both ends? What do these differences imply about income inequality in the United States?

    If rich people are spending their money on titanium toothbrushes, does that mean inequality is harmless or even good by stimulating the economy? Along these lines, is it better or worse for society as a whole if one rich person buys a $5,000 toothbrush meant to last a lifetime, or buys a 10-cent toothbrush for every day of the year, keeping the difference in the bank?

SOCIOLOGY ON THE STREET

Someone you would consider rich may not think of themselves that way. If the label "rich" is relative, what does it mean to be rich? Watch the Sociology on the Street video to find out more: **digital.wwnorton.com /youmayask6core**.

WANT MORE PRACTICE?

Complete the InQuizitive activity for this chapter at digital.wwnorton.com /youmayask6core

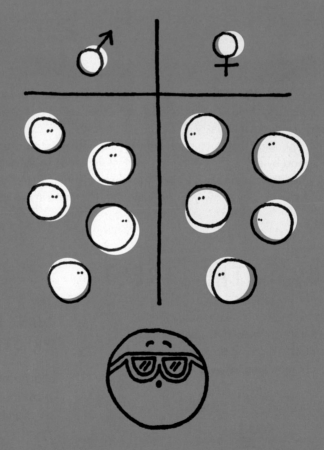

HOW DO WE INVESTIGATE INEQUALITY
BETWEEN MEN AND WOMEN
WITHOUT REINFORCING BINARY
THINKING ABOUT GENDER?

Gender

When most people go to the bathroom, it's because they need to pee. But for one student in Bentonville, Arkansas, nature calling wasn't such a simple affair:

> I got followed into a women's bathroom today at school.
>
> The man who followed me, a teacher, called, "Young man!" in a shocked voice.
>
> I turned, and immediately realized what had happened. Several people, men and women alike, have confronted me in bathrooms at this point. Most apologize after they hear my voice; I look fairly androgynous, but I sound like a Disney princess.
>
> I am a trans man, but in situations like this, people assume that I am a trans woman, which can be dangerous for me.
>
> In most cases, I back down quickly when someone confronts me, for my own safety. Today, though, I saw an opening to fight back a little. I'm on my period, and I was heading into the bathroom to change my pad. I had a new, unopened one in my pocket.
>
> "Sorry," I said to the teacher in my highest pitched voice. I pulled out the pad and held it up. "I was just going to change my pad."
>
> He muttered something about me looking like a boy and backed up a little.
>
> I was secretly flattered, because I *want* to look like a boy, but obviously that isn't a good reason to chase someone into a bathroom. "No worries," I said in my sugarplum fairy voice. "Sometimes it's easy to forget that there are a lot of ways to look like a girl." Then I marched into a stall and shut the door.

The boy's name is Elliot Jackson. In this story, the teacher makes a key assumption: that because Elliot looks like a boy, he shouldn't be going into

a girl's bathroom. But when Elliot turns up the volume on his self-described "sugarplum fairy" voice and brandishes a feminine hygiene product, the teacher becomes confused—Elliot is blurring lines that many often take for granted.

The teacher isn't unlike many Americans, who in recent years have been increasingly exposed to the visibility of transgender people in our society. Television and film are depicting more transgender characters; two notable TV shows are *Transparent* and *Orange Is the New Black*. Transgender rights issues are also in the news, for the very reason Elliot experienced so much discomfort: a series of "bathroom bills," passed in states like North Carolina and considered in others ranging from Florida to Washington, have made the bathroom the site of gender policing.

A protester expresses concern over North Carolina's House Bill 2, also known as the "Bathroom Bill," which requires transgender people to use public restrooms that match the sex on their birth certificate.

But the bathroom is just one of the many places where gender matters. Gender is a powerful force in our lives, determining everything from the toys we play with to our opportunities in school and at work. A key idea we'll encounter throughout this chapter is that that gender is composed of a set of traditions, assumptions, and expectations—and that together, these norms are a key building block of society. As Elliot opened the door to the girls' restroom he violated one of these unspoken rules.

We'll hear more from Elliot later in this chapter, but first we have an important sociological task at hand. What is the difference between gender, sex, and sexuality? While intertwined, these concepts are analytically distinct.

Let's Talk about ~~Sex~~ Gender

SEX

the perceived biological differences that society typically uses to distinguish males from females.

Sex is typically used to describe socially accepted, perceived biological differences that distinguish males from females. There are many biological differences between humans; society gives some of those biological differences gendered labels that, in turn, come to define sex categories such as "male" and "female." The categories are seemingly rooted in biological reality, since these biological differences tend to cluster into two groups according to sex chromosomes a person has. Gender denotes a social position—namely, the set of social arrangements that are built around normative sex categories. Sexuality, meanwhile, refers to desire, sexual preference, and sexual identity and behavior.

No one disputes that many biological differences exist between those who typically identify as men and those who see themselves as women—nor that many of those distinctions can be traced to the difference between having two copies of the X chromosome (XX, female) versus one X and one Y (XY, male). However, what we make of those differences—which themselves are a matter of averages, not absolutes—does not inevitably arise out of the biological. Like many social categories we've discussed, gender is one set of stories we tell each other to get by in the world. It's a collectively defined guidebook that humans use to make distinctions among themselves, to comprehend an otherwise fuzzy mass of individuals. But the gender story can change, and we'll see how it has done so throughout history and across cultures.

And if gender is a human invention, we have to take it as seriously as we would any other institution, as Judith Lorber argues. In *Paradoxes of Gender* (1994), Lorber claims that gender is a social structure that "establishes patterns of expectations for individuals, orders the social processes of everyday life, is built into the major social organizations of society, such as the economy, ideology, the family, and politics, and is also an entity in and of itself." Although gender is a social construction, it matters in the real world, organizing our day-to-day experiences and having profound and unequal consequences for the life chances of people—effects that themselves vary by the other social categories to which an individual belongs. Gender is not just about people—institutions, occupations, and even nouns (in some languages) can all be gendered. In this way gender ultimately embodies power struggles and how they organize daily life, from household economies and wage labor to birth control and babies' names.

GENDER

a social position; behaviors and a set of attributes that are associated with sex identities.

SEXUALITY

desire, sexual preference, and sexual identity and behavior.

Sex: A Process in the Making

We make sense of much variation between men and women by referring to their assumed biological differences. Such differences can range from the behavioral consequences of hormones (such as premenstrual syndrome), relative physical strength of bodies (for example, women's gymnastics emphasize balance, and men's upper-body strength), brain architecture (men supposedly being more left-brain dominant), and chromosomes (XX or XY). But in so doing, we tend to miss a crucial link between nature and nurture. The study of gender boils down to seeing how the two spheres, nature and nurture, overlap, penetrate, and shape each other. The biological world of sex and bodies does not exist outside of a social world, and the social world of human beings is always made up of human bodies. Studying the links between the two allows us to see the social construction of both gender and sex.

SEEING SEX AS SOCIAL: THE CASE OF NONBINARY INDIVIDUALS

Bodies are, so we often think, natural, God- (or evolution-) given, sacred, hardwired. Human babies come equipped with a set of male or female organs, hormones, and chromosomes—what we might call "the plumbing" determined by our DNA. We usually think of sex as an either/or binary. You're either male or female. But in fact, there are some exceptions and blurred lines that have led sociologists to view this model of "natural" sex as more of an approximation than an absolute.

Take, for example, people who are born with both male and female genitalia, neither, or ambiguous ones that do not conform to the gender binary (also included in this category are those whose sex chromosomes do not match up in the normative way to outward sexual appearance). The medical industry used to refer to such children as "intersex"; these days, they refer to a "disorder of sex development." Most doctors today still typically recommend secretive surgery during infancy to make nonbinary children conform to a preconceived notion of what "unambiguous" genitalia should look like. About 90 percent of these surgeries reassign an ambiguous male anatomy into a female one because, in the disquieting (and offensive) phrase of the surgical world, "It's easier to make a hole than build a pole" (Hendricks, 1993).

These practices have lately come under scrutiny, however. Founded in 1993, the Intersex Society of North America (ISNA) was succeeded in 2008 by the Accord Alliance, which has the same mission: to reduce embarrassment and secrecy over intersex conditions. Sociologist Georgiann Davis (2015), who herself learned only as an adolescent that she had male sex chromosomes, has documented rifts within the intersex community. Some members find comfort in the scientific, medicalizing label of "disorder of sex development," while many resist the assertion that their identity is a pathology rather than just part of natural human variation.

The Intersex Society of North America published this satirical postcard to argue that medical standards for "normal" baby genitalia are arbitrary and result in unnecessary surgeries.

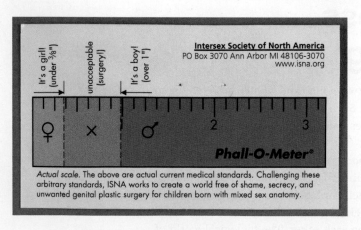

Intersex Society of North America
PO Box 3070 Ann Arbor MI 48106-3070
www.isna.org

It's a girl! (under ⅜")

unacceptable (surgery!)

It's a boy! (over 1")

Phall-O-Meter®

Actual scale. The above are actual current medical standards. Challenging these arbitrary standards, ISNA works to create a world free of shame, secrecy, and unwanted genital plastic surgery for children born with mixed sex anatomy.

In 2017, three former surgeons general of the United States co-authored a report advocating the reconsideration of sex-reassignment surgeries (Elders et al., 2017). They wrote about the emerging medical consensus that "children born with atypical genitalia should not have genitoplasty performed on them absent a need to ensure physical functioning." They also point out that such surgeries became commonplace in the United States only as of the 1950s, and that many other governments and the World Health Organization have already called for a moratorium.

Most people think—if they think about it at all—that the medical construction of sex applies only to a handful of individuals. However, Brown University biology researcher Anne Fausto-Sterling (2000) estimates that the number of deviations from the binary of male or female bodies may be as high as 2 percent of live births, and the number of people receiving "corrective" genital surgery runs between 1 and 2 in every 1,000 births. Just to give you an idea of what 2 in 1,000 looks like, there are now about 2 in 1,000 children born with trisomy 21 (also known as Down syndrome), a biological condition that is more visible than nonbinary sexual identity. Activists argue that social discomfort and fear of difference, rather than medical necessity, may be what pushes parents and surgeons to the operating table.

SEXED BODIES IN THE PREMODERN WORLD

Whereas today we often operate on a mutually exclusive two-sex model of human body types, a lesser known but equally plausible "one-sex" model dominated Western biological thought from the ancient Greeks until the mid-eighteenth century (Laqueur, 1990). In the old one-sex way of thinking, there was only one body (a male body) and the female body was regarded as its inversion—that is, as a male body whose parts were flipped inside rather than hanging on the outside. People believed that women were a lesser but not so radically different version of men—an illustration of how social power relationships can shape scientific belief.

Not until the two-sex model of human bodies gained ground did women and men become such radically different creatures in the popular conception. Incidentally, historian Thomas Laqueur shows us that this differentiation of bodies prompted changes in ideas about the female orgasm. In the one-sex model, it was believed that both a man's orgasm and a woman's were requirements for conception. (There is new medical evidence, in fact, that dual orgasms do increase the chance of conception.) But in the mid-1800s, around the time the two-sex model was gaining ground, female orgasm was considered unnecessary. Whereas seventeenth-century (female) midwives advised would-be mothers that the trick to conceiving lay in an orgasm, nineteenth-century (male) doctors debated whether female orgasm was even possible.

CONTEMPORARY CONCEPTS OF SEX AND THE PARADOXES OF GENDER

The point of all this sex talk is to challenge our tendency to think of bodies as wholly deterministic. This is, to acknowledge that our understandings of, categorizations of, and behavior toward bodies are not set in stone. We're pushing back against essentialist arguments, which explain social phenomena in terms of natural ones. Essentialist thought often relies on biological determinism—which assumes that what you do in the social world is a direct result of who you are in the natural world. If you are born with male parts, essentialists believe, you are essentially and absolutely a man, and you will be sexually attracted to women only, as preordained by nature. As we've seen, medical experts have maintained the ideal of a dimorphic or binary (either male or female) model of sex by tweaking babies who blur the boundaries. The trick is to recognize that the very boundaries separating male and female bodies are themselves contested.

This is not to say that there is no reality to biology. Simply put, rather than believing biology comes before or dictates behavior, sociologists now think of the nature–behavior relationship as a two-way street. Feminist philosopher Elizabeth Grosz (1994) proposes that we view the relationship between the natural and the social (in this case, sex and gender) as akin to a Möbius strip. The Möbius strip is an old math puzzle that looks like a twisted ribbon loop, yet it has just one side and one edge. Biological sex makes up the inside of the strip, whereas the social world—culture, experience, and gender—make up the outside. But as happens in the contours of gender, the inside and outside surfaces are inseparable. In thinking or talking about sex and gender, we often switch from one to the other without even noticing that we've changed our focus (Fausto-Sterling, 2000).

Gender: What Does It Take to Be Feminine or Masculine?

I hope you can now see that biological sex does not exist in the world in some fixed, natural state—medically or historically. Our next step is to trace the different social meanings that humans have taken from sex. There are many historical and cross-cultural meanings, roles, and scripts for behavior that we like to think correspond to more or less fixed biological categories. This complete set of scripts is what sociologists refer to as gender, which,

---

ESSENTIALIST

arguments explaining social phenomena in terms of natural, biological, or evolutionary inevitabilities.

---

broadly speaking, has divided people, behaviors, and institutions into two categories: masculine and feminine.

Much like sex difference, people tend to think that gender difference is a natural cleavage between two static groups. Why do men tend to fight one another, dominate the natural sciences, and outnumber women in the top political and executive offices? Why do women tend to stay more connected with their families and outnumber men in occupations that involve caring for others? The short (essentialist) answer is that men and women are naturally (that is, biologically) different, so they behave differently.

The longer, sociological answer is that gender differences are much more fluid and ambiguous than we may care to admit. Take Elliot, for example. He called himself androgynous—that is, his gender presentation didn't seem obviously masculine or feminine. Elliot's gender cannot be easily explained by appeals to natural differences between the sexes, because it doesn't neatly sync up to a clearly male or clearly female outward appearance or biology (remember the pad he was waving?). And it's not just Elliot who challenges the essentialist argument: There are plenty of people whose behaviors, occupations, and roles don't correspond neatly to essentialist expectations.

But just because gender isn't tied to some fixed biological reality doesn't mean it doesn't have real consequences. Gender establishes patterns of expectations for people, orders our daily lives, and is one of the fundamental building blocks of society—law, family, education, the economy, everything. The process of forming a gendered identity starts before a person is even born, as soon as the sex of a fetus is estimated or even earlier in the case of in vitro fertilization (IVF), where the sex chromosomes can be identified even before implantation into the uterus. Through socialization and personality development—such as being steered to certain games, ways of dressing, and even how they are named—many children acquire a gendered identity that, in most cases, reproduces the attitudes and values of their society. We impose rigid boundaries to maintain a gender order, but if we look at how gender systems vary, we can expose those boundaries as social constructions. Our challenge is to identify those systems and inequalities without unwittingly reproducing the binary thinking we wanted to investigate in the first place.

MAKING GENDER

To start seeing gender sociologically, it helps to look at other cultures. While many Western cultures divide the world neatly between men and women with correspondingly clear male and female bodies, that's not so for the Navajo society of Native Americans. In Navajo tribes, there are not two but three genders: masculine men, feminine women, and the *nadle*. The *nadle* might be born with ambiguous genitalia or they may declare a *nadle* identity

ANDROGYNOUS

neither masculine nor feminine.

Would a member of the *hijra* community define their identity in terms of gender, or something else?

later on regardless of genitalia. The *nadle* perform both masculine and feminine tasks and dress for the moment, according to whatever activity they're doing. Although they are often treated like women, they have the freedom to marry people of any gender, "with no loss of status" (Kimmel, 2000).

Nadles are not the only non-Western examples of nonbinary gender configurations. Serena Nanda (1990) and Gayatri Reddy (2005) studied *hijras* in India, a group that is often included in textbooks like this one to stand in as proof that a binary either/or gender system is not so natural after all. From Reddy we know that "hijras are phenotypic men who wear female clothing and ideally, renounce sexual desire and practice by undergoing a sacrificial emasculation—that is, an excision of the penis and testicles." To our Western ears, then, these men who renounce manhood but who are not women are actively staking a claim for a third gender. But Reddy goes on to develop the definition of what it means to be a *hijra*, which includes behaviors that may have little to do with gender: dedication to the goddess Bedhraj Mata, conferring fertility to newlyweds and newborns; a sometimes reluctant, sometimes quite dedicated entry into prostitution; communal living; self-sacrifice; and poverty. Thus *hijra* identity is a master status, but it is not experienced by the *hijras* as a fight for turf between gender categories. Like the Brazilian *travesti* described by Don Kulick (see page 328), the *hijras* may have few qualms about the balance they've struck between gender and sexuality but more fears about the way their poverty and stigmatization will shape their chances in life.

And as the chapter-opening story about Elliot shows, the growing prominence and social awareness of transgender people in our own society helps

TRANSGENDER

describes people whose gender does not correspond to their birth sex.

us break out of binary thinking about gender, since we tend to assume everyone is cisgender. Amos Mac, a man who was raised as a girl, came to talk to my class about his experience and the magazine *Original Plumbing* he publishes for a transgender audience. When I asked Mac whether he felt more like a man now than he had before he transitioned, he politely corrected my assumptions about a before–after binary gender narrative: "I actually don't even know like what is a man supposed to feel like. You know, I don't really know what a woman is supposed to feel like....I feel comfortable in my skin right now, and I feel comfortable the way people are perceiving me, like strangers on the street. I don't care if people know my history" (Conley, 2014e). He allows that some of his friends may have been "born knowing something [was] terribly wrong...or that they [didn't] feel comfortable in their body," but that there is no common narrative that encompasses what being transgender is like. Mac explained, "I don't think there's a right or wrong way to have this experience, and I don't think it's necessarily a fun one that you would want to go out of your way to have." The search or space for a third gender may be limited by our assumption that gender should be a fixed category. Rather, gender is a spectrum that is constantly changing, along which individuals may change positions over their lifetimes.

DIGITAL.WWNORTON.COM/YOUMAYASK6CORE

To see my interview with Amos Mac, go to
digital.wwnorton.com/youmayask6core

CISGENDER

describes people whose gender corresponds to their birth sex.

GENDER DIFFERENCES OVER TIME

Within a two-gender system, there is enormous variation in what counts as a "good" or "bad" man or woman. For example, how the ideal man or woman looks is itself historically contingent. Specifically, ideal feminine beauty has been a continuous site of change and contestation. Look at the seventeenth-century Rubenesque women, the voluptuous beauties who by today's high-fashion standards are simply overweight. In traditional economies where food was scarce, a plump woman was a sign of good health, wealth, and attractiveness. The long-standing preference for a robust female body has changed as industrialized societies moved from relative scarcity of food to plenitude. Today the cheaper foods are the fattening ones, and

WELCOME TO ZE COLLEGE, ZE

In recent years, colleges have been at the forefront of a startling reconfiguration in our society: to acknowledge difference in sex/gender identities. For example, the State University of New York, one of the nation's largest state university systems, is now allowing students to choose among seven gender identities, including "trans man," "questioning," and "genderqueer." Harvard University and other colleges now have spaces to allow for students to choose gender-neutral pronouns such as "they" or "ze." Other schools are sure to follow.

Some activists and scholars advocate a brand-new pronoun like "ze" or "hir," noting that "Ms." seemed strange when first proposed but now is commonly accepted. Others suggest "they" as a non-gender-binary pronoun since it is already readily used informally when English speakers want to avoid gendering the person to whom they are referring. That's the easy part: Just like choosing how we want someone to address us—by our formal name or nickname or initials—it seems pretty straightforward to allow individuals to select their own pronouns. Elliot Jackson, whose story opened this chapter, for example, asked that I use "he" and "him" pronouns. Easy enough!

But implementing a fluid notion of sex/gender identity for all aspects of the life in the institution we call campus gets trickier. Take campus sports. The policy of Bates College—which is fairly typical of the NCAA—is the following: "A transgender student athlete should be allowed to participate in any sports activity so long as that athlete's use of hormone therapy, if any, is consistent with the National College Athletic Association (NCAA) existing policies on banned medications." So, a female transitioning to male via hormone therapy is no longer allowed to com-

only people with enough disposable income can afford gym memberships and healthy diets. So we can see dominant or "emphasized" definitions of femininity—as embodied by looks—are always undergoing change, from the hysterical Victorian housewife to the sporty working girl of the 1980s to today's heroically perfect-in-all-ways supermom.

This might be an easy point to grasp about femininity, but most people think that masculinity is less subject to such trends and fashions. It is always

pete on women's teams but is eligible for men's sports. Ditto in reverse. The NCAA does not have a mandatory policy for transgender inclusion, but the association claims that inclusion is an "NCAA value" (Kanno-Youngs, 2015).

The final frontier—on college campuses and beyond—has been locker room and bathroom policy (as illustrated by Elliot's story at the start of the chapter). Under the Obama administration, federal agencies ruled that Title VII of the Civil Rights Act bars discrimination based on gender identity (i.e., not just sex), and this includes locker room segregation. The Office of Civil Rights of the US Department of Education rejected the argument that female students needed to be protected from being seen naked by an individual who was biologically still male. Separate facilities for transgender individuals were seen to be akin to the now unacceptable "separate but equal" racial policy of Jim Crow (Phillips, Wagner, and Clifton, 2015).

However, as political winds shift, this is likely not the final word on the matter. For example, the Obama administration interpreted Title IX of the Education Amendments of 1972 as applicable to discrimination based on gender identity, not just gender. This meant that schools that received federal funding (i.e., almost all) were required to make nonseparate restrooms accessible to transgender individuals based on their self-identified gender. However, the Trump administration withdrew that guideline in its first year in office. (To be continued…)

Harvard University student Schuyler Bailar became the first openly transgender swimmer to compete in the NCAA.

harder to denaturalize the dominant category; being the norm, it often is invisible. Among social categories, those who go unquestioned tend to be most privileged. In *Manhood in America* (1996), Michael Kimmel traces the development of hegemonic masculinity in the West and finds that in the eighteenth century, the ideal man was not associated with physical fitness, money-making endeavors, or sports. Business endeavors were the boorish concerns of the rude trade classes, and physical strength undermined one's

HEGEMONIC MASCULINITY

the condition in which men are dominant and privileged, and this dominance and privilege is invisible.

gentlemanly dispositions. Ideal masculinity in the 1700s went hand in hand with kindness and intellect, and preferably a little poetry, a very different image from the modern-day ideal of the "man's man."

Erving Goffman (1963) describes the masculine ideal of mid-twentieth-century America as a young man who is "married, white, urban, northern, heterosexual, Protestant, father, of college education, fully employed, of good complexion, weight and height, and [with] a decent record in sports." Today that definition has blossomed to include more forms of dominant masculine identity, including, for example, the metrosexual male, who, as described by sociologist Kristen Barber (2016), is typically white and wealthy and spends a large amount on grooming activities, such as manicures, that would have been considered feminine in the days Goffman was writing. Likewise, many men form what Tristan Bridges and C. J. Pascoe (2014) call "hybrid masculinities," in which, for example, young white men may try to distance themselves from hegemonic masculinity and instead adopt aspects of, say, African American masculinity. They argue, however, that this cultural practice can often serve to conceal inequalities. That is, when white adolescent males emulate the masculine practices of other marginalized groups, this dynamic serves to mitigate the real power differences and inequalities between the groups, as if race and class were just a style one can adopt or shed as one likes (Aboim, 2016, p. 56; Bridges & Pascoe, 2014).

Over time and from place to place, our ideas about gender are fluid, changing, and context specific. Many of the differences we observe between men and women do not have much to do with individual gender differences at all; instead, the behaviors arise as a result of the different positions men and women occupy. Sociologist Cynthia Fuchs Epstein (1988) calls these "deceptive distinctions," which grossly exaggerate the actual differences between men and women. To illustrate her idea, here's a quick thought experiment: Imagine a doctor and a nurse. Did you assume the doctor was

Kristen Barber would probably call the man on the left "metrosexual": he's getting his eyebrows trimmed in a new men-only section of a beauty salon. The white men on the right exemplify what Tristan Bridges and C. J. Pascoe call "hybrid masculinities": By emulating black hip-hop culture, they erase their own privilege.

a man and the nurse a woman? Tied up in these expectations are other ideas, such as the stereotypes that women are nurturing or men are analytical. The main difference between the doctor and nurse isn't gender, but power and social status.

Baby names provide another striking example of the power of gender. Analysis by Stanley Lieberson, Susan Dumais, and Shyon Baumann (2000) shows that names flow from male to androgynous to female but never in reverse. If you know, for example, a male named Kim, chances are he was born before 1958, the year that Hitchcock's movie *Vertigo* was released, making the actress Kim Novak a household name. The number of boys named Kim dropped to almost zero the next year. Carol, Aubrey, and Lindsay (and many others) started as male names and became feminized. The fact that there is a one-way flow from male to female in terms of child naming tells us something about gender norms and inequality—namely that masculinity dictates culture more than femininity does, or that femininity seeks to emulate masculinity while the masculine sees the feminine as corrupting.

While gender norms can be fluid, one constant across time and culture is that men have held more power than women—a question that has preoccupied social thinkers for a long time.

Theories of Gender Inequality

In applying our sociological imaginations to sex and gender, we will come across many feminist strands of thought. Feminism was at first embraced as a social movement to advocate for women's right to vote; it later became a consciousness-raising movement to get people to understand that gender is an organizing principle of life. One of the central ideas of this "second-wave" of feminism is that gender is important because it structures relations between people. Further, as gender shapes social relations, it does this on unequal ground, meaning gender is not just an identifying characteristic (I'm a guy, you're a girl), but embodies real powers and privileges. Regardless of which "wave" we are talking about, the basic idea behind feminism is that women and men (and other-gendered people) should be accorded equal opportunities and respect.

At the start of the second wave of the feminist movement in the 1960s, theorists scrambled to find an answer to the "woman question": What explains the nearly universal dominance of men over women? What is the root of patriarchy, a system involving the subordination of femininity to masculinity?

FEMINISM

a social movement to get people to understand that gender is an organizing principle in society and to address gender-based inequalities that intersect with other forms of social identity.

PATRIARCHY

a nearly universal system involving the subordination of femininity to masculinity.

A "new feminist bookstore" in 1970. Feminist theorists since the 1960s have attempted to explain the "woman question"—why, in nearly every society across history, do men hold power over women?

RUBIN'S SEX/GENDER SYSTEM

Anthropologist Gayle Rubin was one of the first in a long string of thinkers to argue that the nearly universal oppression of women was in need of an explanation. In the field of anthropology, most scholars studying societies around the world had previously assumed that, since women's subordination occurred everywhere, it must be fulfilling some societal function, and therefore was less interesting than other possible research questions.

Rubin (1975) challenged this notion and proposed the "sex/gender system." In this system, the raw materials of biological sex are transformed through kinship relations into asymmetrical gender statuses. She used the structural perspective of Claude Lévi-Strauss, a French anthropologist, to suggest that because of the universal taboo against incest—fathers and brothers cannot sleep with their daughters and sisters—women who start out belonging to one man (their father, or their brother if Dad dies) must leave their families of origin and go belong to another man (their husband). Women are treated like valuable property whose trade patterns strengthen relations between families headed by men. Indeed, traditional wedding vows in the West involve a commitment to "obey" only on the part of the wife. This "traffic in women" gives men certain rights over their female kin. The resulting sex/gender system, she argued, was not a given; it was the result of human interaction.

Rubin's theories made waves. Feminist thinkers widely agreed that the task at hand was to explain universal male dominance. Why were women—despite a few token examples of matrifocal or women-led tribes in the anthropology books—typically on the bottom of stratification systems?

That is, why did women in almost every society seem to get short shrift? Anthropologist Michelle Rosaldo (1974) answered that it must be women's universal association with the private sphere. Because women give birth and then rear children, they become identified with domestic life, which universally is accorded less prestige, value, and rewards than men's public sphere of work and politics. Meanwhile, anthropologist Sherry Ortner (1974) claimed that women are identified with something that every culture defines as lower than itself—nature. A woman, she reasoned, comes to be identified with the chaos and danger of nature because of bodily functions like lactation and menstruation (what feminist philosopher Simone de Beauvoir [1952] has poetically described as "woman's enslavement to the species").

PARSONS'S SEX ROLE THEORY

Ortner's answer to the woman question sounds like a plausible one. But note the binary logic at work in these anthropological accounts: culture versus nature, private versus public, man versus woman. The world is less rigid, more nuanced, and certainly less easily molded into binary categories than these theories suggest. What underlies these theoretical approaches is structural functionalism. This theoretical tradition dominated early anthropological thought, and it assumed that every society had certain structures (such as the family, the division of labor, or gender) that existed to fulfill some set of necessary functions. Talcott Parsons, an influential American sociologist of the 1950s, offered a widely accepted functionalist account of gender relations. According to Parsons's (1951) sex role theory, the heterosexual nuclear family is the ideal arrangement in modern societies because it fulfills the function of reproducing workers. With a work-oriented father in the public sphere and a domestic-oriented mother in the private sphere, children are most effectively reared to be future laborers who can meet the labor demands of a capitalist system. Sex, sexuality, and gender are taken to be stable and dichotomous, meaning that each has two categories. Each category is assigned a role and a given script that actors carry out according to the expectations of those roles, which are enforced by social sanctions to ensure that the actors do not forget their lines. Women and men play distinctive roles that are functional for the whole of society; a healthy, harmonious society exists when actors stick to their normal roles. Generally speaking, according to Parsons, social structures such as gender and the division of labor are held in place because they work to ensure a stable society.

Functionalism was a hit in the 1950s; after all, it makes intuitive sense (and feels very satisfying) that there must be a good reason for the way things happen to be, some internal logic—at least to those who benefit from those arrangements. But Parsons's sex role theory falls short on several points. For starters, the argument is tautological—that is, it explains the

STRUCTURAL FUNCTIONALISM

theoretical tradition claiming that every society has certain structures (the family, the division of labor, or gender) that exist to fulfill some set of necessary functions (reproduction of the species, production of goods, etc.).

SEX ROLE THEORY

Talcott Parsons's theory that men and women perform their sex roles as breadwinners and wives/mothers, respectively, because the nuclear family is the ideal arrangement in modern societies, fulfilling the function of reproducing workers.

In the 1950s, Talcott Parsons advanced the idea that the nuclear family effectively reared children to meet the labor demands of a capitalist system.

existence of a structure in terms of its function, essentially claiming that things work the way they do because they work. In explaining a phenomenon in terms of its function, functionalists relied on the presumption that the need for the function preexists the phenomenon, a tricky leap of logical faith. Furthermore, they glossed over the possibility that other ways of organizing society than the structures of gender, race, and so on that we currently have could fulfill the same functions, and the question of whether those "functions" are themselves legitimate or ultimately desirable. The end result is a theory that tends to justify or "naturalize" existing forms of social relationships—such as gender wage gaps and the unequal division of labor in housework.

The functionalist sex role story also does not explain how and why structures change throughout history. If traditional husband-and-wife sex roles were so functional, why did they change drastically in the 1970s? In Parsons's account, gender roles appear to be a matter of voluntarism, as if women and men choose, independently of external power constraints, to be housewives and breadwinners, respectively. Of course, this is a myopic view of roles since women of color and immigrant women had always worked outside the home at high rates. Finally—and this is a problem with much early feminist thinking as well—sex and gender are regarded as being composed of dichotomous roles, when these categories are, in fact, fuzzy, flexible, and variable in combination with other social positions, such as race and class.

PSYCHOANALYTIC THEORIES

Where functionalism focuses too much, perhaps, on society as a whole, Freudian theorists have provided an overly individualistic, psychoanalytic account of sex roles. The father of psychoanalysis, Sigmund Freud (1856–1939), famously quipped, "Anatomy is destiny." Although biological determinism plays a major role in Freudian theory, so does the idea that gender develops through family socialization.

According to Freud's developmental psychology, girls and boys develop masculine and feminine personality structures through early interactions with their parents. Boys, so the story goes, have a tormented time achieving masculinity because they must resolve the Oedipal complex. In this stage of development, around age three, every normal boy experiences heterosexual love for his maternal figure. But he soon realizes that he will be castrated (psychologically, not literally) by his father if he continues to fancy his mother. To resolve this Oedipal conflict, the boy rejects his mother, in turn emulating his emotionally distant father and developing rigid ego boundaries.

In Freud's view, girls do not experience quite the same resolution to their analogous "penis envy." When a little girl realizes that she lacks the plumbing to have sexual relations with her mother, she experiences penis envy toward her father, according to Freud. However, she ultimately realizes that one day she too can have a baby, thus providing feminine gratification. Girls end up identifying with their mothers, growing up with less rigid ego boundaries and more easily connecting with others than boys do.

Many thinkers have picked up and used Freudian ideas to theorize differences between men and women. For example, feminist psychoanalyst Nancy Chodorow (1978) modified Freud's theory to answer a particular version of the woman question: Why are women predominantly the caregivers? She reasoned that parents' unequal involvement in child rearing was a partial cause for the universal oppression of women. Her answer was that mothering by women is reproduced in a cycle of role socialization, in which little girls learn to identify as mothers and little boys as fathers (and workers outside the home). Chodorow concluded that egalitarian relations between the sexes would be possible if men shared the mothering.

Before you see a psychoanalyst to discuss your own penis envy, consider the assumptions underlying these psychological accounts. For starters, Freud's theories lacked empirical evidence and assumed a heterosexual, two-parent nuclear family. Since only about 48 percent of American children live in households headed by two married biological parents (US Census Bureau, 2016a), most of you reading this probably did not grow up in the nuclear family setting that Freud and Chodorow describe. Sociologist Carol Stack underscored this point in her ethnography of a poor black community. In *All Our Kin* (1974), she finds that the division between male and female roles and attitudes is not as clear-cut as Freud or Chodorow would have us think. In the community Stack studied, caregiving was a valued responsibility for both men and women. Moreover, these early psychoanalytic theorists took for granted a binary sex/gender system, whereas we now know that those categories are much more fluid in real life.

CONFLICT THEORIES

By the 1980s, another wave of thinkers began to tackle an issue missing from earlier discussions of the woman question: power. Conflict theorists mixed old-school Marxism with feminism to claim that gender, not class, was the driving force of history. Socialist feminists, also known as radical feminists, claimed that the root of all social relations, including relations of production, stemmed from unequal gender relations.

Many conflict theorists argue that patriarchal capitalists benefit through systems that subordinate women. For instance, many cotton mills in the nineteenth and early twentieth centuries hired young, unmarried women and required them to live in company boarding houses in order to regulate their behavior.

Economist Heidi Hartmann (1981) and legal theorist Catharine Mac-Kinnon (1983), for example, both analyzed how capitalism combines with patriarchy to make women economically dependent on men's incomes. This means that in a capitalist society, women have a disadvantaged position in the job market and within the family. Capitalists (that is, men) in turn reap all the benefits of women's subordination. When women are subordinate, men benefit. To radical feminists, gender inequality is first and foremost about power inequalities, and gender differences (as in personality development) emerge from there. However, what lies beneath these conflict theories is yet another variant of essentialism. Radical feminists basically posited that the world is divided into two groups: men and women (red flag number one). These two groups are necessarily pitted against each other in a struggle for resources (red flag number two). Again, men and women are reduced to automatons in a static battle, which women always lose.

"DOING GENDER": INTERACTIONIST THEORIES

A social theory is useful only if it helps you understand the social world in which you live. For example, Candace West and Don H. Zimmerman argue in their influential article "Doing Gender" (1987) that gender is not a fixed identity or role that we take with us into our interactions. Rather, it is the product of those interactions. In this framework, gender is a matter of active doing, not simply a matter of natural being. To be a man or a woman, they argue, is to perform masculinity or femininity constantly. In this social constructionist theory, gender is a process, not a static category.

The "doing gender" perspective is rooted in Erving Goffman's drama-turgical theory, symbolic interactionism, and ethnomethodology. (See discussion of these terms in Chapter 4.) That is, West and Zimmerman argue that people create their social realities and identities through interactions with one another. Unlike the structural functionalists, psychoanalysts, and conflict theorists, however, social constructionists view gender roles as having open-ended scripts. Perhaps individuals come to the stage situated differently according to their place in power hierarchies or personality development, but their lines and gestures are far from being predetermined. Regardless of social location, individuals are always free to act, sometimes in unexpected ways that change the course of the play. For example, the presence of openly transgender people has been said to "undo" or "redo" gender by subverting the binary norms (Connell, 2010; Sawyer et al., 2016). But by and large, as a result of doing gender, people contribute to, reaffirm, and reproduce masculine dominance and feminine submissiveness in the bedroom, kitchen, workplace, and so on.

BLACK FEMINISM AND INTERSECTIONALITY

If gender is a performance, it is much more than a set of neutral scripts that we "voluntarily" follow. Our actions are influenced by structural forces that we might not even be aware of, such as class or race privilege. As Patricia Hill Collins (1990) claims, we "do" a lot more than gender; gender intersects with race, class, nationality, religion, and so forth. Black feminists have made the case that early liberal feminism was largely by, about, and for white middle-class women. In trying to answer the woman question and explain women's oppression, early feminists assumed that all women were in the same oppressed boat. In so doing, they effaced multiple lines of fragmentation and difference into one simple category: woman. For example, by championing women's rights to work outside the home in *The Feminine Mystique* (1997), leading second-wave feminist Betty Friedan ignored the experiences of thousands of working-class and women of color who were already working, sometimes holding down two jobs to support their families. Indeed, a third wave of feminism focuses on how the identities surrounding gender, sex, and sexuality intersect with other meaningful social categories like race or class in a process called intersectionality, as articulated by critical race theorist Kimberlé Williams Crenshaw (1989) and sociologist Beatrice Potter Webb back in 1913.

As illustrated by the concept of intersectionality, "woman" is not a stable or obvious category of identity. Rather, women are differentially located in what Collins calls a matrix of domination. A 40-year-old poor, black, straight, single mother living in rural Georgia will not have the same conception of what it means to be a woman as a 25-year-old professional, white, single, lesbian in Chicago. More fundamentally, Collins (1990) argues, black women face unique oppressions that white women don't. For instance, black women experience motherhood in ways that differ from the white masculinist ideal of the family, as "bloodmothers," "othermothers," and "community

INTERSECTIONALITY

the idea that it is critical to understand the interplay between social identities such as race, class, gender, ability status, and sexual orientation, even though many social systems and institutions (such as the law) try to treat each category on its own.

MATRIX OF DOMINATION

intersecting domains of oppression that create a social space of domination and, by extension, a unique position within that space based on someone's intersectional identity along the multiple dimensions of gender, age, race, class, sexuality, location, and so on.

Patricia Hill Collins criticized feminist leaders such as Betty Friedan (pictured in the red dress), Billie Jean King (in tan pants and a blue shirt on the left), and Bella Abzug (in the gray dress, with the hat) for ignoring the experiences of thousands of working-class women and women of color.

othermothers," thus revealing that white masculine notions of the world do not capture daily lived experiences of many black women.

Gender, sexual orientation, race, class, nationality, ability, and other factors all intersect. Just as some women enjoy privilege by virtue of their wealth, class, education, and skin color, some men are disadvantaged by their lack of these same assets. As bell hooks (1984) noted, if women's liberation is aimed at making women the social equals of men, women should first stop and consider which men they would like to equal. Certainly, not all men are privileged over all women. Making universal comparisons of men to women misses these nuances and implicitly excludes marginally positioned people from the discussion. For example, men of color certainly experience oppression that intersects with their gender identities (for example, black men bearing the brunt of police harassment). Even going beyond demographic groups, men who do not embody or perform the dominant form of masculine behavior often suffer for it, occupying a subordinate status (Connell, 1987). Power comes from many different angles; it doesn't sit evenly on a plane for all women. When the black activist Sojourner Truth (1797–1883) asked, "Ain't I a woman?" (1851), she summed up some elusive philosophical questions: What is woman? Who counts as a woman and why?

POSTMODERN AND GLOBAL PERSPECTIVES

As perspectives have expanded and previously rigid categories have begun to crumble, the validity of the woman question is itself now in question. For instance, anthropologist Oyèrónké Oyěwùmí argues that the woman question is a product of uniquely Western thought and cannot be applied to African societies. In *The Invention of Women* (1997), she presents ethnographic research of Yoruban society in West Africa, which she claims was once a genderless society. Among the Yoruba, before the arrival of anthropologists, villagers did not group themselves as men or women or use body markers at all. Rather, they ranked themselves into strata by seniority. When Western feminist scholars arrived on the scene, presuming the preexistence of gender relations, they, of course, found a system of gender. But this system of categories—distinct males and females—indicates a Western cultural logic, what Oyěwùmí has termed *bio-logic*.

Bio-logic runs deep in our cultural experiences and understandings of gender. It acts as a sort of filter through which all knowledge of the world runs, though there may be different ways of knowing outside such a paradigm. But if "woman" is such an unstable, fragmented category—one that is merely "performed" through discourse, as postmodern and queer theorist Judith Butler (2006) suggests—how are we supposed to study it? Feminists must have some sturdy ground on which to unite, build coalitions, and tackle injustice. Philosopher Susan Bordo (1990) provides a pragmatic buoy by arguing that there are hierarchal and binary power structures out there that do still oppress women and are handy when it comes to addressing issues such as the wage gap, eating disorders, or rape.

Growing Up, Getting Ahead, and Falling Behind

In summer 2017, a Google employee named James Damore circulated a memo to his colleagues questioning the pursuit of gender diversity at the company, in which he claimed the real reason that women were underrepresented among coders and in management was due to "biological causes." Perhaps understandably, the memo caused an uproar within the firm, and once it was leaked to the press, it went viral. Upper-level management at Google disavowed the statement, Damore was forced out of the company, and critics referred to the memo as a blatant example of workplace sexism, while others defended his right to air "un-PC" views on free speech grounds.

Damore's memo shone a light on the underrepresentation of women at Google and companies like it, but the tech industry is just the tip of the iceberg when it comes to gender inequality. Fifty-six percent of college students in 2015 were women (National Center for Education Statistics, 2014; Figure 8.1, next page). However, despite their increased enrollment, women remain

SEXISM

occurs when a person's sex or gender is the basis for judgment, discrimination, or other differential treatment against that person.

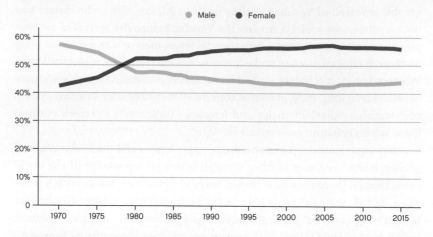

FIGURE 8.1 College Enrollment by Gender, 1970–2015

Male   Female

SOURCE: National Center for Education Statistics, 2017b.

overrepresented in traditionally feminine fields of study: the arts and the humanities. Men also outnumber women at elite colleges, where they are groomed for high-power professions in finance, law, or politics.

Do men disproportionately become computer scientists and financiers and women kindergarten teachers and dental hygienists because they are hardwired to do so? Putting aside Damore's controversial argument, we do know that cis-men and cis-women vary in thousands of major and minor measures. Let's start at the beginning.

GROWING UP WITH GENDER

Gender socialization begins at birth (see Chapter 4) when parents dress their children differently by sex. It continues through school—where boys and girls are treated differently such that boys are called on by teachers more often (Sadker & Sadker, 1994). And it is apparent in family life, where even though women's roles have expanded in the formal labor force, men's have only slightly budged toward more involvement in the domestic sphere (see Chapter 10).

The average male newborn weighs two ounces more than the average female newborn. For infant death rates, however, the disadvantage is tilted against males, who are at higher risk of death than females. Psychologist Carol Gilligan (1982) contends that by adolescence, the disadvantages are stacked against girls, who "lose their voices" as they suffer blows to their self-confidence. Depending on the study, eating disorders may be up to twice as likely to affect adolescent and teen girls compared with their male peers (Pisetsky et al., 2008). More than half of teenaged girls are on diets or think they should be. What's more, girls more frequently report low self-esteem, more girls attempt suicide, and more girls report experiencing some form of sexual harassment in school (Johnson et al., 2016).

Psychologist Christina Hoff Sommers challenges this "girl crisis" in *The War against Boys: How Misguided Feminism Is Harming Our Young Men* (2000). She finds that because they are inadvertently penalized for the short-changing of girls, boys are the ones suffering in education and adolescent health. For example, although females in general attempt suicide twice as frequently as males, boys ages 15 to 19 succeed in killing themselves four times more often than girls. (That's not to mention transgender people, who unfortunately suffer from extraordinarily high rates of suicide. Researchers estimate that between a third and half of trans people attempt suicide at some point in their lives [Virupaksha et al., 2016].) Teenage girls may sneak more cigarettes than boys, but boys are more likely to be involved in crime, alcohol, and drugs and to be suspended from school or drop out. Male teens are also 40 percent more likely to be victimized by violent crime, and that's even taking rape into account (Child Trends Data Bank, 2015).

What accounts for the wide range of statistical differences that exist between men and women? Essentialists might refer to natural sex differences, but as we saw earlier, sociologists are apt to call these same differences "deceptive distinctions," those that arise because of the particular roles individuals come to occupy (Epstein, 1988). Why might some people believe men to be more capable of logic, abstraction, and rationality? Perhaps because their employment possibilities are more likely to include jobs with these demands—a circular argument. Why do women seem to be so good at parenting? Maybe because their weaker employment prospects encourage them to accept domestic roles and rely on a male's salary. Who is more relational and who is more rational? Anthropologists Jean O'Barr and William O'Barr find that the true test of what type of language an individual uses during testimony in court is the witness's occupation, not gender (O'Barr, 1995; Kimmel, 2000). Physicists tend to speak in more abstract terms, teachers in more relational ones, regardless of gender. Once such deceptive distinctions are revealed, it is easy to flip the essentialist rhetoric to see that gender differences may be a product of gender expectations, rather than the cause. Physicists speak in more abstract terms, teachers in more relational ones, because their jobs demand it. The question then shifts to: How and why does gender inequality exist? We need to peel back another layer of the gender onion.

INEQUALITY AT WORK

The past few decades have brought arguably the biggest change in American gender relations—in the world of paid work. Since the 1970s, almost 43 million more women have entered the labor force, from 31.5 million in 1970 to 74.4 million in 2016 (Figure 8.2, next page). Of course, the overall

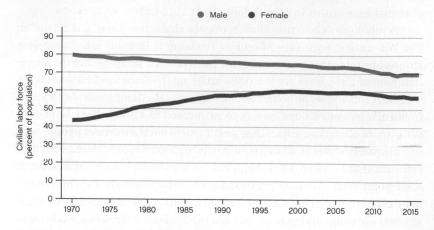

FIGURE 8.2 Increase of Women in the Workforce, 1970–2016

● Male   ● Female

SOURCES: Bureau of Labor Statistics, 2014c, 2017c.

population has increased by 50 percent in that period as well, accounting for part of the rise, but the percentage of women ages 16 to 64 who are in the workforce has risen from approximately 43 to 57 percent in 2016. Because the Great Recession of 2008–9 hit male-dominated industries the hardest (such as construction), it helped accelerate the trend toward gender equality in labor force participation; in fact, there was a brief period at the depth of the economic crisis during which women's totals exceeded those of men in the workforce. However, men have since caught up and reclaimed their majority position: men's labor force participation is at 69 percent compared to women's 57 percent. Because women are disproportionately employed in the public sector, and state governments have yet to recover from the twin challenges of economic woes and ballooning budgets, it may be some time before women once again approach parity (Covert, 2011).

The greater number of women entering the labor force has not catapulted them to equality with their male peers in the workplace. Despite the passage of Title VII of the 1964 Civil Rights Act, which declared it unlawful for employers to discriminate on the basis of a person's race, nationality, creed, or sex, women have continued to fare much worse than their male counterparts in the workforce. Although legally entitled to enter all lines of work, women routinely face sexual harassment, an illegal form of discrimination that runs the gamut from inappropriate jokes on the job to outright sexual assault to sexual "barter," in which sexual favors are extracted from victims under the threat of punishment (Kimmel, 2000). Intended to make people feel uncomfortable and unwelcome, sexual harassment occurs across many settings and in all kinds of relationships; walking down the street and having to listen to "Whooo, baby!" is just one annoying everyday example.

SEXUAL HARASSMENT

an illegal form of discrimination revolving around sexuality that can involve everything from inappropriate jokes to sexual "barter" (where victims feel the need to comply with sexual requests for fear of losing their job) to outright sexual assault.

In the workplace, some argue, sexual harassment is one of the chief ways in which men resist gender equality. In 1982, the US Court of Appeals ruled that sexual harassment is a violation of the Civil Rights Act, because it is a form of discrimination against an individual on the basis of sex. But even though sexual harassment is illegal, it often takes insidious forms not easily detected by everyone or verifiable in court. Most commonly, sexual harassment takes the form of "hostile environments" in which individuals feel unsafe, excluded, singled out, and mocked.

In addition to sometimes having to work in hostile environments, women have consistently been paid less than their male peers, earning about 81 cents to every $1 of a man's wage (Bureau of Labor Statistics, 2017a; Figure 8.3, next page). This overall gap obscures stark racial differences, however. Black women earn 87 cents to the black male dollar (but only 66 cents to the white male dollar). Meanwhile, white women have similar earnings to black men, yet only earn 78 cents to the dollar of white men (Wilson & Rodgers, 2016). (In some large, economically successful cities, however, the gender gap may be shrinking or even reversing among young adults [Bacolod, 2017]). The media touted women's gains in the 1970s and 1980s, when masses of women entered the US labor force. Yet women disproportionately entered the lower rungs of the occupational hierarchy. These feminized jobs, what Louise Howe (1977) calls "pink-collar" jobs, are low-paid service industry jobs. Cleaning buildings, filing papers, and making coffee are hardly what women imagine when they dream of independence. Caring work tends to be feminized as well—think home health aides, nurses, or child-care workers. Meanwhile, "purple-collar" labor—a name giving to occupational niches typically filled by transgender individuals—often involves trans women

It takes a special kind of girl to inspire this kind of trust

As a Stewardess with British Airways, European Division, you'll be in a very responsible position. Our passengers— some very young, some a little apprehensive—will look to you for the attention and reassurance that is such a vital part of your job. A great deal of trust will be placed in you, as the member of the aircrew most in the public eye.

It takes a special kind of girl—perhaps you? You must be aged 19 to 30 and be 5′3″ to 5′9″ tall. Looks are important and you must be physically fit. Adaptability and friendliness are called for as is patience and the ability to work happily with the rest of the crew of a modern airliner.

If you feel you can match these requirements we would like to hear from you— training for these **London** based vacancies starts in early 1975. You'll enjoy an initial

salary of £1517 rising to £1732 after 3 months, but with generous allowances whilst flying, average earnings are in excess of £2200. (Threshold payments are additional to salaries quoted.) The job may be more demanding than most, but offers travel and the freedom from a 9 to 5 rut.

Send us a postcard for an application form to: Assistant Personnel Officer (NV1) (Cabin Services), British Airways European Division PO Box 6, Heathrow Airport London, Hounslow, Middlesex.

(Local interviews will be held at selected centres throughout the UK).

British airways

This British Airways recruiting ad from the 1970s seeks a "very special kind of girl" for a stewardess position: "You must be aged 19 to 30 and be 5′ 3″ to 5′ 9″ tall. Looks are important and you must be physically fit. Adaptability and friendliness are called for as is patience and the ability to work happily with the rest of the crew of a modern airliner."

FIGURE 8.3 Pay Discrepancy Based on Gender

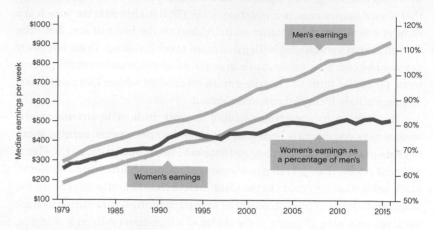

SOURCE: Bureau of Labor Statistics, 2017c.

performing emotional work to reduce tensions in high-stress environments, such as one Philippine call center that was studied by Emmauel David (2015).

How does this occupational segregation happen? Many new women workers find themselves shuffled into occupations dominated by other women, and women who go into a male-dominated field often find that, soon enough, they are surrounded by women. Like names, jobs become feminized when too many women hit the scene. This has happened in fields such as real estate sales, clerical work, pharmacy, public relations, bartending, bank telling, and more recently, the academic disciplines of sociology and biology. In *Job Queues, Gender Queues* (1990), sociologists Barbara Reskin and Patricia Roos argue that women end up in lower-paid jobs because these occupations lose (or have lost) their attractiveness for white men. When a job becomes deskilled and less autonomous—such as secretarial work, which way back when was men's work—earnings decline, routes of upward mobility close off, work conditions in general deteriorate, and men flee to better positions, leaving (typically white) women next in line to shuffle into the ranks of men's cast-off work.

Take book editing, which changed from a "gentlemen's profession" for most of its history to a virtual female ghetto by the end of the twentieth century (Reskin & Roos, 1990). Formerly the high-cultural mission of men with elite academic records, book editing evolved into a more commercial enterprise in the 1960s and 1970s. The result was that editorial work took a downturn in autonomy, job security, and importance. As the job lost its attractiveness to male candidates, the female labor supply was increasing on the crest of second-wave feminism. Thus book-editing jobs, like the other fields that optimists point to as women's inroads into traditionally male-dominated work, became resegregated and ghettoized as women's work. And anything categorized as women's work tends to yield lower pay, prestige, and benefits such as health coverage than men's.

That's not to say that gender norms do not exact a cost on workers in male-dominated jobs as well. Take for example, the construction industry: In an era of weak worker protections and declining male job prospects, one study has found that men who work in construction feel a need to demonstrate their masculinity through sexist, homophobic, and racist comments and through displays of physical strength. The consequences are not just cultural, however. Such displays of masculinity often involve bodily risk-taking with respect to occupational safety that serve to prove workers are "man enough" to perform and keep their jobs in this era of stiff worker competition. The end result is higher rates of worker accidents (and disability), acceleration of the decline of union power, and an overall loss of social status for the workers themselves (Landsbergis et al., 2014).

Glass Ceilings When women do enter more prestigious corporate worlds, they often encounter gendered barriers to reaching the very top: the so-called glass ceiling, which effectively is an invisible limit on women's climb up the occupational ladder. Sociologist Rosabeth Moss Kanter argues in her classic study *Men and Women of the Corporation* (1977) that the dearth of women in top corporate positions results from a cultural conflation of authority with masculinity. In Indesco, the fictitious name of the corporation Kanter studied, she found that most people believed that men and women come to occupy the kinds of jobs for which they are naturally best suited. To the contrary, Kanter showed, the job often makes the person; the person doesn't make the job.

Employee behavior tends to be determined by job requirements and the constraints of the organizational structure of the company. At Indesco, secretarial positions (98.6 percent female) were based on "principled arbitrariness," meaning there were no limits to the (male) boss's discretion as to what his secretary should do (type, fax, pick up his dry cleaning, look after his dog when he's away). Furthermore, secretarial work was characterized by fealty, the demand of personal loyalty and devotion of secretary to boss. Under these conditions, secretaries adopted certain behaviors to get through a day's work, including timidity and self-effacement, addiction to praise, and displays of

GLASS CEILING

an invisible limit on women's climb up the occupational ladder.

Betty Dukes (right) was a longtime greeter for Wal-Mart and became the lead plaintiff in a class-action gender discrimination lawsuit against the discount retailer. She and her counterparts claimed that they were passed over for promotions and paid less than men for the same work. In 2011 the Supreme Court dismissed the case in a controversial 5–4 decision.

emotion—all qualities that Indesco bosses tended to think of as just the way women are.

But what happens when a woman breaks into a top managerial position at Indesco? She becomes a numerical minority, what Kanter calls a token, a stand-in for all women. Because tokens have heightened visibility, they experience greater surveillance and thus performance pressures. Male peers tend to rely on gender stereotypes when interpreting a token's behavior, seeing female managers as "seductresses, mothers, pets," or tough "iron maidens." When a token female manager botches the job, it just goes to show that women can't handle the corporate world and should be kept out of it.

Similarly, Jennifer L. Pierce's (1995) study of law firms showed that sexual stereotypes, as much as organizational structure, are underlying causes of job segregation. Paralegals—who are 86 percent female—are expected to be deferential (that is, they should not stick up for themselves), caring, and even motherly toward the trial lawyers for whom they work. Given the adversarial model of the US legal system, trial lawyers, in contrast, perform what Pierce calls masculine emotional labor. To excel in the job takes aggression, intimidation, and manipulation. Even though almost half of law school graduates are female, women make up only 16 percent of partners at firms (Scharf & Flom, 2010). When a woman joins the higher ranks of trial lawyering, she's likely to face exclusion from informal socializing with her colleagues (no drinks after work), deflation of her job status (frequently being mistaken for a secretary), and difficulty bringing clients into the firm.

Thus female litigators find themselves in a double bind not experienced by their male counterparts. When deploying courtroom tactics of aggression and intimidation—the very qualities that make a male lawyer successful and respected—women litigators find themselves being called "bitches," "obnoxious," and "shrill." But when they fail to act like proper "Rambo litigators" in the courtroom, women are equally chided for being "too nice" and "too bashful." The trade-off between being a good woman and being a successful lawyer adds yet another obstacle for women to make it to the top in male-dominated jobs—and that's not even taking into account the added burden of sexual harassment in the workplace (which women bear the brunt of; see the #MeThree Policy feature on page 334).

Glass Escalators Just as the odds are stacked against female tokens, they tilt in favor of men in female-dominated jobs. In *Still a Man's World* (1995), Christine Williams found that male nurses, elementary school teachers, librarians, and social workers inadvertently maintain masculine power and privilege. Specifically, when token men enter feminized jobs, they enjoy a quicker rise to leadership positions on the aptly named glass escalator. These escalators also operate in law firms, where male paralegals, themselves tokens in the overwhelmingly female semi-profession, reap benefits from

GLASS ESCALATOR

the accelerated promotion of men to the top of a work organization, especially in feminized jobs.

their heightened visibility (Pierce, 1995). Male paralegals are said to enjoy preferential treatment over their female peers, such as promotions and even the simple (but substantial) benefit of being invited to happy hour with the litigators. Recent research finds, however, that the glass escalator is racialized: men of color do not ascend within their occupations at the same rates as their white counterparts (Dill et al., 2016; Wingfield, 2009).

Williams's (2013) new research suggests that women are holding a more equitable share of the top spots in nursing, elementary schools, and libraries, but that a larger problem is emerging that hits both men and women, and their families overall well-being: Wages in these careers have not kept pace with the cost of living. Careers that have traditionally been associated with women are not the only ones that have seen declining wages. Manufacturing, trucking, and warehouse jobs—traditionally dominated by male workers— are also paying less (Mishel & Shierholz, 2013).

In the fashion industry, the escalator reverses the gender filter Williams originally found for women-dominated fields. Aspiring male models are forced off at the second floor while a small number of women models can continue to top model status. I talked to former model Ashley Mears about the wage structure in the modeling industry. She confirmed that "you see a complete inverse" where "women outearn men by two to one, sometimes much more" and that "there are just more jobs and opportunities for women models." She explained that there are two reasons for this, one cultural and one structural. From a cultural perspective, "for a man to do the work of showing his body, displaying his body, it's read as being less than what we fully expect in a hegemonically masculine way. It's read as being effeminate work." From a structural perspective there simply are no stepping-stones to managerial positions or other promotions for models. According to Mears (who became a sociologist), on the models' job escalator, "there's no place to go," because career positions in the fashion industry go to experienced businesspeople and designers who, surprise, tend to be men (Conley, 2013c; Reimer, 2016). So even though male *models* don't enjoy a glass escalator, the wider industry ends up dominated by men all the same.

DIGITAL.WWNORTON.COM/YOUMAYASK6CORE

Go to digital.wwnorton.com/youmayask6core to see more of my interview with Ashley Mears.

Sociology in the Bedroom

Sex. You think it's the most personal, intimate act. The bedroom is surely the one place where the sociological imagination is the last thing you'll need. But just as we've untangled the terms *sex* and *gender*, we'll take a sociological look at sex, which can reveal some startling insights. Connections may be found between the sheets and history, desire and science, how we do it and how it in turn organizes what we do, as well as with whom and with what meanings. As expected, an excavation into the social construction of sexuality divulges a surprising amount of variation in what is considered normal bedroom behavior. By exposing different social patterns of sexuality, throughout history and across cultures, the sociologist can trace unequal gender relations and show how sexuality expresses, represses, and elucidates those relations.

An ancient Greek image of two male lovers. How can comparing social patterns of sexuality across cultures and throughout history help sociologists understand modern sexuality?

SEX: FROM PLATO TO NATO

Among the ancient Greeks, relationships between two men were accepted as normal. Engaging in same-sex acts did not confer a particular identity as the practice varied in frequency across the population among those who did and those who did not also participate in opposite-sex sexual relations. Rather, the socially important distinction revolved around active–passive dichotomy (although historians have shown that many exceptions existed). The active partner was supposed to be older or higher in status than the passive partner. To flip the rules was a violation, and it was considered shameful for a master or noble to be penetrated by a younger man. A more extreme and brutal example of sexual relations founded on power relations can be found today in the social orders in US prisons, where it is easy to see that rape is about power: who is in charge, who is being penetrated, and who is normal versus deviant. Prison rape, like any rape, may not be primarily about seeking sexual gratification; indeed, it is reported that few prison rapists climax during the act. In prison, the same-sex sex act is often seen as something altogether distinct from gay identity.

Or, consider sexual normality among the Sambia, a mountain people in Papua New Guinea. Anthropologist Gilbert Herdt (1981) reported that fellatio played a significant role in a boy's transition into manhood. Young boys are initiated into manhood by a daily ritual of fellatio on older boys and men. By taking in the vital life liquid (semen) of older men, boys prepare themselves to be warriors and sexual partners with women. Fellatio, for Sambia boys, then, is the only way to become "real" men.

Other cultures' attitudes toward male homosexuality are on a whole different level. Both the Siwans of North Africa and the Keraki, also in New Guinea, prefer homosexuality to heterosexuality, for fairly straightforward, practical reasons (Kimmel, 2000). Because every male is homosexual during his adolescence and then bisexual after heterosexual marriage, limiting straight sex keeps down the birthrate. In these cultures with scant resources, homosexuality makes practical sense to limit the chances of teen pregnancy and overpopulation.

BISEXUAL

an individual who is sexually attracted to both genders/sexes.

There is enormous variation in how humans have sex and what it means to them. Mouth-to-mouth kissing, common in Western cultures, is unthinkable among the Thonga and Sirono cultures: "But that's where you put food!" (Kimmel, 2000). American, heterosexual middle-class couples have sex a few nights a week for about 15 minutes a pop; the people of Yapese cultures near Guam engage in sex once a month. Marquesan men of French Polynesia are said to have anywhere from 10 to 30 orgasms a night! Which of these is "normal"?

THE SOCIAL CONSTRUCTION OF SEXUALITY

By treating sexuality as a social construction—that is, as always shaped by social factors—the sociologist would argue that the notion of normal, especially pertaining to what happens behind closed bedroom (or bathroom or car) doors, is always contested. In other words, there is no natural way of doing it. If an essentially right way existed, we wouldn't be able to find such wild and woolly variation throughout history and across cultures. Starting from the view that sex itself is a social creation, sociologists tend to argue that humans have sexual plumbing but no sexuality until they are located in a social environment. The range of normal and abnormal is itself a construction, a production of society. The study of this range can lead the willing sociologist into an exploration of the social relations on which sex is built.

Sexuality and Power Marxist feminists, for instance, argue that sexuality in America is an expression of the unequal power balance between men and women. Catharine MacKinnon (1983) argued that in male-dominated societies, sexuality is constructed as a gender binary, with men on top and women on the bottom (literally and figuratively). To MacKinnon, sexuality is the linchpin of gender inequality, an expression of male control. Some feminists also argue that our experiences of what is titillating are shaped by the fetishization of male power. So being "taken" is exciting and pleasurable to women, revealing that even those things we think of as the most personal of experiences are shaped by social arrangements of power.

Adrienne Rich (1980) called sexuality in America a "compulsory heterosexuality." This "political institution," at least for some, is not a preference but something that has been imposed, managed, organized, and enforced to serve a male-dominated capitalist system in which women's unpaid domestic work is required to support men's paid work outside of the home. Because her work is unpaid, the woman in this situation is unable to leave a bad husband. According to Rich, people come to see heterosexuality as the norm, when it is, in fact, a mechanism integral to sustaining women's social subordination.

Same-Sex Sexuality The connections between sex and power are best exemplified in the social construction of the homosexual in the current West, the social identity of a person who has sexual attraction to and/or relations with people of the same sex. For most of the past century, the dominant view has been that individuals are born either homosexual or heterosexual, gay or straight, queer or normal. Sexual orientation was thought of as a kind of personhood automatically acquired at birth or something carried in our genes (despite documented cases of identical twins in which one is straight and the other gay). But recall the ancient Greeks' homosexual love, the Sambian rites of manhood, or prison rape. Jane Ward (2015), for example, studied straight men who have sex with other men in the contemporary United States, including fraternity members or military personnel who engage in sex acts with their compatriots as part of hazing or men who answer ads to find other men to masturbate with. Men in these examples engage in homosexual acts without adopting a homosexual identity (Silva, 2017). When do acts, behaviors, or desires crystallize into identities?

Before 1850, there was no such thing as a homosexual. Sure, people engaged in same-sex sexual behaviors—but it was not yet an identity in the way we now know it. French philosopher Michel Foucault (1926–1984) led the way to poststructuralist notions of the body and sexuality as historical productions. In *The History of Sexuality* (1978), Foucault made the case that the body is "in the grip" of cultural practices. That is, there is no pre-social or natural body; instead, culture is always already inscribed on our bodies. Foucault further argued that the way in which we know our bodies is linked to power, and knowledge and power go hand in hand. As the population expanded in nineteenth-century Europe, newly formed states and their administrators developed a concern for population management. The rise of scientific ways of thinking at the time led to the creation of what Foucault calls "bio-power," the control of populations by influencing patterns in births, deaths, and illnesses.

Discourses of sexuality had by then surfaced. People talked about sex, scientists studied it, and government officials tried to regulate it, whereas a

---

HOMOSEXUAL

the social identity of a person who has sexual attraction to and/or relations with people of the same sex.

century before, all this sex talk did not exist. With the development of the biological and human sciences in the nineteenth century, doctors wanted to know which kinds of sex were normal and which were deviant, and new attention was paid to policing those differences. By the late 1800s, "Homosexuality appeared as one of the forms of sexuality when it was transposed from the practice of sodomy onto a kind of interior [identity]" (Foucault, 1978). In Foucault's account, the category of homosexuality arises from the efforts of government bureaucrats trying to assert their power (that of the state) over human populations.

Meanwhile, the American Psychiatric Association and the American Psychological Association listed homosexuality on their list of mental disorders until 1973. Homosexuals were regarded as needing to be regulated, observed, studied, and most important, controlled. Because everyone was capable of, or in danger of, sexual deviance, the urgency to confess one's sexuality grew, as did the scientific need for public surveillance. Who was gay? How could a homosexual be detected? Anyone could be a pervert—your professor, your roommate, even you!

Sexual acts became synonymous with the person who performed them. Even our language reflects the way sexual behavior becomes more than a single event among the many other behaviors that make up an individual's life—it often becomes a master status. Here's what I mean. If a man roasts a whole pig—a serious, time-consuming undertaking that requires research, dedication, and planning—he is not known ever after as a "pig roaster" or even as a gourmand. He's still just Bob from down the block who hosted one heck of a barbecue. But if a man has anal sex with another man—something that takes less time and probably less dedication and research than roasting a pig—he becomes a sodomite, a homosexual, gay, or whatever terminology might fit the context.

This was not always so. At one point, what happened in the bedroom (or car or motel or . . .) stayed there. But over the last century what happened behind closed doors was outed, and a whole new person was produced. Today our society places a huge emphasis—our entire selfhood, according to Foucault—on sexuality. Sexuality, in other words, is a prime factor that goes into the construction of gender—a dynamic that may not have been as important in days gone by. Now, what you do (and with whom) defines who you are: not just how others see you, but how you experience yourself.

Consider a widely used argument against lesbian and gay couplings: "It's unnatural," meaning that sex is supposed to be about reproduction.

Homosexuality was considered a mental disorder by the American Psychological Association until 1973. Here, in 1967, Dr. Joseph Wolpe treats a homosexual patient by showing a series of risqué images intended to elicit arousal.

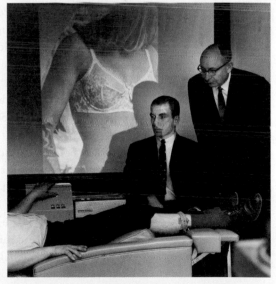

Thousands participate in a pride parade in 2018. How does Foucault show that the rise of the "homosexual" as an identity was a historical process? How might he respond to pride parades and other contemporary assertions of gay identity?

This is a popular argument against same-sex marriages, but it also stigmatizes straight couples who can't or don't have children. Besides, homosexuality occurs widely in nature; one of our closest primate cousins, the bonobo, practices homosexuality (Parker, 2007). Though based on questionable data, Alfred Kinsey's 1948 study *Sexual Behavior in the Human Male* suggested that at least 10 percent of men were homosexual. He challenged the psychiatric model of homosexuals as perverse and abnormal and instead viewed sexuality as a continuum. Most people, he claimed, experience both heterosexual and homosexual desires and behaviors. Kinsey's figures have since been disputed—his sampling was not representative of the US population as a whole—but the basic idea holds. Across the globe and throughout history, there has been more or less the same degree of homosexual behavior.

What is different across time and place, however, is how sexuality is perceived. Gender often plays a large role. For example, Yale historian George Chauncey makes the case in *Gay New York* (1994) that in early-twentieth-century New York, an emerging working-class gay world was split between masculine men (tops) and the effeminate men who solicited them (bottoms). As long as men stuck to their masculine gender scripts, no matter how often they engaged in gay sex, they were not considered abnormal like the more feminine men, who were derided for their effeminacy.

The reverse is true among the *travesti*, transgender prostitutes in Brazil. Anthropologist Don Kulick (1998) conducted an ethnography of the *travesti*, males who adopt female names, clothing styles, hairstyles, cosmetic practices, and linguistic pronouns; ingest female hormones; and inject industrial silicone directly into their bodies to give themselves breasts and round buttocks or *bunda*. They display stereotypically feminine traits, yet they do not self-identify as women. In fact, they think it is both repugnant and impossible for men to try to become women. Kulick argues that in Brazil, gender is determined by sexual practice: "What one does in bed has immediate and lasting consequences for the way one is perceived (and the way one can perceive oneself) as a gendered being. If one penetrates, in this particular configuration of sexuality and gender, one is a 'man.' If one expresses interest in the penis of a male, and especially if one 'gives'—allows oneself to be penetrated by a male—then one is no longer a man." However, the *travesti* do not think they are women; they think of themselves as *travesti*, men who emulate women but are not women. The primacy of penetration as a determinant of gender is common across cultures and challenges Americans to look beyond chromosomes and hormones to practices and cultural context.

CONTEMPORARY SEXUALITIES: THE Q WORD

But it is not just straight men who have sex with other men or bisexuals that complicate the sexuality (and/or gender) binary. There's a reason the term LGBTQIA has seemed to add letters lately—there's a growing recognition that if we look closely and without judgment, we see many forms of gender/sexuality identity that do not conform to categories that dominated discussion since, roughly, the 1850 period that Foucault marked as the start of the era of the "homosexual."

We have already discussed L (lesbian), G (gay [male]), B (bisexual), and T (trans). We have also discussed I (intersex). So let's skip to the last letter in the acronym, "A." Asexuality has, of late, become a more visible sexual identity. Many individuals do not feel sexual attraction; these individuals may identify as asexual (often "ace" for short). They may not experience sexuality-related bodily feelings or actions, but others who identify as asexual may, for example, masturbate to fantasies of fictional characters. The American Psychiatric Association has defined asexuality as a paraphilia—which is an atypical sexual attraction. This lumps it in with medical disorders, whether or not asexuality results in significant distress for the individual or others, such as a partner. Sociologists challenge this definition, pointing out that many ace individuals enjoy sexual fantasies that involve others or involve themselves

How do *travesti* challenge common binary models of understanding gender and sexuality?

in a completely fictional way (that is, younger, kinkier, or even as a superhero) (Scherrer & Pfeffer, 2017). A better way to think about asexuality is as an identity based on membership in a shared sexual community. Hence the "A" was added to the LGBT alphabet as asexual activists have sought to be included in nondominant gender/sexuality social movements, protests, and so on.

What about the Q-word? *Queer* is a derogatory term, reclaimed by those it was originally intended to wound, to become a broader, encompassing term for nonheteronormative sexual behaviors/identities. That is, when someone self-identifies as queer, they may be a lesbian looking for a broader, more inclusive umbrella that is not already overly associated with, say, white middle-class people. Or they may be someone who participates in what might be called "kinky" sexual practices, such as BDSM (a term meant to capture bondage/discipline/submission/sadomasochism and related practices). Or they may simply be someone who does not feel comfortable in and wishes to reject other aspects of heteronormativity. For instance, some queer people reject the notion of same-sex marriage. They reject what they see as assimilation into heteronormative culture, and instead prefer to organize their lives differently in terms of kinship, sexuality, and so on.

One form of nonheteronormativity that has garnered a lot of attention of late is non-monogamy. Non-monogamy is not just having sex with multiple people (because, for example, people can be in monogamous relationships and cheat on their partners). Rather, it is the practice of having multiple sexual (or intimate) relationships (or merely the desire to) with the full knowledge and consent of all the people involved. Polyamory is a particular form of non-monogamy (Sheff, 2014). While there is no agreed on, universal definition of polyamory, it usually involves *intimate* relationships with more than one partner (again, with knowledge and consent of all involved). So a couple who agrees that they can have one-night stands with other people are non-monogamous but not polyamorous. Mimi Schippers (2016) finds that some forms of polyamory subvert gender/sexuality norms more than others. For example, a relationship including a woman and two men challenges heterosexual norms and gender power relations more than a relationship among one man and two women, since the latter more easily allows for male dominance and because female bisexuality is more socially acceptable than male bisexuality.

Fluid notions of acceptable sexual relationships aren't only limited to those under the LGBTQIA umbrella: Indeed, how we think about sex among teen and college-aged people is rapidly changing.

"HEY": TEEN SEX, FROM HOOKING UP TO VIRGINITY PLEDGES

Sometimes the policies meant to achieve one end backfire and cause the opposite outcome, despite the best intentions. Government efforts to

HETERO-NORMATIVITY

the idea that heterosexuality is the default or normal sexual orientation from which other sexualities deviate.

influence healthy teen sexuality provide one example. Recent studies have described the blasé attitude that contemporary American teenagers supposedly bring to their love lives. When sociologist Paula England surveyed students at one midwestern university about their sex lives, half reported that their previous sexual partner was someone they had slept with only once. Here's how she explained hook-up culture, which has replaced dating as the route to romance on college campuses around the country. She did her first surveys at Northwestern, where "before people ever go on a date, they hook up." Lest there be any confusion: a hook-up is not sex…unless it is. A hook-up means "something sexual happens, that 'something sexual' is not always intercourse, often in fact [in] the majority of cases it isn't, and there is no necessary implication that anybody's interested in a relationship, but they *might* be interested." Dating is infrequent, but it is not dead. It is "charged with more meaning now, and it's more likely to be leading to a relationship." Furthermore, "there's a strong norm that relationships should be monogamous and marriages should be monogamous and that people eventually want to get to monogamy and a marriage. They're just putting it off a lot longer. That's really what's changed, I think" (Conley, 2009d).

Why so many short-lived hook-ups instead of longer-term relationships? For one thing, the students are not looking to begin families anytime soon. For another, relationships are just more work—for the men, at least. England found that heterosexual males are not necessarily expected to sexually satisfy the women they hook up with, but the opposite is true when they are in relationships (Armstrong et al., 2012). This has led to what they call "the orgasm gap." England and co-authors' analysis shows "that specific sexual practices [such as oral sex or genital stimulation], experience with a particular partner, and commitment all predict women's orgasm and sexual enjoyment"— all conditions less present in hook-ups than in relationships. For example, with respect to oral sex, England explains that in a hook up "it's much more likely that the woman is servicing the guy than vice versa. So it seems like the hook-ups are prioritizing male pleasure" (Conley, 2009d).

In a hook-up culture, what might determine whether a pair continues to hook up or moves on to new partners? England has a theory that goes like this:

DIGITAL.WWNORTON.COM/YOUMAYASK6CORE

To see an interview with Paula England about hook-up culture, go to digital.wwnorton.com/youmayask6core

Partner-specific investment may well be important. In other words, ... our ability to please partners may not just be generally what have we learned from a lifetime of experience but what have we actually learned about this particular partner and what works for them and what they like and what they need.... If it's hook-up number two, she's more likely to have an orgasm than if it's one, and if it's three more likely, and if it's four even more likely. (Conley, 2009d)

She also noted that these hook-ups rarely produce babies—but nearly 14 percent of students still did not use any method to prevent pregnancy the last time they had sex.

Here are more numbers: Fewer than 40 percent of American high-school students tell surveyors that they have had sexual intercourse (Figure 8.4), but this rate is not growing; it has remained more or less steady since 2003 (Centers for Disease Control and Prevention, 2016a). Boys probably lie more often about the extent of their sexual experience (citing the proverbial girlfriend in a different state), and girls possibly downplay their sexual activity. Many of those who have not had intercourse are still sexually active in other ways. Teenagers' romantic interludes last about 15 months on average, leaving them plenty of time before marriage (median age of 27.9 for women and 29.9 for men in 2016) to have many partners (US Census Bureau, 2017m). To top it off, most adolescents with a sexually transmitted disease (STD) don't know that they are infected. All of these factors combine to put American teenagers at high risk for STDs, which have been increasing dramatically since the 1970s.

FIGURE 8.4 Percentage of High-School Students Who Have Had Sex, 1991–2015

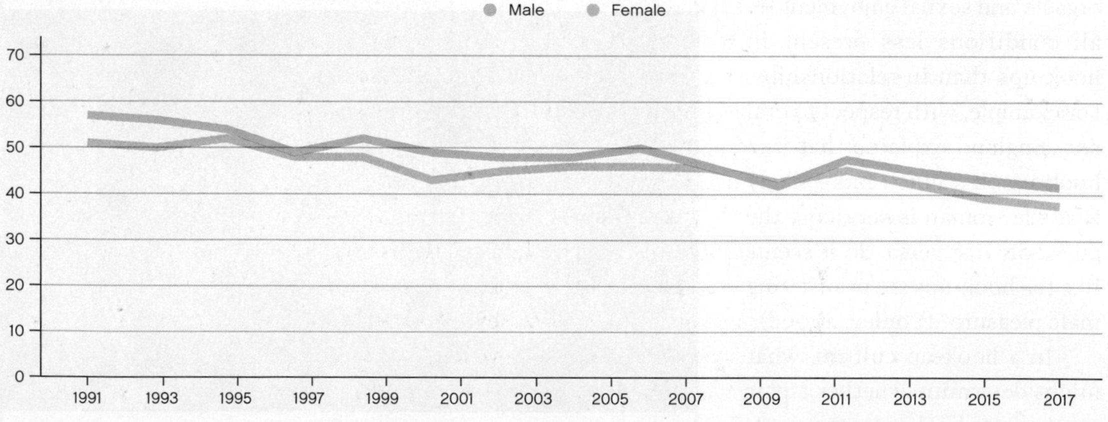

SOURCE: Centers for Disease Control and Prevention, 2016a.

So what's a public health officer to do? During its years in office, the Bush administration (2001–9) advocated a "virginity pledge" and other abstinence policies, and the abstinence advocacy group True Love Waits says that 2.5 million young people have made the virginity pledge since 1993 (Herbert, 2011). As it turns out, the pledge does, on average, delay the onset of sexual activity as well as reduce a teenager's number of sexual partners, according to the National Longitudinal Study of Adolescent Health (Brückner & Bearman, 2005)—particularly when it occurs in a school context where a sizable number of students take the pledge, creating a meaningful identity of sorts through collective abstinence (Bearman & Brückner, 2001). That said, most pledgers (about 60 percent) break their pledge. And when sex happens, it's much more likely to come in a rush of surprisingly strong, unfettered desire. Among the pledgers, only 40 percent ended up using a condom during sex—why would they have a condom if they were planning to be virgins until marriage?—compared with 60 percent of teens who had not pledged. The end result was a net higher rate of HPV infection and pregnancy among the pledgers (Paik et al., 2016).

What is for certain is that teens—like their adult counterparts—are navigating new terrain in terms of romantic and sexual relations thanks to technology. One way we can see this is in the debate over opening lines in online dating apps. Three million messages per day on the Tinder app begin with "Hey." Does such a generic approach represent laziness (that is, not having read someone's profile in order to say something specific)? Or is it just a universal human way to say "hi" without concocting some phony pickup line (even if it is not as vulnerable online as it would be in person at a party or a bar)? One thing is for sure—it probably beats the most searched GIF on Tinder: Joey from the show *Friends*, asking "How you doing?" (Verdier, 2016).

Figuring out the right way to approach someone online is not the only challenge that teens and adults have in the evolving world of Tinder, Grindr, and OkCupid. A new phenomenon of the "define-the-relationship" (DTR) conversation is emerging. In a world where a new potential partner is only a swipe away and where people are inventing new ways of being romantic or sexual and where one cannot assume a single narrative arc of how such relationships are meant to progress, the DTR talk is a way that people can clarify expectations, hopes, fears, and so on after getting to know each other. Will we be monogamous or friends with benefits? If monogamous, do we delete our dating apps? Will we be public on Facebook as a couple? Is this just something fun for now, or do we mutually envision a longer-term relationship between us? These are just some of the questions the DTR conversation may address. As to when to have (or avoid) that conversation, just as there are myriad relationship types nowadays, there is no clear norm on what's too soon (or too late) to discuss or define one's relationship. Stay tuned: Norms can evolve fast in the face of technological innovation.

#METHREE

The #MeToo moment that began in 2017 unleashed a tsunami of sexual harassment and assault allegations that had built up over decades while predatory behavior went on unabated. It has prompted many policy makers to ask what can be done to prevent such a buildup of allegations in the future. That is, how do we get harassment or assault victims to step forward in real time? Or better still, get perpetrators fearful enough of being called out with consequences for bad behavior so that they don't perpetrate in the first place! One thing is for sure: What organizations from colleges to corporations to Congress are doing presently does not work. Namely, watching videos or even engaging in live training with HR professionals does not seem to have the effects on actual workplace or school harassment and assault that it is intended to. It can even backfire by reinforcing gendered roles, according to sociologist Justine Tinkler (2013), by depicting men exclusively as harassers and women only as victims. But all this doesn't mean *nothing* works and we should give up hope of a civil public space where individuals can feel safe and respected.

The ideas that have been offered include rewarding managers when reports of harassment go up, focusing training on getting bystanders (as opposed to the victims themselves) to intervene in some way, and promoting more women into leadership roles (which often changes the norms of what is acceptable behavior). One particularly novel notion is the formation of "information escrows"—secret repositories where complaints can be reported anonymously. The idea is this:

Many people want to give others the benefit of the doubt. They don't want to start a huge stir and possibly ruin someone's career if it is truly a case of a one-off mistake or a misunderstanding. They also don't want to risk the cost to their own reputation (or career) in coming forward in what may end up being a he said–she said (or he said–he said, or…) situation that ends up inconclusive and uncomfortable for all involved. But if the action—say a misplaced hand or inappropriate humor or someone who doesn't take no for an answer when asking out a co-worker—is part of a larger pattern of behavior, then as most would agree, it should have consequences. That's where the information escrow comes into the picture.

The escrow is managed by someone in human resources or could even be set up to be totally computerized. If I experience something that may qualify as harassment, I can lodge a complaint confidentially, detailing the alleged infraction. But I can then specify the conditions under which I want my grievance to trigger action. For example, I might say, as long as nobody else has complained about our shift manager pinching them, my allegation should remain secret and dormant. But the moment someone else also files a report to the information escrow of inappropriate touching (or perhaps other infractions), my testimony becomes an official complaint too. That way, it is no longer one person's word against another's but an established pattern of behavior documented independently—in other words, a much more powerful case. Likewise, if the behavior was a one-off and never occurred again, I might chose to let it go. Of course, the presence of the information escrow option doesn't (and isn't meant to) preclude someone from reporting someone's misbehavior straightaway through formal channels.

As proposed by Ian Ayres and Cait Unkovic (2012), such allegation escrows work best when infractions are underreported and when they typically involve more than one victim. Such harassment escrows are a particular case of a broader class of possible information escrows that can cover everything from bids in a trade (where the trade occurs and bids are revealed only if the seller and buyer overlap in price) to revelation of mutual romantic crushes. In fact, some modern dating apps like Tinder and Bumble essentially act as escrows: Only if we both swipe right does our mutual interest get revealed. Imagine a similar app for deterring harassers. In an era where one camp is worried about the potential for false accusations to ruin someone's life without due process while the other side is concerned with victims who fear going public alone, it may be worth a try.

Conclusion

Starting with our friend Elliot, we've learned that what seems to be normal or natural often turns out to be fluid and contingent. The stuff we build up around biological plumbing—roles, expectations, psyches, institutions—is not essential. These are socially constructed facts, built less on the biological and more on the existing social structures of power.

We've learned a lot about how these norms of gender can create inequality, but what happens when gender can become something empowering? Elliot concludes his story on an uplifting note:

Today, I felt gender euphoria.

Euphoria is the opposite of dysphoria. I see my reflection, and I love the guy I see. No matter what I'm wearing, I feel nothing but confidence in how I look and who I am. I am brave enough to confront rude people who follow me into bathrooms, to wave menstrual hygiene products at them. I feel like myself. I feel like an Elliot.

Days like today were almost nonexistent before I realized I was trans and began socially transitioning. When I was pretending to be a girl, even the highest joys and deepest sadnesses in my life were missing something, a certain depth, a certain extra grounding in reality. I'm not a girl. No matter how I try, dressing like a girl, acting like one, will always ring hollow, always feel wrong. That isn't who I am.

I'm a boy. Maybe nobody but me sees or feels a difference when I present male, but I feel better. I feel real, alive. I feel every bit of emotion the world has to offer me.

That, no matter how many self-appointed bathroom guardians I have to face, will always be worth it.

We've traced how our intricate system of sexes, genders, and sexualities has evolved. Can we "undo gender," or are we stuck in the paradox of reproducing the binaries we started with, even if we recognize their inequalities? If we humans have constructed gender as a way to organize, simplify, and control a messy social world, then indeed we can deconstruct it. As Judith Lorber (1994) argued, only when we stop using gender as a basis for dividing up the world—in terms of which jobs people hold, what rights they exercise, how much money they earn, how much control they have over their bodies, with whom they can have sex, and yes, what bathroom they use—will we find true equality.

QUESTIONS FOR REVIEW

1. What is binary thinking about sex and gender? How do individuals who challenge these binaries (such as transgender or intersex people) help us understand that sex and gender are social, not natural, categories?

2. Women represent a minority group in the military. Men are in the minority as nurses and paralegals. How are women and men treated differently in these positions, and what does this difference suggest about the way gender structures social relations?

3. How do the cases of the *hijras* and the *travesti*, as described in this chapter, challenge our understanding of sex and gender?

4. What might the idealized man of the mid-twentieth century as described by Erving Goffman think of more recent forms of masculinity, such as Kristen Barber's metrosexual male or the "hybrid masculinities" described by Tristan Bridges and C. J. Pascoe? Despite these differences, how are most forms of dominant masculine identity similar?

5. Names such as Carol and Aubrey used to be boys' names but today are almost exclusively female. Careers such as nursing and secretarial work used to be associated with men but these days are overwhelmingly performed by women. Why do names or careers become feminized, and what does this say about masculinity?

6. How does Talcott Parsons describe the role of men and women in his "sex role theory"? Explain how conflict theories can be seen as a critique of structural functionalism, and describe some limitations of each approach.

7. What is the difference between homosexual "acts" and a homosexual "identity"? How did the historical development of the latter, according to Michel Foucault, affect how one sees oneself?

8. More differences seem to exist among boys and girls than between them. Nonetheless, we tend to think of them as different. What are "deceptive distinctions," and how do they create gender differences? Use an example from Rosabeth Moss Kanter's work to support your answer.

MEASURING MANSPLAINING

Sites like **arementalkingtoomuch.com,** the "Time to Talk" app, and others have sprung up recently after it became widely reported that men tend to dominate conversations and talk over women. This is something that sociologists have known since Carol Stack and Don Zimmerman famously found that men overwhelmingly interrupt women in everyday conversations (1975). More recent research has continued to support this finding: For example, a recent study by the *Harvard Business Review* used 15 years of oral argument transcripts to show that male Supreme Court justices interrupt each other three times less often as they interrupt their female peers (Jacobi & Schweers, 2017). And it's not just interrupting: compared to women, men also tend to take up more conversational air time in social settings from the classroom to the board room.

TRY IT!

Observe a class discussion and measure how much time men speak versus how much time women speak. You can try any of the apps or websites mentioned above, or simply use a stopwatch. It would be ideal to sit in on a class you're not taking (but be sure to check with the instructor first). And don't spill the beans about your project—if your research subjects know what you're measuring, you'll probably get lousy results.

While you're measuring female versus male airtime, also note how many times a male student interrupts a female student and vice versa.

THINK ABOUT IT

Once you have your results, compare the amount of airtime to the proportion of men and women in the class. Did men disproportionately dominate the conversation? Did men interrupt women more often than women interrupted men? Did these gender dynamics vary intersectionally—that is, for example, by the race of the would-be speakers?

Whether or not the results of your study supports the finding that men are socialized to talk over women, reflect on the assumptions you made in performing the experiment. Did a two-category scorecard force you to make any assumptions about the gender of individuals—assumptions that you might not have made without discrete categories?

SOCIOLOGY ON THE STREET

Internet dating is a major resource for people looking for potential partners, with a seemingly unlimited pool. What are the similarities and differences between dating online compared to dating in person? Watch the Sociology on the Street video to find out more: **digital.wwnorton.com /youmayask6core**.

WANT MORE PRACTICE?

Complete the InQuizitive activity for this chapter at digital.wwnorton.com /youmayask6core

RACE AS WE KNOW IT HAS NO DETERMINISTIC BIOLOGICAL BASIS: ALL THE SAME, RACE IS SO POWERFUL THAT IT CAN HAVE LIFE-OR-DEATH CONSEQUENCES.

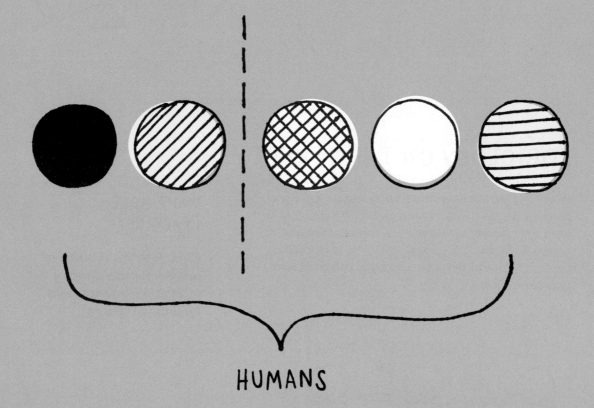

HUMANS

Race

"I found my baby sister!" I declared to my mother, wheeling a carriage around for her to see the newest member of our family, whom I had just kidnapped. I was not quite three years old, and the toddler was only a few months younger than that, with cornrows braided so tightly on her little head that they pulled the brown skin of her face tautly upward. I remember that she was smiling up at me, and I must have taken her smile as permission to swipe the unattended stroller from the courtyard of our housing complex.

"No, you haven't!" my mother gasped, putting a hand over her open mouth. The child was quickly returned to her frantically searching mother, despite my tearful protests.

This story fascinates me today: I wanted a baby sister so badly that I kidnapped a black child, not realizing that race is a primary way we divide families. How could I be so oblivious to the meaning of race—something that years later feels so natural, so innate? To my childhood self, race was neither a meaningful category nor a too obvious one. In the largely minority housing project where I grew up in the 1970s, race was not something mutable, like a freckle or hairstyle; it defined who looked like whom, who was allowed to be in the group and who wasn't. But for my sister and me, as whites, race was turned inside out. We had no idea that we belonged to the majority group, the privileged one. We just thought we didn't belong.

Is race real? You might consider this a paradoxical question. After all, how is a huge part of your identity *not* real? The sociological study of race treats it as a social phenomenon that seems natural but isn't. That is, race is a real social distinction, and people around the world and throughout history have drawn sharp lines between "us" and "them" on the basis of race. But as a biological, genetic, geographic, or cultural category, race has fluid and changeable boundaries. In this sense, race is constructed in the interests of groups that wish to maintain power and social exclusion. To the sociologist, understanding racial differences—including by income, educational attainment, crime rates, and teen pregnancy—means treating differences not just as personal matters but as pieces of a larger social picture. Having grown up as a "honky" in the housing projects of New York City's Lower East Side, a fish out of water, I've been looking at race with the sociological imagination ever since my failed kidnapping. Now it's your turn.

The Myth of Race

Perhaps you have heard claims that race is fake, that it's "just a myth." Race refers to a group of people who share a set of characteristics—typically, but not always, physical ones—and are said to share a common bloodline. People obviously have different physical appearances, including eye color, hair texture, and skin color, so it's perhaps puzzling to hear that (biological) racial differences somehow do not exist. To speak of the myth of race is to say that it is largely a social construction, a set of stories we tell ourselves to organize reality and make sense of the world, rather than a fixed biological or natural reality. In this sense, it resembles the socially constructed notions of childhood and adolescence that we discussed in Chapter 4. We tell the set of stories over and over and, collectively, believe in it and act on it, therefore making it real through practices such as largely separate marriage and reproductive communities. But we could organize our social distinctions a different way (for example, based on foot size or hair color), and indeed, throughout history, we have told this set of stories in myriad ways.

Take, for example, the following passage from an 1851 issue of *Harper's Weekly Magazine*, in which the author describes the physiognomy of a certain racial group. Try to guess which race the author is describing:

> [They are] distinctly marked—the small and somewhat upturned nose, the black tint of the skin.... [They] are ignorant, and as a consequence thereof, are idle, thriftless, poor, intemperate, and barbarian.... Of course they will violate our laws, these wild bisons leaping over the fences which easily restrain the civilized domestic cattle, will commit great crimes of violence, even capital offences, which certainly have increased as of late.

Most people would guess that the minority group in question here is African Americans. This passage was written, in fact, about Irish immigrants, who in late-nineteenth-century America struggled to assimilate amid fierce and widespread racism (Knobel, 1986). It was believed that the Irish were a distinct category of people who carried innate differences in their blood, differences that made them permanently inferior to their white American neighbors.

When the term *race* comes up in America today, we usually think in two colors: black and white. But, at the turn of the twentieth century, Americans categorized themselves into anywhere from 36 to 75 different races that they organized into hierarchies, with Anglo-Saxon at the top followed by Slav, Mediterranean, Hebrew, and so on down the list (Jacobson, 1998). Even though the United States was a nation of immigrants, many Americans doubted whether "ethnic stock," such as the Irish, were fit for self-governance in the new democracy.

In 1790, Congress passed the first naturalization law, limiting the rights of citizenship to "free white persons." This law strikes us today as both restrictive and inclusive. It was restrictive because it granted naturalization only to free whites, thereby coloring American citizenship. Yet it also set up an initially broad understanding of "whiteness," an umbrella term that in common parlance could include not just Anglo but also Slavic, Celtic, and Teutonic (German) Europeans. However, as millions of immigrants surged to the shores of America—25 million European immigrants arrived between 1880 and World War I—the notion of "free white persons" was reconsidered. With an Irish-born population of more than 1 million in 1860, Americans began to theorize racial differences within the white populace. Questions arose in the popular press and imagination, such as "Who should count as white?" "Whom do we want to be future generations of Americans?" "Who is fit for self-governance?" The inclusiveness of "white persons" splintered into a range of Anglos and "barbarous" others, and Americans began to distinguish among Teutons, Slavs, Celtics, and even the "swarthy" Swedes. The Immigration Act of 1924 formalized the exclusive definition of whiteness by imposing immigration restrictions based on a quota system that limited the yearly number of immigrants from each country. The law set an annual ceiling of 18,439 immigrants from eastern and southern Europe, following the recommendation of a report stating that northern and western Europeans were of "higher intelligence" and thus ideal "material for American citizenship" (Jacobson, 1998).

A racist ideology can be seen in this early line of thinking about whiteness. Racism is the belief that members of separate races possess different and unequal traits coupled with the power to restrict freedoms based on those differences. Racist thinking is characterized by three key beliefs: that humans are divided into distinct bloodlines and/or physical types; that these

THE USUAL IRISH WAY OF DOING THINGS.

An anti-immigration cartoon from an 1871 issue of *Harper's Weekly*. How have attitudes about race changed over the course of American history?

RACISM

the belief that members of separate races possess different and unequal traits.

bloodlines or physical traits are linked to distinct cultures, behaviors, personalities, and intellectual abilities; and that certain groups are superior to others.

European immigration slowed during World War I and essentially came to a halt as a result of the 1924 National Origins Act, while internal African American migration from the rural South to the industrial North skyrocketed. These shifts, along with the solidification of the one-drop rule (see later in the chapter), shifted national attention away from white–nonwhite relations toward white–black relations. Whites of "ethnic stock" were drawn back into the earlier, broad category of white, thereby reuniting Anglos and other Europeans. Public horror at Nazi crimes following the conclusion of World War II further strengthened the idea of whiteness as an inclusive racial category.

Today being of Irish descent is a matter of ethnic—not racial—identification, a reason to celebrate on St. Patrick's Day. I know this firsthand: Being one-eighth Irish and having an Irish-WASP (white Anglo-Saxon Protestant) name like Dalton Conley entitles me to free drinks on the Irish holiday. Irish American is no longer considered a restrictive racial category, as once was the case. Whiteness today is something we take for granted, a natural part of the landscape. But dig just a hundred years back into the unnatural history of race, and you might not even recognize it. Not only have groups of people been categorized differently over time, showing that there is nothing natural about how we classify groups into races today, but the very concept of what a race is has changed over time.

The Concept of Race from the Ancients to Alleles

The idea of race, some scholars have claimed, did not exist in the ancient world (Fredrickson, 2002; Hannaford, 1996; Smedley, 1999; Snowden, 1983). Well, it did and it didn't. It did in the sense that the ancients recognized physical differences and grouped people accordingly. In ancient Egypt, for example, physical markers were linked to geography. Believing that people who looked a certain way came from a certain part of the world, the Egyptians spoke for instance of the "pale, degraded race of Arvad," whereas their darker-skinned neighbors were designated the "evil race of Ish." The Chinese also linked physical variation to geography, as laid out in a Chinese creation myth. As the ancient tale goes, a goddess cooked human beings in an oven. Some humans were burned black and sent to live in Africa. The underdone ones turned out white and were sent to Europe. Those humans cooked just right, a perfect golden brown, were the Chinese.

However, in the ancient worlds of Greece, Rome, and early Christendom, the idea of race did not exist as we know it today, as a biological package of traits carried in the bloodlines of distinct groups, each with a separate way of being (culture), acting (behavior), thinking (intelligence), and looking (appearance). The Greek philosopher Hippocrates, for instance, believed that physical markers such as skin color were the result of different environmental factors, much as the surface of a plant reflects the constitution of its soil and the amount of sunlight and water it receives. To be sure, the Greeks liked the looks of their fellow Greeks the best, but the very notion of race goes against Aristotle's principle of civic association, on which Greek society was based. The true test of a person was to be found in his (women were excluded) civic actions. Similarly, the Romans maintained a brutal slavery system, but their slaves, as well as their citizens, represented various skin colors and geographic origins. The ancients may have used skin color to tell one person from the next—they weren't color-blind—but they didn't discriminate in the sense of making judgments about people on the basis of their racial category without regard to their individual merit (Hannaford, 1996). The notion has been so thoroughly displaced by racialized thinking that to us moderns, it is impossibly idealistic to imagine a society without our race concept.

A medieval illustration depicting Ham as a black man. How did Europeans use the biblical story to defend colonialism?

RACE IN THE EARLY MODERN WORLD

Modern racial thinking developed in the mid-seventeenth century in parallel with global changes such as the Protestant Reformation in Europe, the Age of Exploration, and the rise of capitalism. For example, European colonizers, confronted with people living in newly discovered lands, interpreted human physical differences first with biblical and later with scientific explanations, and race proved to be a rather handy organizing principle to legitimate the imperial adventure of conquest, exploitation, and colonialism. To make sense of what they considered the "primitive" and "degraded" races of Africa, Europeans turned to a biblical story in the book of Genesis, the curse of Ham. According to this obscure passage, when Noah had safely navigated his ark over the flood, he got drunk and passed out naked in his tent. When he woke from his stupor, Noah learned that his youngest son, Ham, had seen him naked, whereas

his other sons had respectfully refused to behold the spectacle. Noah decided to curse Ham's descendants, saying, "A slave of slaves shall he be to his brothers" (Gourevitch, 1998). European Christians and scientists interpreted this tale to mean that Ham was the original black man, and all black people were his unfortunate, degraded descendants. For an expanding Europe and America, the Hamitic myth justified colonialism and slavery.

When the divine right of conquest lost its sway, science led the way as an authority behind racial thinking, legitimating race by scientific mandate. Scientific racism, what today we call the nineteenth-century theories of race, brought a period of feverish investigation into the origins, explanations, and classifications of race. In 1684, François Bernier (1625–1688) proposed a new geography based not on topography or even political borders but on the body, from facial lineaments to bodily configurations. Bernier devised a scheme of four or five races based on the following geographic regions:

- *Europe* (excluding Lapland), *South Asia, North Africa,* and *America:* people who shared climates and complexions

- *Africa proper:* people who had thick lips, flat noses, black skin, and a scanty beard

- *Asia proper:* people who had white skin, broad shoulders, flat faces, little eyes, and no beard

- *Lapps* (small traditional communities living around the northern regions of Finland and Russia): people who were ugly, squat, small, and animal-like

Scientific racism sought to make sense of people who were different from white Europeans—who constituted the norm, according to the French scientist Comte de Buffon (1707–1788). This way of thinking, called ethnocentrism, the judgment of other groups by one's own standards and values, has plagued scientific studies of "otherness." In Buffon's classification schemes, anyone different from Europeans was a deviation from the norm. His pseudoscientific research, like all racial thinking of the time, justified imperial exploits by automatically classifying nonwhites as abnormal, improper, and inferior.

With the publication of *On the Natural Varieties of Mankind* in 1775, Johann Friedrich Blumenbach (1752–1840), widely considered the founder of anthropology, cataloged variation by race, including differences in head formation, a pseudoscience called phrenology. Blumenbach's aim was to classify the world based on the different types of bumps he could measure on people's skulls. Based on these skull measurements, he came up with five principal varieties of humans: Caucasian, Mongoloid, Ethiopian, American, and Malay. Caucasians (named after the people who live on the southern

SCIENTIFIC RACISM

nineteenth-century theories of race that characterize a period of feverish investigation into the origins, explanations, and classifications of race.

ETHNOCENTRISM

the belief that one's own culture or group is superior to others and the tendency to view all other cultures from the perspective of one's own.

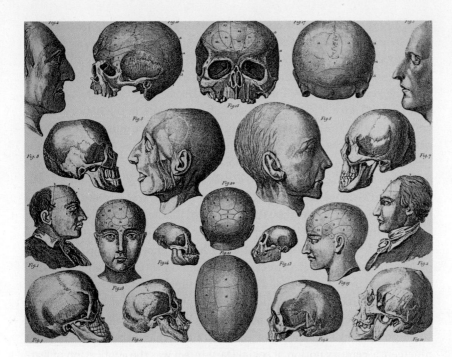

Charts like this one helped phrenologists interpret the shapes of human skulls. How did nineteenth-century theorists use this sort of pseudoscience to justify racism?

slopes of the Georgian region of eastern Europe), he decided, were the superlatives of the races based on their excellent skull qualities.

Another eighteenth-century thinker, the Swiss theologian Johann Caspar Lavater (1741–1801), popularized physiognomy, which correlated outside appearances to inner virtues. Not surprisingly, light skin and small features signified high intellect and worthy character. Political philosophers were also on board with racial thinking. Charles de Secondat, Baron de Montesquieu, had already made the connection between climate and certain forms of government in *The Spirit of the Laws* (1748/1750).

Race was now considered not just a set of physical traits but something that comes with social implications. Immanuel Kant (1724–1804) argued for a link between inner character and outside physiognomy and further claimed that these individual markers were also imprinted on an entire nation's moral life.

However, racial differences were still believed by many to be the product of climate (and therefore not immutable or innate to the soul). In fact, in 1787 the Reverend Minister Samuel Stanhope Smith, who was president of what is now Princeton University, wrote an essay in which he proposed that dark skin should be thought of as a "universal freckle." Differences in skin shade, he maintained, were really just like different levels of suntans. It was his belief that if an African from the sub-Sahara were transplanted to Scandinavia, his dark-brown skin would turn lighter over the course of generations (and perhaps the underlying social and cognitive characteristics associated with race would change as well). Notice how Smith's pliable view

ONTOLOGICAL EQUALITY

the philosophical and religious notion that all people are created equal.

of race captures a spirit of ontological equality: We are all born the same deep beneath our skin; it just so happens that some of us have been out in the sun a bit longer than others. Ontological equality is the philosophical and religious notion that all people are created equal.

The Reverend Smith's line of thinking also demonstrates the scientific influence of Lamarckism, now a discredited footnote in the history of scientific thought. The basic tenet of Lamarckism is that acquired traits can be passed down across generations. For example, an acquired attribute such as flexibility, language skill, or sun exposure can be passed down to a person's offspring, affecting generations to come.

Lamarckism was debunked by Charles Darwin, who in 1859 published his theory of natural selection. Darwin argued that acquired attributes could not be transmitted; instead, change can occur only through the positive selection of mutations. Darwin's theory had an enormous impact on how people thought of race. In effect, it called into question the popular belief that climate influenced racial difference and instead offered an account in which racial lineages were much more deeply rooted and long-standing. What's more, humankind was now seen as being on a trajectory in which some groups have advanced (or evolved) more than others. The popular nineteenth-century notion of social Darwinism was the application of Darwinian ideas to society—namely, the evolutionary "survival of the fittest." Social theorist Herbert Spencer (1820–1903) promulgated the idea that some people, defined by their race, are better fit for survival than others and are therefore intended by nature to dominate inferior races. A new puzzle arose with Darwinian ideas: What, if not inherited climate change, could explain the development of humans along such radically different lines?

SOCIAL DARWINISM

the application of Darwinian ideas to society—namely, the evolutionary "survival of the fittest."

Herbert Spencer coined the term *survival of the fittest*. How did Spencer draw from the work of Darwin to justify racism?

EUGENICS

Scientists now had to arrive at a new explanation of physical difference among humans, and the scientific community confronted a growing debate: monogenism versus polygenism. The debate turned on the origins of the various races of humans. Were humans a united species, or did we come from separate origins? Monogenists, including religious traditionalists, believed that humans were one species, united under God. Polygenists believed that different races were, in fact, distinct species. Darwin sided with the monogenists, claiming that the notion of different species was absurd. (Politics, it is said, makes for strange bedfellows. It certainly did in this case, as Darwinists and religious traditionalists, usually opposed, became allies in arguing that all humans were one species.)

Even though the monogenists won the debate, the notion of separate roots and distinct reproductive genetic histories has had a lasting impact on how we think of human difference. Under the model of natural selection, human difference must have evolved over tens or hundreds of thousands

Howard Knox (center), assistant surgeon at Ellis Island from 1912 to 1916, examined immigrants for diseases and disabilities, including mental illnesses.

of years (if not millions), not just over a few generations in relative sun or shade. Such a vast time frame was used as evidence that races were very different (and not simply superficially so).

A new movement, eugenics, took the idea of very distant origins and ran with it. Eugenicists, led by Sir Francis Galton (1822–1911), claimed that each race had a separate package of social and psychological traits transmitted through bloodlines. Eugenics literally means "well born"; it is the pseudoscience of genetic lines and the inheritable traits they pass on from generation to generation. Everything from criminality and feeblemindedness to disease and intelligence, Galton asserted, could be traced through bloodlines and selectively bred out of or into populations. One of his followers, the American psychologist H. H. Goddard (1866–1957), applied eugenic thinking to generalize findings from his intelligence tests in America. He tested a handful of immigrants arriving at Ellis Island in the early twentieth century and generalized their test scores to whole populations, claiming— and garnering many believers too—that around 70 percent of the immigrants sailing from eastern and southern Europe were, in his phraseology, "morons" who posed a serious threat to the good of the nation. Goddard supported the immigration exclusion acts that in 1924 largely blocked non-Anglos from immigrating and were intended to improve the "stock" of the nation. This concern about the new and objectionable stock of immigrants, as opposed to "native," more desirable immigrants of an earlier epoch, was the crux of nativism, the movement to protect and preserve indigenous land or culture from the allegedly dangerous and polluting effects of new immigrants. Madison Grant (1865–1937), an influential writer, epitomized the spirit of nativism when he argued that not restricting the immigration of southern and eastern Europeans was "race suicide" for the white race (1916/1936).

The problem with race, for eugenicists, scientists, and politicians, has always been that if race is such an obvious, natural means of dividing the

EUGENICS

literally meaning "well born"; a pseudoscience that postulates that controlling the fertility of populations could influence inheritable traits passed on from generation to generation.

NATIVISM

the movement to protect and preserve indigenous land or culture from the allegedly dangerous and polluting effects of new immigrants.

Bhagat Singh Thind.

world, why does no foolproof way of determining racial boundaries exist? According to the social historian Ian Haney López (1995), the US Supreme Court grappled with this question in the late nineteenth and early twentieth centuries. In a landmark case in 1923, for example, Dr. Bhagat Singh Thind, a Sikh from India, was denied American citizenship. The Supreme Court ruled that he did not qualify as a "free white person," despite being the first Indian Sikh to be inducted into the US Army in World War I. In previous cases, the Court relied on a combination of scientific evidence and "common knowledge" to decide who counted as white. But the Thind case posed a particular challenge because leading anthropologists at the time uniformly classified Asian Indians as members of the Caucasian race. The very notion of whiteness was at stake: If the anthropologists were right, then the commonly accepted conception of whiteness would be radically changed to include dark-skinned immigrants like Thind. The Court therefore decried science as failing to distinguish human difference sufficiently, relying on common knowledge alone to deny Thind's claims to whiteness. As the Court put it, "The words 'free white persons' are words of common speech, to be interpreted in accordance with the understandings of the common man" (Haney López, 1995).

TWENTIETH-CENTURY CONCEPTS OF RACE

The judges in the Thind case were not the only people who attempted to define whiteness and nonwhiteness in the absence of a stable scientific taxonomy of race. In Nazi Germany, for example, race posed certain key questions: How can Jewishness be detected? Are Jews a race or a religious group? Both, actually: They are a religious group that has been racialized. Scholars have pointed out that the seeds of racism may be traced to anti-Judaism among early Christians, who forced Jews to convert. Anti-Semitism grew in the eleventh century and was based on the belief that getting rid of Jews was preferable to converting them. But Jewishness was still a social identity at this point—a matter of having religious beliefs that differed from the norm. Anti-Semitism did not turn into racism until the idea took hold that Jews were intrinsically inferior, having innate differences that separated them from their Christian neighbors (Fredrickson, 2002; Smedley, 1999). In Nazi Germany, where Jews were believed to have such innate and inherited differences, the problem remained: How can a person be identified as Jewish? This became an obsession during the Nazis' program of racial purification. They devised a "scientific" way to detect Jewishness by measuring ratios of forehead to nose size to face length, but they had little luck in nailing down a reliable strategy for making such a determination (hence, Jews in Nazi-occupied countries were forced to wear a yellow Star of David as a marker of their identity).

In the 1960s, many whites in rural parts of America similarly failed in trying to distinguish themselves from their mixed Native American and black neighbors (known as "tri-racial isolates") by searching for distinguishing signs on the body, such as differences in fingernails, feet, gums, and lines on the palm of a hand. In exasperation, some whites reported having to rely on good, old-fashioned "instinct" to distinguish themselves from nonwhites (Berry, 1963).

One means (but still not foolproof) of drawing sharp racial boundaries in America was the one-drop rule, asserting that just "one drop" of black blood makes a person black. The rule developed out of the laws passed in many US states forbidding miscegenation, or interracial marriage. By 1910, most whites in the United States had accepted this doctrine. The one-drop rule was integral in maintaining the Jim Crow system of segregation upheld in the 1896 Supreme Court decision in *Plessy v. Ferguson*. In the American South, it was clear that anyone of black lineage fell on the unfortunate side of the racial divide, and the rule essentially cleaved America into two societies: one black, one white. This meant again clumping together all "white ethnics" into one united category. As F. James Davis notes (1991), the one-drop rule was highly efficient, not least because it completely erased stratification *within* the black community that had previously been based on skin tone.

Scientific racial thought slowly passed out of vogue as theories of cultural difference gained momentum among American intellectuals from the 1920s to the 1940s. Anthropologist Franz Boas dismissed the biological bases of discrete races, and sociologists such as Robert Park advanced new ideas about culture's importance in determining human behavior. Race, these thinkers argued, was less about fixed inherited traits than about particular social circumstances. Furthermore, when World War II exposed the kind of atrocities to which scientific racism could lead, it became socially and scientifically inappropriate to discuss race in biological terms, and eugenics came to be considered a dangerous way of thinking. With the decline of scientific racism and the shift toward cultural theories of race and ethnicity, the Immigration Act of 1924 was gradually chipped away at starting in 1959 and then completely repealed in 1965. Don't let the formal denouncement of racial thinking fool you, however. Cultural explanations of race often reflect a disguised racist ideology just as much as biological ones do.

Despite the ideas of scholars such as Boas and Park, the old idea of fixed, biological racial differences remains alive and well today, although in modified form. The search for racial boundaries continues in the twenty-first century with the rise of molecular genetics. DNA research now allows us to look deeper than the bumps on our heads, deeper than skin tone or palm lines. Today you can find a number of DNA testing companies offering you an inside look at your heritage. For as little as $70 or so, a testing kit arrives in the mail, you swab your mouth according to the directions, and then

ONE-DROP RULE

the belief that "one drop" of black blood makes a person black, a concept that evolved from US laws forbidding miscegenation.

MISCEGENATION

the technical term for interracial marriage, literally meaning "a mixing of kinds"; it is politically and historically charged—sociologists generally prefer *exogamy* or *outmarriage*.

↑

Marion West (center) embraces Vy Higgensen (right). The distant cousins discovered their relationship after the results of separate DNA tests were entered into a database.

send the swab to the company. Your "real" identity comes back from the lab in about seven weeks.

Wayne Joseph, a 53-year-old Louisiana high-school principal with Creole roots, did just that. Born and raised black, but having light skin, Joseph was mildly curious about the ancestry in his veins. He received some unexpected results: His genetic makeup is 57 percent Indo-European, 39 percent Native American, 4 percent East Asian, and zero percent African (Kaplan, 2003). Despite the findings, Joseph continues to embrace his ethnic identity as black. As he put it to reporters, "The question ultimately is, are you who you say you are, or are you who you are genetically?" Rachel Dolezal, a Washington woman whose parents say she is Caucasian, maintains that "from my earliest memories I have awareness and connection with the black experience, and that's never left me. It's not something that I can put on and take off." Her controversial self-identification as a black woman caused her to lose her local leadership position in the Spokane chapter of the National Association for the Advancement of Colored People (NAACP) and her job teaching in the Africana Studies department at Eastern Washington University (Samuels, 2015).

Cells, alleles, and gene sequences have become the new tools of science that promise to reveal our racial truths, but the old idea hasn't much changed—that there is a biological and social package of traits inside our bodies that can be traced through our lineage—despite our knowledge that humans are biologically one species. There is no doubt that there exists genetic variation that corresponds to the general geographic origins of what we call race, but the amount of variation is nowhere near as great as most people believe. Further, the relationship between genes and complex social behavior (i.e., intelligence) is not very well understood. Yet when the 2004 General Social Survey (GSS) asked respondents why, on average, African Americans have worse jobs, income, and housing than white people, about 80 of the 888 respondents, or 9 percent, responded that blacks "have less in-born ability to learn." (Moreover, a whopping 49.7 percent of respondents believed that blacks are worse off because they "just don't have the motivation or will power to pull themselves up out of poverty.") Even over a decade later, the newest data available from the GSS in 2015 show that these numbers remain largely the same (Barry-Jester, 2015). The historical search for difference affects our belief system today.

Racial Realities

The biological validity of discrete racial categories—be it the bumps on your skull or the DNA in your blood—may be debatable, but in social life, race is real, with real consequences. Just ask someone of the Burakumin race in Japan. Today making up anywhere from 1 to 3 percent of the Japanese population, depending on the estimate, the Burakumin originated as a group of displaced people during fourteenth-century feudal wars (Hankins, 2014). With no connection other than being Japanese, the Burakumin suddenly shared something undesirable—they were homeless, destitute, and forced to wander the countryside together. Imagine all the homeless people today in Miami or Los Angeles suddenly uniting. The Burakumin formed a distinct social category, with complete social closure, their own reproductive pool, their own occupational pool, and so on, although they were not a distinct group genetically. Today, however, it is commonly believed that the Burakumin "are descendants of a less human 'race' than the stock that fathered the Japanese nation as a whole" (De Vos & Wagatsuma, 1966). Six hundred years later, the Burakumin still display no physical distinctions from their fellow Japanese citizens. For those people in Japan wishing to avoid interrelations with the Burakumin, this lack of distinctiveness poses a dilemma. So for a hefty price, there are private investigators for hire who will confirm the pedigree of your prospective employee, tenant, or future son-in-law.

In Japan, the Burakumin live in ghettos, called *burakus*, and score lower on health, educational achievement, and income compared with their fellow Japanese citizens. Yet when Japanese and Burakumin immigrate to America, the scoring gap narrows dramatically. The distinction between Burakumin and other Japanese is meaningless outside of the significance bestowed on it in their home country. Again, we see that race is not necessarily just about physical or biological differences.

To take an example of racial realities closer to home, consider the consequences of being Arab—or perhaps I should say being perceived as Muslim—in post-9/11 America. In an interview for this book, Jen'nan Read used the image of a Venn diagram to explain that "Arab is an ethnicity; being a Muslim is a religious categorization. In the US context a lot of Muslims and Arabs seem to be the same group. In fact, most Arabs in the United States are Christians who immigrated prior to World War One. And most Muslims are

The Burakumin are a minority who can be distinguished from the rest of the Japanese population only by genealogical detectives. Prejudice against this group often leads to homelessness.

actually not Arab…one-third are South Asian, one-third are Arab, and about a quarter are African Americans who have converted to the religion" (Conley, 2009i). In the United States, Muslims are often identified with Islamic terrorists. Followers of Islam these days are often lumped into a fixed racial category as a dangerous and undemocratic "other," seen as separate from, and inferior and hostile to, Christians.

In the first year after the events of 9/11, the number of anti-Muslim hate crimes shot up 1,600 percent (Read, 2008). And though reported crimes dropped over the following decade and a half, such backlash has resurged after the 2016 presidential campaign of Donald Trump, which some see as linked to his tendency to retweet incendiary remarks or memes. In the first six months of his administration, anti-Islamic hate crimes rose by a quarter from the prior year (which was already higher than the previous period) (Council on American-Islamic Relations, 2017). Most notable were mosque torchings, cemetary vandalism, and an incident in Bloomington, Minnesota, when a bomb was thrown into a mosque in an act Governor Mark Dayton called terrorism (Bromwich, 2017).

In the wake of terrorist anxiety, and several years into the war on terror, Muslims in America have undergone what scholars call racialization, the formation of a new racial identity by drawing ideological boundaries of difference around a formerly unnoticed group of people. These days, any brown-skinned man with a beard or woman with a headscarf is subject to threats, violence, and harassment. And men with turbans bear some of the worst discrimination, although nearly all men who wear turbans in the United States are Sikh, members of one of the world's largest religious groups, which originated in India. Four days after 9/11, Balbir Singh Sodhi, a Sikh living in Mesa, Arizona, was shot five times and killed in the gas station he owned. He was the first victim of an anti-Muslim epidemic, and he wasn't even Muslim. In one Harvard study, 83 percent of Sikhs interviewed said that they or someone they knew personally had experienced a hate crime or incident, and another 64 percent felt fear or danger for their family and themselves (Han, 2006). Even more striking is what happens to Caucasian Americans who convert to Islam. One woman, despite having fair

RACIALIZATION

the formation of a new racial identity by drawing ideological boundaries of difference around a formerly unnoticed group of people.

skin and green eyes, has been categorized by people as Palestinian when she wears the Muslim headscarf, called the *hijab*. She's even been told, "Go back to your own country," although she was born and raised in California (Kuruvil, 2006; Spurlock, 2005). The frequency of such incidents is likely to increase as more people convert to Islam in the United States (Read, 2008).

The racialization of Muslims operates on several flawed assumptions. First, people make stereotyped assumptions based on appearance (turban = Osama bin Laden), even if in their own personal experience they know better (most people wearing turbans in the United States are Sikhs, not Muslim extremists). Second, making snap judgments about Muslims requires a gross caricaturization of Islam's followers. Remember, two-thirds of Arabs in the United States are not Muslim but Christian (Pew Research Center, 2011a).

As it turns out, American Muslims are both a highly diverse population and a very mainstream one. Muslims have been in North America since the seventeenth century, when they were transported from Africa as slaves, and about 37 percent of Muslims were born in America (Pew Research Center, 2011a). As a group, they are assimilated with the mainstream, having income and education levels similar to those of the rest of the population. By and large, they hold fast to the ideas that education is important and that hard work pays off in a successful career. Doesn't sound too radical, does it? That's because the overwhelming majority of Muslims in America and throughout the world strongly disagree with Islamic extremism (Pew Research Center, 2011a). Of course, no racial boundaries are drawn along accurate lines or real differences, but again, once racialized, a group faces real social consequences.

A group of Sikhs protest after the murder of Balbir Singh Sodhi.

Race versus Ethnicity

What's the difference between race and ethnicity anyway? Some books use the terms interchangeably; most subsume *race* under the umbrella label of "ethnicity." Today's understanding of race is that it is

- *Externally imposed:* Someone *else* defines you as black, white, or other.

- *Involuntary:* It's not up to you to decide to which category you belong; someone else puts you there.

- *Usually based on physical differences:* Those unreliable bumps on your head.

- *Hierarchal:* Not white? Take a number down the ranks.

- *Exclusive:* You don't get to check more than one box.

- *Unequal:* It's about power conflicts and struggles.

This last point is important for making sense of the Burakumin and Muslim Americans. Racial groupings are about domination and struggles for power. They are organizing principles for social inequality and a means of legitimating exclusion and harassment.

Ethnicity, one's ethnic affiliation, is by contrast

- *Voluntary:* I choose to identify with my one-eighth Irish background (it makes me feel special, so why not?).

- *Self-defined:* It is embraced by group members from within.

- *Nonhierarchal:* Hey, I'm Irish, you're German. *Great!*

- *Fluid and multiple:* I'm Irish *and* German. *Even better!*

- *Cultural:* Based on differences in practices such as language, food, music, and so on.

- *Planar:* Much less about unequal power than race is.

Ethnicity can be thought of as a nationality, not in the sense of carrying the rights and duties of citizenship but in the sense of identifying with a past or future nationality. For Americans, Herbert Gans (1979b) called this identification symbolic ethnicity. Symbolic ethnicity today is a matter of choice for white middle-class Americans. It has no risk of stigma and confers the pleasures of feeling like an individual. For example, from 2000 to 2010 the second-fastest-growing racial group in the United States was Native Americans. According to the US Census, the American Indian and Alaska Native (alone or multiracial) population grew by 1.1 million, about 39 percent, whereas the entire US population grew during that same time by 9.7 percent (US Census Bureau, 2012a). These numbers are not the outcome of migration or a Native American baby boom, but instead reflect a growing interest in claiming one's heritage, so long as it's not too stigmatizing and brings just the right amount of uniqueness. Native Americans are eligible for numerous types of federal assistance, ranging from health-care services to preferential admission rates at colleges (though a 2004 Civil Rights Commission report found that the federal government spends more per capita for prisoners' health care than for Native Americans' [Ho, 2009]).

Massachussetts senator Elizabeth Warren provides one high-profile

---

ETHNICITY

one's ethnic quality or affiliation. It is voluntary, self-defined, nonhierarchal, fluid and multiple, and based on cultural differences, not physical ones per se.

SYMBOLIC ETHNICITY

a nationality, not in the sense of carrying the rights and duties of citizenship but of identifying with a past or future nationality. For later generations of white ethnics, something not constraining but easily expressed, with no risks of stigma and all the pleasures of feeling like an individual.

example of such ethnic claims. At certain times in her career she claimed Native American heritage. In the dog-eat-dog world of politics this symbolic ethnic claim came back to bite her as opponents challenged her to prove her ancestry. Most infamously, President Trump has mockingly called her "Pocahontas," using the term even during a meeting with Native American leaders. One lesson from the Warren controversy is that while symbolic ethnic assertions of Native American status are rising, formally claiming American Indian ancestry is more than a matter of checking a box. Membership criteria vary from tribe to tribe but most require genealogical proof.

Crowds line the street at the St. Patrick's Day Parade in New York City. How is this an example of symbolic ethnicity?

These differences between race and ethnicity underscore the privileged position of whites in America, who have the freedom to pick and choose their identities, to wave a flag in a parade, or to whip up Grandma's traditional recipe and freely show their ethnic backgrounds. The surge of ethnic pride among white Americans today implies a false belief that all ethnic groups are the same, but in the very way that symbolic ethnicity is voluntary for white ethnics, it is not so for nonwhite ethnic Americans such as Latinos and Asians. As soon as someone classifies you as different on the basis of your phenotypical (racial) features, you lose the ability to choose your ethnic identity. It becomes racialized—subsumed under a forced identifier, label, or racial marker of "otherness" that you cannot escape. Thus, although it is common to use the term *ethnicity* across the board to refer to Latino, black, Asian, or Irish backgrounds, being Irish in America is something that a person can turn on or off at will. You can never *not* be Asian or black: Your body gives away your otherness, no matter how much you want to blend in.

To be black in America is to be just that—black. Some scholars argue that this is the fundamental issue about race in America. Blacks were considered, until recently, a monolithic group. They were unique among racial groups in that their ethnic (tribal, language group, and national) distinctions were deliberately wiped out during the slave trade in order to prevent social organization and revolt (Eyerman, 2001). Alex Haley's landmark novel *Roots* (1976) raised an awareness among African Americans about tracing their ethnic rather than racial identity. This is changing now as more African Americans trace their roots to specific places in Africa through genealogical research or DNA testing. Likewise, immigration from Africa and the Caribbean has created distinctly recognizable national groups of origin among the

US black population. Finally, the presidential campaign of Barack Obama led, perhaps, to a symbolic expansion of who counts as black within the African American community: even the son of a white, Kansan mother and Kenyan father who has half-Asian and fully African half-siblings. Though some African Americans were initially uncomfortable with embracing Obama and his mixed racial heritage as "black like us," their attitudes would have had little impact on the wider perception of Obama as a black man because of the continuing significance of the one-drop rule in America. Indeed the white nationalist backlash that followed Obama's inauguration, which has continued through the campaign and election of Donald Trump, suggests that the one-drop rule is alive and well. That is, the fact that Obama was biracial and not a descendant of US slaves did not seem to matter to Americans who cheered his ascent or those who wished for his downfall.

Ethnic Groups in the United States

The United States is home to countless ethnicities today. It has such a heterogeneous population, in fact, that it is on its way to having no single numerically dominant group. Until the mid-nineteenth century, ethnic diversity was minimal because immigration rates were relatively low until about 1820. Early in the country's formation, white Anglo-Saxon Protestants dominated over Native Americans as well as black slaves. From that point forward, Anglos secured their place at the top of the cultural, political, and economic hierarchy, prevailing over other immigrant populations. As exemplified in the following snapshot of ethnic groups in America today, these historical hierarchies have remained relatively stable and intact despite drastically changing demographics.

NATIVE AMERICANS

According to archaeological findings, the original settlers of the North American continent arrived anywhere between 12,000 and 50,000 years ago from northeast Asia, traveling by foot on glaciers. This is disputed by some tribes, such as the Ojibwe, who believe that their ancestors came from the east, not the west. Before European explorers arrived in significant numbers for extended periods in the fifteenth century (there is evidence that the Vikings had reached North America before then), the indigenous population was anywhere between 10 and 100 million. The tribes living here when Europeans showed up were geographically, culturally, and physically diverse, but they were categorically viewed as a single uncivilized group by arriving Spanish, French, and British explorers. Even today, "Indians" in America are part of 280 distinct cultural groups (Gray & Nye, 2001). Foreshadowing the lack of respect to come, Columbus called all of the people he met Indians, despite their clear cultural differences, because he thought he was in India. Confronted with foreign diseases and unfamiliar military technology, the Indians were quickly dominated by white invaders. In Central and South America, the Spanish brutally enslaved them as labor for the mining industry. In the northern parts of North America, French colonialists nurtured their relationships with the Indians in order to cultivate a profitable fur trade. The British, chiefly concerned with acquiring land, settled colonies with the long-term goal of expanding the British state, dispossessing and "civilizing" America's indigenous population in the process (Cornell, 1988).

The Oglala Nation Powwow and Rodeo parade is a bright spot for the Pine Ridge Indian Reservation, which, like many federal reservations, is plagued by problems linked to low socioeconomic status.

American Indians' way of life was completely obliterated by the European settlements, as vital land was taken from them and their communal infrastructure was disrupted. Most devastating were the newly imported diseases, such as smallpox and cholera, against which the Indians, having no native immunity, were virtually defenseless. There were also grueling forced marches from native lands to dedicated reservations. By the end of the 1800s, the Native American population had dwindled to approximately 250,000 (US Census Bureau, 1993). The Indian Bureau (later called the Bureau of Indian Affairs) was established as part of the War Department in 1824 to deal with the remaining "Indian problem," and its chief means was "forced assimilation." This involved removing Indian children from their families and putting them in government-run boarding schools that taught the superiority of Anglo culture over "primitive" Native culture and religion. Children who refused to adopt Western dress, language, and religion met with harsh physical and emotional punishment (Cornell, 1988). (A similar project was undertaken in British-ruled Australia with the native Aboriginal tribes.) Despite this poor treatment, Navajo men served the United States in World War II, in which 29 "code talkers" used the Navajo language in lieu of cryptography to protect the secrecy of American communications. The work of the code talkers was critically instrumental in the American victory at Iwo Jima and many other battles throughout the war (Bixler, 1992).

Today, people claiming at least some Native American ancestry number about 5.6 million (US Census Bureau, 2017a). Only about one-fifth of Native Americans live in a designated American Indian area. The largest reservation, Navajoland or Diné Bikéyah, covers approximately 16 million acres of

Arizona, New Mexico, and Utah. Reservations are generally impoverished and rife with health problems, domestic abuse, substance abuse, poor infrastructure, and high crime. In fact, Native Americans as a whole are plagued by the lowest average socioeconomic status. They rank among the worst in terms of high-school dropout rates and unemployment, which go hand in hand with poor health outcomes such as alcoholism, suicide, and premature death. Suicide rates are more than double that of the general population and are the second-leading cause of death for Native Americans from ages 15 to 34 (Centers for Disease Control and Prevention, 2012a). Around 33 percent of Native Americans die before age 45, compared with 11 percent of the US population as a whole (Garrett, 1994).

Though facing many social problems, Native Americans are becoming more politically organized and active as a bloc. For example, sustained protests against the Dakota Access Pipeline crossing sacred ground—and the reservation's water source—in North Dakota led to a delay and rerouting of the oil conduit (McKenna, 2016). And recent lawsuits brought by Native Americans in Colorado have successfully challenged the gerrymandering that had concentrated their numbers into a single state district, thereby diluting their power (Turkewitz, 2018).

AFRICAN AMERICANS

The first black people in North America arrived not as slaves but as indentured servants contracted by white colonialists for set periods. They were in the same circumstances as poor, nonfree whites such as those from Ireland or Scotland (Franklin, 1980). The system of slavery, however, evolved to meet colonial labor needs, and the slave trade was a fixed institution by the end of the seventeenth century. African Americans have been, all in all, on the bottom of the racial hierarchy ever since.

Just before the American Revolution, slaves made up more than 20 percent of the colonial population (Dinnerstein et al., 1996). Today about 12.7 percent of the American population is black (US Census Bureau, 2017b). Like the Japanese Burakumin, this minority group has high rates of poverty, health problems, unemployment, and crime. According to the US Census Bureau, the median income of African Americans as a group is roughly 62.8 percent that of whites (US Census Bureau, 2017c, 2017d). In 2015, more than 6.7 million people in the United States were on probation or parole, in jail, or in prison—that's 2.7 percent of all US adult residents, or 1 in every 37 adults (Kaeble & Glaze, 2016). Among men ages 25 to 39, blacks are imprisoned 2.5 times and 6 times as often as Hispanics and whites, respectively;. overall, just 0.5 percent of white men are imprisoned (Carson, 2014).

Sociologists and demographers today are beginning to study how new black immigrants are fracturing the holistic conception of "African American." For the first time, more Africans are entering the country than during

Wiley College student body president Kabamba Kiboko dances at a pep rally. Her family came to the United States from the Congo. For the first time, more Africans are entering the country than during the slave trade.

the slave trade. More than 9 percent of the black population is foreign born (US Census Bureau, 2012b). Afro-Caribbeans such as Cubans, Haitians, and Jamaicans resent being unilaterally categorized as African American, because each of these immigrant groups enjoys a unique history, culture, and language that do not correspond to the American stereotypes of black skin. For this reason, new black immigrant groups would rather not assimilate, but instead retain their distinctive immigrant status, setting themselves apart from the lowest status group in America, blacks (Greer, 2006).

LATINOS

Latino, like the term *Hispanic* (the two are often used interchangeably), refers to a diverse group of people of Latin or Hispanic origin. In 2016, the majority of Latinos in the United States were from Mexico (about 63.2 percent), Puerto Rico (about 9.5 percent), Cuba (3.9 percent), and the Dominican Republic (3.3 percent) (US Census Bureau, 2017e). (Figure 9.1, next page, provides a breakdown of the US Hispanic population by region of origin.) They are a huge and rapidly expanding segment of the American population; in 2012 they made up approximately 17 percent of the population, surpassing African Americans (US Census Bureau, 2012c). Latinos also live in a wide array of locations, although they have clustered on the West Coast and in the South and Midwest (US Census Bureau, 2011).

Hispanics are often called an "in between" ethnic group because of their intermediate status, sandwiched between Caucasians and African Americans. Unlike African Americans, the majority of Latinos in America today

FIGURE 9.1 US Hispanic Population by Region of Origin, 2016

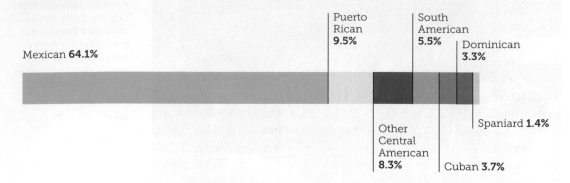

Mexican **64.1%**

Puerto Rican **9.5%**

South American **5.5%**

Dominican **3.3%**

Other Central American **8.3%**

Spaniard **1.4%**

Cuban **3.7%**

SOURCE: US Census Bureau, 2017e.

Framingham, Massachusetts, is home to the nation's highest concentration of Brazilian American immigrants. With recent immigration policies, how do you think these kinds of communities will be affected?

have come by way of voluntary immigration, particularly during the last four decades of heavy, second-wave immigration. Puerto Ricans are the exception, because they have been able to travel freely to the United States since 1917, when Puerto Rico became an American territory and its inhabitants US citizens. The chief motivation for Latino immigration is economic because of America's high demand for labor in the service, agriculture, and construction industries.

Mexicans, Puerto Ricans, and Cubans all have diverse phenotypical traits that make racial distinctions of a unified Latino type nearly impossible. Mexicans are generally classified as "mestizos," a term referring to a racially mixed heritage that combines Native American and European traits. And for instance, Puerto Ricans are often a mixture of African, European, and Indian backgrounds. So ambiguous is the Latino label that at various times the US Census has classified them as part of the white race and as a separate race. At the moment, being Hispanic or Latino is considered an ethnic identity, not a racial identity.

Most Cubans consider themselves Hispanic whites, although their immigration status has changed drastically in recent years. Following the Communist revolution led by Fidel Castro, the first large wave of Cuban immigrants arrived in southern Florida in the 1960s. These immigrants were upper or middle class, were educated, and were perceived as the victims of a Communist regime

that came to power during the Cold War. As such, they were welcomed enthusiastically to this country, and their assimilation started smoothly (Portes, 1969). By 1995, however, that warm welcome had faded. The US federal government terminated its 35-year open-door policy toward Cuban refugees, and the heavy media coverage of Cubans arriving since then in small boats has led to stereotypes of a desperate "wetback" invasion. Arrivals since the 1970s—who generally are from lower socioeconomic backgrounds in Cuban society—have met more resistance from their host society, consequently experiencing higher rates of unemployment, low-wage work, and dependence on welfare and charity than native whites and previous Cuban émigrés (Portes et al., 1985). More recently, with the normalization of relations between the United States and Cuba, the special status that Cubans had with respect to immigration (essentially a fast track to American citizenship) has been terminated, and as a result, deportations have increased. Some 37,000 Cubans face deportation orders as of 2018 (Torres, 2017). The Trump administration is revising Obama's policies so limbo and uncertainty may continue for a while.

ASIAN AMERICANS

Like *Latino*, the term *Asian American* is very broad, encompassing diverse and sometimes clashing peoples from China, Korea, Japan, and Southeast Asia. The first wave of Asians to arrive in the United States in the mid-nineteenth century were predominantly laborers of Chinese, then Japanese, then Korean and Filipino origin. A second large wave of immigration is currently under way, mostly made up of well-educated and highly skilled people from all over Asia.

Early Asian immigrants were perceived as a labor threat and therefore met with extreme hostility. The Chinese Exclusion Act of 1882, which led to a ban against the Chinese in 1902, marked the first time in American history in which a group was singled out and barred entry. Urban "China-towns" developed out of ghettos in which marginalized Chinese workers, mostly men, were forced to live. Japanese immigrants faced similar hostilities and were formally barred entry by the Oriental Exclusion Act of 1924. By 2010, Asian and mixed-Asian US residents amounted to 5.9 percent of the population. This was a 46 percent increase over a decade, making Asian Americans the fastest-growing racial group (US Census Bureau, 2014). They are most heavily concentrated in California, Hawaii, New York, Illinois, and Washington.

Asian Americans are unique among ethnic minorities because of their high average socioeconomic status, surpassing that of most other ethnic minorities as well as most whites in terms of educational attainment. For example, the median family income for the US population as a whole in 2013 was $71,062, whereas for Asian Americans it was $92,260 (US Census

Bureau, 2017f). That said, despite the overall success of Asian Americans, certain groups—notably Cambodians and Hmong (from Laos and Vietnam)—experience very high poverty rates.

Furthermore, Asian Americans overall find it a bit more difficult to be reemployed once they lose a job. In 2013, the Asian American unemployment rate for those with bachelor's degrees or higher was 0.1 percentage point higher than the rate for whites (Bureau of Labor Statistics, 2014b), and Asian Americans and blacks suffer from the longest average duration of unemployment spells. In 2013, unemployed Asian Americans experienced a median 20.5 weeks before they found work again. For blacks that year, the figure was 21.5 weeks. (Whites and Latinos had 14.5- and 15.4-week median durations between jobs, respectively, in 2013.) However, these black–Asian and white–Latino similarities obscure different reasons for the long duration for the various groups. Whereas blacks typically face barriers to employment, Asians have more family support and savings and thus are able to wait out a bad labor market for the best possible job.

In recent years, Asians have been applauded for their smooth assimilation as the "model minority," the implication being that if only other ethnic groups could assimilate so well, America would have fewer social problems. Such a view, however, effaces the rather unsmooth history Asian immigrants have faced in this country as well as the continuing poverty and discrimination faced by some Asian ethnics. Furthermore, "positive" stereotypes of high achievement are not always beneficial and can place enormous pressure on Asian youths to measure up to an impossibly high ideal.

Not only can the model minority myth be damaging to Asians themselves but it can be used by those with an extremist agenda. For example, the president of the National Policy Institute, a white supremacist think tank, Richard Spencer, routinely cites data claiming that East Asians have the highest IQs on average, followed by those of European descent. This placement of whites second to another group paints a patina of scientific objectivity onto the claims that so-called alt-right race rabble-rousers like himself are merely "race realists." How biased can he be, after all, if he's not even listing his own group first? Of course, the real point is to pave the way for claims that other nonwhite groups are intellectually inferior. Even back in the 1960s, the case of Asian American assimilation and upward mobility was used to blame other groups for their failures to achieve success—ignoring the circumstances minority groups faced and inherited (Lim, 2018).

MIDDLE EASTERN AMERICANS

Middle Easterners come from places as diverse as the Arabian Peninsula, North Africa, Iran, Iraq, and the Palestinian territories. They established communities in the United States as far back as the late 1800s, but their numbers have swelled since the 1970s as part of the rising tide of non-European immigration. Middle Easterners in this second wave of immigration often arrive from politically tumultuous areas to seek refuge in the United States—such as the many refugees who have recently fled from war-torn Syria.

Today about 2 million Americans report Arab ancestry, and even more Americans have a Middle Eastern heritage, because not all Middle Easterners are Arab (US Census Bureau, 2017g), despite the fact that most Americans regard anyone from the Middle East as Arab and Muslim. In fact, the largest Middle Eastern population in the United States today is from Iran, and they are Persians, not ethnic Arabs, and do not generally speak Arabic. Similarly, although the majority of new Middle Eastern Americans are Muslim, many of them are Christian, and a small number are Jewish (Bozorgmehr et al., 1996).

Widespread misunderstandings about Middle Easterners derive, in part, from their negative stereotyping in the mainstream media. In one study of television portrayals of Arabs, researchers found four basic myths that continue to surround this ethnic group. First, they are often depicted as fabulously wealthy—as sultans and oil tycoons. Second, they are shown as uncivilized and barbaric. Third, they are portrayed as sex-crazed, especially for underage white sex slaves. Fourth, they are said to revel in acts of terrorism, desiring to destroy all things American (Shaheen, 1984). Little has changed since this study came out over 30 years ago, although after 9/11, the emphasis shifted away from stereotypes of Arabs as extremely rich and toward one of Middle Easterners as terrorists.

The Importance of Being White

We've seen some of the trajectories of various ethnic and racial groups in America, but what about the largest racial population, whites? Scholars have begun to pay more attention to what it means to be a white person. Every year on the first day of my introduction to sociology class, I ask my 200 or so students to write down the five social categories that best describe who they are. Black students almost always put their race at or near the top of the list. Latino and Asian American students usually list their ethnicity as well. Until recently, I could be fairly confident that whites would not list their race. Some might identify Polish, German, or another ethnic or national origin, but not one white student would write down Caucasian, white, or even Euro-American. And that was the point of my experiment.

We have already seen how the category of whiteness is socially constructed—first inclusively defined as all "free white persons" in 1790, then restrictively defined as only northern and western European whites in the early twentieth century, and reformulated back to an umbrella category by the mid-twentieth century. We know now that this category, which seems so natural and innate, is actually a flexible label that has expanded over time to include many formerly nonwhite groups such as Jews, Irish, and Italians. Today most white people have little awareness of the meaning of whiteness as a category. As Nell Irvin Painter, the author of *The History of White People*, says, "The foundation of white identity is that there isn't any. You're just an individual" (Schackner, 2002).

Whiteness, argues Peggy McIntosh (1988), is an "invisible knapsack of privileges" that puts white people at an advantage, just as racism places nonwhites at a disadvantage. In her now classic essay "White Privilege: Unpacking the Invisible Knapsack" (1989), McIntosh catalogs more than 50 "Daily Effects of White Privilege," ranging from the mundane to the major. Here are just a few McIntosh notices:

- I can log into Netflix and count on finding movies and TV featuring people of my race represented, go into a supermarket and find the staple foods which fit with my cultural traditions, and walk into a hairdresser's shop and find someone who can cut my hair.

- I can arrange to protect my children most of the time from people who might not like them.

- I do not have to educate my children to be aware of systemic racism for their own daily physical protection.

- I am never asked to speak for all the people of my racial group.

- I am not made acutely aware that my shape, bearing, or body odor will be taken as a reflection on my race.

- I can choose blemish cover or bandages in "flesh" color and have them more or less match my skin. (pp. 3–5)

Whiteness, then, is about not feeling the weight of representing an entire population with one's successes or failures. It's about not having to think about race much at all.

In recent years, however, awareness of whiteness has been on the rise, as evidenced by the profusion of scholarship on whiteness, the goal of which is to call attention to the social construction and ensuing privilege of the category. Calling attention to whiteness helps whites understand how slanted the playing field really is. It also helps rectify something wrong with the way we study race in America: By traditionally focusing on minority groups, the implicit message that scholarship projects is that nonwhites are "deviant," to borrow from the Comte de Buffon, and that's why we study them. Even popular culture has caught on with memes like "stuff white people like," "columbusing," and "if black people said the stuff white people say," which offer humorous yet pointed critiques of privileges that affluent white people tend to have in America.

But white consciousness may have another, more troubling side. In 1980, before white studies got under way in universities, the white supremacist David Duke left his position as the grand wizard of the Knights of the Ku Klux Klan and founded the National Association for the Advancement of White People (NAAWP), attempting to sugarcoat his racist movement with a seemingly more politically correct approach. In this new framework, Duke presented whites as a besieged minority, writes sociologist Mitch Berbrier (2000), defining the NAAWP's mission as a pro-white heritage movement as opposed to an antiblack one. Sociologist Abby Ferber has analyzed the clever appropriation of civil rights language in Duke's white supremacist discourse. For example, in an article by Duke in the *White Patriot*, Ferber finds the rhetoric of reverse discrimination, victimhood, and the right to cultural difference:

↑

Former Klansman David Duke stands under a Confederate battle flag. How was Duke able to cast the NAAWP as a pro-white movement instead of a racist organization?

> [O]ur race and all others should have the right to determine their own destiny through self-determination and rule.... [E]very people on this planet must have the right to life: the continued existence of its unique racial fabric and resulting culture. (*White Patriot*, no. 56, p. 6, quoted in Ferber, 1999)

Sounds reasonable, right? That's because the new language of white supremacy allows racists to move away from explicitly racist language (of biological inferiority, for example). Duke's NAAWP also co-opts civil rights discourse, as in the organization's original mission statement: "The NAAWP is a not for profit, nonviolent, civil rights educational organization, demanding equal rights for whites and special privileges for none."

Such examples demonstrate one possible outcome of the emergence of white studies: to politically empower extremists by giving them a legitimate language for their racist ends. Note, however, that this rhetoric does not acknowledge the advantages whites typically enjoy. The power of whiteness studies is that it exposes the social construction of a seemingly natural (and neutral) category, giving a sense of the unequal footing beneath the labels "white" and "black." (That is, whites can embrace ethnic identity, but blacks are stuck in a racial category.) But like any new "technology," such discourse can be used for many ends.

Minority–Majority Group Relations

What are the *social* consequences of race? Scholars have defined four broad forms that minority–majority group relations can take: assimilation, pluralism, segregation, and conflict.

In the 1920s, sociologist Robert Park began to wonder what, on the one hand, held together the diverse populations in major American cities and, on the other hand, sustained their cultural differences. He came up with a race relations cycle of four stages: contact, competition, accommodation, and assimilation. His model, called straight-line assimilation, was at first accepted as the universally progressive pattern in which immigrants arrive, settle in, mimic the practices and behaviors of the people who were already there, and achieve full assimilation in a newly homogenous country.

Milton Gordon (1964) tweaked Park's model by suggesting multiple kinds of assimilation outcomes. For Gordon, an immigrant population can pass through (or stall in) seven stages of assimilation: cultural, structural, marital, identificational, attitude receptional, behavior receptional, and civic assimilation (Table 9.1).

With Park and Gordon in mind, let's do a thought experiment. Imagine yourself as a Polish immigrant arriving at Ellis Island in 1900. You don't have much money, and you've come to America in search of opportunity; this is the land of plenty, so you've been told. You settle into a Polish enclave of

STRAIGHT-LINE ASSIMILATION

Robert Park's 1920s universal and linear model for how immigrants assimilate: they first arrive, then settle in, and achieve full assimilation in a newly homogenous country.

TABLE 9.1 Gordon's Stages of Assimilation

STAGES OF ASSIMILATION	CHARACTERISTIC
Cultural assimilation	Change of cultural patterns to those of host society
Structural assimilation	Large-scale entrance into cliques
Marital assimilation	Large-scale intermarriage
Identification assimilation	Development of sense of collective identity based exclusively on host society
Attitude reception assimilation	Absence of prejudice
Behavior reception assimilation	Absence of discrimination
Civic assimilation	Absence of value and power conflict

Manhattan, where you connect with friends and maybe some family. You do your best to learn English. You buy a pair of riveted denim pants, popular among American workmen. You secure work in a factory thanks to your connections in the Polish community. After an initial period of tension and conflict stemming from job competition and housing constraints, you eventually are accepted by your Anglo-American neighbors, first by being allowed to join the workers' union and then—and this probably only happens to your children—by being allowed to marry into an Anglo family. By this time, you think of yourself as an American. Congratulations, you have reached Milton Gordon's final stage of civic assimilation.

Harold Isaacs (1975) noticed something that these theories of assimilation could not explain: People did not so easily shed their ethnic ties. Ethnic identification, among white ethnics and everyone else, persisted even after a group attained certain levels of structural assimilation. Clifford Geertz (1973) explained this persistence as a matter of primordialism—that is, the strength of ethnic ties resides in deeply felt or primordial ties to one's culture. Ethnicity is, in a word, *fixed*. If not biologically rooted, it's rooted in some other intractable source that Geertz reasoned must be culture.

The flip side of this argument came from Nathan Glazer and Daniel P. Moynihan in *Beyond the Melting Pot* (1963). Far from being a deeply rooted structure that kept people bonded to their culture, ethnic identification, they reasoned, persisted because it was in an individual's best interest to maintain it. They saw ethnic groups as miniature interest groups—individuals uniting for instrumental purposes, such as fending off job competition. Glazer and Moynihan believed that ethnicity was fluid and circumstantial. More recently, scholars have posited that ethnic identification is both a deeply felt attachment and an instrumental position that can change according to circumstance (Cornell & Hartmann, 1998).

PLURALISM

For most people, however, assimilation into American society is not so easy, and acceptance varies systematically. At times a pressure cooker has been invoked as a more appropriate metaphor than a melting pot. Park's model was useful in shifting attention away from essentialist explanations of the so-called innate differences among immigrants, but it suffers from several shortcomings. Most obviously, it does not apply to nonwhite immigrants, many of whom are not fully accepted into all areas of American society. Park's model also does not apply to involuntary immigrants, notably African Americans and some refugees. Ernest Barth and Donald Noel (1972) noted that assimilation is not necessarily the end result for immigrants. On the contrary, other outcomes such as exclusion, pluralism, and stratification are possibilities. And as others point out (Lieberson, 1961; Massey, 1995; Portes & Zhou, 1993), some immigrants assimilate more easily than

PRIMORDIALISM

Clifford Geertz's term to explain the strength of ethnic ties because they are fixed and deeply felt or primordial ties to one's homeland culture.

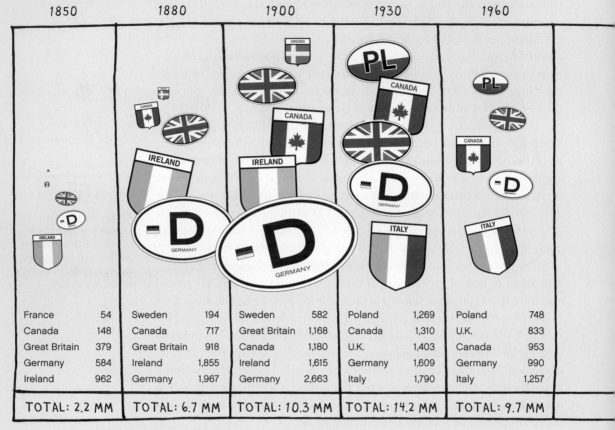

FIGURE 9.2 Five Largest Foreign-Born Populations in the United States, 1850 to 2016 (in thousands)

1850		1880		1900		1930		1960	
France	54	Sweden	194	Sweden	582	Poland	1,269	Poland	748
Canada	148	Canada	717	Great Britain	1,168	Canada	1,310	U.K.	833
Great Britain	379	Great Britain	918	Canada	1,180	U.K.	1,403	Canada	953
Germany	584	Ireland	1,855	Ireland	1,615	Germany	1,609	Germany	990
Ireland	962	Germany	1,967	Germany	2,663	Italy	1,790	Italy	1,257
TOTAL: 2.2 MM		**TOTAL: 6.7 MM**		**TOTAL: 10.3 MM**		**TOTAL: 14.2 MM**		**TOTAL: 9.7 MM**	

NOTE: The total numbers include all foreign-born populations, not just those in the top five.
SOURCES: Gibson & Jung 2006; US Census Bureau 2017g.

others, depending on a variety of structural factors, like migration patterns; differences in contact with the majority groups; demographics including fertility, mortality rates, and age structure; and ultimately, power differentials among groups. This is the case for the "new immigration," which, in comparison with the earlier era of European immigration (1901–30), is a large-scale influx of non-European immigration that began in the late 1960s and continues to the present (Figure 9.2).

The new immigrants, largely from Hispanic and Asian countries, are racialized as nonwhites—even though Asians are widely considered a "model minority." They therefore are subject to a different set of conditions for assimilating and face greater obstacles to their upward mobility (Massey, 1995). For example, Portes and Zhou (1993) report that the children of Haitian

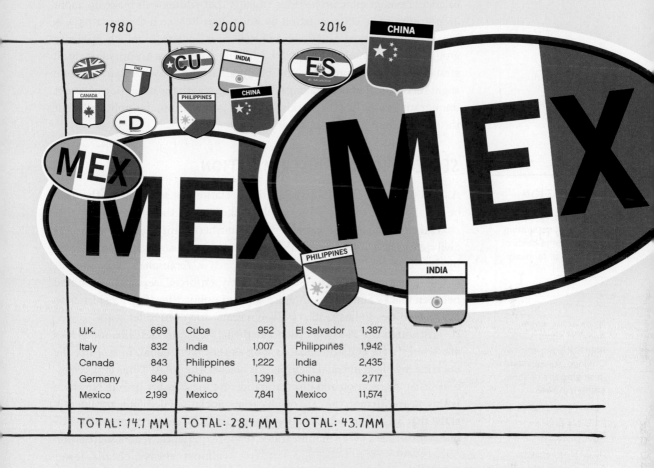

1980		2000		2016	
U.K.	669	Cuba	952	El Salvador	1,387
Italy	832	India	1,007	Philippines	1,942
Canada	843	Philippines	1,222	India	2,435
Germany	849	China	1,391	China	2,717
Mexico	2,199	Mexico	7,841	Mexico	11,574
TOTAL: 14.1 MM		TOTAL: 28.4 MM		TOTAL: 43.7MM	

immigrants living in Miami's Little Haiti are at high risk for downward mobility because they face social ostracism from their own ethnic community if they choose to adopt the outlook and cultural ways of native-born Americans. Among Haitian youth, a common message is the devaluation of mainstream norms, and anyone who excels at school or abides by mainstream rules runs the risk of being shunned for "acting white." This picture of assimilation is much more complex than Park's initial formulation.

A society with several distinct ethnic or racial groups is said to exhibit pluralism, meaning that a low degree of assimilation exists. A culturally pluralistic society has one large sociocultural framework with a diversity of cultures functioning within it. This is the premise of multiculturalism in America. Statistically speaking, in a pluralist country, no single group

PLURALISM

the presence and engaged coexistence of numerous distinct groups in one society.

commands majority status. Switzerland, with its three linguistic groups—German, French, and Italian—is a striking example of ethnic autonomy and balance. Demographic projections suggest that whites will make up about 46 percent of the US population by 2065 (Pew Research Center, 2015c). A broader definition of pluralism, however, is a society in which minority groups live separately but equally. Imagine America with no substantial stratification, no oppression, and no domination. In Switzerland, despite slight income differences among ethnic groups (with the exception of recent immigrant groups like the Turkish), no one group dominates politically, but the same cannot be said for America.

SEGREGATION AND DISCRIMINATION

SEGREGATION

the legal or social practice of separating people on the basis of their race or ethnicity.

A third paradigm for minority–majority relations is segregation, the legal or social practice of separating people on the basis of their race or ethnicity. An extreme case of segregation was the southern United States before the civil rights movement. Under the Jim Crow system of segregation, reinforced by the Supreme Court's 1896 ruling in *Plessy v. Ferguson*, a "separate but equal" doctrine ruled the South. Strictly enforced separation existed between blacks and whites in most areas of public life—from residence to health facilities to bus seats, classroom seats, and even toilet seats.

Although the *Plessy* decision ruled that separate facilities for blacks and whites were constitutional as long as they were equal, in real life the doctrine legalized unequal facilities for blacks. The NAACP has long recognized that segregation and discrimination are inescapably linked. Nowhere is this clearer than in the case of education. Social science data consistently show that an integrated educational experience for minority children produces advantages over a nonintegrated school experience. School segregation almost always entails fewer educational resources and lower quality for minority students.

As Anthony Marx (1998) has noted, concern over segregation grew during World War II as America was caught in the embarrassing contradiction of espousing anti-racist rhetoric against its Nazi foes while upholding an egregiously racist doctrine at home. America emerged from the war as a global force with heightened stakes for its world reputation; this new status,

Black actor, singer, and civil rights leader Paul Robeson leads Oakland dockworkers in singing the national anthem in 1942.

Elementary school students in Ft. Myer, Virginia, face each other on the first day of desegregation.

along with growing public dissent, perhaps helped motivate the Supreme Court's landmark 1954 decision in *Brown v. Board of Education*. The Court's majority opinion that legally segregated public schools were "inherently unequal" is considered the ruling that struck down the "separate but equal" doctrine. It was also the spark that ignited the civil rights movement of the 1960s.

However, school desegregation has been under fire since several Supreme Court decisions in the 1990s (*Dowell*, 1991; *Pitts*, 1992; *Jenkins*, 1995). Two 2007 cases in Louisville, Kentucky, and Seattle, Washington, came the closest to overturning the spirit (if not the letter) of *Brown*. Earlier, former presidents Richard Nixon and Ronald Reagan had both openly attacked desegregation initiatives, especially busing. In 1981, President Reagan's attorney general, William Bradford Reynolds, flatly proclaimed that the "compulsory busing of students in order to achieve racial balance in the public schools is not an acceptable remedy" (Orfield, 1996). Today most US schools are only marginally less segregated than they were in the mid-1960s. It seems that 1988 was the least segregated year in US schools. Namely, researchers at the University of California, Los Angeles, found the number of "hyper-segregated schools, in which 90% or more of students are minorities, grew since 1988 from 5.7% to 18.4%" (Orfield et al., 2016). School segregation is invariably linked to poverty, which is perpetuated by residential segregation, and perhaps this issue is the one we should be addressing if we are concerned about mitigating racial disparities.

In 1968, under President Lyndon B. Johnson's initiative, the Kerner Commission reported that despite the civil rights movement sweeping the nation, America was split into two societies: "one black, one white—separate and unequal." The main reason for the fissure was residential segregation, what sociologist Lawrence Bobo (1989) has termed the "structural linch-pin of American racial inequality." Residential segregation, scholars argue, maintains an urban underclass in perpetual poverty by limiting its ties to upwardly mobile social networks, which connect people to jobs and other opportunities. When you live in the ghetto, your chances of landing a good job through your social network are indeed slim (Wilson, 1987).

It has also been suggested that residential segregation inflicts poverty through a "culture of segregation" (Massey & Denton, 1993). According to this argument, you live in a ghetto that's extremely isolated from the outside world—no family restaurants like the Olive Garden, no mainstream bank branches, not even a chain grocery store that sells fresh vegetables. You're surrounded daily by the ills that accompany poverty: poor health, jobless-ness, out-of-wedlock children, welfare, educational failure, a drug economy, crime and violence, and in general, social and physical deterioration. In the ghetto, the most extreme form of residential segregation, you come to believe that this is all there is to life; the social ills become normative. It's no big deal to sell drugs, drop out of school, depend on welfare, or run with a gang. You slide into the very behaviors that, in turn, reproduce the spiral of decline of your neighborhood.

Whether you buy this line of thought or not (and this viewpoint has been criticized as being overly deterministic), consider how a segregated neigh-borhood got that way in the first place. It didn't just pop up out of nowhere, nor was it always there. As Douglass Massey and Nancy Denton (1993) have argued, the ghetto was deliberately and systematically constructed by whites to keep blacks locked into their (unequal) place. Before 1900, blacks faced job discrimination but relatively little residential segregation. Blacks and whites lived side by side in urban centers, as the index of dissimilarity numbers in Figure 9.3 shows. The index of dissimilarity, the standard measure of segregation, captures the degree to which blacks and whites are evenly spread among neighborhoods in a given city. The index tells you the percentage of nonwhites who would have to move in order to achieve residential integration.

Various structural changes—industrialization, urbanization, the influx to the North of South-ern blacks who competed with huge waves of European immigrants—led to increased hostility and vio-lence toward blacks, who found

In 1942, a race riot broke out in Detroit, Michigan, during an attempt by white residents to force African Americans out of the neighborhood.

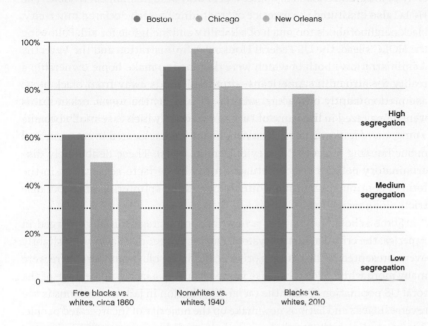

FIGURE 9.3 Index of Racial Dissimilarity

Legend: ● Boston ● Chicago ● New Orleans

High segregation
Medium segregation
Low segregation

- Free blacks vs. whites, circa 1860
- Nonwhites vs. whites, 1940
- Blacks vs. whites, 2010

NOTE: The index of dissimilarity measures the degree of segregation in a given city. A value under 30 is low, one between 30 and 60 is medium, and one over 60 is high.
SOURCES: Massey & Denton, 1993; Brookings Institution, 2010.

themselves shut out of both white jobs and white neighborhoods. The color line, previously more flexible and fuzzy, hardened into a rigid boundary between black and white.

The black ghetto was manufactured by whites through a set of deliberate, conscious practices. Boundaries separating black neighborhoods were policed by whites, first with the threat of violence and periphery bombings in the 1920s and then with "neighborhood associations" that institutionalized housing discrimination. Property owners signed secret agreements promising to not allow blacks into their domain. When a black family did move to a neighboring block, whites often adopted the strategy of flight instead of fight, and this process of racial turnover yielded the same result: black isolation. Even today, when a black family moves into a white neighborhood, the property value declines slightly, in subtle anticipation of the process of white flight, which leaves behind a run-down, undesirable black neighborhood—a veritable vicious circle. On the other hand, the formerly black neighborhood of Harlem in New York City is now no longer majority black, a consequence of the recent in-migration of whites and Hispanics (Roberts, 2010).

Also helping create the black ghetto have been specific government policies, such as the Home Owners' Loan Corporation (HOLC), which in the early 1930s granted loans to homeowners who were in financial trouble. The HOLC also instituted the practice of "redlining," which declared inner-city, black neighborhoods too much of a liability and ineligible for aid. Following the HOLC's lead, the US Federal Housing Administration and the Veterans Administration—both of which were designed to make home ownership a reality for struggling Americans—funneled funds away from black areas and predominantly into white suburbs. Finally, by the 1950s, urban slums were being razed in the name of "urban renewal," which essentially became a program of removal, as African Americans were relocated to concentrated public housing projects (Massey & Denton, 1993). These deliberately discriminatory policies are perpetuated today by de facto segregation in the form of continued suburban white flight and the splintering of school districts along racial lines.

Some scholars argue that a new form of segregation has emerged in America: the criminal justice system. During the 1960s, blacks were slightly overrepresented in the nation's prisons, but in absolute numbers there were many more white felons, because whites made up a larger percentage of the total US population. Today the racial distribution in jails and prisons is the reverse. Blacks and Latinos now make up the majority of incarcerated people. Recent studies estimate that nearly half of all black men will be arrested before the age of 23 (Brame et al., 2014). Is imprisonment just another means of confining the black population away from whites? Several scholars make this case, based on changes in drug laws that seem to affect minorities disproportionately.

Ethnic Tutsi troops overlook a pile of skulls that will be reburied in a memorial to approximately 12,000 Tutsi massacred by Hutu militias.

RACIAL CONFLICT

The final paradigm of race relations is conflict relations, when antagonistic groups within a society live integrated in the same neighborhoods, hold the same jobs, and go to the same schools. This was the volatile scenario in Rwanda in 1994, when roughly 800,000 Tutsi were murdered by Hutu, mostly by machete, in the span of 100 days. That's about 333.3 killings per hour, or five and a half lives every second (Gourevitch, 1998). The killings, as well as the maiming and systematic rape of Tutsi women, were the culmination of more than a century of racial hostility that began with the Belgian colonization of Rwanda.

Belgian explorers, immersed in discourses of scientific racism, confronted two Rwandan tribes, the Hutu and the Tutsi, who for ages had been living and working together and intermarrying. Because of all their shared social, cultural, and genetic heritages, scientists today cannot distinguish Hutu and Tutsi into separate biological populations. But in the late nineteenth century, the Belgians gave preferential treatment to the Tutsi, whom they believed to be superior to the Hutu. What followed was a brutal system of oppression in which the Tutsi dominated the Hutu.

When Rwanda was granted independence in 1962, after a century of hatred brewing between the two groups, the Hutu took power under a dictatorship masked as a democracy, and their long-standing animosity simmered into an explosion in April 1994 after three years of failed crops. The result was genocide, the mass killing of a particular population based on racial, ethnic, or religious traits. The genocide, backed by the government and media, turned neighbors into murderers overnight: Friends killed friends, teachers killed students, and professionals killed co-workers. The Rwandan genocide was a stark reminder that when we speak of the myth or fiction of race, we cannot deny its reality in social life.

GENOCIDE

the mass killing of a group of people based on racial, ethnic, or religious traits.

Group Responses to Domination

There are several forms of response to oppression, four of which are briefly outlined in this section: withdrawal, passing, acceptance, and resistance. Although we tend to think of minority groups as being oppressed by majority groups, keep in mind that sometimes the majority are the oppressed group, as in South African apartheid, where 4.5 million white Afrikaners and British dominated 19 million indigenous people.

WITHDRAWAL

An oppressed group may withdraw, as the Jewish population did after Nazi persecution in Poland. Before World War II, Jews in Poland numbered 3.3 million, the second-largest Jewish population in the world. Eighty-five percent of Polish Jews died in the Holocaust, leaving roughly 500,000. After World War II, violence against Jews continued, and many moved. These conditions, plus the bitter taste of Polish complicity in the Holocaust itself, caused many Jews to leave for good. By 1947, Poland was home to just 100,000 Jews.

Another case of withdrawal was the Great Migration of the mid-twentieth century in the United States. Blacks streamed from the Jim Crow rural South in search of jobs and equality in the industrialized urban North and West; an estimated 1.5 million African Americans left per decade between 1940 and

1970. The North opened opportunities to blacks that previously had been violently denied them in the South, including economic and educational gains as well as the cultural freedom manifested in the Harlem Renaissance. But leaving the South did not always lead to immediate improvements. In their search for a better life, many African Americans found cramped shantytowns on the edge of urban centers, exploitation by factory owners looking for cheap black labor, and increasing hostility from white workers. Racialized competition for housing and employment sometimes led to violent clashes, such as the East St. Louis riots in the summer of 1917. The riots, principally involving white violence against blacks, raged for nearly a week, leaving nine whites and hundreds of African Americans dead. An estimated 6,000 black citizens, fearing for their lives, fled the city, another stark example of withdrawal.

PASSING

Another response to racial oppression is passing, or blending in with the dominant group. For example, during his early adulthood, Malcolm X attempted to look more like white men through the painful process of chemically straightening or "conking" his Afro. A more recent example was the pop star Michael Jackson. And an even more recent, interesting case is that of Rachel Dolezal, who was of white, European ancestry but tried to (and did for a while) pass as African American by altering her hair and skin tone. Her efforts were so successful for a time that she became president of the Spokane, Washington, chapter of the NAACP. Once it was revealed that she did not have any African ancestry, she was dubbed the "undisputed heavyweight champion of racial appropriation" (Vestal, 2017). Passing is not necessarily about physical changes, though. One of the most common ways people have tried to pass has been to change their surnames. The single largest ethnic group in the United States today is German Americans. Not English, but German. Where, do you ask, are all the Schmidts and Muellers? They now go by Smith and Miller, after a huge wave of name changing among German Americans during the world wars—if not during the first one, then often by the second.

ACCEPTANCE VERSUS RESISTANCE

Another response is acceptance, whereby the oppressed group feigns compliance and hides its true feelings of resentment. In Erving Goffman's (1959) terms, members of this group construct a front stage of acceptance, often using stereotypes to their own advantage to "play the part" in the presence of the dominant group (see Chapter 4). Backstage, however, privately among their subaltern or oppressed group, they present a very different self. Sociologist Elijah Anderson refers to this as "code-switching," a strategy used by

SUBALTERN

a subordinate, oppressed group of people.

African Americans in the presence of dominant white society. In Anderson's ethnography of a black neighborhood in Philadelphia, blacks learn two languages, one of the street and one of mainstream society, and daily survival becomes a matter of knowing which one to speak at the right time. For an inner-city youth, an act of code-switching could be as simple as putting on a leather jacket and concealing his textbook beneath it for the walk home from school (Anderson, 1999). (Code-switching is akin to the double consciousness that W. E. B. Du Bois ascribed to African Americans who maintain two behavioral scripts; see Chapter 1.)

A more overt form of resistance, the fourth paradigm of group responses to domination, would be collective resistance through a movement such as revolution or genocide or through nonviolent protest as in the US civil rights movement.

Prejudice, Discrimination, and the New Racism

On the individual level, racism can manifest itself as prejudice or discrimination. Prejudice refers to thoughts and feelings about an ethnic or racial group, whereas discrimination is an act. Robert Merton (1949a) developed a diagram for thinking about the intersections of prejudice and discrimination (Figure 9.4). One who holds prejudice and discriminates is an "active bigot." This is the prototypical racist who puts his money (or burning cross) where his (or her) mouth is. Those who are neither prejudiced nor discriminatory are "all-weather liberals," not only espousing ideologies of racial equality or talking the talk, but also walking the walk when faced with real choices.

COLLECTIVE RESISTANCE

an organized effort to change a power hierarchy on the part of a less-powerful group in a society.

PREJUDICE

thoughts and feelings about an ethnic or racial group, which lead to preconceived notions and judgments (often negative) about the group.

DISCRIMINATION

harmful or negative acts (not mere thoughts) against people deemed inferior on the basis of their racial category, without regard to their individual merit.

FIGURE 9.4 Merton's Chart of Prejudice and Discrimination

ACTIVE BIGOT (PREJUDICED-DISCRIMINATES)	TIMID BIGOT (PREJUDICED)
FAIR-WEATHER LIBERAL (DISCRIMINATES)	ALL-WEATHER LIBERAL

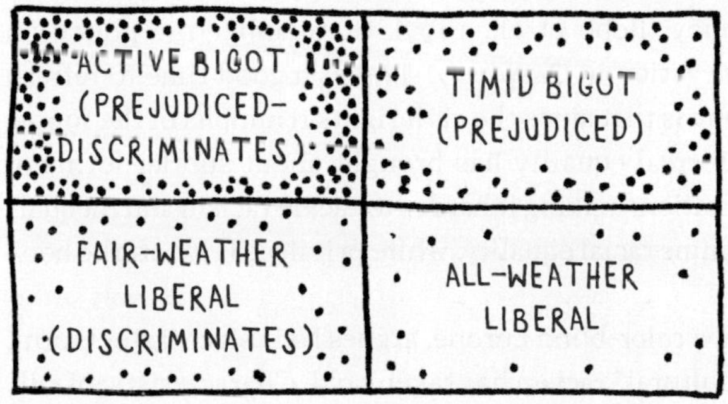

SOURCE: Merton, 1949a.

Many people fall between these two stances. A "timid bigot" is one who is prejudiced but does not discriminate—a closet racist perhaps, who backs down when confronted with an opportunity for racist action. Conversely, one who is not prejudiced but does discriminate is termed a "fair-weather liberal." Despite the fair-weather liberal's inner ideological stance in favor of racial equality, he or she still discriminates, perhaps without knowing it. For instance, a couple who consider themselves open-minded about race relations may feel compelled to sell their home as soon as they are confronted with new black neighbors (white flight, as discussed earlier). Of course, they do not cite residential integration as their motivation for leaving, but instead offer an excuse that has embedded racist reasons, such as the differences in school districts. In fact, the prime time for families to move out of integrated neighborhoods happens to be around the time when the eldest child of the family turns five and begins school.

Active bigots are rarer to come by, because prejudicial viewpoints are largely unacceptable in most of the ostensibly antiracist West. But don't be fooled into thinking that race doesn't matter anymore. It does, and racism is still alive and going strong, although it's veiled in different terms. You just have to know how to look for it. Old-fashioned, overt racism tended to convey three basic ideas: Humans are separable into distinct types; they have essential traits that cannot be changed; and some types of people are just better than others. Not many people openly make such claims in America today, as they tend to be frowned on.

COLOR-BLIND RACISM

the view that racial inequality is perpetuated by a supposedly color-blind stance that ends up reinforcing historical and contemporary inequities, disparate impact, and institutional bias by "ignoring" them in favor of a technically neutral approach.

As this old kind of racism declines, race scholars are beginning to find traces of a new kind of racism gaining ground. Eduardo Bonilla-Silva calls this "color-blind racism." According to Howard Winant (2001), such new racial hegemony comes with "race neutral" rhetoric and relies more on culture and nationality to explain differences between nonwhites and whites than immutable physical traits. The new line of thinking replaces biology with culture and presumes that there is something fixed, innate, and inferior about nonwhite cultural values. In America, this kinder, gentler antiblack ideology is characterized by the persistence of negative stereotyping, the tendency to blame nonwhites for their own problems, and resistance to affirmative action policy (Bobo et al., 1997). (Remember the policy on class-based affirmative action in Chapter 7? This is a good time to refresh your memory.) The irony is that since the civil rights triumphs of the 1960s, the official stance of formal equality has brought about subtler forms of prejudice and discrimination, making it harder to tackle racism and inequality. When a state proclaims racial equality, white privilege is let off the hook and goes unnoticed.

Likewise, in the new color-blind Europe, argues Neil MacMaster (2001), a "differentialist" or "cultural" racism has taken hold. Characteristic of cultural racism is the call to protect national (white) identity from "criminal" and polluting cultural outsiders, constructing an image of "fortress Europe."

Right-wing National Democratic Party members protest immigration and refugee policies in Germany. How is this an example of cultural racism?

Antirefugee commentary is one example of the new racist ideology. For example, in a televised speech in 1978, Margaret Thatcher made the following appeal to cultural purity in regard to 4 million immigrants from recently decolonized countries: "Now that is an awful lot, and I think it means that people are really rather afraid that this country might be swamped by people with a different culture." (MacMaster, 2001). In France, for example, the government does not collect information about race, but social scientists have conducted experiments by sending in job applications with identical résumés other than the names being changed. They show that, with all else equal, those with Arab (Adida et al., 2010) or North African (Pierné, 2013) sounding names get fewer callbacks. Racism can exist even when race does not officially exist.

How Race Matters: The Case of Wealth

Nonwhites, especially African Americans, Latinos, and Native Americans, lag behind whites on a number of social outcomes, from income and educational attainment to crime rates and infant mortality rates. For example, blacks are half as likely as whites to graduate from college or hold a professional or managerial job and are twice as likely to be unemployed and to die before their first year of life. As striking as the figures are, if there is one statistic that captures the persistence of racial inequality in the United States, it is net worth. If you want to determine your net worth, all you have to do is add up everything you own and subtract from this figure the total amount of your outstanding debt. When you do this for nonwhite and white families, the differences are glaring. So while most of America felt the impact of the housing market crash and 2008–09 recession, African Americans and Latinos felt the blow far more sharply than whites. Median wealth for whites fell 16 percent, yet African Americans and Latinos experienced 53 and 66 percent losses, respectively (Pew Research Center, 2011b).

Latinos are a varied group but largely reflect African Americans, who had an average household net worth of $17,100 in 2016, on wealth measures. The median Latino family in 2016 had about $20,600 in assets, and about one-third of both black and Latino households had zero or negative net worth. Compare those figures to the $171,000 in household wealth of the average white family (Kochhar & Cilluffo, 2017). We know considerably less about Native Americans because reliable data are lacking, but given that they have a poverty rate of 26.2 percent (compared with just 11.6 percent for whites), their wealth is not likely to be high (US Census Bureau, 2017h). Asian Americans, however, have low rates of poverty (11.8 percent) (US Census Bureau, 2017i) and high rates of home ownership, at about 60 percent (US Census Bureau, 2017j).

This "equity inequality" has grown in the decades since the civil rights progress of the 1960s. What's more, the wealth gap cannot be explained by income differences alone. That is, the asset gap remains large even when we compare black and white families at the same income levels. For many among the growing black and Latino middle classes, the lack of assets may mean living from paycheck to paycheck, being trapped in a job or neighborhood that is less beneficial in the long run, and not being able to send one's kids to college. Parents' wealth is also a strong predictor of children's teenage and young adult outcomes—everything from teenage premarital childbearing to educational attainment to welfare dependency (Conley, 1999).

Equity inequality captures the historical disadvantage of minority groups and the way those disadvantages accrue over time. The institutional barriers to blacks acquiring property, discussed earlier, were only one such mechanism. These included redlining by banks (whereby loans, especially mortgages, were not given in predominantly black neighborhoods seen as higher risk), racially restrictive covenants (whereby owners had to agree to sell their homes to only whites), and blacks' disproportionate exclusion from government benefits such as FHA and VA mortgages and Social Security retirement pensions (agricultural and domestic workers, who were largely black, were initially excluded from these programs; Truman corrected this).

Similar processes and policies have decimated the wealth of Native Americans, who went from living off the land (the entire US territory) to being disproportionately impoverished and dispossessed over the course of a century by exploitative US policies. One of the most telling examples of this sort of institutionalized dispossession happened to Japanese Americans. As skilled farmers, Japanese immigrants accrued enough wealth in the early twentieth century to attract resentment, culminating in the 1924 Alien Land Act, which prohibited noncitizens from owning land. Japanese immigrants then found success in business, running nurseries and selling cut flowers, and amassed considerable wealth by 1941, about $140 million cumulatively (Lui, 2004). When World War II broke out and panic spread over the possibility of a treacherous Japanese population in America, the Roosevelt administration mandated a program of internment by Executive Order 9066. Japanese American citizens were placed into camps in the western part of the United States. They were given only a week to dispose of all their assets, forcing them to sell their homes and businesses to whites at scandalously low prices (Lui, 2004). The result was a huge forced transfer of wealth from Japanese to whites under discriminatory government policy.

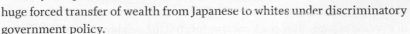

What were the consequences of the Japanese internment camps? How are they an example of equity inequality?

Japanese Americans still alive at the time were paid reparations in 1988. Some have argued for a similar policy to compensate African Americans for the institution of slavery (and racially unequal asset policies since Emancipation). Each year, until his retirement in 2017, Congressman John Conyers introduced a bill to study the issue; so far, even the creation of a commission to assess the issue has been blocked. Meanwhile, the tax reform bill known as the Tax Cuts and Jobs Act of 2017 raised the amount of an estate not subject to taxes to 11.2 million dollars. Because the number of African Americans and Hispanics who would benefit is extremely small, it will serve to widen the already large racial wealth gap.

In summary, policies intending to address disparities between non-whites and whites must take into account the extreme wealth gap and its historical trajectory. Affirmative action aimed at improving wages and increasing job openings for blacks, Hispanics, and Native Americans can address only a piece of a larger cycle of wealth inequality. Income from work provides for the day-to-day, week-to-week expenses; wealth is the stuff long-term upward mobility is made of.

INSTITUTIONAL RACISM

Another cause of the asset inequality discussed in this chapter is the simple fact that property in black neighborhoods doesn't accrue value at the same rate as that in mostly white areas. Property reflects only the value accorded to it in the marketplace. In this vein, when a neighborhood's housing values precipitously decline as the proportion of black residents rises, the price changes provide a record of the economic value of "blackness." And it does not take any "active bigots" or otherwise racist individuals to generate this phenomenon. Namely, aside from any personal ideology, whites have an economic incentive to sell when they sense a neighborhood starting to integrate, as evidence shows that once a neighborhood reaches somewhere between 5 and 20 percent black, it quickly becomes predominantly black due to a rash of selling (with an accompanying drop in values) (Card et al., 2008). This expectation means that it is rational for whites to be the first to sell as blacks increase in numbers in a given area in order to prevent a loss of equity. This, of course, becomes a self-fulfilling prophecy and a vicious circle linking race and property values—to the disadvantage of blacks. (Meanwhile, property value does not follow the same clear-cut pattern for other ethnic minorities, although minority enclaves generally have lower real estate values than exclusively white neighborhoods.)

INSTITUTIONAL RACISM

institutions and social dynamics that may seem race neutral but actually disadvantage minority groups.

The case of race and property values is an example of institutional racism—institutions and social dynamics that may seem race neutral but actually end up disadvantaging minority groups. Another example is provided by sentencing laws for dealing or consuming cocaine. At the height of a period of panic over crack cocaine infiltrating neighborhoods, the United States passed the Anti-Drug Abuse Act of 1986. This law declared that for sentencing purposes, 1 gram of crack cocaine was equivalent to 100 grams of powdered cocaine. The result was a mandatory minimum sentence of five years for a first-time possession charge for a typical crack user. Because crack was cheaper and more prevalent in low-income, predominantly black communities, the result was a huge racial disparity in drug sentences by race. President Obama addressed this issue in 2010 by signing the Fair Sentencing Act, which reduced the sentencing ratio between powder cocaine and crack from 100-to-1 to 18-to-1. This ratio, while an improvement, still leaves a significantly disparate impact by race (Davis, 2011).

Given the role of drug prosecutions as a lynchpin of institutional racism in the criminal justice system, one might think that legalization would be a way to eliminate such disparate impact. However, if we look at the case of marijuana, we can see that institutional racism can even color decriminalization efforts. Blacks and whites use marijuana at similar rates, but blacks are up to 8 times more likely to be arrested for possession thanks to who gets targeted by police and where and how the drug is consumed (American Civil Liberties Union, 2013). Such disparities continue into an age of semi-legalization. That is, to be legal, marijuana must be consumed in your home. But those who live in public housing or rental apartments that do not permit it tend to consume the drug outside, where they are vulnerable to arrest. In fact, in Washington, D.C., arrests for posession of marijuana almost tripled in the year after it was legalized there. Ditto for arrests for selling: Those affected are primarily black residents. Meanwhile, when licenses were awarded to legal growers in nearby Maryland, black companies were shut out since the commission awarding the contracts sought "geographic diversity." So the profits in what is likely to become a massive industry will be unequally distributed by race too (Milloy, 2017).

Yet another example of institutional racism can be found in the case of hiring patterns by employers. With limited information about job applicants, employers may be rational to use social networks to recruit employees since informal ties (i.e., references) can provide more reliable information about individuals than can be gleaned from paper job applications. And because whites tend to hold more managerial positions, and social networks tend to be segregated by race, this need for additional information on the part of employers also perpetuates racial disparities with no racially explicit motivation.

A related dynamic is called statistical discrimination, where firms use race as a shorthand proxy for having attended poorer schools and having experienced other disadvantages that would lead to less productive performance. Although this dynamic is not completely color-blind, it is different from overt racism in that the motivation is not about race per se but about underlying characteristics that tend to be associated with race. Finally, institutional racism can even be encoded into the educational system through test construction. There has been much debate about cultural bias in testing. Beyond this issue, however, is the effect of stereotypes on performance. The psychologists Claude Steele and Joshua Aronson, for example, have shown that they can drive black students' test scores down just by priming them with negative stereotypes before they sit for the exam (Steele & Aronson, 1998). These are just a few of the ways race can continue to disadvantage certain groups even in an age when overt racial animus may have waned in significance.

The Future of Race

This brief overview of the history of race and its present-day ramifications allows us to make some guesses about the future of race. For starters, racial and ethnic diversity in America will tend to increase. The 2010 Census data show a 134 percent increase in Americans who identify as multiracial—that is, 9 million people (Pew Research Center, 2011). The number of foreign-born people in the United States surpassed 37 million (US Census Bureau, 2010a). And according to National Research Council projections, by the year 2060, largely thanks to the most recent wave of immigration (along with differential fertility rates), America's Latino and Asian populations will triple, making up about 28.6 percent and 11.7 percent of the US population, respectively (Colby & Ortman, 2015). No longer black and white, America is now a society composed of multiple ethnic and racial groups with an ever-shifting color line marking fuzzy boundaries (Figure 9.5, next page).

FIGURE 9.5 An Ethnic Snapshot of America in 2016

White, not Hispanic
62%

SOURCE: US Census Bureau, 2017k.

In 1996, there was a Multiracial March on Washington in which multiracial activists demanded a separate census category to bolster their political claims and recognition. Although the movement did not result in a multiracial identity category, for the first time ever, the 2000 Census allowed respondents to check off more than one box for racial identity. The resulting multiracial population is currently estimated to be about 9 million, and those are just the self-identified people who checked more than one race box in 2010. The latest census also asks separate questions about race and ethnicity, which means that census data can now be used to examine some of the racial diversity within the Hispanic population, as well as some ethnic diversity among African American and white populations (Figure 9.6).

This so-called browning of America brings us to a new crossroads. The white–black divide may become the white–nonwhite divide or the black–nonblack divide. Sociologist Jennifer Lee has looked at just this question, and her research shows that the experience of first- and especially second-generation Asians and Latinos indicates the new color line is black–nonblack. What does this mean? In the simplest terms, it means that the biggest differences in demographic characteristics like income, educational attainment, and interracial marriage will be between blacks and all other groups while the distinctions between these other groups continue to narrow. Lee notes,

FIGURE 9.6 Race Questions from the 2010 US Census

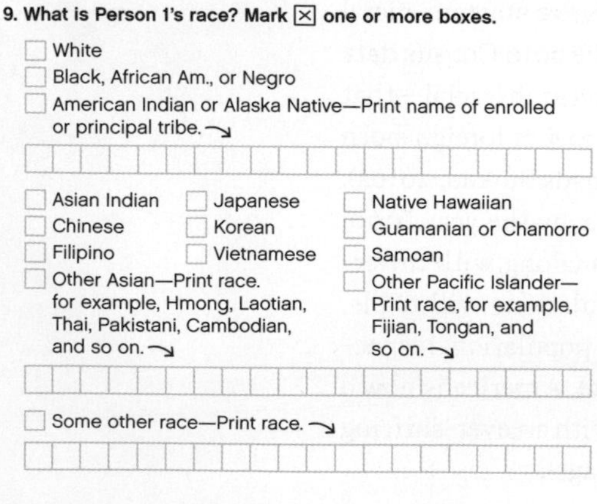

9. What is Person 1's race? Mark ☒ one or more boxes.

☐ White
☐ Black, African Am., or Negro
☐ American Indian or Alaska Native—Print name of enrolled or principal tribe. ↳

☐ Asian Indian ☐ Japanese ☐ Native Hawaiian
☐ Chinese ☐ Korean ☐ Guamanian or Chamorro
☐ Filipino ☐ Vietnamese ☐ Samoan
☐ Other Asian—Print race. for example, Hmong, Laotian, Thai, Pakistani, Cambodian, and so on. ↳ ☐ Other Pacific Islander— Print race, for example, Fijian, Tongan, and so on. ↳

☐ Some other race—Print race. ↳

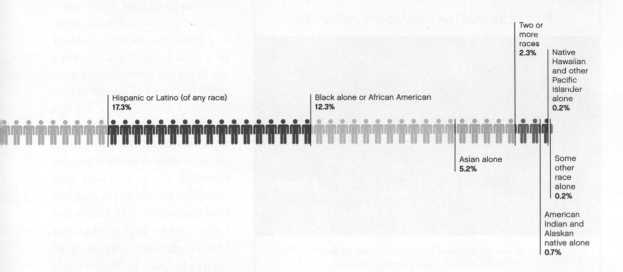

Hispanic or Latino (of any race)
17.3%

Black alone or African American
12.3%

Two or more races
2.3%

Native Hawaiian and other Pacific Islander alone
0.2%

Asian alone
5.2%

Some other race alone
0.2%

American Indian and Alaskan native alone
0.7%

If you look at rates of educational attainment or interracial marriage or multiracial reporting, Asians and Latinos, their disadvantage stems from their immigrant histories, and so with each generation you see outcomes improving. For African Americans, you see there's some progress, but it's not nearly at the same speed, and so what I see is [that] the divide is really blacks and everyone else.... The one caveat I would add is that those Latinos who have darker skin, who are more mistaken for African Americans or black Americans...will probably fall on the black side of the divide. In part because of their skin color, but also because their socioeconomic status is not on par with some of the other lighter skin Latino groups like Cubans. (Conley, 2009j)

Lee points out that the biggest losers in this new configuration of race in America are likely to be—no surprise—blacks, who may be blamed for their own poverty when compared with successful Latino and Asian immigrants. Lee cautions, "Someone will look at African Americans who have been here longer and who aren't attaining certain levels of mobility and say, why can't they make it? If race isn't an issue for Asians and Latinos, it shouldn't matter for blacks. What I argue is that the stigma attached to blackness because of the history of blackness is so different; they are not immigrants. To compare African Americans to immigrants is never fair."

How close are we to this new black–nonblack color line? At least four states (California, Texas, Hawaii, and New Mexico) are deemed "majority minority" states, where whites are not the majority of the population within major metropolitan areas (US Census Bureau, 2010b). But let's not forget

DIGITAL.WWNORTON.COM/YOUMAYASK6CORE

To see my interview with Jennifer Lee, go to
digital.wwnorton.com/youmayask6core

that racial categories are social constructions, not static entities. To claim a multiracial identity presupposes the existence of a monoracial identity, when we know no scientifically pure or distinct race of people exists. Similarly, the notion of a white minority presumes that whiteness is a fixed racial category, whereas whiteness has expanded to include groups that were considered nonwhite in the past and may continue to fold Asians and Latinos into a kind of whiteness that emphasizes symbolic ethnicity. Indeed, as Warren and Twine (1997) point out, as long as blacks are present, a back door is open for nonblacks to slip under the white umbrella. For example, the Asian success story as the "model minority" is made possible by Asians' ability to blend in with whites, because they are unequivocally not black. We are at a point in American history when whiteness may be expanding again, with blacks, as always, serving as the counterweight.

Another consequence of the "browning of America" is the adoption by white nationalists of the rhetoric of minority status. Though it will not be until 2044 that whites are no longer a majority—at which time they will still be the single biggest racial group at 49.9 percent of the population (Colby & Ortman, 2015)—that has not stopped many alt-right politicos from using the fears of being lost in the melting pot to bolster their political power very effectively. Once we actually do arrive at a population distribution in which whites are a minority—though if history holds, likely a wealthy and politically powerful one—it will be interesting to observe what kind of dynamics emerge. There are cases worldwide of what Amy Chua (1998) calls "market dominant minorities," such as whites in South Africa, Chinese in Indonesia, and Indians in some Caribbean nations. When the minority group is the economically and socially dominant group, it is a delicate situation because blame and backlash can flare up when that country's economic prospects falter. For the time being, however, it is (some) whites themselves who seem to be playing the part of aggreived, oppressed minority group—at least at the ballot box.

POLICY

DNA DATABASES

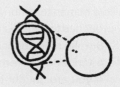

The #BlackLivesMatter social movement is a product of many social forces. There is, of course, explicit and implicit racial bias among law enforcement officers that causes racialized police violence. But there is an important role of technology in explaining how the issue of police violence became so big so quickly. First, there is the social media that allow what might have been local incidents to be strung together into a national awareness—not to mention acting as an organizing tool for actual protests. Perhaps the most important technological innovation contributing to the concern about racialized police brutality, however, is the cell phone camera. Urban theorist Jane Jacobs (1961) once talked about the "eyes on the street" as the primary means by which order was kept in cities. That is, there simply were not enough police to enforce good behavior at all times in all places. Informal watching of our neighbors along with social sanctions is what really kept order, according to Jacobs. But today it is more likely that the cell phones on the street matter most—at least when it comes to policing the policers. Has racialized police violence gotten worse or are we simply more aware of it now that almost every citizen has a handy-dandy camera at hand? Technology reveals—and may ultimately help resolve—a racial problem. One of the most debated policies around police behavior is the implementation of body cameras on police to keep records of interactions.

Seems like a no-brainer, right? But what might be the racialized consequences of this technological fix? Police, for instance, may avoid interactions with African American residents, fearful that any misstep might cost them their career if caught on camera. The potential result? Even less attention might be paid to segregated black communities on the part of police. Some have argued that the huge spike in murders among African Americans in Baltimore in the two months after the anti-police riots of 2015 was the result of a police pullback reaction.

The point is that as long as race matters, seemingly "neutral" technology will be inflected with racialized power relations. Take another example: forensic DNA databases—the stuff of TV courtroom drama. Today, when people are taken into custody by law enforcement in most jurisdictions, a sample of DNA is taken from them. This DNA is then available to compare with all other criminal cases that contain DNA data. Many of the stories in the media are about DNA exonerating—often years later—wrongly convicted individuals. The Innocence Project—started in 1992 at Cardozo Law School—pioneered the movement to initiate appeals of flimsy convictions based on often-overlooked biological samples. And the success of the Innocence Project—along with its heart-wrenching tales of wrongful imprisonment seems to suggest that civil libertarians should welcome this new era of forensic science. This would seem to have a racial-equality-producing effect by countering other forms of bias (say by juries) in the criminal justice system.

On the side of prosecutors, DNA has also been a boon where it has been available at a crime scene or obtained from a rape kit. Now suspects could be definitively matched to the scene of the crime via blood or semen, for instance. At first blush, then, it seems like DNA in the courtroom is an unalloyed good: It reduces the error term in an otherwise error-ridden system that relies on fallible human testimony and other less "scientific" approaches to establishing perpetrator identities.

But like most technologies, forensic genetics has a tendency to reproduce existing inequalities. If you are caught because you have a prior conviction and your bodily fluids at the scene of a new crime match those that are on file, that is unfortunate for you, but it is hard to make a case that it is inherently or systematically unfair. People who have committed prior crimes are more likely to be caught than those who are not in the system. However, if your brother or mother has been genotyped by law enforcement—even if you have led a squeaky-clean life until now—then you are also more likely to be fingered by your genetic fingerprint. That is, your DNA will identify you as a first-degree relative of your sibling. And that information, plus a little detective work, is almost as good as you, yourself, appearing in the dataset. DNA fingerprinting can even identify cousins or grandchildren—although with less certainty. So if your relatives are more likely to have been registered in the database, you are more likely to be located, even if you yourself have no priors. Of course, in this case, the unfairness lies not in the fact that you were tripped up by your DNA but rather in the fact that the person who committed a crime who comes from a more advantaged background and thus does not have relatives in the database gets away with murder (literally or figuratively). Add in the sort of class or race stratification in the criminal justice system that is thought to exist, and you have a perfect storm by which DNA amplifies existing inequalities. In the United Kingdom, for example, 1 in 4 black children over the age of 10 have their DNA in a database, while only 1 in 10 of white children do (Doward, 2009). Multiply that out to their relatives, and boom, a technology that gets a lot of press for exonerating wrongly convicted minorities suddenly does more to add to the disproportionality in the system.

Conclusion

When you look at race using the sociological imagination, you'll recognize that it's hardly a cut-and-dried issue. You'll see the historical social construction of ideas and identities. You'll see the present-day realities—sometimes monstrosities—of an aspect of our lives so often called a myth or fiction. And you'll be equipped to look at the changing nature of race and race relations that will affect your future.

QUESTIONS FOR REVIEW

1. What does the childhood anecdote from the beginning of the chapter, about not yet having learned the meaning of race, teach us regarding the nature of race? Would a nonwhite three-year-old have brought home a white "baby sister"? How does this question bring up the "invisible knapsack of privileges" that puts white people at an advantage?

2. What does real estate value have to do with school segregation? With this link in mind, how have inequalities in wealth contributed to long-term inequality between blacks and whites in the United States?

3. Although the validity of "race" is debatable, why do sociologists study race as it relates, for example, to the likelihood of going to prison? What does this mean about what is "real," the way people understand the world, and what sociologists should study?

4. How has science been informed by culture (including racist beliefs), and in turn, how has science fueled racism?

5. As the saying goes, "You can't judge a book by its cover." How do eugenics and physiognomy contradict this saying (in regard to people)? Are the principles behind these pseudosciences still with us today? If so, in what capacity?

6. What is "racialization," and how has it differed between Muslims and the Irish?

7. How is stating your ethnicity more similar to stating that you like the Beatles than describing your race?

8. Which of the four forms of minority–majority group relations has recently been most prevalent in the United States and Rwanda, respectively? How have the minority groups in these countries responded quite differently to domination?

9. Thinking about the history of race, what do you predict for the future of "race" and "ethnicity" as social categories? Will they stay the same? What do demographic trends and history lessons suggest might happen in the coming decades in the United States?

HOW SEGREGATED ARE YOU?

In recent decades, the United States has undergone what some scholars call "the Big Sort," whereby we have segregated ourselves by political views (think red state versus blue state) and are now clustered into communities of like-minded people. But as we separate into different communities by lifestyle and belief systems, does that mean that we are also living in more racially homogenous social worlds? There is some debate about the overall trends when it comes to racial segregation. By some measures, America today is actually less integrated across racial lines than it was 30 years ago; by others it is more so. It all depends on how we measure the phenomenon.

TRY IT!

Put your hometown's zip code into American Fact Finder at **factfinder.census.gov**. (If you aren't from the United States, use the zip code where your college is based or pick your favorite five-digit number.) From the list of data sets, choose a recent estimate of racial and ethnic* origin—a good one is the American Community Survey's "Demographic and Housing Estimates."

Then compare the racial/ethnic diversity of your community to the country overall. Follow the same instructions above, except instead of your zip code, search for "United States" and pull up the same data set you used to measure the diversity in your community. I'll plug in my NYC zip code:

RACE / ETHNICITY	PERCENTAGE
WHITE	65.0
BLACK OR AFRICAN AMERICAN	9.0
AMERICAN INDIAN AND ALASKA NATIVE	0.4
ASIAN	16.8
NATIVE HAWAIIAN AND OTHER PACIFIC ISLANDER	0.0

RACE / ETHNICITY	PERCENTAGE
SOME OTHER RACE	5.0
TWO OR MORE RACES	3.8
HISPANIC OR LATINO (OF ANY RACE)*	17.3

*Remember that the US Census Bureau does not consider "Hispanic" or "Latino" as racial categories. That's why most data sets include a breakdown of racial origin (e.g., white, African American, Asian), and a separate breakdown of Latino or Hispanic origin.

THINK ABOUT IT

Is your hometown more racially homogenous than the national average? How do you think the results would change if you zoomed into your specific neighborhood or zoomed out to a wider view such as your metropolitan area or state?

The American Community Survey includes data over five years. Compare the diversity of your community today to five years ago. Is yours a community in transition?

SOCIOLOGY ON THE STREET

Decades after real estate segregation (or "redlining") became illegal, many Americans still live in communities that are highly segregated by race and/or socioeconomic status. How does unofficial segregation occur? Watch the Sociology on the Street video to find out more: **digital.wwnorton.com/youmayask6core**.

WANT MORE PRACTICE?

Complete the InQuizitive activity for this chapter at digital.wwnorton.com /youmayask6core

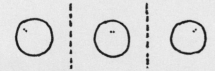

10

WE THINK OF THE FAMILY
AS A HAVEN IN A HARSH WORLD,
BUT IN FACT, INEQUALITY
BEGINS AT HOME.

ELDEST

MIDDLEBORNS

YOUNGEST

Family

In many ways, Ozzie and Harry typify the American ideal of a family. They met while Harry was vacationing in Italy, and it was love at first sight. Harry was 31 and successful; Ozzie was 24, talented, and full of promise. The two quickly initiated a passionate monogamous relationship. Eight years later the couple owned a house in the Los Angeles area, where they lived with their two daughters. Harry worked as a literary agent, supporting Ozzie, a full-time homemaker, and the kids. Theirs is a fairy tale come true.

Yet the story of Ozzie and Harry brings the dominant concept of family into a modern perspective. The two could not come from more dissimilar backgrounds. Harry is white, Jewish, and Ivy League educated; Ozzie is Afro-Brazilian, Catholic, and less formally schooled. Harry is well-off; Ozzie grew up in a poor family and was raised without a father. The couple is "transracial, transnational, cross-class, [and] interfaith" (Stacey, 2004). In fact, they share only one demographic characteristic: They are both male.

Not only is the couple an against-all-odds success story, but their story illustrates just how strong our preconceptions of family are. Even in this most progressive of arrangements, the two assume the historically gendered roles that we often associate with family units. Where do these normative notions of the family come from? To examine this, let's start with the original Ozzie and Harriet: the Nelsons.

There is something inevitably comforting about watching family television shows from the 1950s. It is a glimpse into a foreign yet consoling world where growing boys like Ricky and David Nelson may be viewed dashing into a kitchen, with schoolbooks in hand and a cheerful "Hi, Mom. Hi, Pop!" Their loving parents, Ozzie and Harriet, together listen to their children's problems and together steer them on the right path to adulthood. *The Adventures of Ozzie and Harriet*, the situation comedy that first aired on ABC in 1952, typifies a familiar, lionized form of family that today evokes nostalgia and fuels political debates: the male breadwinner (although no one knows

Why did the idea of the nuclear family—embodied here by the Nelsons from the television show *The Adventures of Ozzie and Harriet*—emerge after World War II?

for sure what Ozzie did for a living); the kind, female homemaker; and their nurtured, well-socialized children.

The scenes of 1950s family life can overwhelm the unprepared viewer with a surge of sentiments—everything from the warm fuzzies to bleak loneliness to bitter cynicism. At first blush, the quality of family life today, with widespread divorce, out-of-wedlock births, dual incomes, and ever-increasing work hours, seems to have taken a nosedive. But if the domestic world of Ozzie and Harriet seems too good to be true, that's because, of course, it is. Anyone watching at the time readily acknowledged that such television shows did not reflect how Americans actually lived. Rather, they portrayed American families as people wanted to see them. Millions of viewers turned to the idealized life presented in *The Adventures of Ozzie and Harriet* to be entertained by the dramatization of mundane details of suburban life and soothed by the easy resolution of the day's problems by the end of the episode.

In the postwar era, the nuclear family—the idealized model of a male breadwinner, a female homemaker, and their dependent children—emerged as the dominant, normative, and mythical model for domestic life. It was dominant, because indeed many Americans took to it; in the 1950s, 86 percent of children lived in two-parent families, and 60 percent of children were born into homes with a male breadwinner and a female homemaker (Coontz, 2001). It was normative, because it has been hailed, then and now, as the proper form for families, the way things ought to be. Finally, the taken-for-grantedness of the nuclear family is mythical, because its existence represents not a natural, timeless, or universal approach to family arrangements but one that appeared in specific historical contexts.

In the sociological imagination, this version of the family looks less like a universal norm and more like an ideological construct, and an unusual one at that. Arising out of unique social structural conditions of the 1950s, characters such as the Nelsons are exceptional, not just for their special Christmas decor and family picnics but also because they are, in fact, unique. That is, their family model deviates from both its predecessors and its successors. Yet many of us still idealize this model and miss it terribly. By studying cross-cultural variation in family forms, we'll see that this allegedly "normal" family is not always the dominant model. By unpacking the historical development of the nuclear family model and its subsequent breakdown, we'll come to recognize that this idea of family is not a universal fact. And regardless of what kind of family ties you may deem worth forging, worth loving, and worth defending, as a sociologist your job will be to study all family forms without prejudgment.

Family Forms and Changes

Why do we fall for the people we do? At first thought this might seem like a simple question. You start by envisioning the personal qualities of your lover, perhaps something about his or her physique, personality, sense of humor, or charm. If you take a step back and really try to think about it, you might admit to more practical reasons as well—physical proximity, shared language, financial security—those everyday factors vital to functioning relationships. If you take yet another step back, you face the larger question of mate selection: the phenomenon of human partnering that is patterned by history, culture, and law.

I know this is an unromantic way to view your love life, but consider the host of cultural and legal codes that prohibit you from partnering with, say, your 14-year-old first cousin. In Victorian England or a modern-day Muslim tribe, however, a man's 14-year-old first cousin might be a perfectly good match. Historically, interracial partnerships have also been prohibited. In 1961, when President Barack Obama's parents were married, it was still illegal in 22 different states for a white and a nonwhite to be married. Then in *Loving v. Virginia* (1967) the US Supreme Court unanimously struck down America's antimiscegenation laws. (See Chapter 9 for a discussion of the one-drop rule.) Similar laws, aimed at maintaining white "racial purity,"

existed in South Africa during apartheid as well as in Nazi Germany. In the contemporary world, ethnically mixed couples are hardly shocking, but 50 years ago in the United States they were almost unthinkable. When a culture maintains either legal or normative sanctions against people marrying outside their race, class, or caste, we call it a rule of endogamy, meaning marriage from within. To some extent, we all practice endogamy—that is, on some level people tend to hook up with similar people, because it makes for easier relations if you and your partner share a social group. In other parts of the world, such as India, historically people haven't had much of a choice: The caste system of India is based on a rigid adherence to endogamy (see Chapter 7).

In much of the contemporary Western world, exogamy, or marriage to someone outside one's social group, is legally possible, if not always culturally acceptable. Consider the following unlikely pair: the African American son of a steel mill worker in Illinois and the only child and sole heir of a mega-millionaire, white, chart-topping rock-and-roll star in Tennessee. The 1994 union of Michael Jackson and Lisa Marie Presley lasted less than two years, but it is an example of a couple who transgressed social, class, and racial groups. However, what is also telling is that by the time they met and fell for each other, they were already both celebrities and social equivalents, despite the gulf between their backgrounds. Total exogamy—when people from completely different social categories get together—is rare.

In addition to codes of endogamy and exogamy, another basic social structure governs your love life: how many partners you're expected to have. In societies that practice monogamy a person can partner with only one other. In societies that practice polygamy people have more than one sexual partner or spouse at a time. (See Chapter 8 for a more thorough discussion of a wide range of sexual practices.) This can take the form of a woman having several husbands, called polyandry, as happens in some rural areas of Asia. Or a man can take several wives, called polygyny, the most common form of polygamy, practiced in many contemporary Islamic and African cultures. Although the Mormon Church outlawed polygamy in 1890, some Mormon splinter groups, called fundamentalists, continue the practice of one man supporting several families at the same time (Lee, 2006).

The next time someone asks why you fell for your special someone, keep in mind that the answer could be quite long and sociologically involved. Cultural norms and state regulations play a fundamental, if invisible, role

The Lovings embrace at a press conference the day after the Supreme Court ruled in their favor in *Loving v. Virginia*, June 13, 1967.

ENDOGAMY

marriage to someone within one's social group.

EXOGAMY

marriage to someone outside one's social group.

MONOGAMY

the practice of having only one sexual partner or spouse at a time.

in your love life. These factors first set the limits as to who is even available on your romantic horizon. Once you find that person (or those persons, for you polygamists), your next move will be to form a relationship with him or her, probably in the shape of a family. But just how might that family look?

MALINOWSKI AND THE TRADITIONAL FAMILY

The normative family model is one in which a heterosexual couple lives with their dependent children in a self-contained, economically independent household. (It is called "traditional" by some, but as we will see later, this is a fairly recent historical development.) This family is typically a patriarchal one, governed by a male head with a dependent wife and children. In 1913 Bronislaw Malinowski put to rest a long-standing family debate among anthropologists about the universal existence of families. Scholars had argued that traditional tribal societies couldn't possibly have family units because of their egregious nondiscriminating sexual promiscuity. Malinowski, based on his research of Australian Aboriginals (who seemed to have sex with everyone), suggested that these natives did, in fact, form ties indicative of familial arrangements. The Aboriginals, he found, recognized family relations and kept them distinct from other forms of social connection. They maintained a central place, the equivalent of a hearth and home, where the family gathered. And they bestowed feelings of love, affection, and care on each family member. Thus the dispute was settled: The family was accepted as a universal human institution for most of the twentieth century, and this notion continues to endure today. It is so strong that scientists look for Malinowski's family ties even among rodents. Neurogeneticists have discovered a genetic basis for monogamous coupling in one species of prairie voles and are hard at work trying to determine whether humans share the same neurological mechanisms for partner preference (Fink et al., 2006; Young & Hammock, 2007; Young & Wang, 2004).

Malinowski argued that the family, in addition to being a universal phenomenon, was a necessary institution for fulfilling the task of child rearing in society. The influential structural-functional sociologist Talcott Parsons would expand on this notion in the 1950s. As we learned in Chapter 8, according to Parsons, the traditional nuclear family, consisting of a mother and father and their children, was a functional necessity in modern industrial society, because it was most compatible with fulfilling society's need for productive workers and child nurturers. And so the nuclear family reigned supreme, timeless, and universal. (Note that although the nuclear family is a particular form of the traditional model, in this chapter we will use the two terms synonymously.)

The problem with functionalist arguments such as these is that even if one social institution seems to perform some function for society, the

POLYGAMY
the practice of having more than one sexual partner or spouse at a time.

POLYANDRY
the practice of having multiple husbands simultaneously.

POLYGYNY
the practice of having multiple wives simultaneously.

NUCLEAR FAMILY
familial form consisting of a father, a mother, and their children.

Families come in all different forms. What are some of the ways that family groups can differ from culture to culture? For instance, how does an extended family in the United States (top) differ from the Na in China (middle) or Zambian families in Africa (bottom)?

function is not necessarily performed only by that particular institution. A functional need is not sufficient cause for the development of an institution, especially not when other kinds of institutions can perform the function just as well. But when you have Western blinders on, as most social scientists following Malinowski did, it becomes difficult to see anything other than the kind of family life you expect to see. When properly fitted with your sociological lens, however, you can make the familiar strange, and you'll find some colorful variation among familial arrangements.

Take the Munduruku villagers of South America, for instance. They give new meaning to the hearth-and-home feature of Malinowski's universal family. Munduruku men and women live in separate houses at different ends of their village; they eat separate meals; they sleep apart. In fact, they meet up with each other only to have sex (Collier et al., 1997). Among the Na people of southern China, who have managed to retain their distinctive culture despite war, Communism, and the onset of quasi-capitalism, the institution of marriage doesn't exist. The Na do not have any practice like it, nor do they put much thought into fatherhood. Children grow up with uncles rather than fathers as the primary males in the home. Sex occurs in the middle of the night, when men visit women for anonymous and spontaneous encounters. There are no social rules restricting who can partner with whom. Yet the Na manage just fine, reproducing from generation to generation and maintaining a stable economy.

If the Na take hands-off fatherhood to the extreme, women in present-day Zambia implode our notions of caring motherhood. Zambian mothers don't nurture their daughters in the way Westerners would expect. When a Zambian girl needs advice, she is expected to seek out an older female relative as a confidante in preference to her mother (Collier et al., 1997). Mothers' "essential" nature has always been defined in the West as nurturing and connected to offspring, presumably ordained by women's biological birthing functions. However, the way motherhood is practiced in Zambian culture suggests that a mother's unconditional warmth, despite the rhetoric, is not necessarily a biological given.

THE FAMILY IN THE WESTERN WORLD TODAY

Even the modern Western family comes in a variety of forms. The typical American family today is not likely to resemble the Nelsons. In addition to nuclear families, one needs to count the extended family—kin networks that extend outside or beyond the nuclear family. Families with no children also exist (for instance, couples that become "empty nesters" after their children move out), as do two-wage-earner families (dual-income families), single-parent families, blended stepfamilies, and adopted families, to name just a few possibilities. Although the traditional family was the dominant model for a majority of people living in the 1950s, it never described home

EXTENDED FAMILY

kin networks that extend outside or beyond the nuclear family.

life for all Americans. And it's increasingly losing its edge today as families take on new shapes and sizes.

In 2016, less than 10 percent of families consisted of two married parents with children and with the husband as the sole earner (Bureau of Labor Statistics, 2017e). Only one-fifth of American households consist of a married couple with their own children, and 62 percent of households have only one or two people in them (US Census Bureau, 2016b).

In the face of soaring divorce rates, the Ozzie and Harriet way of life is indeed becoming a historical artifact. Nowadays, although the exact figure varies considerably from study to study, approximately 40 percent of all marriages end in divorce (Hurley, 2005). This statistic doesn't seem to discourage too many people, though, because three of four divorced men and two of three divorced women try their hand at marriage again. But alas, the second time around has about the same rate of failure as the first, and those marriages are even a bit more likely to hit the skids (Coontz, 2010). It's commonly thought that divorce has skyrocketed since the 1950s, but actually the divorce rate has been steadily rising since the nineteenth century, as divorce became less and less of a social and religious taboo (see Figure 10.4). Divorce rates seem to have stabilized over the past decade or so, with about 90 percent of marriages making it to the 5-year mark and 70 percent making it to 15 years; then, another 1 percent of marriages dissolve each year after that, whether by divorce or death.

But rising divorce rates in the United States have not put marriage in any danger of extinction. Eventually, 86 percent of men and 89 percent of women are projected to get married, but men are more likely to remarry after a divorce (Vespa et al., 2013).

Given the increased incidence of divorce and remarriage, families are taking on new shapes and sizes, such as single-parent and blended arrangements (Figure 10.1). As of 2013, about 15 percent of children in America lived in a blended family (Pew Research Center, 2014), which reconfigures the aftermath of divorce into step-relations. (We'll have more to say about the effect of divorce on children later.) Cohabitation, living together in an intimate relationship without formal legal or religious sanctioning, has also emerged as a socially acceptable arrangement; roughly 24 percent of never-married Americans between 25 and 34 years old live together without marriage. Within three years, though, 58 percent will be married, 19 percent will have broken up, and 23 percent will still be cohabiting (Copen et al., 2013). Many cohabiters believe that living together is an effective means of curtailing future divorce. It's the commonsense notion that living together provides a sample of what married life is like and will inform and improve the couple's future marriage. On the opposite side of the debate, conservative supporters of "family values" point to studies showing a higher divorce rate among couples who cohabit before marriage than those who don't (Meckler, 2002). Conservative Christian counselors advise that "living in sin" sets up

COHABITATION

living together in an intimate relationship without formal legal or religious sanctioning.

FIGURE 10.1 Work-Family Living Arrangements of Children, 1960 and 2015, Ages 0–14

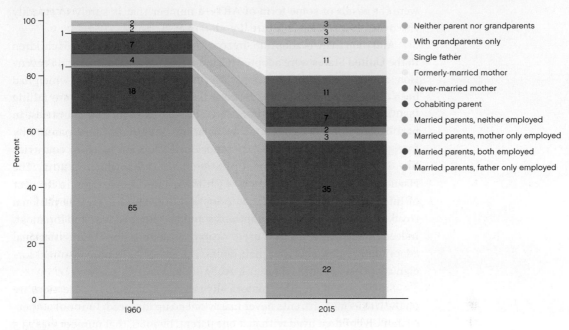

SOURCE: Cohen, 2017.

a rockier road for a marriage, but the actual reason for the higher divorce rate probably has to do more with selection bias: Most people who cohabit are open to premarital sex, which also means they are less likely to see divorce as sinful.

Single women are having their days as well. Twenty-three percent of American households are headed by single moms (they have children, but are not married), and 15 percent of women remain child-free (Livingston, 2015). Women who do have children are doing so at a later age than ever before—now averaging 26 years, up from 21 in the 1970s (Martin et al., 2017). However, women have been delaying marriage for even longer, tying the knot at 27.9 years old, on average (US Census Bureau, 2017n). The age when a woman has her first baby varies by subgroup: White professional women, on average, postpone motherhood even longer. Another interesting new twist in American families is the rapid rise in the number of twins. Between the years 1980 and 2016, the number of twin births has nearly doubled, from 68,339 to 133,155 births, and triplet and other higher-order multiple births increased by a staggering 3.1 times, from 1,337 to 4,123 (Martin et al., 2017). This rise took place almost entirely among white women. As professional white women increasingly delay childbearing, they are biologically more likely to give birth to twins, and this trend is compounded by the more wide-spread use of fertility treatments, which increase the likelihood of a multiple birth (Martin et al., 2010). Indeed, the rate of use of Assisted Reproductive Technology (ART) has risen rapidly. As of 2015, 1.7 percent of all live births were the result of some form of ART—a number that is surely to steadily rise over the foreseeable future (Sunderam et al., 2018).

And then there is adoption. In 2010, approximately 1 percent of children in the United States were adopted (Centers for Disease Control and Prevention, 2011b; Chandra et al., 2005). International adoption rates, which had been climbing since the 1980s, peaked at 22,000 in 2004 and are falling fast. In 2016 only 5,370 children were adopted from foreign countries, in part because of more rigorous qualifications required of new parents by foreign countries, as well as UN-sanctioned adoption bans on countries like Guatemala (US Department of State, 2018). Meanwhile, in 2012, The Hague Convention on Protection of Children and Co-operation in Respect of Intercountry Adoption (Hague Adoption Convention), an international treaty, stated that domestic adoptions should be pursued first and foremost. Indeed, many countries that used to serve as sources for US international adoptions are now shutting their doors. For example, Ethiopia banned foreign adoptions in 2018 (Hosseini, 2018).

Single-parent families, historically the result of death or desertion, are on the rise as more parents never marry or end up divorced. In 1960, 9.1 percent of US children lived with just one parent; by 1986, that number was 23.5 percent; by 2013, it was 27 percent, and 83.9 percent of those children lived in single-mother families (US Census Bureau, 2017o). Single-parent families are also becoming more common worldwide; currently, approximately 14 percent of all families around the globe are headed by women (Chamie, 2016).

Keeping It in the Family: The Historical Divide between Public and Private

Given the enormous cross-cultural and even within-cultural variation among family forms, you may be wondering how and why the so-called traditional family ever came to be the standard marker of normalcy. The short answer is that the development of the family was tied to the development of modernity, state formation, and the rise of the modern economy. For starters, *traditional family* is a misleading term; this family structure is not characteristic of all tradition, just that of the 1950s. The nuclear model of stay-at-home mom and working father is itself deviant compared with American family arrangements both 50 years before it and 50 years after. It was a historically specific model in a unique era, in which many white city-dwellers moved to the suburbs, married young, and raised three or four children. It was a prosperous era, too: In the 1950s real wages grew more in any single year than they did in the 1980s as a whole. With just a high-school education, an average 30-year-old white man could earn enough at his manufacturing job to buy a median-priced home on 15 percent of his salary (Coontz, 2001). Today, with just a high-school education, an average 30-year-old man is likely to be tenuously positioned in the service sector, making about $15 an hour and probably lacking employer-provided health insurance. But it's not just money that set the 1950s apart. As Stephanie Coontz notes in *The Way We Never Were: American Families and the Nostalgia Trap* (1992), people long for the sense of simplicity, wholesomeness, and ease that 1950s families like the Nelsons are perceived to have enjoyed.

PREMODERN FAMILIES

In the preindustrial family of the nineteenth century, each household unit operated like a small business—that is, as a miniature family economy. It was a site for both production and consumption, where work was done in the home and the home was a working unit. Families made and used their own food, clothes, and goods, and there was little if any surplus wealth.

Preindustrial families, such as these settlers, operated like a small business. The home was a site for work, and the entire family was involved.

KINSHIP NETWORKS

strings of relationships
between people
related by blood and
co-residence (i.e.,
marriage).

Families tended to live near their kin and thus could get help and support from their kinship networks, strings of relationships between people related by blood and co-residence (i.e., marriage). Preindustrial communities didn't have huge savings banks, insurance companies, payday lenders, or government agencies to help in hard times; that was the role of family. For example, a down-and-out uncle might have a failed crop of wheat one summer. He could call in an IOU from his luckier cousin across the village, borrowing some of his crop and setting off a reciprocal exchange of food, clothing, and child care. Such families would have a grapevine structure where more lateral kinship ties endured than vertical ones (usually, no more than three generations of one family were alive at a time). Communities cooperated in a noncash economy, using a barter system to swap goods. There was no significant accumulation of goods, no savings, no wealth, and not much beyond what individuals needed to survive. There was also minimal division of labor between the sexes, such that men were involved in child care, and women and children often performed the same manual labor as men. Indeed, in the preindustrial family, children were thought of as "small adults," as Philippe Ariès argued in 1962. They didn't warrant any special treatment or consideration, and childhood was hardly the nurturing period we think of today. In fact, some scholars argue that the whole notion of childhood, or children's special needs, is a relatively recent invention, as is that of motherhood as a full-time occupation. Both have emerged only since the middle of the nineteenth century and have been made possible through industrialization, the rise of the cash economy, formal schooling, and the establishment of privacy in the family.

THE EMERGENCE OF THE MALE BREADWINNER FAMILY

With the Industrial Revolution, the realms of the public and private—previously intertwined as one in the working family—split into separate spheres of work and home, as men left household production for wage work in factories. Families stopped being productive mini-economies for the barter system and instead became strictly sites for consumption. Out went household spinning and weaving, and in came the factory-made sweater. Food, clothing, furniture, decorations, appliances, cosmetics, pharmaceuticals: You name it, the family was there to consume it. Specifically, women made the choices of what and how much their families would consume.

Furthermore, "women's work"—tending the home and raising children—became relegated to the private, domestic sphere, where it went unpaid as a woman relied on a man's wages, at least among the middle classes. All the extra money in the new wage economy passed through the hands of women, who remained in charge of running the home and doing the shopping. Such functions might seem like a move toward women "wearing the pants" in the family, but they, in fact, provided a point of emergence for a new form of gender inequality. Women were in charge of spending the family's money, not earning it, and in our society, work for money is the most highly valued form of labor. Women's work was the unpaid management of the home, and this distinction established unequal positions for men and women in society. At the very least, a man had the chance to spend his wages before they ever reached the other members of his family.

The results of these structural changes were far-reaching. First, a gendered division of labor arose in the household, where women now were in exclusive charge of maintaining the home and rearing children. Second, as the mobility of families searching for paid labor opportunities increased, they became separated from their kinship networks. Family structures changed from grapevine forms to "beanpole" families, in which kinship ties are vertical. Because people didn't live near their siblings, aunts, and uncles, they could depend only on their children and parents who lived with them. As life expectancies increased, up to five generations may have been alive at any given time, but lateral ties (such as those existing between cousins) weakened because of the longer distances separating these kin.

The beanpole family structure is particularly taxing on women today, who, because of delayed childbirth, are likely to find that their parents need assistance just as their own children are leaving home. As a result of declining fertility and longer life expectancies, the US population is growing older, so the problem of elder care (and its disproportionate impact on women) will only get worse. In 2014 the median age in the United States was 37.9 years, with 10 states having a median age of 40 or older (US Census Bureau, 2016c, 2016d).

A third fallout of the new cash economy and the resulting split between public and private realms was the creation of the cult of domesticity, the notion that true womanhood centers on domestic responsibility and child rearing. During the first half of the twentieth century, ideas sprang up surrounding woman's true nature—ideas meant to support her newly created role as housewife. These included the notion that women, more than men, are endowed with the innate emotional qualities required to provide warmth and comfort. According to this ideology, not only are women better suited

This nineteenth-century painting illustrates the Victorian feminine domestic ideal. How did the Industrial Revolution transform the division of labor between men and women?

CULT OF DOMESTICITY

the notion that true womanhood centers on domestic responsibility and child rearing.

for home life, but also their domesticity comes to be seen as necessary for the survival of society. Women, so the argument goes, are needed to ensure that the home remains a safe haven in the otherwise cold seas of capitalist enterprise. As breadwinners, men struggle in the dog-eat-dog commercial world, whereas women provide the emotional shelter that anchors private life in human sentiment, emotion, care, and love.

FAMILIES AFTER WORLD WAR II

By the 1950s the model nuclear family had already come to be idealized, although it was mostly attainable only by white middle- and upper-class families (see the discussion of racial and ethnic differences in family structure later in the chapter). Most men's earnings were simply not great enough to afford a stay-at-home wife and dependent children. The gap between ideal and real narrowed as real wages (wages adjusted for inflation) increased in the 1950s, making the patriarchal tradition of a male breadwinner and a female homemaker a feasible arrangement for a greater number of families. Still, many nonwhite, non-middle-class families and individuals were excluded from the prosperity of the 1950s (see Chapter 9), and women worked outside the home throughout history, especially nonwhite women.

During the post–World War II economic boom, the manufacturing economy thrived with unionized jobs, real living wages, government housing subsidies, and job-training programs. It was a period of great optimism for good reasons: The Depression was over; America and its allies had won the war; and the United States was now dominant on the world's economic stage. Americans moved en masse to the suburbs—often to homes that

Textile factory workers in 1949. Why did many women leave their jobs after World War II?

required little to no down payment. Meanwhile, many women quit their jobs, if they had them, and returned to cultivating domesticity full-time; the divorce rate dropped to just over one in four marriages; and fertility rates soared during this "baby boom." The Ozzie and Harriet Nelson type of family was in its prime.

As family scholars are quick to point out, however, these trends in families were atypical. Although the divorce rate dipped in the 1950s, it had been on the rise since the end of the nineteenth century, so modern appeals to return to an age when divorce did not exist ring somewhat hollow. Furthermore, the fertility boom following World War II represented an unusual spike in family size, which had otherwise been on the decline since well before the turn of the century. What's more, the 1950s were also an era of rampant teen pregnancies: The teenage birthrate was twice as high in 1957 as in the 1990s. The difference was that pregnant teens in the postwar era had less access to abortion, and when they bore children, they more often got married. In fact, the 1950s marked the twentieth century's youngest national average for age at time of marriage—an average of 20 and 22 years old for women and men, respectively. Finally, the decline in women's workforce participation represented a dip in an otherwise upward trend, especially given women's labor power during World War II (when women were hired as temporary replacements for men serving in the armed forces).

Although the period was an anomaly in many ways, we tend to look back fondly (and sometimes bitterly) to the 1950s family as the normal and right way for families to be. But Stephanie Coontz found that while children's well-being and family economic security were at an all-time high toward the end of the 1960s, it was also a time of tumultuous struggle for racial and sexual equality (Coontz, 2001). What, then, do we really miss?

Family and Work: A Not-So-Subtle Revolution

Since the 1970s, American men and women have been caught up in what Kathleen Gerson (1985) calls a "subtle revolution" in our way of organizing work and home, even as we fantasize about an imagined time gone by. Women's participation in the labor force has soared, whereas fertility rates have plummeted. By the end of the 1970s, more women were in, rather than out of, the labor force for the first time. In 1950, approximately 30 percent of women worked outside the home; that number more than doubled to 60

FIGURE 10.2 Women in the Labor Force, 1970–2015

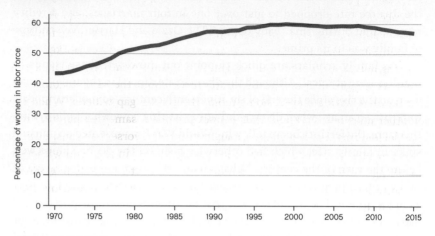

SOURCE: Bureau of Labor Statistics, 2016a.

percent by 1999 (Francis, 2005). In 2015, the labor force participation rate for women was 56.7 (Bureau of Labor Statistics, 2016a; Figure 10.2).

At the same time, birthrates have dropped, sinking from a baby boom to a baby bust. In the mid-1950s the birthrate was about 106.2 births per 1,000 women. That rate had dropped to just 62.9 in 1980, and as of 2014 it was at that point after dropping to 62.5 in 2016 (Martin et al., 2017). The National Center for Health Statistics reports that marriage rates since the 1960s have also declined, and many adults are postponing marriage until later in life. With the rising divorce rate, marriage is looking less and less like the stable force in women's lives that it was in the postwar era. In fact, by 2007, more Americans age 15 and older were single than married for the first time in a century. The result is that the cult of domesticity is out of the picture for most women, and daughters since the 1970s have increasingly departed from their mother's paths.

The gender revolution in work and family is probably here to stay, and with notable effects on children. Research study after research study has hinged on one basic question: Is maternal employment good, bad, or neutral for kids? There's no shortage of hypotheses. A study published in *Child Development* (Brooks-Gunn et al., 2003) suggests that having a working mother in one's early years can result in lower cognitive achievement and increased behavioral problems for a child. Yet another study published in the prestigious journal *Science* (Chase-Lansdale et al., 2003) claims the opposite: For moms with lower income levels, leaving kids in day care to enter the workforce is beneficial. Hundreds of studies fall on both sides of the argument, with mothers' employment being either disastrous or advantageous to their children. Family situations are just too complex and diverse to generalize. Perhaps, then, this mishmash of results is a consequence of asking the wrong question. Maybe we should instead ask, "How does maternal employment affect children differently within the family?"

In *The Pecking Order* (2004), I found that when a mother worked while raising her children, the adult daughters and sons eventually attained jobs that were more or less equivalent and made about the same income. But in families with a stay-at-home mom, the gender gap widens. When the mother does not work and thus cannot provide a same-sex parental role model in terms of career choices, daughters fare far worse than sons, earning roughly $8,000 less a year than their brothers later in life. And without a working-mother role model, daughters are 15 percent less likely to graduate from college than their brothers (whereas in the general population, more women than men earn college degrees).

A Feminist "Rethinking of the Family"

As noted in Chapter 4, the family is the primary unit of socialization for most of us. If family life is structured by larger forces in the social order, such as wage policy and gender inequality at work, then it is also simultaneously a site for the reproduction of these relations. As we have just seen, whether or not a mother works outside the home may affect the outcomes of her sons and daughters differently. Sarah Fenstermaker Berk (1985) has characterized the family as a "gender factory" of sorts, where women and men learn to take on distinct roles paralleling the divide between public and private spheres. The family is where people first learn how to "do gender" in conformity with social rules. Sociologist Marjorie Duvall (1991) shows in *Feeding the Family* that family ties are constructed through women's acts of shopping, preparing meals, and serving them. In everything from the planning of menus to accommodating a child's dislike of lima beans, women are doing gender. As Barbara Risman (1998) notes, "It is at home that most people come to believe that men and women are and should be essentially different." In the act of traditional marriage itself, rituals are gender-stratified: Brides are given away by their fathers, and they take the last name of a husband in patrilineal custom.

The idealized cult of domesticity, although a historical and cultural anomaly among many possibilities, has had powerful and lasting effects on many women's family ideals. Feminist consciousness, raised in the 1960s, led many women to question the ideology of domesticity, finding it "stultifying, infantilizing, and exploitative" of women (Stacey, 1987). In 1963, the feminist writer Betty Friedan led the assault against the limits of being a homemaker with her classic *The Feminine Mystique*. She writes in the introduction:

> The problem lay buried, unspoken, for many years in the minds of American women. It was a strange stirring, a sense of dissatisfaction, a yearning that women suffered in the middle of the twentieth century in the United States. Each suburban housewife struggled with it alone. As she made the beds, shopped for groceries, matched slipcover material, ate peanut butter sandwiches with her children, chauffeured Cub Scouts and Brownies, lay beside her husband at night—she was afraid to ask even of herself the silent question—"Is this all?" (1997, p. 15)

Sociologists elaborated on Friedan's invocation, claiming that what really distinguishes the head of the family is power. The family, as they see it, is a battleground for the power to make collective decisions on everything from who does the laundry to what neighborhood to live in to what college fund to invest in. This is the case, argues sociologist Jessie Bernard in *The Future of Marriage* (1972), even in the comfortable living rooms of traditional nuclear families such Ozzie and Harriet Nelson's. Take, for instance, the tensions that arise from the different ways that household members spend money. In studies of welfare recipients, women have been known to spend a greater portion of their welfare benefits on children than fathers do. And in a study of Japanese American families, working wives tended to keep their earnings separate from their husbands' income, and even secret, to preserve their autonomy to spend it how they—and not their husbands—chose (Glenn, 1986).

Betty Friedan in 1967, advocating for the addition of an equal rights clause for women in the New York state constitution.

Furthermore, money is not all the same within the family; rather, husbands', wives', and children's incomes are earmarked in distinctive ways. Women's earnings tend to be spent on extras, so-called luxury items, or otherwise nonessential expenses. Her money is devalued as "fun money." Men's earnings, on the other hand, are earmarked for essentials. When sociologist Margaret Nelson (1990) interviewed working wives, one woman explained, "What he makes there is mainly like insurance, taxes, and all that. What I make usually goes for food, clothing—whatever I find necessary...or just to take the kids once a week to Middlebury and blow it." So even a household's income is gendered. As Viviana Zelizer (2005) notes, the distinctions between household incomes help protect a man's status and sense of masculinity when a wife also takes on a breadwinner role in dual-income families. In this way, feminist rethinking of the family urges us not to view the family as a necessarily cohesive, unified whole but to recognize that the family is not immune from socially structured gender relations. And as gender relations change, so too do family forms. As women become less financially dependent on men, for example, more equality slowly develops in other areas of their relationships. Or does it?

When Home Is No Haven: Domestic Abuse

Family life is not always the warm, nurturing environment it appears to be on prime-time television, where moms and dads stand ready with hugs, siblings crack jokes, and grandma is strict but well intentioned. Abuse, neglect, and manipulation happen across all familial relationships: Sibling "rivalry" becomes physically aggressive, husbands and wives with powerful tempers hit and control one another, adult children plunder their elderly parents' life savings. The most frequent form of family violence is sibling on sibling (Eriksen & Jensen, 2006). But before you assign all the blame for your current low self-esteem on that time your brother walloped you upside the head with a shoe, note that the strongest two predictors of sibling-on-sibling violence are dads with short tempers and moms who get physical when it comes to punishing the kids (Eriksen & Jensen, 2006). So is it your brother's fault that he learned how to solve problems by mimicking dad's anger management strategy and piggybacking on mom's heavy hand of authority? This monkey-see, monkey-do explanation has had a long history in the study of domestic violence. Studies have found that abuse transmission is not genetic and that only about one-third of people who are abused as children go on to have a seriously neglectful or abusive relationship with their own children (Oliver, 1993). Broad social factors like poverty, single-parent households, and low levels of educational attainment are associated with higher levels of all types of domestic abuse.

When violence occurs within a couple it's often called IPV, intimate partner violence. More than one-third of women and one-quarter of men have been raped, physically attacked, and/or stalked by an intimate partner at some point in their lives (Centers for Disease Control and Prevention, 2011c). In a study of 10,018 homicides for women over 18. "Over half of all homicides (55.3%) were IPV-related; 11.2% of victims of IPV-related homicide experienced some form of violence in the month preceding their deaths, and argument and jealousy were common precipitating circumstances" (Petrosky et al., 2017).

Elder abuse is a relatively new field of study that investigates physical, verbal, and financial abuse intentionally or unintentionally perpetrated against people (usually family members) who are at least 57 years old. Estimates suggest elder abuse is not particularly widespread; one-year prevalence (that is, a statistical measure of how frequently these problems occur within a given twelve-month period) was 4.6 percent for emotional abuse, 1.6 percent for physical abuse, 0.6 percent for sexual abuse, 5.1 percent for potential neglect, and 5.2 percent for current financial abuse by a family member (Acierno et al., 2010). The impact of mistreatment in intimate relationships can have serious and long-lasting financial, health, and emotional consequences for individuals, families, and communities.

The Chore Wars: Supermom Does It All

One of the main ways that gender is enacted within the family is with respect to the unpaid labor that needs to be done at home. In the case of housework and the chore wars, gender remains a salient social force that shapes family life. The cult of domesticity lingers on, such that even though 56.7 percent of women participated in the workforce in 2015, domestic duties, such as housework and child care, still fall disproportionately on their shoulders (Bureau of Labor Statistics, 2015a; Francis, 2005). Women return from the office to take up what Arlie Hochschild (1989) calls the second shift: Women take responsibility for housework and child care, which includes everything from cooking dinner to doing laundry, bathing children, reading bedtime stories, and sewing Halloween costumes. Despite women's gains in the public realm of work, the revolution at home has, as Hochschild described it, stalled.

Within a two-career household, parents are likely to spend their at-home time on separate—and unequal—tasks. One study from 1965 to 1966 found that working women averaged 3 hours each day on housework, whereas men put in a meager 17 minutes. When it comes to leisure activities, however, men surpass their working wives. Working fathers watch an hour more of television per day than working mothers. They also sleep a half hour longer (Hochschild, 2003). The resulting "leisure gap" can brew hostilities between exasperated, exhausted wives and their unresponsive husbands.

Using national studies on time use from the 1960s and 1970s, Hochschild counted the hours that women and men put in on the job in addition to their time spent on housework and child care. She determined that women worked roughly 15 hours longer each week than men (Hochschild, 2003). After 52 weeks, this added up to 780 hours more that women work than men—that's an extra month! This gap seems to have stabilized (Figure 10.3). The Pew Research Center found that as of 2016, women spent roughly 14 hours more per week on housework and child care combined (Parker & Livingston, 2018).

SECOND SHIFT

women's responsibility for housework and child care—everything from cooking dinner to doing laundry, bathing children, reading bedtime stories, and sewing Halloween costumes.

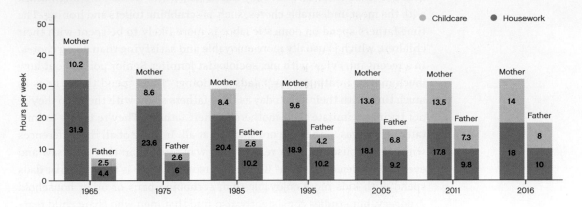

FIGURE 10.3 Trends in Housework and Child Care since 1965

● Childcare   ● Housework

Hours per week

1965: Mother 10.2 / 31.9; Father 2.5 / 4.4
1975: Mother 8.6 / 23.6; Father 2.6 / 6
1985: Mother 8.4 / 20.4; Father 2.6 / 10.2
1995: Mother 9.6 / 18.9; Father 4.2 / 10.2
2005: Mother 13.6 / 18.1; Father 6.8 / 9.2
2011: Mother 13.5 / 17.8; Father 7.3 / 9.8
2016: Mother 14 / 18; Father 8 / 10

SOURCES: Pew Research Center, 2013; Parker & Livingston, 2018.

Recent research shows how and when this gendered division of labor arises. Before the birth of the first child, men exhibit higher total work hours than women: They tend to work more paid hours and also do a relatively equitable share of housework. But after the birth of children, the division of labor in the home changes, both because men do less of the child-care work and because, at the same time, they reduce their non-child-care housework. In total, before becoming parents, married men spend about three additional hours doing all forms of work (jobs and housework) compared to women—but after the birth of a child, women do about six hours more than men (Yavorsky et al., 2015).

The division of labor in the home refers to not just who does how much but also who does what. As Viviana Zelizer (2005) shows, housework tasks are patterned around gender. When men do housework, their contribution is referred to as "helping out around the house," and their wives zealously thank them. Women who have "helping" husbands in the home consider themselves extremely lucky. Many men, however, never report feeling lucky or extremely thankful when their wives do the housework. And women *run* the household; they don't simply "help out."

In addition, men are typically in charge of outdoor, stereotypically masculine activities, whereas women disproportionately do the work inside. Even cooking is broken down by gender: Men more often cook the meat for a meal when they share work in the kitchen. Men also have more control over when they help out, because they are more likely to be in charge of chores that don't require completion on a daily basis, such as changing the oil in the family car or mowing the lawn. Women are more likely to be locked into a fixed and harried housework routine: Pick up the kids from day care, cook dinner, wash the dishes, administer the bath, and put the baby to bed (Zelizer, 2005).

Yet another male advantage in the division of household labor is that men get to do more of what they like to do, while women are often stuck with the most undesirable chores, such as scrubbing toilets and ironing. The time fathers spend on domestic labor is more likely to be spent with their children, which is usually more enjoyable and satisfying than doing chores. In a recent interview with me, sociologist Jennifer Senior pointed out how much more parenting today's dads are doing: "Dads spend three times as much time with their kids today as their fathers spent with them. So they're not trying to imitate their mothers or their fathers. They're trying to imitate their wives...they are not slackers at all. In spite of all the differences within the house, they are really doing way more than they ever saw and ever had modeled for them" (Conley, 2015b). Not only is the time that dads spend with kids more enjoyable than scrubbing pans or other household drudgery, but studies consistently also find that men who share child rearing and housework have happier marriages, better health, and longer lives (Kimmel, 2000). As best-selling author and pediatrician Dr. Spock (1998) put it, "There is no reason why fathers shouldn't be able to do these jobs as well as mothers." Furthermore, studies show that when left to their own devices, men are perfectly fine housekeepers, cooks, and primary parents (Gerson, 1993).

Neoclassical economists typically look at a member's power in the family as a direct expression of that member's utility to the family unit, and that tends to be measured in terms of his or her income. In such a formulation, money talks, and whoever earns the bacon doesn't have to cook it. This is not so for women, whose income goes only so far in increasing their power in the family. The more a wife contributes to the household income, the more her husband is likely to share, or "help out" with, the second shift. That seems to fit the bacon theory. But as soon as the wife's earnings start to overtake her husband's, he drops out of the shared domestic equation, leaving the entire second shift up to her (Bittman et al., 2003). Furthermore, when women earn more than their husbands, the perceived insult to masculinity may make for a tense home environment. Men, it seems, are not willing to let anyone else wear the proverbial pants in the household.

DIGITAL.WWNORTON.COM/YOUMAYASK6CORE

To see an interview with Jennifer Senior, go to digital.wwnorton.com/youmayask6core

With all the advances in home cooking and cleaning technology over the past few decades, you might wonder why the second shift can't be handled a little more efficiently than in a whole extra month of labor a year. But as Ruth Schwartz Cowan reports in *More Work for Mother: The Ironies of Household Technology from the Open Hearth to the Microwave* (1983), all those time-saving devices like the vacuum cleaner and the washing machine have paradoxically increased the amount of time women spend on housework. As Windex, dishwashers, Swiffers, and domestic media perfectionists like Martha Stewart present sparkling everyday homes, standards of cleanliness and household design rise, and even more housework is necessary than before. That said, as women have increased their hours in the formal labor market, the gender gap has declined in recent years: Women now do only 1.8 hours of household work for every hour a man completes, compared with the sixfold difference in the 1960s.

The entrepreneur Don Aslett, who built a business around domestic cleaning services, told one sociologist, "The whole mentality out there is that if you clean, you're a scumball" (Ehrenreich, 2001). Although not always so harshly spoken of, housework certainly is overlooked as a worthy or meaningful activity in America. For one thing, it's unpaid labor, and in a capitalist economy, it seems pretty, well, worthless. For another, it has been construed historically as women's work. A housewife, measured against modern criteria for success such as salaries, pensions, benefits, bonuses, and promotions, is just a housewife. Feminists in the 1960s argued that housework should be fairly compensated, because without a woman's labor, the American workforce would not be able to keep going, an argument that is being reviewed in Italy, the United Kingdom, and the United States (Ellen, 2014; Scalise, 2014; Shulevitz, 2016). This "reproductive labor" argument stressed the value of social reproduction: All the activities and tasks that women performed daily kept their working husbands' lives running smoothly.

This movement, you've probably figured out, did not catch on immediately. In fact, it is taking more than four decades and one tragedy. Zelizer (2005) showed that women's unpaid work commanded a price tag in the distribution of monetary rewards for surviving family members of those killed in the September 11, 2001, terrorist attacks. In the aftermath of 9/11, friends and family grappled with the shock and grief of losing loved ones. Some relatives of victims filed wrongful death suits against the airlines of the hijacked planes. To curb individual lawsuits and to spare the airlines from insurmountable litigations, the US government established the September 11 Victim Compensation Fund. Attorney Kenneth Feinberg had the job of administering the $7 billion fund to settle 2,880 death claims and 2,680 personal injury claims (Chen, 2004). It was up to Feinberg to decide, among those injured and the families of those killed, who received what.

As Zelizer (2005) notes, at first the fund was distributed to surviving family members in the amount of the deceased's estimated future earnings

↑

Many women juggle full-time jobs with caring for their children and running their home with little help from their spouses. According to Arlie Hochschild, what are the consequences of the supermom strategy?

and made no provision for compensating unpaid household work. Feminists organized and lobbied Feinberg on the issue, arguing that "ignoring the unpaid work performed by full-time workers raises sex discrimination concerns.... Women victims, especially mothers, are much more likely to have expended significant time on unpaid work." In the end, the Victim Compensation Fund Final Rule allowed for a case-by-case review of compensation claims on the lost economic value of household services that would have been provided by the deceased. Still, the small nod to the value of some women's valuable but unpaid domestic work has not yet translated into broader policy changes in the way housework functions in the US economy where the government still offers no guarantee of paid maternity (or paternity) leave.

It also takes marginalization to appreciate the center, as Christopher Carrington's study of gay couples in San Francisco demonstrates. Blumstein and Schwartz (1983) have suggested that an egalitarian division of labor is much more likely to exist in the households of gay and lesbian couples. But as Carrington (1999) notes, they also pay much more care and attention to their household chores, spending about two times as long on cleaning as heterosexual couples. The explanation? Homosexual couples see housework as a means of legitimating their households; thus it is a more central, validating activity for them than it is for the typical working wife. Even as same-sex couples have become more societally accepted, their more egalitarian division of household labor has appeared to persist (Bauer, 2016; Brewster, 2015). This suggests that egalitarianism may not just be an effort to gain legitimacy but, rather, is a reflection of the attitudes of those who decide to enter into a same-sex marriage or is a result of the psychological effect of being in such a household or, perhaps, simply reflects the more equal labor market positions of such spouses.

If, as sociologists argue, the burden of the second shift falls on working mothers in dual-income families, we might expect some nasty consequences. Indeed, Hochschild found that in response to the stalled revolution, some women tried to do it all. This is the supermom strategy. She cooks, she cleans, she climbs the career ladder, all while being a devoted parent and loving partner. Supermoms, of course, don't really exist, but there is a perpetual myth that some women can do it all, and if other women can't, it's a result of some personal flaw.

Women who buy into the supermom myth burn out quickly, and typically their marriages absorb the shock. In one study, Hochschild (1989) found that married women are more likely than men to think about divorce (30 percent of married women have considered divorce versus 22 percent of men). Women are also likely to give a more comprehensive list of reasons for wanting a divorce. In her interviews of dual-income families, Hochschild encountered bitter women, fed up with doing all the work, and puzzled men, confused over their wives' hostilities or bitter themselves at begrudgingly having to help out more.

Because women still earn less than men, about $0.81 for every dollar that a man earns, a disproportionate financial shock hits women after a divorce (Hegewisch & Williams-Baron, 2017). Given the rampant threat of divorce, some women bite the second-shift bullet, just letting the hostilities simmer. Other women manage by cutting household corners where they can, allowing the dust to pile up and skimping on the children's evening story time. For them, hostilities linger beneath a growing sense of parental guilt.

By contrast, in households in which men and women genuinely share the second shift, marriages are much more likely to be stable and happy ones. Studies (e.g., that conducted by Michael S. Kimmel in 2000) have found that when men share the housework, working wives experience less stress. Barbara Risman (1998) calls these "fair families," where husbands and wives equally split the roles of breadwinner and homemaker. In fair families, couples honor an ideology of gender equality, and both men and women benefit from such an arrangement. Women win the self-respect that comes with being economically independent, and they respond by regarding marriage less as a necessity and more as a voluntary, love-based relationship. Men, by sharing the breadwinner burden, are less stressed and feel freer to find jobs they enjoy. Perhaps most important, Risman speculates about a uniqueness in these sharing couples: They are more often close friends.

Swimming and Sinking: Inequality and American Families

AFRICAN AMERICAN FAMILIES

For millions of American households, the idealized traditional family in *The Adventures of Ozzie and Harriet* never even came close to a lived reality. For African American families, who throughout history have combined work and family, the split between the public and private spheres has never made much sense. Neither has the ideal of exclusive, "isolationist" motherhood made sense for women lacking the resources that allow for full-time homemaking. More often, black and poor women have come to rely on extra-familial female networks in order to manage child-care and work responsibilities (Stack & Burton, 1994). If the average American woman has two shifts, then the typical black American mother has three, because she is more often the primary or only breadwinner. This greater importance of women in the black family has unfortunately been conflated with "backward" female domination or matriarchy.

Social scientists in the 1960s made heavy use of the matriarchal thesis to explain social problems in the African American community. In *The Negro Family: The Case for National Action* (1965), Daniel Patrick Moynihan found that 25 percent of black wives outearned their husbands, versus only 18 percent of white wives. This "pathological" matriarchy, Moynihan argued, undercuts the role of the father in black families and leads to all sorts of problems later in life, such as domestic violence, substance abuse, crime, and degeneracy. You name it, the matriarch caused it. The root of matriarchy, Moynihan asserted, dated back to the days of slavery, which reversed roles for men and women and continued to haunt African Americans up to the time of his report.

According to the Moynihan report and similar arguments, the matriarch is a stereotypically bad black mother. She's domineering and unfeminine, always wearing the pants in the family. She's hefty and gruff. She spends so much time away from the home that she can't supervise her own kids. And when she is at home, she's too bossy and strong; thus she emasculates her man and drives him away. If only the matriarch could be a little more feminine, a little less strong, a little *whiter*, suggested the report, the black family could lift itself up out of poverty. The image of the matriarch makes it easier to lay the blame on African American families for their own problems—social ills such as neighborhood decay, stagnating wages, single motherhood, higher divorce rates, and lower school achievement. The matriarch is also a powerful reminder to all women of just what can go wrong when women challenge the patriarchal decree that they be submissive, dependent, and feminine. The Moynihan report's prescription for the social problems of black people was just this: Black women should aspire to the cult of true (white) womanhood.

Scholars like W. E. B. Du Bois had argued all along, however, that African American female-headed families were the outcome, rather than the cause, of racial oppression and poverty. And in 1987 William Julius Wilson graphed a "marriageable Black male index," which highlighted the scarcity of employed, un-incarcerated black men (fewer than 50 marriageable black men per 100 black women back then). In such a context, if black women didn't work to pay the rent, put food on the table, and take care of the kids, who would? As Elaine Kamplain has pointed out, African American mothers were "damned if they worked to support their families and damned if they didn't" (Hochschild, 2003).

Matriarchy is one of many misreadings that social critics have imposed on the black family. But when critics hold fast to the idealized concept of the nuclear family, they are likely to see any model that differs from the traditional yardstick as deviant. The sociologist's trick is to view the traditional family, in which a heterosexual couple lives together with their dependent children in a self-contained, economically independent household, as just one of many potential family forms. After all, this particular kind of family evolved from the socially constructed separation of home and work—a separation rooted in the upper classes—and the experiences of most African American families stray from that norm. In fact, asserts Patricia Hill Collins (1990), women of color have never fit this model. Collins looks at the family with a different set of lenses on—what she calls an Afrocentric worldview—which allows for alternative concepts of family and community. One instance of this worldview is more of a collective effort with strong neighborhood support: the "It takes a village" model.

Many African American parents rely on extrafamilial community members to help raise children. What are some of the criticisms of the matriarchal thesis?

Identifying the traditional family as a specific historical phenomenon that has rarely applied to black families enables us to analyze the black community's unique characteristics in a less judgmental manner. African American communities tend to have an expanded notion of kinship, as denoted by the all-encompassing use of the words *brother* and *sister*. This may have developed out of blacks' historical experience with slavery. Because slaves were separated from their biological families at the auction block by white owners, blacks adapted by expanding the definition of *family* from immediate bloodlines to racial ones. Under slavery, black men and women occupied indistinct work roles, often performing the same labor side by side in plantation fields. The legacy of their shared labor was incompatible with a nuclear family, a model predicated on the differentiation between male and female spheres (Collins, 1990).

As a result of segregation throughout the twentieth century, the particular African American notions of community and family continued to endure. But after World War II, the manufacturing economy took a downturn and many black workers, employed by shrinking industries, found themselves at risk of economic marginalization. Marital rates among blacks have been in decline since the 1960s, and female-headed households have been on the rise among both the poor and middle classes. Kathryn Edin's ongoing research in poor African American communities has found a consistent interest in the elements that constitute family life, such as having children and getting married, though poor-quality schools, lack of jobs, and (sometimes) substance abuse make it difficult to undertake a traditional progression of dating, love, marriage, and childbearing. Childbearing often precedes marriage, but unlike common stereotypes of indifferent cads interested primarily in sex, fathers-to-be are often excited about becoming parents and tend to strengthen their commitment to the baby's mother. Unfortunately, parenting the same child (or children) may not be enough to hold these couples together until they feel financially secure enough to marry. Black children (53 percent) and Hispanic children (29 percent) were more likely to live with a single parent than non-Hispanic white children (22 percent) or Asian children (12 percent) (US Census Bureau, 2017p). This statistic reflects, in part, the high rates of unemployment and incarceration and low rates of education among black men.

LATINO FAMILIES

According to the US Census Bureau (2016e), Latinos are the largest minority group in America, making up 17.8 percent of the US population. The Census Bureau (2011c) defines Latino as "a person of Cuban, Mexican, Puerto Rican, South or Central American, or other Spanish culture or origin regardless of race." Latinos also live all over the United States, although they have clustered on the West Coast and in the South and Midwest (Hernandez et al., 2007). Given the diverse origins and geography of Latinos, how can we make general claims about the Latino family? Sociologists don't all agree that such a thing exists or that a generic Latino identity is even imaginable (Santa Ana, 2004). But there are a few common threads that run through many Latino families. Their family ties are strong, and family is a top priority to many Latinos—so important, in fact, that individual Latinos often define their self-worth in terms of their family's image and accomplishments (Ho, 1987). Like African American families, Latino families act as safety nets, and members take seriously their responsibilities to help one another, even in long chains of needy extended relations (Skogrand et al., 2004). They have a strong sense of community and allegiance to it (Hurtado, 1995). For example, so many Latino immigrants send remittances, or money, to family members back home that remittances are now one of the largest sources of cash to the Mexican and many other Central American economies.

Traditional rules of gender and authority also loom large in Latino culture. Women listen to their men, and children to their elders, in a clear-cut hierarchy. This should make sense to anyone who's taken a Spanish language class (or studied any other Romance language). The heavy use of titles and ranks in the everyday vocabulary shows that respect and formality are crucial (DeNeve, 1997). Most Latino families are also shaped by a tradition of devout Catholicism. Combine high religiosity with the importance of family honor, and you soon recognize a few tendencies: high rates of marriage and relatively low rates of divorce, high rates of marriage at a young age, and a lot of babies born out of wedlock. This third outcome may seem contradictory, but think about it: If a young Latina does have premarital sex and becomes pregnant without a marriage prospect in sight, she's less likely than her African American or white peers to have an abortion, a worse breach of Catholic culture than an out-of-wedlock birth. Add to that a tight-lipped stance on discussing sex at home, the dwindling of preventive sex education, and the difficulty (not to mention the embarrassment) teens face trying to get their hands on effective and affordable birth control, and you end up with a lot of babies born to single Latina moms. In 2015, about 53 percent of all Latino babies were born to unwed mothers, compared with 29.2 percent of non-Hispanic white babies (those numbers are still lower than the nearly 71 percent of African American babies born outside of marriage) (Martin et al., 2017).

FLAT BROKE WITH CHILDREN

Single mothers and poverty often go hand in hand in America, and these twin challenges plague nonwhite communities to a greater extent than white Americans. In a shrinking welfare state with widening economic gaps between the rich and the poor, single mothers are faced with tough trade-offs between mothering and work. Single mothers rely on a combination of welfare, family, lovers, luck, and creativity to make ends meet. At any given point in the past few decades, half of all mothers raising children by themselves have relied on welfare to get by. In contrast to the media myth of the "lazy welfare mother" who watches television and buys filet mignon with her food stamps, many of these mothers work hard (often off the books) and desperately want to escape poverty. Welfare critics might be surprised to read *Making Ends Meet: How Single Mothers Survive Welfare and Low-Wage Work* (1997), in which sociologists Kathryn Edin and Laura Lein find that all single mothers prefer self-reliance to welfare dependency. In fact, a majority get off the welfare rolls in two years, and hardly any stay on welfare for more than eight years.

In their study of 379 mothers across four US cities, Edin and Lein traced what happens when poor single mothers are faced with the choice between welfare and work. The word *choice* here is misleading: It implies freedom, as though these women can easily move from dependency to self-sufficiency. However, because single mothers are often unskilled or semiskilled and have less education than average, their employment options are limited to low-wage work that rarely provides benefits (the kinds of jobs listed in the paper Reagan had flashed).

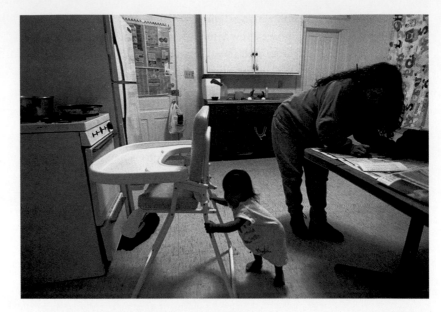

Jenni McGlaun talks on the phone while her 15-month-old son, Jesse, plays in their Milwaukee home. McGlaun has been on and off welfare for five years. What does Kathryn Edin and Laura Lein's research on single mothers reveal about the choices these women face between welfare and work?

Edin and Lein found that mothers on welfare could cover about three-fifths of their expenses. In low-wage jobs, they faced a larger gap between earnings and expenses, in part to cover the costs of transportation, child-care arrangements, increased rent, and fewer food stamps. (Both the food stamp program and federal housing program consider income when determining benefits; even a father's child support can translate into a rent increase for a single mother.) This system makes a savings account virtually impossible for welfare recipients to maintain and is arguably the real culprit behind the trap of dependency. In fact, leaving welfare for work substantially increases these women's expenses, such that they can cover only two-thirds of those expenses on low wages alone. One working mother voiced her frustrations:

> Ask any politician to live off my budget. Live off my minimum wage job and just a little bit of food stamps—how can he do it? I bet he couldn't. I'd like him to try it for one month. Come home from work, cook dinner, wash clothes, do everything, everything, get up and go to work the next day, and then find you don't have enough money to pay for everything you need. (Edin & Lein, 1997, p. 149)

Barbara Ehrenreich, a sociologist and writer, tried to make ends meet with low-wage work in 1998. That was a year in which the Preamble Center for Public Policy estimated that the odds against a typical welfare recipient's landing a job at a living wage were about 97 to 1. Nationwide, it took an average hourly wage of $8.89 to afford a one-bedroom apartment. Currently, the federal minimum wage is $7.25, which amounts to a full-time yearly salary of $15,080. Only approximately 4.3 percent of the labor force makes the minimum wage or less, but about 20 percent of the workforce plugs away for $9 an hour or less, and Ehrenreich wanted to figure out just how they did it (Bureau of Labor Statistics, 2014a; Ehrenreich, 2001; Greenhouse, 2013).

Ehrenreich traveled the country working for $6 or $7 an hour as a waitress, hotel maid, and Walmart sales clerk. At these low wages, she had to pay high rates for rent by the week (because she lacked the required one-month security deposit). She had to eat for less than $9 a day, a budget that at best included canned beans, fast food, or noodle soup microwaved at a convenience store, and she supplemented this diet by sponging junk food from charities. All the while, Ehrenreich prayed for her health to keep up (low-wage jobs are often physically taxing), gave up clothes shopping, and juggled two or more jobs. She still could not afford to live off of her wages:

> I grew up hearing over and over, to the point of tedium, that "hard work" was the secret of success: "Work hard and you'll get ahead" or "It's hard work that got us where we are." No one ever said that you could work hard—harder than you ever thought possible—and still find yourself sinking ever deeper into poverty and debt. (2001, p. 220)

Most welfare recipients live with few extras, often in neighborhoods with high poverty concentration and elevated crime rates. Yet politicians and voters alike worry about the cycle of dependency, the so-called welfare trap, and taxpayers deeply resent any person, real or perceived, who gets a free ride. Welfare critic Charles Murray, author of the conservative classic *Losing Ground* (1984), believes that welfare moms are staining the very moral fabric that holds the country together. As he sees it, all economic support should be pulled from under the feet of single mothers, and that would keep poor women and teens from having babies for whom they cannot care.

The 1996 Personal Responsibility and Work Opportunity Reconciliation Act, the national welfare reform enacted during the Clinton administration, didn't go as far as Murray would have liked, but it did make into policy the sentiment common among taxpayers that it was time for lazy welfare moms to get their act together, learn self-sufficiency, and take responsibility for themselves. But responsibility can take on several, often conflicting, meanings for mothers. Faced with either welfare or low-wage work, single mothers find that the government's definition of responsibility is narrowly defined as wage work. In *Flat Broke with Children: Women in the Age of Welfare Reform* (2003), Sharon Hays shows that many of these women are forced to take low-paying jobs with no future and little career stability. As a result, they are frequently driven to marry for financial support. All too often, Hays finds, single mothers (and their children) end up in poverty, homeless, or seeking alternatives to get by, including illicit sources of income.

Jill Lawrence is a single mother in South Dakota. Like many of the single mothers that Sharon Hays has written about in *Flat Broke with Children*, she struggles to raise her son and make ends meet on her own.

Single mothers' paid labor often requires that they sacrifice responsible parenthood. By forcing welfare moms to enter the workforce while not providing adequate child care, the government encourages women to abandon their children in an unsupervised home. To the single mothers Hays interviewed, real responsibility meant taking care of and supervising their children, looking after their health, and providing them with a safe home environment. According to Hays, for the most desperate women in America, being a responsible worker requires being an irresponsible mother. To manage this tension between being a good worker or a good mother, single mothers may shoot for the nearly impossible strategy of self-reliance, but in reality they often end up depending on cash assistance from family or boyfriends. They barter with sisters, cousins, and kindly neighbors. They also take side jobs (again hiding their income to avoid rent increases or food stamp ineligibility) or go to relief agencies, although this last option proves too humiliating for most to bear. As a last resort, some single mothers turn to criminal activity such as prostitution or drug dealing, where the earnings are high but the moral costs "rob them of the self-respect they gained from trying to be good mothers" (Edin & Lein, 1997).

The Pecking Order: Inequality Starts at Home

By now, you can see that there's no universal family form and that plenty of gender inequality exists in American families. But you might still think that, at least when it comes to the children, the home (no matter who makes up a family) is a haven of equality, altruism, and infinite love in an otherwise harsh world. In *Haven in a Heartless World* (1977), Christopher Lasch paints a rosy portrait of domestic life. To him, the home really is a haven where workers seek refuge from the cold winds of a capitalist public sphere. The family is sacred because it provides intimate privacy, managed by a caring woman who shields male workers from "the cruel world of politics and work." Building on the socially constructed split between the public and private spheres, this idea of the family as insulated from the outside world is deeply entrenched in America.

However, as you might have guessed by now, such ideas give a misleading sense of the family as a harmonious unit, its members altruistically sacrificing for one another to forge a healthy, hearty, and loving private world. As I found while researching my book *The Pecking Order* (2004), a compelling but largely invisible struggle also takes place in the home. In each American family, there exists a pecking order among siblings—a status hierarchy, if you will—that can ignite the family with competition, struggle, and resentment. Parents often abet such struggles, despite their protestations that they "love all their children equally."

Let me illustrate with the story of a future president, William Jefferson Blythe IV, born to a 23-year-old widow named Virginia. Childhood was rough for young Bill, especially after his mother married Roger Clinton, a bitter alcoholic who physically abused his wife. Bill cites the day that he stood up to his stepfather as the most important day in his transition to adulthood and perhaps his entire life. In 1962, when Bill was 16, Virginia finally left Roger, but by then there was another Roger Clinton in the family, Bill's younger half-brother.

Despite the fact that Bill despised his stepfather, he went to the Garland County courthouse and changed his last name to Clinton so that he would have the same surname as the younger brother he cherished. Although they were separated by 10 years, were only half-siblings, and ran in very different circles, the brothers were close. The younger Roger probably hated his father more than Bill did, but he nonetheless took on some of the old man's traits as he came of age, most notably substance abuse. By age 18, he was heavily into marijuana. During Bill's first (unsuccessful) congressional run in 1974, Roger spent much of his time stenciling signs while smoking joints at campaign headquarters.

As Bill's political fortunes rose, Roger's prospects first stagnated and then sank. He tried his hand at a musical career, worked odd jobs, and eventually began dealing drugs—and not just pot. In 1984 the Arkansas state police informed then-Governor Bill Clinton that his brother was a cocaine dealer under investigation. After a sting operation, which the governor did not obstruct, Roger was arrested. He was beside himself in tears, threatening suicide for the shame he had brought on his family and especially his brother, the successful politician.

You might be wondering, as Clinton biographer David Maraniss (1995) has, "How could two brothers be so different: the governor and the coke dealer, the Rhodes scholar and the college dropout?" To be sure, a pair of brothers who are, respectively, a former president and an ex-con is a fairly extreme example. But the basic phenomenon of sibling differences in success that the Clintons represent is not all that unusual. In fact, in explaining economic inequality in America, sibling differences represent more than half of all the differences between individuals.

What do sibling disparities as large as these indicate? If asked to explain why one brother succeeded while the other failed miserably, most people would point to different individual characteristics, such as work ethic, responsibility, personal motivation, and discipline. To account for a sibling's failures, people also tend to cite personal reasons, such as "a bad attitude," "poor emotional or mental health," or most commonly, "lack of determination" (Conley, 2004).

Other people will grapple for an explanation by suggesting the role of birth order. The commonly held idea is that firstborns are naturally more driven and successful, if just because they are most favored by their parents. But this is still a form of individual explanation—something unique to the biology or psychology of the sibling—that fails to take into account the role of sociological factors. For example, in families with two kids, birth order doesn't matter that much. Firstborns don't have too much advantage over one younger sibling. For example, slightly fewer than one-fourth of US presidents were firstborns, about what we would expect from chance. Birth position matters only in the context of larger families and limited resources. When family resources are stretched thin, love really does become a pie, as they say. The children born first or last into a large family seem to fare better socioeconomically than those born in the middle. Middle kids feel the effects of a shrinking pie, as they tend to be shortchanged on resources like money for college and parental attention.

Taken as a whole, the facts about intrafamily stratification present a much darker portrait of American family life than we are accustomed to. Sure, we want to think of the home as a haven in a heartless world, but the truth is that inequality starts at home. These statistics also pose problems for media stories and politicians concerned with the erosion of the idealized nuclear family. In fact, they hint at a trade-off between economic opportunity and stable, cohesive families. The family is, in short, no shelter from the cold winds of capitalism; rather, it is part and parcel of that system.

A pecking order emerges during the course of childhood. It both reflects and determines siblings' positions in the overall status ordering that occurs within society. It is not just the will of parents or the "natural" abilities of children (or lack thereof); the pecking order is conditioned by the swirling winds of society, which in turn envelop the family. Furthermore, sibling disparities are much more common in poor families and single-parent homes than in rich, intact families. In fact, when families have limited resources, the success of one sibling often generates a negative backlash among the others. As the parents unwittingly put all their eggs—all their hopes and dreams—in just one basket, the other siblings inevitably are left out in the cold. Americans like to think that their behavior and destiny remain solely in their own hands. But the pecking order, like other aspects of the social fabric, ends up being shaped by social forces.

The Future of Families, and There Goes the Nation!

DIVORCE

Throughout this chapter we have examined the idealization of monogamous marriage as a key characteristic of American culture. Sociologist Andrew Cherlin writes in *The Marriage-Go-Round: The State of Marriage and the Family in America Today* (2009) about how the idealized conception of marriage is currently helping delay age at first marriage and otherwise shape contemporary American marital patterns. He finds a paradox. On the one hand, Americans value marriage very highly—85 to 90 percent of us will eventually get married. On the other hand, America has the highest divorce rate of any comparable Western country. How does Cherlin explain this love–hate relationship? He suggests that both the drive to get married and the desire to divorce are rooted in our collective past,

> all the way to the Colonial days when marriage was the nexus of civil society that the early settlers established. It's very important here.... We want to be married. At the same time though, we're very individualistic, and that has deep cultural roots too. Think about the saying "Go West, young man." Think about the rugged individual. We're individualists. And so we want to be married, but we evaluate our marriages in very personal terms. Am I getting the personal growth that I need out of my marriage? And if you think the answer is no, you feel justified in leaving. (Conley, 2009k)

Social science research over the past couple of decades has struggled to establish just how high rates of divorce alter the social fabric. Figure 10.4 illustrates the divorce rate in America since 1920. Adding a layer to the American marriage paradox, it turns out that the most politically and religiously conservative states have the highest rates of divorce: "The states with the top ten divorce rates, eight out of ten of those voted for John McCain in the 2008 election. All ten voted for George Bush in the 2004 election" (Conley, 2009k). Cherlin explains that personal economics, not personal values, may end up contributing to rocky marriages. The states with high divorce rates are relatively poor; a sizable number of their citizens struggle to find jobs with decent wages, which compromises the ideal vision of family as a haven protected from crass financial concerns. Couples who cannot provide themselves a middle-class lifestyle may begin to question the utility of marriage in the first place. They face all of the responsibility of looking out for each other without the means to live the ideal lifestyle to which they aspire. Cherlin also notes that younger people have even more difficulty finding good jobs because they have less work experience, and this may delay their marriage plans (Conley, 2009k). Young couples may live together, even have children together, holding off on long-term personal relationships until after they have established long-term financial relationships with reliable employers.

Much of the debate about divorce follows the children of divorced parents through their educational careers and into their adult relationship choices. The moral and political debate surrounding the long-term consequences of divorce has largely treated its effects as uniform for all offspring. In *The Pecking Order* (2004), I paint a more nuanced picture, in which circumstances and context widely vary the impact of divorce on kids—even those in the same family. Everything from the timing of the divorce to a parent's hostility can influence a kid's educational attainment, future earnings, and socioeconomic success. Blanket condemnations of divorce are therefore dangerously naive, as are those who say that it's no big deal.

For example, in their best seller *The Unexpected Legacy of Divorce* (2000), Judith Wallerstein and her colleagues claim that divorce almost universally damages children's self-esteem and developmental trajectories. Based on interviews with about 50 offspring of divorced parents, they conclude that adult children of divorce suffer from higher rates of depression, endure low self-esteem, and have difficulties forming fulfilling, lasting relationships of their own.

Sociologist Linda Waite and columnist Maggie Gallagher echo this view in *The Case for Marriage: Why Married People Are Happier, Healthier, and Better Off Financially* (2000). (The title, in this case, says it all.) They claim that

FIGURE 10.4 US Divorce Rate, 1920–2016

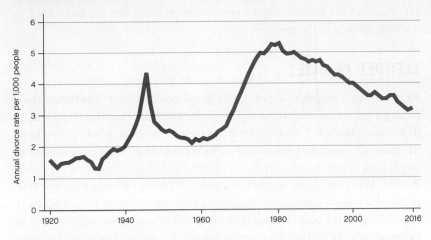

SOURCES: US Census Bureau, 2010c; Centers for Disease Control and Prevention, 2015f; National Center for Health Statistics, 2017.

married parents provide better homes for their children than divorced ones because they have more money and time to spend on the children, enjoy stronger emotional bonds with them, have more social capital (connections) that will be helpful to their children's chances for success, and are physically and mentally healthier. By contrast, they say, divorced families are likely to manifest more child abuse, neglect, and delinquency, and the children will probably attain less education.

On the flip side of the debate are research findings that suggest a less calamitous future for the children of single-parent families, indicating that children from divorced households generally do not do much worse than other kids. In *For Better or for Worse: Divorce Reconsidered* (2002), psychologists E. Mavis Hetherington and John Kelly argue that the children of divorce, for the most part, do adjust well to the new reality. Under some conditions, but certainly not all, divorce produces in kids more stress and depression, and lower future socioeconomic success. Continued parental conflict and role reversals in which children "play parent," for example, make divorce potentially destructive. In other cases, however, parents who leave a high-conflict marriage probably spare their children from ongoing family feuds, hostilities, or abuse. At least one study shows that kids from high-conflict marriages that stay together may do worse than kids from high-conflict marriages that break up (Morrison & Coiro, 1999).

Too often, social science research is carelessly picked up in political debates as simple sound bites. Saying kids from divorced families fare worse than kids from intact families is one thing; saying that divorce caused those worse outcomes is quite another matter. That is, we can't really know for sure that had those parents stayed together, the kids would have been better off. We can only say that high levels of parental conflict are bad—with or without divorce in the picture.

BLENDED FAMILIES

As should be clear by this point, the dynamics involved in maintaining a family (both as a whole and in terms of the individual relationships within it) are complicated. This complexity increases drastically when two family units are integrated. As stated on **helpguide.org**, "To a child who does not belong to one, stepfamily may suggest Cinderella's family or the Brady Bunch." However, if you are part of a blended family or know someone who is (there's a good chance you do, given that one-third of the children in the United States now fall into this category), you understand that neither extreme reflects reality. As divorce becomes more common, so too does the blended family form. Two people meet, fall in love, get married, have kids—we'll call them family A. Then they decide to split. Another pair does the same (family B). Some time later, Mom A encounters Dad B, and the process repeats. The result is a blended family or stepfamily, with stepparents, stepchildren, and sometimes stepsiblings. Blended families are the result of not only divorce but also death. When either partner dies, remarriage is not uncommon. According to the 1990 US Census, 3.6 percent of grooms and 10 percent of brides had previously lost a spouse (Kreider & Ellis, 2011).

In fact, in his book, Cherlin estimates that more than a quarter of American children today experience at least two maternal partner changes, and more than 8 percent experience three or more. Some developmental psychologists worry about this dynamic, as it impacts children's ability to form trusting, stable ties with adult figures in their lives. Others see children as robust and more or less adaptable to most changes that come their way.

GAY, LESBIAN, AND TRANSGENDER FAMILIES

Marriage, divorce, remarriage. Stepfamilies, blended families. Another family arrangement on the rise, as we saw at the beginning of the chapter, is same-sex couples. At present, same-sex marriages are legal in an increasing number of countries, including the United States and about half of Europe. As for gay parenting, research consistently finds that lesbian and gay parents are at least as successful as heterosexuals in producing well-adjusted, successful offspring (Stacey, 1997).

However, until recently homosexual unions faced fierce opposition. In 1996 President Clinton signed into federal law the Defense of Marriage Act (DOMA), which stipulated that federally recognized marriages had to be heterosexual and that states where gay marriage was previously illegal were not required to recognize same-sex marriages conducted in states where it was legal. But DOMA was struck down in 2013 and the Supreme Court effectively legalized same-sex marriage across the United States in 2015.

An old joke goes, "Gays and lesbians getting married—haven't they suffered enough?" That is, if homosexual couples want to marry, why not just let them do it? Opponents of gay marriage—many of whom are strongly religious, "social values" conservatives—argue that nontraditional familial arrangements will wreak social havoc. Let gays get married, so the argument goes, and you effectively destroy the family. Once the family goes, look out—the next step is chaos, no moral order. Furthermore, opponents of gay marriage claim that the purpose of a family is to procreate and raise children into functioning adults, and this is determined by biology. Because gay marriage does not produce the biological children of both parents, such people believe that it is not functional, right, or natural.

According to the Pew Research Center, views are steadily shifting. In 1996, 64 percent of Americans were opposed to gay and lesbian marriage. By 2017, those numbers have flipped, where 62 percent supported gay marriage while 32 percent are opposed. Young people are the strongest supporters of same-sex marriage, with 74 percent of the millennial generation (born since 1980) supporting same-sex marriage, while 45 percent of people over 65 oppose same-sex marriage (Pew Research Center, 2017c).

Families with one or more transgender parents are another group that has challenged conservative notions of how a family "should" look. Transgender individuals lag behind gays and lesbians in terms of societal acceptance (see Chapter 8) but the trend lines are similarly toward tolerance. There is, as of yet, no reliable data on how many families include transgender parents, but best estimates put the figure at less than 1 percent.

Two married women greet their eight-year-old daughter after school in Ridgefield, Washington in 2017. How have attitudes toward gay and lesbian families changed?

MULTIRACIAL FAMILIES

MISCEGENATION

the technical term for interracial marriage, literally meaning "a mixing of kinds"; it is politically and historically charged—sociologists generally prefer *exogamy* or *outmarriage*.

Gays have not been the only ones who have faced obstacles to legal marriage. The technical term for interracial marriage is miscegenation, which literally means "a mixing of kinds." This term, however, is politically and historically charged and, as such, should be used with a degree of caution. Sociologists and other social scientists generally prefer the term *exogamy* or *outmarriage*.

Considering the history of race relations in the United States (see Chapter 9), it is not surprising that the idea (and even legality) of interracial marriages and families has been cause for controversy. Although interracial marriage was never illegal at the federal level, from 1913 to 1948, 30 states enforced antimiscegenation laws. Many of these laws lasted until 1967, when the Supreme Court finally declared them unconstitutional, reversing the restrictions in 16 remaining states.

We have undoubtedly come a long way since 1967. One year after the Supreme Court ruling (1968), a Gallup poll revealed that 80 percent of Americans were opposed to blacks and whites getting married. By 2009 that number had dropped to 17 percent (Wang, 2012). However, attitudes toward exogamy in the United States vary greatly by location, religion, and a host of other factors. And, of course, such polls capture only expressed attitudes, subject to social desirability bias, not actual behavior (i.e., actual intermarriage rates).

In 2015, 17 percent of new marriages were between people of different races or ethnic groups. Asian Americans had the highest rate of outmarriage at 29 percent, followed by 27 percent of Hispanics, 18 percent of blacks, and 11 percent of whites (Bialik, 2017). These percentages significantly change when gender is taken into account. For example, 36 percent of Asian American

American families are increasingly looking like the family pictured here. How are attitudes about multiracial families changing?

women marry someone outside of their race, unlike Asian American men who have only a 21 percent rate of outmarriage. Black men have a 24 percent rate of outmarriage versus black women's rate of 12 percent.

IMMIGRANT FAMILIES

Immigrants—today arriving mostly from Latin America and Asia—face a unique set of challenges in forming and sustaining families. Aside from the difficulty of the immigration process itself, through which it can take years to reunify spouses and siblings through visa sponsorships, families who migrate to America tend to be poorer, larger, and hold less educational credentials. For instance, about 9.1 percent of people born in the United States have less than a high-school degree, while the same is true for 28.8 percent of people born abroad (Ryan & Bauman, 2016). What's more, the gaps between natives and immigrants who have not obtained citizenship or who are undocumented are even larger. There are more than four times as many noncitizens who have less than a high-school degree (or equivalent) compared to those born in the United States, while those native households make about $14,000 (or about 25 percent) more in earnings (Ryan & Bauman, 2016). These socioeconomic disparities are exacerbated by the reality that an overwhelming proportion of immigrants live in expensive cities on the East and West Coasts, like Los Angeles and New York (López & Bialik, 2017). Because of these resource gaps, undocumented immigrants face a particularly daunting set of challenges in providing for their families and obtaining safe, stable, and affordable housing.

Most Americans nowadays delay forming families until they receive college degrees and earn more money precisely because of the difficulties of having children with fewer resources. Yet though we might expect immigrant families to form and stay together less often because of these challenges, the opposite is true. Immigrants tend to start families regardless of their educational attainment or income, with the poor marrying and having children at a similar rate compared to their peers with higher socioeconomic status (Qian, 2013). About a quarter of immigrants have never been married, while that's true for more than a third of native-born populations (Ryan & Bauman, 2016). And in terms of keeping the family unit intact, the gaps are similarly big—about 79 percent of foreign-born children live with married parents, compared to 64 percent of natives (US Census Bureau, 2017p).

Why do families who face greater socioeconomic, geographic, and linguistic challenges manage to form and sustain themselves more than US natives? While some propose a different set of cultural norms—for example, a greater emphasis on family ties and the necessity of social support to get ahead in America—immigration policy itself likely contributes to the structure of immigrant families. Even though the process is difficult, current immigration policy favors family-based reunification rather than education level or labor market skills (Kandel, 2018). So while the Republican Party made immigration a key issue in the 2016 election (and beyond)—and ending family-based immigration policy, in particular (not to mention separating detainee families from their parents for a period)—foreign-born populations actually show a greater tendency to live out the kinds of mid-twentieth-century nuclear family arrangements that many political conservatives often express a desire to return to.

POLICY

 ## EXPANDING MARRIAGE

When the poet Elizabeth Barrett Browning wrote, "How do I love thee? Let me count the ways," little did she know that a century and a half later the list would extend into the hundreds—at least for married couples. As of 2004, the US General Accounting Office had identified more than 1,000 legal rights and responsibilities attendant to marriage (Shah, 2004). The era of big government is clearly not over when it comes to family policy.

These rights and responsibilities range from the continuation of water rights upon the death of a spouse to the ability to take funeral leave. And that's just the federal government. States and localities have their own marriage provisions. New York State, for example, grants a spouse the right to inherit a military veteran's peddler's license. And Hawaii extends to spouses of state residents lower fees for hunting licenses. No wonder gay and lesbian activists put such a premium on access to marriage rights. Some gay marriage advocates want all spousal rights immediately

and will settle for nothing less. Others take an incremental approach, aiming to secure first the most significant domestic partner rights, such as employer benefits like health care.

But with all the divorces and blended families, it's unclear if our current legal system is best suited to the task of serving families' needs. Perhaps it is time to consider if we'd be better off by breaking down or unbundling the marriage contract into its constituent parts. Then, applying free-market principles, we could allow each citizen to assign the various rights and responsibilities now connected to marriage as he or she sees fit. In addition to employer benefits, some of the key marital rights include the ability to pass property and income back and forth tax-free, spousal privilege (i.e., the right not to testify against one's husband or wife), medical decision-making power, and the right to confer permanent residency to a foreigner, just to name a few. The mutual responsibilities of marriage include parenthood—a

Jeanne Fong (left) and Jennifer Lin, a same-sex couple from San Francisco, celebrate during a marriage equality rally in Washington, D.C.

husband is the legal father of any child born to his wife regardless of biological paternity—and shared tort liability. Why not allow people to parcel out each of these marital privileges?

Take my own first marriage as a case in point: My ex-wife is a foreigner who applied for citizenship, and her marriage to me made that possible. Marrying a foreigner is pretty common—7 percent of married couples in the United States are made up of a US-born spouse and a foreign-born spouse (US Census Bureau, 2013). As the law now stands, I, as a straight American man, theoretically could become a green card machine. As long as I can convince the overstretched Department of Homeland Security that my penchant for falling in love with foreign women is genuine, I can divorce and remarry as many times as I like, obtaining permanent residency for each of my failed loves along the way. Is this fair to the many Americans who can't sponsor aging grandparents or, in some

cases, even parents? Or, is it even fair to the vast majority of Americans who are happy marrying other Americans and don't want to see the country fill up with my romantic baggage?

Instead, why not give all Americans the right to sponsor one person in their lifetime—a right that they could sell, if they so desire? This would mean that if I wanted to marry a Kenyan after divorcing an Australian, I could, but I would need to purchase, perhaps on eBay, the right to confer citizenship from someone else who didn't need it. Similarly, why not let all Americans name one person (other than their lawyer, priest, or therapist) who can't be forced to testify against them in court? This zone of privacy could be transferred over the course of a lifetime, perhaps limiting such changes to once every five years.

While we are at it, how about allowing each of us to choose someone with whom we share our property, with all the tax (and liability) implications that choice would imply? We might even allow parenthood to become contractual and in multiples greater than two by letting people name individuals they want to be the co-parents to their biological children, even if that means a kid ends up with three or four parents, none of whom is married to any of the others. This would help grandparents, for instance, have the legal right to act on behalf of the grandchildren in their care.

We could go down the long list of rights and responsibilities embedded in the marriage contract. Ideally, most people would choose one person in whom to vest all these rights, but everyone would have the freedom to decide how to configure these domestic, business, legal, and intimate relationships in the eyes of the law.

Other people have proposed changing marriage by making it more flexible. Some queer activists continue to argue that the law should recognize a variety of households, from conjugal couples to groups of elderly people living collectively, as "legal families." But they have not specified the rights that should go along with such arrangements or addressed the problems and paradoxes that might arise if legal rights and responsibilities were extended to groups of more than two. Unbundling marital rights would achieve the same end without creating new inequities based on group size.

It might also take some of the vitriol out of the marriage debate. Marriage itself could stay in church (or in Las Vegas, as the case may be). And each couple could count their own ways to love.

Conclusion

By now you should be wary of any social institution that is hailed supreme because it is "more natural." You should be skeptical of any family arrangement that is deemed more functional than another, and you should hold the traditional family at a critical distance, especially considering the experiences of women, African Americans, gays and lesbians, the poor, the mainstream, and the marginalized.

Under the "postmodern family condition," as Judith Stacey (1996) calls it, clear rules no longer exist in our complex, diversified, and sometimes messy postindustrial society. Gone are the ruling days of the normative Nelsons. Families today take on many shapes and sizes that best fit their members' needs, and they are defined not by blood ties but by the quality of relationships. Let us count the ways...

QUESTIONS FOR REVIEW

1. The case of Ozzie and Harry at the beginning of the chapter brings to mind the variety of family arrangements. Describe the "nuclear family" and three other family forms. Does sociological research suggest that one such arrangement is necessarily the right one?

2. Think about the discussion regarding rights for same-sex couples. How do the functionalist arguments by Bronislaw Malinowski (1913) and Talcott Parsons (1951) help explain the changes (or lack thereof) in the institution of marriage?

3. How do historical cultural ideals relate to current rates of marriage and divorce in the United States? Name one reason that the likelihood of being currently married might be lower among those with low incomes.

4. Sometimes what's considered "normal" is far from what's most prevalent currently or historically. How does this statement relate to perceptions about the "traditional" family?

5. Describe how a gendered division of labor arose after the Industrial Revolution. How was this change tied to kinship networks?

6. What is the cult of domesticity? How has it changed over the last century?

7. What is the second shift, and how does it relate to a leisure gap between husbands and wives?

8. Are mothers on welfare lazy? Use the findings by Kathryn Edin and Laura Lein (1997) to help answer this question.

9. What is a pecking order, and what does the term mean for children in a family? According to this concept, does your birth position in the family or number of siblings matter to your life chances for success in school and beyond?

MAKING INVISIBLE LABOR VISIBLE

Building on the ideas introduced by Arlie Hochschild's book *The Second Shift* (1989), In recent years commentators have talked about "invisible" domestic labor—work that happens at home that is often unacknowledged and unpaid.

TRY IT!

Think about all the work a parent or caretaker did when you were growing up, and consider the economic value of that labor. Caretakers often chauffeur kids like an Uber driver, act as a health aid, or function as a short-order cook—how much was that worth per week? Here are some estimates to get you started:

TASK	NATIONAL AVERAGE
UBER DRIVER	$4 PER 15-MINUTE RIDE
BABYSITTER / CARETAKER	$12 PER HOUR
SHORT-ORDER COOK	$5 PER MEAL
HOUSE CLEANER	$10 PER ROOM
HOMEWORK TUTOR	$20 PER HOUR

WORK PROVIDED BY PARENT OR CARETAKER	AMOUNT WORTH PER WEEK
_____	_____
_____	_____
_____	
(For simplicity's sake, you could use minimum wage in your state for any tasks not defined above.)	**TOTAL AMOUNT** $ _____

THINK ABOUT IT

So, have you broken out your checkbook to pay your mother, father, or other primary caregiver back for all those years of uncompensated labor on your behalf? Think about your family and others you know well: Who typically does this unpaid labor? Is it equally borne by all adults in the family? Does it break down along gender lines? Or age? Or by task (Dad is chauffeur and Mom is short-order cook, for example)? How do you feel about the social division of unpaid labor in your family—as it compares to the typical American household?

SOCIOLOGY ON THE STREET

It seems like everyone has an opinion on whether women with children—especially young children—should be in the workforce. Why is the conversation over "having it all" still mainly about mothers? Watch the Sociology on the Street video to find out more: **digital.wwnorton.com /youmayask6core**.

WANT MORE PRACTICE?

Complete the InQuizitive activity for this chapter at digital.wwnorton.com /youmayask6core

Glossary

A

achieved status a status into which one enters; voluntary status.

altruistic suicide suicide that occurs when one experiences too much social integration.

androgynous neither masculine nor feminine.

anomic suicide suicide that occurs as a result of insufficient social regulation.

anomie a sense of aimlessness or despair that arises when we can no longer reasonably expect life to be predictable; too little social regulation; normlessness.

ascribed status a status into which one is born; involuntary status.

association *see* correlation.

B

biological determinism a line of thought that explains social behavior in terms of who you are in the natural world.

bisexual an individual who is sexually attracted to both genders/sexes.

bourgeois society a society of commerce (modern capitalist society, for example) in which the maximization of profit is the primary business incentive.

bourgeoisie the capitalist class.

broken windows theory of deviance theory explaining how social context and social cues impact whether individuals act deviantly; specifically, whether local, informal social norms allow deviant acts.

C

case study an intensive investigation of one particular unit of analysis in order to describe it or uncover its mechanisms.

caste system a religion-based system of stratification characterized by no social mobility.

causality the notion that a change in one factor results in a corresponding change in another.

cisgender describes people whose gender corresponds to their birth sex.

class system an economically based hierarchical system characterized by cohesive, oppositional groups and somewhat loose social mobility.

code switch to flip fluidly between two or more languages and sets of cultural norms to fit different cultural contexts.

cohabitation living together in an intimate relationship without formal legal or religious sanctioning.

collective resistance an organized effort to change a power hierarchy on the part of a less-powerful group in a society.

color-blind racism the view that racial inequality is perpetuated by a supposedly color-blind stance that ends up reinforcing historical and contemporary inequities, disparate impact, and institutional bias by "ignoring" them in favor of a technically neutral approach.

comparative research a methodology by which two or more entities (such as countries), which are similar in many dimensions but differ on one in question, are compared to learn about the dimension that differs between them.

conflict theory the idea that conflict between competing interests is the basic, animating force of social change and society in general.

conformist individual who accepts both the goals and the strategies that are considered socially acceptable to achieve those goals.

consumerism the steady acquisition of material possessions, often with the belief that happiness and fulfillment can thus be achieved.

content analysis a systematic analysis of the content rather than the structure of a communication, such as a written work, speech, or film.

contradictory class locations the idea that people can occupy locations in the class structure that fall between the two "pure" classes.

corporate crime a particular type of white-collar crime committed by the officers (CEOs and other executives) of a corporation.

correlation or association when two variables tend to track each other positively or negatively.

crime the violation of laws enacted by society.

cult of domesticity the notion that true womanhood centers on domestic responsibility and child rearing.

cultural relativism taking into account the differences across cultures without passing judgment or assigning value.

cultural scripts modes of behavior and understanding that are not universal or natural.

culture the sum of the social categories and concepts we embrace in addition to beliefs, behaviors (except instinctual ones), and practices; everything but the natural environment around us.

culture jamming the act of turning media against themselves.

culture lag the time gap between appearance of a new technology and the words and practices that give it meaning.

culture shock doubt, confusion, or anxiety arising from immersion in an unfamiliar culture.

D

deductive approach a research approach that starts with a theory, forms a hypothesis, makes empirical observations, and then analyzes the data to confirm, reject, or modify the original theory.

dependent variable the outcome the researcher is trying to explain.

deterrence theory philosophy of criminal justice arising from the notion that crime results from a rational calculation of its costs and benefits.

dialectic a two-directional relationship following a pattern in which an original statement or thesis is countered with an antithesis, leading to a conclusion that unites the strengths of the original position and the counterarguments.

discrimination harmful or negative acts (not mere thoughts) against people deemed inferior on the basis of their racial category, without regard to their individual merit.

divide et impera the role of a member of a triad who intentionally drives a wedge between the other two actors in the group.

double consciousness a concept conceived by W. E. B. Du Bois to describe the two behavioral scripts, one for moving through the world and the other incorporating the external opinions of prejudiced onlookers, which are constantly maintained by African Americans.

dramaturgical theory the view (advanced by Erving Goffman) of social life as essentially a theatrical performance, in which we are all actors on metaphorical stages, with roles, scripts, costumes, and sets.

dyad a group of two.

E

egoistic suicide suicide that occurs when one is not well integrated into a social group.

elite–mass dichotomy system a system of stratification that has a governing elite, a few leaders who broadly hold power in society.

embeddedness the degree to which social relationships are reinforced through indirect ties (i.e., friends of friends).

endogamy marriage to someone within one's social group.

equality of condition the idea that everyone should have an equal starting point.

equality of opportunity the idea that everyone has an equal chance to achieve wealth, social prestige, and power because the rules of the game, so to speak, are the same for everyone.

equality of outcome the idea that each player must end up with the same amount regardless of the fairness of the "game."

essentialism a line of thought that explains social phenomena in terms of natural ones.

essentialist arguments explaining social phenomena in terms of natural, biological, or evolutionary inevitabilities.

estate system a politically based system of stratification characterized by limited social mobility.

ethnicity one's ethnic quality or affiliation. It is voluntary, self-defined, nonhierarchal, fluid and multiple, and based on cultural differences, not physical ones per se.

ethnocentrism the belief that one's own culture or group is superior to others, and the tendency to view all other cultures from the perspective of one's own.

ethnomethodology literally "the methods of the people"; this approach to studying human interaction focuses on the ways in which we make sense of our world, convey this understanding to others, and produce a shared social order.

eugenics literally meaning "well born;" a pseudo-science that postulates that controlling the fertility of populations could influence inheritable traits passed on from generation to generation.

exchange mobility mobility resulting from the swapping of jobs.

exogamy marriage to someone outside one's social group.

experimental methods methods that seek to alter the social landscape in a very specific way for a given sample of individuals and then track what results that change yields; they often involve comparisons to a control group that did not experience such an intervention.

extended family kin networks that extend outside or beyond the nuclear family.

F

face the esteem in which an individual is held by others.

fatalistic suicide suicide that occurs as a result of too much social regulation.

feminism a social movement to get people to understand that gender is an organizing principle in society and to address gender-based inequalities that intersect with other forms of social identity.

feminist methodology a set of systems or methods that treat women's experiences as legitimate empirical and theoretical resources, that promote social science for women (think public sociology, but for a specific half of the public), and that take into account the researcher as much as the overt subject matter.

formal social sanctions mechanisms of social control by which rules or laws prohibit deviant criminal behavior.

free rider problem the notion that when more than one person is responsible for getting something done, the incentive is for each individual to shirk responsibility and hope others will pull the extra weight.

functionalism the theory that various social institutions and processes in society exist to serve some important (or necessary) function to keep society running.

G

gender a social position, behaviors, and set of attributes that are associated with sex identities.

gender roles sets of behavioral norms assumed to accompany one's status as male or female.

generalizability the extent to which we can claim our findings inform us about a group larger than the one we studied.

generalized other an internalized sense of the total expectations of others in a variety of settings—regardless of whether we've encountered those people or places before.

genocide the mass killing of a group of people based on racial, ethnic, or religious traits.

glass ceiling an invisible limit on women's climb up the occupational ladder.

glass escalator the accelerated promotion of men to the top of a work organization, especially in feminized jobs.

H

hegemonic masculinity the condition in which men are dominant and privileged, and this dominance and privilege is invisible.

hegemony a condition by which a dominant group uses its power to elicit the voluntary "consent" of the masses.

heteronormativity the idea that heterosexuality is the default or normal sexual orientation from which other sexualities deviate.

historical methods research that collects data from written reports, newspaper articles, journals, transcripts, television programs, diaries, artwork, and other artifacts that date back to the period under study.

homosexual the social identity of a person who has sexual attraction to and/or relations with people of the same sex.

hypothesis a proposed relationship between two variables, usually with a stated direction.

I

I one's sense of agency, action, or power.

ideology a system of concepts and relationships, an understanding of cause and effect.

income money received by a person for work, from transfers (gifts, inheritances, or government assistance), or from returns on investments.

independent variable a measured factor that the researcher believes has a causal impact on the dependent variable.

inductive approach a research approach that starts with empirical observations and then works to form a theory.

informal social sanctions the usually unexpressed but widely known rules of group membership; the unspoken rules of social life.

in-group another term for the powerful group, most often the majority.

innovator social deviant who accepts socially acceptable goals but rejects socially acceptable means to achieve them.

institutional racism institutions and social dynamics that may seem race neutral but actually disadvantage minority groups.

intersectionality the idea it is critical to understand the interplay between social identities such as race, class, gender, ability status, and sexual orientation,

even though many social systems and institutions (such as the law) try to treat each category on its own.

isomorphism a constraining process that forces one unit in a population to resemble other units that face the same set of environmental conditions.

K

kinship networks strings of relationships between people related by blood and co-residence (i.e., marriage).

L

labeling theory the belief that individuals subconsciously notice how others see or label them, and their reactions to those labels over time form the basis of their self-identity.

large group a group characterized by the presence of a formal structure that mediates interaction and, consequently, status differentiation.

M

macrosociology a branch of sociology generally concerned with social dynamics at a higher level of analysis—that is, across the breadth of a society.

master status one status within a set that stands out or overrides all others.

material culture everything that is a part of our constructed, physical environment, including technology.

matrix of domination intersecting domains of oppression that create a social space of domination and, by extension, a unique position within that space based on someone's intersectional identity along the multiple dimensions of gender, age, race, class, sexuality, location, and so on.

me the self as perceived as an object by the "I"; the self as one imagines others perceive one.

mechanical or segmental solidarity social cohesion based on sameness.

media any formats, platforms, or vehicles that carry, present, or communicate information.

mediator the member of a triad who attempts to resolve conflict between the two other actors in the group.

microsociology a branch of sociology that seeks to understand local interactional contexts; its methods of choice are ethnographic, generally including participant observation and in-depth interviews.

middle class a term commonly used to describe those individuals with nonmanual jobs that pay significantly more than the poverty line—though this is a highly debated and expansive category, particularly in the United States, where broad swaths of the population consider themselves middle class.

midrange theory a theory that attempts to predict how certain social institutions tend to function.

miscegenation the technical term for interracial marriage; literally meaning "a mixing of kinds"; it is politically and historically charged—sociologists generally prefer *exogamy* or *outmarriage*.

monogamy the practice of having only one sexual partner or spouse at a time.

N

narrative the sum of stories contained in a set of ties.

nativism the movement to protect and preserve indigenous land or culture from the allegedly dangerous and polluting effects of new immigrants.

nonmaterial culture values, beliefs, behaviors, and social norms.

norms how values tell us to behave.

nuclear family familial form consisting of a father, a mother, and their children.

O

one-drop rule the belief that "one drop" of black blood makes a person black, a concept that evolved from US laws forbidding miscegenation.

ontological equality the philosophical and religious notion that all people are created equal.

operationalization how a concept gets defined and measured in a given study.

organic solidarity social cohesion based on difference and interdependence of the parts.

organization any social network that is defined by a common purpose and has a boundary between its membership and the rest of the social world.

organizational culture the shared beliefs and behaviors within a social group; often used interchangeably with *corporate culture*.

organizational structure the ways in which power and authority are distributed within an organization.

other someone or something outside of oneself.

out-group another term for the stigmatized or less powerful group, the minority.

P

panopticon a circular building composed of an inner ring and an outer ring designed to serve as a prison in which the guards, housed in the inner ring, can observe the prisoners without the detainees knowing whether they are being watched.

participant observation a qualitative research method that seeks to uncover the meanings people give their social actions by observing their behavior in practice.

party a group that is similar to a small group but is multifocal.

patriarchy a nearly universal system involving the subordination of femininity to masculinity.

polygamy the practice of having more than one sexual partner or spouse at a time.

polygyny the practice of having multiple wives simultaneously.

population an entire group of individual persons, objects, or items from which samples may be drawn.

positivist sociology a strain within sociology that believes the social world can be described and predicted by certain observable relationships (akin to social physics).

postmodernism a condition characterized by a questioning of the notion of progress and history, the replacement of narrative within pastiche, and multiple, perhaps even conflicting, identities resulting from disjointed affiliations.

prejudice thoughts and feelings about an ethnic or racial group, which lead to preconceived notions and judgments (often negative) about the group.

primary deviance the first act of rule breaking that may incur a label of "deviant" and thus influence how people think about and act toward you.

primary groups social groups, such as family or friends, composed of enduring, intimate face-to-face relationships that strongly influence the attitudes and ideals of those involved.

primordialism Clifford Geertz's term to explain the strength of ethnic ties because they are fixed and deeply felt or primordial ties to one's homeland culture.

proletariat the working class.

Q

qualitative methods methods that attempt to collect information about the social world that cannot be readily converted to numeric form.

quantitative methods methods that seek to obtain information about the social world that is already in or can be converted to numeric form.

R

race a group of people who share a set of characteristics—typically, but not always, physical ones—and are said to share a common bloodline.

racialization the formation of a new racial identity by drawing ideological boundaries of difference around a formerly unnoticed group of people.

racism the belief that members of separate races possess different and unequal traits.

rebel individual who rejects both traditional goals and traditional means and wants to alter or destroy the social institutions from which he or she is alienated.

recidivism when an individual who has been involved with the criminal justice system reverts to criminal behavior.

reference group a group that helps us understand or make sense of our position in society relative to other groups.

reflection theory the idea that culture is a projection of social structures and relationships into the public sphere, a screen onto which the film of the underlying reality or social structures of a society is projected.

reflexivity analyzing and critically considering our own role in, and effect on, our research.

reliability the likelihood of obtaining consistent results using the same measure.

research methods approaches that social scientists use for investigating the answers to questions.

resocialization the process by which one's sense of social values, beliefs, and norms are reengineered, often deliberately, through an intense social process that may take place in a total institution.

retreatist one who rejects both socially acceptable means and goals by completely retreating from, or not participating in, society.

reverse causality a situation in which the researcher believes that A results in a change in B, but B, in fact, is causing A.

ritualist individual who rejects socially defined goals but not the means.

role the duties and behaviors expected of someone who holds a particular status.

role conflict the tension caused by competing demands between two or more roles pertaining to different statuses.

role strain the incompatibility among roles corresponding to a single status.

S

sample the subset of the population from which you are actually collecting data.

scientific method a procedure involving the formulation, testing, and modification of hypotheses based on systematic observation, measurement, and/or experiments.

scientific racism nineteenth-century theories of race that characterize a period of feverish investigation into the origins, explanations, and classifications of race.

scientific revolution see paradigm shift.

second shift women's responsibility for housework and child care—everything from cooking dinner to doing laundry, bathing children, reading bedtime stories, and sewing Halloween costumes.

secondary deviance subsequent acts of rule breaking that occur after primary deviance and as a result of your new deviant label and people's expectations of you.

secondary groups groups marked by impersonal, instrumental relationships (those existing as a means to an end).

segmental solidarity *see* mechanical solidarity.

segregation the legal or social practice of separating people on the basis of their race or ethnicity.

self the individual identity of a person as perceived by that same person.

sex the perceived biological differences that distinguish males from females.

sex role theory Talcott Parsons's theory that men and women perform their sex roles as breadwinners and wives/mothers, respectively, because the nuclear family is the ideal arrangement in modern societies, fulfilling the function of reproducing workers.

sexism occurs when a person's sex or gender is the basis for judgment, discrimination, or other differential treatment against that person.

sexual harassment an illegal form of discrimination revolving around sexuality that can involve everything from inappropriate jokes to sexual "barter" (where victims feel the need to comply with sexual requests for fear of losing their job) to outright sexual assault.

sexuality desire, sexual preference, and sexual identity and behavior.

small group a group characterized by face-to-face interaction, a unifocal perspective, lack of formal arrangements or roles, and a certain level of equality.

social cohesion social bonds; how well people relate to each other and get along on a day-to-day basis.

social construction an entity that exists because people behave as if it exists and whose existence is perpetuated as people and social institutions act in accordance with the widely agreed-on formal rules or informal norms of behavior associated with that entity.

social control mechanisms that create normative compliance in individuals.

social Darwinism the application of Darwinian ideas to society—namely, the evolutionary "survival of the fittest."

social deviance any transgression of socially established norms.

social equality a condition in which no differences in wealth, power, prestige, or status based on nonnatural conventions exist.

social institution a complex group of interdependent positions that, together, perform a social role and reproduce themselves over time; also defined in a narrow sense as any institution in a society that works to shape the behavior of the groups or people within it.

social integration how well you are integrated into your social group or community.

social mobility the movement between different positions within a system of social stratification in any given society.

social network a set of relations—essentially, a set of dyads—held together by ties between individuals.

social regulation the number of rules guiding your daily life and, more specifically, what you can reasonably expect from the world on a day-to-day basis.

socialization the process by which individuals internalize the values, beliefs, and norms of a given society and learn to function as members of that society.

sociological imagination the ability to connect the most basic, intimate aspects of an individual's life to seemingly impersonal and remote historical forces.

sociology the study of human society.

status a recognizable social position that an individual occupies.

status hierarchy system a system of stratification based on social prestige.

status set all the statuses one holds simultaneously.

status-attainment model approach that ranks individuals by socioeconomic status, including income and educational attainment, and seeks to specify the attributes characteristic of people who end up in more desirable occupations.

stigma a negative social label that not only changes others' behavior toward a person but also alters that person's own self-concept and social identity.

straight-line assimilation Robert Parks's 1920s universal and linear model for how immigrants assimilate: they first arrive, then settle in, and achieve full assimilation in a newly homogenous country.

strain theory Robert Merton's theory that deviance occurs when a society does not give all of its members equal ability to achieve socially acceptable goals.

stratification the hierarchical organization of a society into groups with differing levels of power, social prestige, or status and economic resources.

street crime crime committed in public and often associated with violence, gangs, and poverty.

strength of weak ties the notion that relatively weak ties often turn out to be quite valuable because they yield new information.

structural functionalism theoretical tradition claiming that every society has certain structures (the family, the division of labor, or gender) that exist to fulfill some set of necessary functions (reproduction of the species, production of goods, etc.).

structural hole a gap between network clusters, or even two individuals, if those individuals (or clusters) have complementary resources.

structural mobility mobility that is inevitable from changes in the economy.

subaltern a subordinate, oppressed group of people.

subculture the distinct cultural values and behavioral patterns of a particular group in society; a group united by sets of concepts, values, symbols, and shared meaning specific to the members of that group distinctive enough to distinguish it from others within the same culture or society.

survey an ordered series of questions intended to elicit information from respondents.

symbolic ethnicity a nationality, not in the sense of carrying the rights and duties of citizenship but of identifying with a past or future nationality. For later generations of white ethnics, something not constraining but easily expressed, with no risks of stigma and all the pleasures of feeling like an individual.

symbolic interactionism a micro-level theory in which shared meanings, orientations, and assumptions form the basic motivations behind people's actions.

T

tertius gaudens the member of a triad who benefits from conflict between the other two members of the group.

theory an abstracted, systematic model of how some aspect of the world works.

tie the connection between two people in a relationship that varies in strength from one relationship to the next; a story that explains our relationship with another member of our network.

total institution an institution in which one is totally immersed and that controls all the basics of day-to-day life; no barriers exist between the usual spheres of daily life, and all activity occurs in the same place and under the same single authority.

transgender describes people whose gender does not correspond to their birth sex.

triad a group of three.

U

upper class a term for the economic elite.

V

validity the extent to which an instrument measures what it is intended to measure.

values moral beliefs.

Verstehen German for "understanding." The concept of *Verstehen* comes from Max Weber and is the basis of interpretive sociology in which researchers imagine themselves experiencing the life positions of the social actors they want to understand rather than treating those people as objects to be examined.

W

wealth a family's or individual's net worth (i.e., total assets minus total debts).

white-collar crime offense committed by a professional (or professionals) against a corporation, agency, or other institution.

Bibliography

Aboim, S. (2016). *Plural masculinities: The remaking of the self in private life.* New York: Routledge.

Abrutyn, S., & Mueller, A. S. (2014). Are suicidal behaviors contagious in adolescence? Using longitudinal data to examine suicide suggestion. *American Sociological Review, 79*(2), 211–27.

Acemoglu, D., & Restrepo, P. (2018, January). Artificial intelligence, automation and work (NBER Working Paper No. 24196). National Bureau of Economic Research. Retrieved from www.nber.org/papers/w24196.

Acemoglu, D., & Robinson, J. A. (2006). *Economic origins of dictatorship and democracy.* New York: Cambridge University Press.

Acemoglu, D., et al. (2001, December). The colonial origins of comparative development: An empirical investigation. *The American Economic Review, 91*(5), 1369–401.

Acierno, R., et al. (2010). Prevalence and correlates of emotional, physical, sexual, and financial abuse and potential neglect in the United States: The National Elder Mistreatment Study. *American Journal of Public Health, 100*(2), 292–97.

Ad Council. (2018). Texting and driving prevention. Retrieved from www.psacentral.org/campaign/Texting_and_Driving_Prevention.

Adida, C. L., et al. (2010). Identifying barriers to Muslim integration in France. *Proceedings of the National Academy of Sciences USA, 107*(52), 22384–90.

Aguirre, B. E., et al. (1998). A test of the emergent norm theory of collective behavior. *Sociological Forum, 13*(2), 301–20.

Alexander, K., et al. (2001). Schools, achievement and equality: A seasonal perspective. *Educational Evaluation and Policy Analysis, 23*(2), 171–91.

Allen, F. A. (1981). *The decline of the rehabilitative ideal: Penal policy and social purpose.* New Haven, CT: Yale University Press.

Allen, J. T. (1973). Designing income maintenance systems: The income accounting problem (paper no. 5). *Studies in Public Welfare.* Washington, DC: Government Printing Office.

Amenta, E. (2006). *When movements matter: The Townsend Plan and the rise of Social Security.* Princeton, NJ: Princeton University Press.

American Civil Liberties Union. (2013). The war on marijuana in black and white. Retrieved from www.aclu.org/files/assets/aclu-thewaronmarijuana-rel2.pdf.

Ammerman, N. T. (2005). *Pillars of faith: American congregations and their partners.* Berkeley: University of California Press.

Anderson, E. (1999). *Code of the street: Decency, violence, and the moral life of the inner city.* New York: Norton.

Anderson, J. L. (2009). The most failed state: Letter from Mogadishu. *The New Yorker, 85*(41), 64.

Ariès, P. (1965). *Centuries of childhood: A social history of family life.* New York: Vintage. (Original work published 1962.)

Armstrong, E. A., & Bernstein, M. (2008). Culture, power, and institutions: A multi-institutional politics approach to social movements. *Sociological Theory, 26*(1), 74–99.

Armstrong, E. A., et al. (2012). Accounting for women's orgasm and sexual enjoyment in college hookups and relationships. *American Sociological Review, 77*(3), 435–62.

Arnold, M. (1869). *Culture and anarchy.* London: Smith, Elder.

Arum, R. (2003). *Judging school discipline: The crisis of moral authority.* Cambridge: Harvard University Press.

Arum, R., & Shavit, Y. (1995). Secondary vocational education and the transition from school to work. *Sociology of Education, 68*(3), 187–204.

Asch, S. E. (1956). Studies of independence and conformity: A minority of one against a unanimous majority. *Psychological Monographs, 70*(9, Whole No. 416).

Attanasio, O., et al. (2013). The evolution of income, consumption, and leisure inequality in the US, 1980–2010. In C. Carroll et al. (Eds.), *Improving the measure of consumer expenditures* (pp. 100–140). Chicago: University of Chicago Press.

Auletta, K. (1981, November 16). I—the underclass. *The New Yorker,* p. 63.

Australian Catholic University. (2008). Principles of inclusive curriculum. Retrieved from www.acu.edu.au/__data/assets/pdf_file/0019/103735/Principles_of_Inclusive_Curriculum.pdf.

Australian Human Rights Commission. (1997, April). *Bringing them home: Report of the national inquiry into the separation of Aboriginal and Torres Strait Islander children from their families.* Retrieved from www.hreoc.gov.au/social_justice/bth_report/report/index.html.

Autor, D. H., et al. (2013). The China syndrome: Local labor market effects of import competition in the United States. *American Economic Review, 103*(6), 2121–68. Retrieved from www.aeaweb.org/articles?id=10.1257/aer.103.6.2121.

Ayres, I., & Unkovic, C. (2012). Information escrows. *Michigan Law Review, 111*(2), 145. Retrieved from https://repository.law.umich.edu/mlr/vol111/iss2/1.

Babbie, E. R. (2007). *The practice of social research* (11th ed.). Belmont, CA: Wadsworth.

Bacolod, M. (2017). Skills, the gender wage gap, and cities. *Journal of Regional Science, 57*(2), 290–318.

Bailey, M. J., & Dynarski, S. M. (2011). Inequality in postsecondary education. In G. J. Duncan & R. J. Murnane (Eds.), *Whither opportunity? Rising inequality, schools, and children's life chances* (pp. 117–32). New York: Russell Sage Foundation.

Baker, D. C. (2005, May 26). Church hosts "drive-thru" Sunday services. *Quad-City Times.* Retrieved from www.qctimes.com/articles/2005/05/26/local/export93493.txt.

Baker, J. O. (2015). *American secularism: Cultural contours of nonreligious belief systems.* New York: NYU Press.

Bakshi, A. (2003). Potential adverse health effects of genetically modified crops, part B. *Journal of Toxicology and Environmental Health, 6,* 211–55.

Baldus, D. C., et al. (1998). Racial discrimination and the death penalty in the post-*Furman* era. *Cornell Law Review, 83,* 1638. Retrieved from www.lawschool.cornell.edu/research/cornell-law-review/upload/baldus.pdf.

Bales, K. (2016). *Blood and earth: Modern slavery, ecocide, and the secret to saving the world.* New York: Spiegel & Grau.

Baltzell, E. D. (1958). *Philadelphia gentlemen.* New York: Free Press.

Banerjee, N. (2005, December 30). Going to church to find a faith that fits. *The New York Times,* pp. 1a, 18. Retrieved from www.nytimes.com/2005/12/30/us/teenagers-mix-churches-for-faith-that-fits.html.

Banfield, E. (1970). *The unheavenly city.* Boston: Little, Brown.

Barry-Jester, A. M. (2015, June 23). Attitudes toward racism and inequality are shifting. Retrieved from https://fivethirtyeight.com/features/attitudes-toward-racism-and-inequality-are-shifting.

Barth, E. A. T., & Noel, D. L. (1972). Conceptual frameworks for the analysis of race relations: An evaluation. *Social Forces, 50,* 333–48.

Bauer, G. (2016). Gender roles, comparative advantages and the life course: The division of domestic labor in same-sex and different-sex couples. *European Journal of Population, 32*(1), 99–128.

Bear, C. (2008, May 13). American Indian school a far cry from the past. *Morning Edition.* National Public Radio.

Bearman, P. S., & Brückner, H. (2001). Promising the future: Virginity pledges and the transition to first intercourse. *American Journal of Sociology, 106*(4), 859–912.

Bearman, P. S., et al. (2004). Chains of affection: The structure of adolescent romantic and sexual networks. *American Journal of Sociology, 110*(1), 44–91.

Beauvoir, S. de. (1952). *The second sex.* New York: Knopf.

Beck, U. (1992). *Risk society: Towards a new modernity* (Mark Ritter, Trans.). London: Sage.

Becker, H. S. (1953, November). Becoming a marihuana user. *American Journal of Sociology, 59,* 235–43.

Becker, H. S. (1963). *Outsiders: Studies in the sociology of deviance.* New York: Free Press.

Belley, P., & Lochner, L. (2007, October). *The changing role of family income and ability in determining educational achievement* (NBER Working Paper No. 13527). National Bureau of Economic Research. Retrieved from www.nber.org/papers/w13527.

Bender, C. J. (2007). Touching the transcendent: Rethinking religious experience in the sociological study of religion. In N. Ammerman (Ed.), *Everyday religion: Observing modern religious lives* (pp. 201–19). New York: Oxford University Press. Retrieved from www.oxfordscholarship.com/oso/private/content/religion/9780195305418/p050.html#acprof-9780195305418-chapter-11.

Benedict, R. (1934). *Patterns of culture.* New York: Mentor Books.

Berbrier, M. (2000). The victim ideology of white supremacists and white separatists in the United States. *Sociological Focus, 33,* 175–91.

Berger, P. L. (1967). *The sacred canopy: Elements of a sociological theory of religion.* Garden City, NY: Anchor Books.

Berk, S. F. (1985). *The gender factory: The apportionment of work in American households.* New York: Plenum Press.

Bernard, J. S. (1972). *The future of marriage.* New York: World Publishers.

Berry, B. (1963). *Almost white.* New York: Macmillan.

Besecke, K. (2007). Beyond literalism: Reflexive spirituality and religious meaning. In N. Ammerman (Ed.), *Everyday religion: Observing modern religious lives* (pp. 169–86). New York: Oxford University Press. Retrieved from www.oxfordscholarship.com/oso/private/content/religion/9780195305418/p009.html.

Bialik, K. (2017, June 12). Key facts about race and marriage, 50 years after *Loving v. Virginia.* Pew Research Center. Retrieved from www.pewresearch.org/fact-tank/2017/06/12/key-facts-about-race-and-marriage-50-years-after-loving-v-virginia.

Bird, W., & Thumma, S. (2011). A new decade of megachurches. Retrieved from http://hirr.hartsem.edu/megachurch/New-Decade-of-Megachurches-2011Profile.pdf.

Bittman, M., et al. (2003). When does gender trump money? Bargaining and time in household work. *American Journal of Sociology, 109,* 186–214.

Bixler, M. T. (1992). *Winds of freedom: The story of the Navajo code talkers of World War II.* Darien, CT: Two Bytes.

Black megachurches surge. (1996). *Christian Century, 113*(21), 686–87. Retrieved from www.christiancentury.org.

Blanchard, R., & Bogaert, A. F. (1996). Homosexuality in men and number of older brothers. *American Journal of Psychiatry, 153,* 27–31.

Blanton, C. K. (2000). "They cannot master abstractions, but they can often be made efficient workers": Race and class in the intelligence testing of Mexican Americans and African Americans in Texas during the 1920's. *Social Science Quarterly, 81*(4), 1014–26.

Blau, P. M., & Duncan, O. D., with collaboration of A. Tyree. (1967). *The American occupational structure.* New York: Wiley.

Blum, B. (2018, June). The lifespan of a lie? *Medium.* Retrieved from http://medium.com/s/trustissues/the-lifespan-of-a-lie-d869212b1f62.

Blumer, H. (1969). *Symbolic interactionism: Perspective and method.* Englewood Cliffs, NJ: Prentice-Hall.

Blumstein, P. W., & Schwartz., P. (1983). *American couples: Money, work, and sex.* New York: Morrow.

Bobo, L. (1989). Keeping the linchpin in place: Testing the multiple sources of opposition to residential integration. *Revue Internationale de Psychologie Sociale, 2,* 306–23.

Bobo, L., et al. (1997). Laissez-faire racism: The crystallization of a kinder, gentler, antiblack ideology. In S. A. Tuch & J. K. Martin (Eds.), *Racial attitudes in the 1990s: Continuity and change* (pp. 15–42). Westport, CT: Praeger.

Bogaert, A. F. (2006). Biological versus nonbiological older brothers and men's sexual orientation. *Proceedings of the National Academy of Sciences USA, 103*(28), 10771–74.

Bogin, B. (1995). Plasticity in the growth of Mayan refugee children living the United States. In C. G. N. Mascie-Taylor & B. Bogin (Eds.), *Human variability and plasticity* (pp. 46–74). Cambridge: Cambridge University Press.

Bogle, K. A. (2008). *Hooking up: Sex, dating and relationships on campus.* New York: NYU Press.

Bonczar, Thomas P. (2003). Prevalence of imprisonment in the U.S. population, 1974–2001. Bureau of Justice Statistics Special Report. Retrieved from www.bjs.gov/content/pub/pdf/piusp01.pdf.

Booth, B. (2015, April 16). How much would you pay to get your kid into Harvard? CNBC. Retrieved from www.cnbc.com/2014/11/10/is-a-college-planner-really-worth-it.html.

Bordo, S. (1990). Feminism, postmodernism, and gender skepticism. In L. J. Nicholson (Ed.), *Feminism/postmodernism* (pp. 133–56). New York: Routledge.

Bourdieu, P. (1977). Cultural reproduction and social reproduction. In J. Karabel & A. H. Halsey (Eds.), *Power and ideology in education* (pp. 487–511). New York: Oxford University Press.

Bowen, W. G., & Bok, D. (1998). *The shape of the river: Long-term consequences of considering race in college and university admissions.* Princeton, NJ: Princeton University Press.

Bowles, S., & Gintis, H. (1976). *Schooling in capitalist America: Educational reform and the contradictions of economic life.* New York: Basic Books.

Bozorgmehr, M., et al. (1996). Middle Easterners: A new kind of immigrant. In R. Waldinger & M. Bozorgmehr (Eds.), *Ethnic Los Angeles* (pp. 347–78). New York: Russell Sage Foundation.

Bradley, M. B., et al. (1992). *Churches and church membership in the United States, 1990.* Atlanta: Glenmary Research Center.

Brame, R., et al. (2014). Demographic patterns of cumulative arrest prevalence by ages 18 and 23. *Crime and Delinquency, 60,* 471–86.

Brandon, R. (2015, December 1). Mark Zuckerberg and Priscilla Chan to donate 99 percent of their Facebook fortune. The Verge. Retrieved from www.theverge.com/2015/12/1/9831554/mark-zuckerberg-charity-45-billion.

Brandow, M. (2008). *New York's poop scoop law: Dogs, the dirt and due process.* West Lafayette, IN: Purdue University Press.

Brewster, M. E. (2015). Lesbian women and household labor division: A systematic review of scholarly research from 2000 to 2015. *Journal of Lesbian Studies, 21*(1): 47–69.

Bridges, T., & Pascoe, C. J. (2014). Hybrid masculinities: New directions in the sociology of men and masculinities. *Sociology Compass, 8*(3), 246–58.

Briody, D. (2003). *The iron triangle: Inside the secret world of the Carlyle Group.* Hoboken, NJ: Wiley.

Brookings Institution. (2010). 2010 Racial and ethnic distributions of 100 largest metropolitan and 2000–2010 change. Retrieved from www.brookings.edu/wp-content/uploads/2016/06/0831_census_race_appendixa.pdf.

Brooks-Gunn, J., et al. (2003). Maternal employment and child cognitive outcomes in the first three years of life: The NICHD study of early child care. *Child Development, 73*(4), 1052–72.

Bromwich, J. E. (2017, August 7). Minnesota governor calls mosque attack a "criminal act of terrorism." *The New York Times.* Retrieved from www.nytimes.com/2017/08/07/us/minnesota-mosque-explosion.html.

Brubaker, R. (1992). *Citizenship and nationhood in France and Germany.* Cambridge: Harvard University Press.

Brückner, H., & Bearman, P. S. (2005). After the promise: The STD consequences of adolescent virginity pledges. *Journal of Adolescent Health, 36,* 271–78.

Bryk, A., et al. (1993). *Catholic schools and the common good.* Cambridge: Harvard University Press.

Buchmann, C., et al. (2006). The growing female advantage in higher education: The role of family background and academic achievement. *American Sociological Review, 71*(4), 515–41.

Budge, S. L., et al. (2013). Anxiety and depression in transgender individuals: The roles of transition status, loss, support, and coping. *Journal of Consulting and Clinical Psychology, 81*(3), 545–57. Retrieved from www.ncbi.nlm.nih.gov/pubmed/23398495.

Bureau of Justice Statistics. (2014). US Correctional population declined by less than 1 percent for the second consecutive year. Retrieved from www.bjs.gov/content/pub/press/cpus13pr.cfm.

Bureau of Labor Statistics, US Department of Labor. (2007). Working and work-related activities done by men and women in 2007. *American Time Use Survey.* Retrieved from www.bls.gov/tus/current/work.htm#a1.

Bureau of Labor Statistics, US Department of Labor. (2013). Table 19: Persons at work in agricultural and non-agricultural industries by hours of work. Retrieved from www.bls.gov/cps/cpsaat19.htm.

Bureau of Labor Statistics, US Department of Labor. (2014a). Characteristics of minimum wage workers, 2013. Retrieved from www.bls.gov/opub/reports/minimum-wage/archive/minimumwageworkers_2013.pdf.

Bureau of Labor Statistics, US Department of Labor. (2014b). Labor force characteristics by race and ethnicity, 2013. Retrieved from www.bls.gov/cps/cpsrace2013.pdf.

Bureau of Labor Statistics, US Department of Labor. (2014c). Women in the labor force: A databook. Retrieved from www.bls.gov/opub/reports/womens-databook/archive/women-in-the-labor-force-a-databook-2014.pdf.

Bureau of Labor Statistics, US Department of Labor. (2015a). Highlights of women's earnings in 2014. Retrieved from www.bls.gov/opub/reports/cps/highlights-of-womens-earnings-in-2014.pdf.

Bureau of Labor Statistics, US Department of Labor. (2015b). Table 19. Persons at work in agricultural and non-agricultural industries by hours of work. Current Population Survey. Retrieved from www.bls.gov/cps/cpsaat19.htm.

Bureau of Labor Statistics, US Department of Labor. (2016a). Labor force participation rate: women. Retrieved from https://fred.stlouisfed.org/series/LNS11300002#0.

Bureau of Labor Statistics, US Department of Labor. (2016b). Tables 4–6. Employment characteristics of families—2015 (News Release, USDL-16-0795). Retrieved from www.bls.gov/news.release/pdf/famee.pdf.

Bureau of Labor Statistics, US Department of Labor. (2016c). Working mothers issue brief. Retrieved from www.dol.gov/wb/resources/WB_WorkingMothers_508_FinalJune13.pdf.

Bureau of Labor Statistics, US Department of Labor. (2017a, August). Highlights of women's earnings in 2016. Retrieved from www.bls.gov/opub/reports/womens-earnings/2016/pdf/home.pdf.

Bureau of Labor Statistics, US Department of Labor. (2017b). College enrollment and work activity of 2016 high school graduates. Economic news release. Retrieved from www.bls.gov/news.release/hsgec.nr0.htm.

Bureau of Labor Statistics, US Department of Labor. (2017c). Women in the labor force: A databook. Retrieved from www.bls.gov/opub/reports/womens-databook/2017/pdf/home.pdf.

Bureau of Labor Statistics, US Department of Labor. (2017d). Unemployment rates and earnings by educational attainment, 2016. Retrieved from www.bls.gov/emp/ep_chart_001.htm.

Bureau of Labor Statistics, US Department of Labor (2017e, April 19). Employment characteristics of families summary. Retrieved from www.bls.gov/news.release/famee.nr0.htm.

Bureau of Labor Statistics, US Department of Labor. (2017f, April 27). TED: Employment in families with children 2016. Retrieved from www.bls.gov/opub/ted/2017/employment-in-families-with-children-in-2016.htm.

Bureau of Labor Statistics, US Department of Labor. (2017g). May 2016 national occupational employment and wage estimates United States. Retrieved from https://www.bls.gov/oes/2016/may/oes_nat.htm#00-0000.

Bureau of Labor Statistics, US Department of Labor. (2018a). CPI inflation calculator. Retrieved from www.bls.gov/data/inflation_calculator.htm.

Bureau of Labor Statistics, US Department of Labor. (2018b, January 9). Union members—2017 [News release]. Retrieved from www.bls.gov/news.release/pdf/union2.pdf.

Bureau of Labor Statistics, US Department of Labor. (2018c). Average hours employed people spent working on days worked by day of week, 2017 annual averages. Retrieved from www.bls.gov/charts/american-time-use/emp-by-ftpt-job-edu-h.htm.

Bureau of Labor Statistics, US Department of Labor. (n.d.) Consumer price index calculator. Retrieved from: http://data.bls.gov/cgi-bin/cpicalc.pl?cost1=1&year1=1940&year2=2015.

Burt, R. (1992). *Structural holes*. Cambridge: Harvard University Press.

Burton, C. (2014). Suburban student raises $25K in 24 hours for college tuition. ABC Eyewitness News. Retrieved from http://abc7chicago.com/education/suburban-student-raises-$25k-in-24-hours-for-college-tuition/176216.

Calhoun, C. (2002). *Dictionary of the social sciences*. New York: Oxford University Press.

Callan, T., et al. (1993). Resources, deprivation and the measurement of poverty. *Journal of Social Policy*, 22, 141–72.

Cantril, H. (1941). *The psychology of social movements*. New York: Wiley.

Card, D., et al. (2008). Tipping and the dynamics of segregation. *Quarterly Journal of Economics*, 123(1), 177–218.

Carlyle, T. (1971). *Chartism*. In *Selected writings*. A. Shelston (Ed.). Harmondsworth: Penguin Books. (Original work published 1839.)

Carnevale, A. P., et al. (2014). The college payoff: Education, occupations, lifetime earnings. Washington, DC: Georgetown University, Center on Education and the Workforce. Retrieved from https://cew-7632.kxcdn.com/wp-content/uploads/2014/11/collegepayoff-complete.pdf.

Carpiano, R. M., et al. (2006). Social inequality and health: Future directions for the fundamental cause explanation for class differences in health. Paper presented at the Russell Sage Foundation conference Social Class: How Does It Work?, New York.

Carrington, C. (1999). *No place like home: Relationships and family life among lesbians and gay men*. Chicago: University of Chicago Press.

Carson, E. A. (2014). Prisoners in 2013. Washington, DC: US Department of Justice, Bureau of Justice Statistics. Retrieved from www.bjs.gov/content/pub/pdf/p13.pdf.

Case, A., & Deaton, A. (2015). Rising morbidity and mortality in midlife among white non-Hispanic Americans in the 21st century. *Proceedings of the National Academy of Sciences USA, 112*(49), 15078–83. Retrieved from www.pnas.org/content/112/49/15078.full.

Centers for Disease Control and Prevention. (2010). Surveillance summaries: Youth risk behavior surveillance. Retrieved from www.cdc.gov/mmwr/pdf/ss/ss5905.pdf.

Centers for Disease Control and Prevention. (2011a). National Survey of Family Growth. 2006–2010. Retrieved from www.cdc.gov/nchs/nsfg.htm.

Centers for Disease Control and Prevention. (2011b). Key statistics from the National Survey of Family Growth, Table 1. Retrieved from www.cdc.gov/nchs/nsfg/key_statistics/a.htm#adoption.

Centers for Disease Control and Prevention. (2011c). National Intimate Partner and Sexual Violence Survey: 2010 summary report. National Center for Injury Prevention and Control, Division of Violence Prevention.

Centers for Disease Control and Prevention. (2011d). Understanding intimate partner violence: Fact sheet 2011. Retrieved from www.cdc.gov/violenceprevention/pdf/ipv_factsheet-a.pdf.

Centers for Disease Control and Prevention. (2012a). Suicide: Facts at a glance. Retrieved from www.cdc.gov/violenceprevention/pdf/suicide-datasheet-a.pdf.

Centers for Disease Control and Prevention. (2012b). Reported cases and deaths from vaccine preventable diseases, United States, 1950–2011. Retrieved from www.cdc.gov/vaccines/pubs/pinkbook/downloads/appendices/g/ases&deaths.pdf.

Centers for Disease Control and Prevention. (2015a). Health, United States, 2014: With special feature on adults aged 55–64. Retrieved from www.cdc.gov/nchs/data/hus/hus14.pdf#016.

Centers for Disease Control and Prevention. (2015b). Understanding literacy & numeracy. Retrieved from www.cdc.gov/healthliteracy/learn/understandingliteracy.html.

Centers for Disease Control and Prevention. (2015c). Reported cases and deaths from vaccine preventable diseases, United States, 1950–2013. Retrieved from www.cdc.gov/vaccines/pubs/pinkbook/downloads/appendices/E/reported-cases.pdf.

Centers for Disease Control and Prevention. (2016a). 1991–2015 high school youth risk behavior survey data. Retrieved from http://nccd.cdc.gov/youthonline.

Centers for Disease Control and Prevention. (2016b). Fact sheet: Today's HIV/AIDS epidemic. Retrieved from www.cdc.gov/nchhstp/newsroom/docs/factsheets/todaysepidemic-508.pdf.

Centers for Disease Control and Prevention. (2017). Health, United States, 2016: With chartbook on long-term trends in health. Retrieved from www.cdc.gov/nchs/data/hus/hus16.pdf#053.

Centers for Disease Control and Prevention. (2018). Youth risk behavior survey data summary & trends report 2007–2017. Retrieved from www.cdc.gov/healthyyouth/data/yrbs/pdf/trendsreport.pdf.

Centers for Medicare & Medicaid Services. (2018, January 8). National health expenditure data: Historical. Retrieved from www.cms.gov/Research-Statistics-Data-and-Systems/Statistics-Trends-and-Reports/NationalHealthExpendData/NationalHealthAccountsHistorical.html.

Cesarini, D., et al. (2016). Wealth, health, and child development: Evidence from administrative data on Swedish lottery players. Quarterly Journal of Economics, 131(2), 687–738. Retrieved from https://academic.oup.com/qje/article/131/2/687/2606947.

Cesarini, D., et al. (2017). The effect of wealth on individual and household labor supply: Evidence from Swedish lotteries. American Economic Review, 107(12), 3917–46.

Chamie, J. (2016, October 15). 320 million children in single-parent families. Inter Press Service. Retrieved from www.ipsnews.net/2016/10/320-million-children-in-single-parent-families.

Chandra, A., et al. (2005, December). Fertility, family planning, and reproductive health of U.S. women: Data from the 2002 National Survey of Family Growth. Vital and Health Statistics, ser. 23, no. 25. Retrieved from www.cdc.gov/nchs/data/series/sr_23/sr23_025.pdf.

Chase-Lansdale, P. L., et al. (1997). Neighborhood and family influences on the intellectual and behavioral competence of preschool and early school-age children. In J. Brooks-Gunn et al. (Eds.), Neighborhood poverty. Vol. 1 (pp. 79–118). New York: Russell Sage Foundation.

Chase-Lansdale, P. L., et al. (2003). Mothers' transitions from welfare to work and the well-being of preschoolers and adolescents. Science, 299(5612), 1548–52.

Chauncey, G. (1994). Gay New York: Gender, urban culture, and the makings of the gay male world, 1890–1940. New York: Basic Books.

Chaves, M. (2002, summer). Abiding faith. Context Magazine, 1(2).

Chaves, M., & Higgins, L. M. (1992). Comparing the community involvement of black and white congregations. Journal for the Scientific Study of Religion, 31(4), 425–40.

Chen, D. W. (2004, November 18). $7 billion for the grief of Sept. 11. The New York Times. Retrieved from www.nytimes.com/2004/11/18/nyregion/7-billion-for-the-grief-of-sept-11.html.

Cherlin, A. J. (2009). The marriage-go-round: The state of marriage and the family in America today. New York: Knopf.

Chetty, R., & Hendren, N. (2015). The impacts of neighborhoods on intergenerational mobility: Childhood exposure effects and county level estimates. Retrieved from www.equality-of-opportunity.org/images/nbhds_paper.pdf.

Chetty, R., Hendren, N., & Katz, L. F. (2015). The effects of exposure to better neighborhoods on children: New evidence from the moving to opportunity experiment. Retrieved from www.equality-of-opportunity.org/images/mto_paper.pdf.

Child Trends Data Bank. (2015). Violent crime victimization: Indicators of child and youth well-being. Retrieved from www.childtrends.org/wp-content/uploads/2015/12/71_Violent_Crime_Victimization.pdf.

Chirot, D. (1985). The rise of the West. American Sociological Review, 50, 181–95.

Chodorow, N. (1978). Reproduction of mothering: Psychoanalysis and the sociology of gender. Berkeley: University of California Press.

Chorev, N. (2007). Remaking U.S. trade policy: From protectionism to globalization. Ithaca, NY: Cornell University Press.

Chua, A. L. (1998). Markets, democracy, and ethnicity: Toward a new paradigm for law and development. Yale Law Journal, 108(1), 1–107.

Clark, A. (2005, January 7). Frequent flyer miles soar above sterling. The Guardian. Retrieved from www.theguardian.com/money/2005/jan/08/business.theairlineindustry.

Clark, G. (2014). The son also rises: Surnames and the history of social mobility. Princeton, NJ: Princeton University Press.

Clifton, R. A., et al. (1986). Effects of ethnicity and sex on teachers' expectations of junior high school students. Sociology of Education, 59(1), 58–67.

Cloward, R., & Ohlin, L. (1960). Delinquency and opportunity. New York: Free Press.

Cohen, P. N. (2017). Families are changing—And staying the same. Educational Leadership, 75(1), 46–50. Retrieved from www.terpconnect.umd.edu/~pnc/EdLead17.pdf.

Colby, S. L., & Ortman, J. M. (2015, March). Projections of the size and composition of the U.S. population: 2014–2060. Current Population Reports. Retrieved from www.census.gov/content/dam/Census/library/publications/2015/demo/p25-1143.pdf.

Coleman, J., & Hoffer, T. (1987). Public and private high schools. New York: Basic Books.

Coleman, J., et al. (1966). Equality of educational opportunity. Washington, DC: US Government Printing Office.

Coleman, J., et al. (1982). High school achievement: Public, Catholic and private schools compared. New York: Basic Books.

College Board. (2018). Average published undergraduate charges by sector and by Carnegie classification, 2017–18. Retrieved from https://trends.collegeboard.org/college-pricing/figures-tables/average-published-undergraduate-charges-sector-2017-18.

Collier, J., et al. (1997). Is there a family? New anthropological views. In R. N. Lancaster & M. di Lionardo (Eds.), *The gender/sexuality reader: Culture, history, political economy* (pp. 71–81). New York: Routledge.

Collins, P. H. (1990). Work, family, and black women's oppression. In *Black feminist thought: Knowledge, consciousness and the politics of empowerment* (pp. 45–68). New York: Routledge.

Collins, R. (1971). Functional and conflict theories of educational stratification. *American Sociological Review*, 36(6), 1002–19.

Collins, R. (1979). *The credential society: A historical sociology of education and stratification.* New York: Academic Press.

Conger, R., et al. (1992). A family process model of economic hardship and adjustment of early adolescent boys. *Child Development*, 63, 526–41.

Conger, R., et al. (1994). Economic stress, coercive family process and developmental problems of adolescence. *Child Development*, 65, 541–61.

Conley, D. (1999). *Being black, living in the red: Race, wealth, and social policy in America.* Berkeley: University of California Press.

Conley, D. (2001). Capital for college: Parental assets and postsecondary schooling. *Sociology of Education, 74,* 59–72.

Conley, D. (2004). *The pecking order: Which siblings succeed and why.* New York: Pantheon Books.

Conley, D. (2009a, October 8). Interview with Mitchell Duneier. New York.

Conley, D. (2009b, October 8). Interview with Duncan Watts. New York.

Conley, D. (2009c, August 8). Interview with C. J. Pascoe. 104th Annual Meeting of the American Sociological Association. San Francisco.

Conley, D. (2009d, August 8). Interview with Paula England. 104th Annual Meeting of the American Sociological Association. San Francisco.

Conley, D. (2009e, August 9). Interview with Victor Rios. 104th Annual Meeting of the American Sociological Association. San Francisco.

Conley, D. (2009f, October 8). Interview with Devah Pager. New York.

Conley, D. (2009g, October 8). Interview with Jeffrey Sachs. New York.

Conley, D. (2009h, August 9). Interview with Michael Hout. 104th Annual Meeting of the American Sociological Association. San Francisco.

Conley, D. (2009i, August 9). Interview with Jen'nan Read. 104th Annual Meeting of the American Sociological Association. San Francisco.

Conley, D. (2009j, August 9). Interview with Jennifer Lee. 104th Annual Meeting of the American Sociological Association. San Francisco.

Conley, D. (2009k, August 8). Interview with Andrew Cherlin. 104th Annual Meeting of the American Sociological Association. San Francisco.

Conley, D. (2009l, August 8). Interview with Stephen Morgan. 104th Annual Meeting of the American Sociological Association. San Francisco.

Conley, D. (2009m, August 9). Interview with Frances Fox Piven. 104th Annual Meeting of the American Sociological Association. San Francisco.

Conley, D. (2009n, October 8). Interview with Alondra Nelson. New York.

Conley, D. (2009o, August 8). Interview with John Evans. 104th Annual Meeting of the American Sociological Association. San Francisco.

Conley, D. (2009p, August 8). Interview with Doug McAdam. 104th Annual Meeting of the American Sociological Association. San Francisco.

Conley, D. (2010, January). Interview with Kate Rich. E-mail.

Conley, D. (2011a, August 21). Interview with Allison Pugh. 106th Annual Meeting of the American Sociological Association. Las Vegas, NV.

Conley, D. (2011b, August 21). Interview with Annette Lareau. 106th Annual Meeting of the American Sociological Association. Las Vegas, NV.

Conley, D. (2011c, August 21). Interview with David Grusky. 106th Annual Meeting of the American Sociological Association. Las Vegas, NV.

Conley, D. (2011d, August 21). Interview with Shamus Khan. 106th Annual Meeting of the American Sociological Association. Las Vegas, NV.

Conley, D. (2011e, August 21). Interview with Nitsan Chorev. 106th Annual Meeting of the American Sociological Association. Las Vegas, NV.

Conley, D. (2013a, August 12). Interview with Julia Adams. 108th Annual Meeting of the American Sociological Association. New York.

Conley, D. (2013b, August 12). Interview with Michael Gaddis. 108th Annual Meeting of the American Sociological Association. New York.

Conley, D. (2013c, August 12). Interview with Ashley Mears. 108th Annual Meeting of the American Sociological Association. New York.

Conley, D. (2013d, August 12). Interview with Mario Luis Small. 108th Annual Meeting of the American Sociological Association. New York.

Conley, D. (2013e, August 12). Interview with Matthew Desmond. 108th Annual Meeting of the American Sociological Association. New York.

Conley, D. (2013f, August 12). Interview with Robb Willer. 108th Annual Meeting of the American Sociological Association. New York.

Conley, D. (2013g, August 12). Interview with Susan Crawford. 108th Annual Meeting of the American Sociological Association. New York.

Conley, D. (2014a, October 1). Interview with danah boyd. New York.

Conley, D. (2014b, November 19). Interview with Shamus Khan. New York.

Conley, D. (2014c, September 17). Interview with Fadi Haddad. New York.

Conley, D. (2014d, September 24). Interview with Marc Ramirez. New York.

Conley, D. (2014e, November 5). Interview with Amos Mac. New York.

Conley, D. (2014f, November 26). Interview with Delores Malaspina. New York.

Conley, D. (2014g, October 8). Interview with Adeel Qalbani. New York.

Conley, D. (2014h, October 22). Interview with Andy Bichlbaum. New York.

Conley, D. (2015a, February 25). Interview with Asha Rangappa. New York.

Conley, D. (2015b, February 2). Interview with Jennifer Senior. New York.

Conley, D. (2015c, March 5). Interview with Adam Davidson. New York.

Conley, D. (2015d, February 4). Interview with Jennifer Jacquet. New York.

Conley, D. (2015e, April 8). Interview with Zephyr Teachout. New York.

Conley, D. (2015f, March 9). Interview with Stephen Duncombe. New York.

Conley, D., & Bennett, N. G. (2000). Is biology destiny? Birth weight and life chances. *American Sociological Review*, 65(3), 458–67.

Conley, D., & Glauber, R. (2006). Parental educational investment and children's academic risk: Estimates of the impact of sibship size and birth order from exogenous variation in fertility. *Journal of Human Resources*, 41(4), 722–37.

Conley, D., & Heerwig, J. (2011). The war at home: Effects of Vietnam-era military service on postwar household stability. *American Economic Review*, 101(3), 350–54.

Conley, D., et al. (2003). *The starting gate: Birth weight and life chances.* Berkeley: University of California Press.

Conley, D., et al. (2007). *Africa's lagging demographic transition: Evidence from exogenous impacts of malaria ecology and agricultural technology* (NBER Working Paper No. 12892). National Bureau of Economic Research.

Connell, C. (2010). Doing, undoing, or redoing gender? Learning from the workplace experiences of transpeople. *Gender & Society*, 24(1), 31–55.

Connell, R. (1987). *Gender and power.* Stanford, CA: Stanford University Press.

Cookson, P. W., Jr., & Persell, C. H. (1985). *Preparing for power.* New York: Basic Books.

Cooley, C. H. (1909). *Social organization: A study of the larger mind.* New York: Scribner's.

Cooley, C. H. (1922). *Human nature and the social order.* New York: Scribner's. (Original work published 1902.)

Coolidge, C. (2011, September 16). Christian goods big seller for Wal-Mart. *ABC News.* Retrieved from http://abcnews.go.com/Business/story?id=86290&page=1#.Tt-3dXPZuoA.

Coontz, S. (Ed.). (2001). *Historical perspectives on family diversity. Shifting the center: Understanding contemporary families.* Mountain View, CA: Mayfield.

Coontz, S. (2010). How to make it work this time. *The New York Times*, Room for Debate section. Retrieved from www.nytimes.com/roomfordebate/2010/12/19/why-remarry/how-to-make-a-second-marriage-work.

Copen, C. E., et al. (2013, April 4). First premarital cohabitation in the United States: 2006–2010 National Survey of Family Growth. *National Health Statistics Reports*, 64. Retrieved from www.cdc.gov/nchs/data/nhsr/nhsr064.pdf.

Cornell, S. (1988). *The return of the native: American Indian political resurgence.* New York: Oxford University Press.

Cornell, S., & Hartmann, D. (1998). Fixed or fluid? Alternative views of ethnicity and race. In *Ethnicity and race: Making identities in a changing world* (pp. 39–71). Thousand Oaks, CA: Pine Forge Press.

Costello, E. J., et al. (2003). Relationships between poverty and psychopathology: A natural experiment. *Journal of the American Medical Association*, 290, 2023–29.

Council on American-Islamic Relations. (2017). Civil rights data quarter two update: Anti-Muslim bias incidents April–June 2017. Retrieved from www.cair.com/images/pdf/Civil-Rights-Data-Quarter-Two-Update-Anti-Muslim-Bias-Incidents-April--June-2017.pdf.

Covenant House. (n.d.). Sleep out to end youth homelessness. Retrieved from www.covenanthouse.org/helping-homeless/sleep-out.

Covert, B. (2011, March 8). Prepare for a possible "womancession." *National Public Radio.* Retrieved from www.npr.org/2011/03/08/134357162/the-nation-prepare-for-a-possible-womancession.

Cowan, R. S. (1983). *More work for mother: The ironies of household technology from the open hearth to the microwave.* New York: Basic Books.

Cox, D., & Jones, R. P. (2017). America's changing religious identity. Public Religion Research Institute. Retrieved from www.prri.org/research/american-religious-landscape-christian-religiously-unaffiliated.

Credit Suisse Research Institute. (2017, November). Global wealth databook 2017. Retrieved from http://publications.credit-suisse.com/tasks/render/file/index.cfm?fileid=FB790DB0-C175-0E07-787A2B8639253D5A.

Croucher, K., & Wendelin, R. (2007). Inclusivity in teaching practice and the curriculum. *Guides for Teaching and Learning in Archaeology*, 6. Retrieved from www.heacademy.ac.uk/system/files/number6_teaching_and_learning_guide_inclusivity.pdf.

Crouse, J., & Trusheim, D. (1988). The case against the SAT. *Public Interest*, 93, 97–110.

Currie, D. H. (1999). *Girl talk: Adolescent magazines and their readers.* Toronto: University of Toronto Press.

Dart, J. (2001, February 28). Hues in the pews: Racially mixed churches an elusive goal. *Christian Century.* Retrieved from http://hirr.hartsem.edu/cong/articles_huesinthepews.html.

David, E. (2015). Purple-collar labor: Transgender workers and queer value at global call centers in the Philippines. *Gender & Society*, 29(2), 169–94.

Davidson, J. D., et al. (1995). Persistence and change in the Protestant establishment, 1930–1992. *Social Forces*, 74(1), 157–75.

Davis, F. J. (1991). *Who is black? One nation's definition.* University Park: Pennsylvania State University Press.

Davis, G. (2015). *Contesting Intersex: The Dubious Diagnosis.* New York: NYU Press.

Davis, G. F. (2003). American cronyism: How executive networks inflated the corporate bubble. *Contexts*, 2(3), 34–40.

Davis, K. (1940). Extreme social isolation of a child. *American Journal of Sociology*, 45, 554–65.

Davis, K., & Moore, W. E. (1944). Some principles of stratification. *American Sociological Review*, 10(2), 242–49.

Davis, L. (2011). Race, gender, and class at a crossroads: A survey of their intersection in employment, economics, and the law. *Journal of Gender, Race & Justice*, 14(2).

Dawkins, R. (2006). *The god delusion*. Boston: Houghton Mifflin.

Death Penalty Information Center. (2015). Innocence: List of people freed from death row. Retrieved from www.deathpenaltyinfo.org/innocence-list-those-freed-death-row.

Death Penalty Information Center. (2017). Execution list 2016. Retrieved from https://deathpenaltyinfo.org/execution-list-2016.

Death Penalty Information Center. (2018a). Execution list 2017. Retrieved from https://deathpenaltyinfo.org/execution-list-2017.

Death Penalty Information Center. (2018b). Facts about the death penalty. Retrieved from https://deathpenaltyinfo.org/documents/FactSheet.pdf.

Death Penalty Information Center. (2018c). Innocence and the death penalty. Retrieved from https://deathpenaltyinfo.org/innocence-and-death-penalty#race.

DellaCava, F. A., et al. (2004). Adoption in the U.S.: The emergence of a social movement. *Journal of Sociology and Social Welfare, 31*(4), 141–60.

DeNeve, C. (1997, winter). Hispanic presence in the workplace. *The Diversity Factor*, 14–21.

DeParle, J. (2004). *American dream: Three women, ten kids and a nation's drive to end welfare*. New York: Viking.

Desmond, M. (2016). *Evicted: Poverty and profit in the American city*. New York: Crown.

De Vos, G., & Wagatsuma, H. (1966). *Japan's invisible race: Caste in culture and personality*. Berkeley: University of California Press.

Dill, J. S., et al. (2016). Does the "glass escalator" compensate for the devaluation of care work occupations? The careers of men in low- and middle-skill health care jobs. *Gender & Society, 30*(2), 334–60.

Dillon, M., & Wink, P. (2003). Religiousness and spirituality: Trajectories and vital involvement in late adulthood. In M. Dillon (Ed.), *Handbook of the sociology of religion* (pp. 179–89). Cambridge: Cambridge University Press.

DiMaggio, P. J., & Powell, W. (1983). The iron cage revisited: Institutional isomorphism and collective rationality in organizational fields. *American Sociological Review, 48*, 149.

Dinnerstein, L., et al. (1996). *Native and strangers: A multicultural history of Americans*. New York: Oxford University Press.

Doran, K. M., et al. (2013, December 19). Housing as health care—New York's boundary-crossing experiment. *New England Journal of Medicine, 369*, 2374–77.

Doucouliagos, C., & Laroche, P. (2003). What do unions do to productivity? A meta-analysis. *Industrial Relations, 42*(4).

Dove. (2005). *Real women have real curves*. Retrieved from www.campaignforrealbeauty.com/flat3.asp?id=2287&src=InsideCampaign_firming.

Doward, J. (2009, August 8). "Racist bias" blamed for disparity in police DNA database. *The Guardian*. Retrieved from www.theguardian.com/politics/2009/aug/09/police-dna-database-black-children.

Downey, D. B. (1995). When bigger is not better: Family size, resources, and children's educational performance. *American Sociological Review, 60*(5), 746–61.

Doyle, D. H. (1977). The social functions of voluntary associations in a nineteenth-century American town. *Social Science History, 1*(3), 333–55.

DPA. (2013, May 16). Palestinian smugglers deliver KFC to Gaza. *Haaretz*. Israel. Retrieved from www.haaretz.com/news/middle-east/palestinian-smugglers-deliver-kfc-to-gaza-1.524370.

Drape, J. (2005, October 30). Increasingly, football's playbooks call for prayers. *The New York Times*. Retrieved from www.nytimes.com/2005/10/30/sports/football/increasingly-footballs-playbooks-call-for-prayer.html.

Du Bois, W. E. B. (1903). *The souls of black folk*. Paris: McClurg.

Duncan, G., et al. (1994). Economic deprivation and early childhood development. *Child Development, 65*(2), 296–318.

Duncan, G. J., et al. (2005). Peer effects in drug use and sex among college students. *Journal of Abnormal Child Psychology, 33*(3), 375–85. Retrieved from www.gse.uci.edu/person/duncan_g/docs/11peerdrugs.pdf.

Duncan, G. (2016, December 19). When a basic income matters most. Economic Security Project. Retrieved from https://medium.com/economicsecproj/when-a-basic-income-matters-most-d90d093458a3.

Duneier, M. (1999). *Sidewalk*. New York: Farrar Straus & Giroux.

Durkheim, É. (1951). *Suicide: A study in sociology* (G. Simpson & J. A. Spaulding, Trans.). New York: Free Press. (Original work published 1897.)

Durkheim, É. (1972). The forced division of labor. In A. Giddens (Ed.). *Selected writings* (p. 181). Cambridge: Cambridge University Press. (Original work published 1893.)

Durkheim, É. (1995). *The elementary forms of religious life* (K. E. Fields, Intro. & Trans.). New York: Free Press. (Original work published 1917.)

Durkheim, É. (1997). *The division of labor in society* (L. A. Coser, Intro.; W. D. Halls, Trans.). New York: Free Press. (Original work published 1893.)

Duvall, M. L. (1991). *Feeding the family: The social organization of caring as gendered work*. Chicago: University of Chicago Press.

Dyer, G. (1985). *War*. New York: Crown.

Easterly, W., & Levine, R. (2002). *Tropics, germs, and crops: How endowments influence economic development* (NBER Working Paper No. W9106). National Bureau of Economic Research.

Eck, D. L. (2006). From diversity to pluralism. In *On common ground: World religions in America* [CD-ROM]. Cambridge: The Pluralism Project.

Economist, The (2018a, February 3). All must have degrees: Going to university is more important than ever for young people. Retrieved from www.economist.com/news/international/21736151-financial-returns-are-falling-going-university-more-important-ever.

Economist, The. (2018b, January 10). The youth of today: Teenagers are better behaved and less hedonistic nowadays. Retrieved from www.economist.com/international/2018/01/10/teenagers-are-better-behaved-and-less-hedonistic-nowadays.

Edgell, P., et al. (2006). Atheists as "other": Moral boundaries and cultural membership in American society. *American Sociological Review, 71*(2), 211–34.

Edgerton, R. (1995). "Bowling alone": An interview with Robert Putnam about America's collapsing civic life. *American Association for Higher Education Bulletin, 48*(1).

Edin, K., & Lein, L. (1997). *Making ends meet: How single mothers survive welfare and low-wage work.* New York: Russell Sage Foundation.

Edwards, B., & McCarthy, J. D. (2004). Resources and social movement mobilization. In David A. Snow, Sarah A. Soule, and Hanspeter Kriesi (Eds.), *The Blackwell companion to social movements* (pp. 116–52). Malden, MA: Blackwell Publishing.

Ehrenreich, B. (2001). *Nickel and dimed: On (not) getting by in America.* New York: Metropolitan Books.

Ehrenreich, B., & Hochschild, A. R. (Eds.). (2004). *Global woman: Nannies, maids and sex workers in the global economy.* New York: Macmillan Holt.

Eisenberg, D., et al. (2013, November). Peer effects on risky behaviors: New evidence from college roommate assignments. *Journal of Health Economics, 33,* 126–38.

Eisenbrey, R. (2007, June 20). *Strong unions, strong productivity: Snapshot for June 20.* Economic Policy Institute. Retrieved from www.epi.org/content.cfm /webfeatures_snapshots_20070620.

Elder, G., et al. (1995). Linking family hardship to children's lives. *Child Development, 56,* 361–75.

Elders, M. J., et al. (2017, June). Re-thinking genital surgeries on intersex infants. Palm Center Report. Retrieved from www.palmcenter.org/wp-content/uploads/2017/06 /Re-Thinking-Genital-Surgeries 1.pdf. Accessed January, 2018.

Ellen, B. (2014). Paid housework? No one'll clean up from that idea. *The Guardian.* Retrieved from www.theguardian.com /commentisfree/2014/mar/08/paying-for-housework -domestic-women-men.

Engels, F. (1878, May–July). *Herrn Eugen Dühring's Umwälzung des Sozialismus* in the supplement to *Vorwärts.*

Entwisle, D. R., & Alexander, K. L. (1992, February). Summer setback: Race, poverty, school composition, and mathematics achievement in the first two years of school. *American Sociological Review, 57*(1), 72–84.

Environmental Performance Index. (2018). 2018 EPI results. Retrieved from https://epi.envirocenter.yale.edu/epi -topline?country=&order=field_epi_rank new&sort=asc.

Epistemological modesty: An interview with Peter Berger. (1997, October 29). *The Christian Century.* Retrieved from www.christiancentury.org/article/epistemological -modesty.

Epstein, C. F. (1988). *Deceptive distinctions: Sex, gender, and the social order.* New Haven, CT: Yale University Press.

Eriksen, S., & Jensen, V. (2006). All in the family? Family environment factors in sibling violence. *Journal of Family Violence, 21*(8), 497–507.

Erikson, K. (2005). *Wayward Puritans: A study in the sociology of deviance* (rev. ed.). Boston: Pearson/Allyn & Bacon. (Original work published 1966.)

Eskenazi, B., et al. (2002). The association of age and semen quality in healthy men. *Human Reproduction, 18*(2), 447–54.

Espenshade, T. J., & Chung, C. Y. (2005). The opportunity cost of admission preferences at elite universities. *Social Science Quarterly, 86*(2), 293–305.

Espenshade, T. J., et al. (2004). Admission preferences for minority students, athletes, and legacies at elite universities. *Social Science Quarterly, 85*(5), 1422–46.

Essman, E. (2007). The American people: Social classes. *Life in the USA.* Retrieved from www.lifeintheusa.com/people /socialclasses.htm.

Estabrook, B. (2009, March). Politics of the plate: The price of tomatoes. *Gourmet.* Retrieved from http:// politicsoftheplate.com/wp-content/uploads/2009/11 /tomatoes.pdf.

Evans, J. (2002). *Playing God? Human genetic engineering and the rationalization of public bioethical debate.* Chicago: University of Chicago Press.

Eyerman, R. (2001). *Cultural trauma: Slavery and the formation of African American identity.* Cambridge: Cambridge University Press.

Fadiman, A. (1997). *The spirit catches you and you fall down: A Hmong child, her American doctors, and the collision of two cultures.* New York: Farrar, Straus, & Giroux.

Fagan, J., et al. (2003, March 31). Reciprocal effects of crime and incarceration in New York City neighborhoods. *Fordham Urban Law Journal, 30,* 1551–602.

Falk, A. (2005, June 30). Mom sells face space for tattoo advertisement. *Deseret News.* Retrieved from www .deseretnews.com/article/600145187/Mom-sells-face -space-for-tattoo-advertisement.html?pg=all.

Farmer, P. (1999). *Infections and inequalities: The modern plagues.* Berkeley: University of California Press.

Farmer, P. (2011). *Haiti after the earthquake.* New York: PublicAffairs.

Faughnder, R. (2016, March 26). Faith-based films are building followings at the box office. *Los Angeles Times.* Retrieved from www.latimes.com/entertainment /envelope/cotown/la-et-ct-faith-based-movies -20160325-story.html.

Fausto-Sterling, A. (2000). *Sexing the body: Gender politics and the construction of sexuality.* New York: Basic Books.

Federal Bureau of Investigation. (2003). *Facts and figures 2003.* Retrieved from www.fbi.gov/libref/factsfigure/wcc.htm.

Federal Bureau of Investigation. (2014a). Financial crimes report to the Public: Fiscal years 2010–2011. Retrieved from www.fbi.gov/stats-services/publications /financial-crimes-report-2010-2011.

Federal Bureau of Investigation. (2014b). Crime in the United States 2013: Property crime, tables 1 and 23. Retrieved from www.fbi.gov/about-us/cjis/ucr/crime-in-the-u.s /2013/crime-in-the-u.s.-2013/property-crime/property -crime-topic-page/propertycrimemain_final.

Federal Bureau of Investigation. (2015). 2014 Crime in the United States: Table 1. Retrieved from www.fbi.gov /about-us/cjis/ucr/crime-in-the-u.s/2014/crime-in -the-u.s.-2014/tables/table-1.

Federal Bureau of Investigation. (2017). 2016 Crime in the United States: Table 11. Retrieved from https://ucr.fbi.gov /crime-in-the-u.s/2016/crime-in-the-u.s.-2016/tables /table-11.

Federal Reserve Bank of St. Louis. (2018, August 30). Personal saving rate. Retrieved from https://fred .stlouisfed.org/series/PSAVERT.

Federal Reserve Board. (2018). *Report on the Economic Well-Being of US Households in 2017.* Retrieved from www.federalreserve.gov/publications/files/2017-report -economic-well-being-us-households-201805.pdf.

Ferber, A. L. (1999). *White man falling: Race, gender, and white supremacy.* Lanham, MD: Rowman & Littlefield.

Ferran, L. (2009, December 31). Megachurch asks for nearly $1M in 48 hours: Pastor Rick Warren of Saddleback Church asks members for $900,000 before New Year. *Good Morning America. ABC News.* Retrieved from http://abcnews.go.com/GMA/church-asks-mil-48-hours/story?id=9455589.

Fetter, J. H. (1995). *Questions and admissions: Reflections on 100,000 admissions decisions at Stanford.* Stanford, CA: Stanford University Press.

Figlio, D. N. (2005). *Boys named Sue: Disruptive children and their peers* (NBER Working Paper No. W11277). National Bureau of Economic Research. Retrieved from http://palm.nber.org/papers/w11277.

Figlio, D. N. (2007, September). Boys named Sue: Disruptive children and their peers. *Education Finance and Policy, 2*(4), 376–94.

Fillinger, K. (2013). Megachurches by the numbers. *Christian Standard.* Retrieved from http://christianstandard.com/2013/05/megachurches-by-the-numbers.

Fink, S., et al. (2006). Mammalian monogamy is not controlled by a single gene. *Proceedings of the National Academy of Sciences USA, 103*(20), 10956–60. Retrieved from www.pnas.org/content/103/29/10956.

Finke, R., & Stark, R. (1992). *The churching of America, 1776–1990: Winners and losers in our religious economy.* New Brunswick, NJ: Rutgers University Press.

Finn, J. D., et al. (2005). Small classes in the early grades, academic achievement, and graduating from high school. *Journal of Educational Psychology, 97*(2), 214–23.

Fischer, C., & Hout, M. (2006). How Americans prayed: Religious diversity and change. In C. Fischer & M. Hout (Eds.), *A century of difference: How America changed in the last one hundred years* (pp. 186–211). New York: Russell Sage Foundation.

Fischer, C. S., et al. (1996). *Inequality by design: Cracking the bell curve myth.* Princeton, NJ: Princeton University Press.

Fischer, P. M., et al. (1991, December). Brand logo recognition by children aged 3 to 6 years: Mickey Mouse and old Joe the Camel. *Journal of the American Medical Association, 266,* 3145–48.

Fisher, D. R., et al. (2017). Intersectionality takes it to the streets: Mobilizing across diverse interests for the Women's March. *Science Advances, 3*(9), eaao1390.

Flora, C. (2013). *Friendfluence: The surprising ways friends make us who we are.* New York: Anchor Books.

Foer, J. S. (2009). *Eating animals.* New York: Little, Brown.

Folbre, N. (1987). *A field guide to the U.S. economy.* New York: Pantheon Books.

Forbes. (2018) Profile: Bill Gates. Retrieved from www.forbes.com/profile/bill-gates.

Ford, H., & Crowther, S. (1973). *My life and work, by Henry Ford.* New York: Arno Books. (Original work published 1922.)

Fordham, S., & Ogbu, J. (1986). Black students' school success: Coping with the "burden of acting white." *Urban Review, 18,* 176–206.

Foss, R. J. (1994). The demise of homosexual exclusion: New possibilities for gay and lesbian immigration. *Harvard Civil Rights–Civil Liberties Law Review, 29,* 439–75.

Foster, D. (2006, February 19). Mind over splatter. *The New York Times.* Retrieved from www.nytimes.com/2006/02/19/opinion/mind-over-splatter.html.

Foucault, M. (1977). *Discipline and punish: The birth of the prison.* New York: Pantheon Books.

Foucault, M. (1978). *The history of sexuality.* New York: Pantheon Books.

Foucault, M. (1980). *Power/knowledge: Selected interviews and other writings, 1972–1977.* C. Gordon (Ed.). New York: Pantheon Books.

Fox News. (2017, May 10). Top 13 highest grossing faith films. Retrieved from www.foxnews.com/entertainment/2017/05/10/top-13-highest-grossing-faith-films.html.

Francis, D. R. (2005, November). Changing work behavior of married women. *NBER Digest.* Retrieved from www.nber.org/digest/nov05/w11230.html.

Frank, R. H. (2007). *Falling behind: How rising inequality harms the middle class.* Berkeley: University of California Press.

Franklin, J. H. (1980). *From slavery to freedom* (5th ed.). New York: Knopf.

Frederick, C. (2010). A crosswalk for using pre-2000 occupational status and prestige codes with post-2000 occupation codes. Center for Demography and Ecology, University of Wisconsin–Madison. Data files retrieved from www.ssc.wisc.edu/cde/cdewp/OccCodes.zip.

Fredrickson, G. M. (2002). *Racism: A short history.* Princeton, NJ: Princeton University Press.

Freedman, J. O. (2003). *Liberal education and the public interest.* Iowa City: University of Iowa Press.

Freeman, C. E. (2004, November 19). *Trends in educational equity of girls and women: 2004* (NCES 2005-016). US Department of Education, National Center for Education Statistics. Washington, DC: US Government Printing Office. Retrieved from http://nces.ed.gov/pubsearch/pubsinfo.asp?pubid=2005016.

Freeman, R. B. (2007, February 22). *Do workers still want unions? More than ever* (Briefing Paper No. 182). Economic Policy Institute, Agenda for Shared Prosperity. Retrieved from www.sharedprosperity.org/bp182.html.

Friedan, B. (1997). *The feminine mystique.* New York: Norton. (Original work published 1963.)

Friedman, M. (1970, September 13). A Friedman doctrine—The social responsibility of business is to increase its profits. *The New York Times Magazine.* Retrieved from http://select.nytimes.com/gst/abstract.html?res=F10F11FB3E5810718EDDAA0994D1405B808BF1D3&scp=1&sq=The+social+responsibility+of+business+is+to+increase+its+profits&st=p.

Fryer, R. G., Jr. (2010, August 10). *Racial inequality in the 21st century: The declining significance of discrimination* (NBER Working Paper No. 16256). National Bureau of Economic Research. Retrieved from www.nber.org/papers/w16256.

Fuchs, V. (1967). Redefining poverty and redistributing income. *Public Interest, 8,* 88–95.

Gamm, G., & Putnam, R. D. (1999). The growth of voluntary associations in America, 1840–1940. Patterns of social capital: Stability and change in comparative perspective, Part II. *Journal of Interdisciplinary History, 29*(4), 511–57.

Gamoran, A., & Mare, R. D. (1989). Secondary school tracking and educational inequality: Compensation, reinforcement or neutrality? *American Journal of Sociology, 94*(5), 1146–86.

Gans, H. (1979a). *Deciding what's news: A study of* CBS Evening News, NBC Nightly News, Newsweek *and* Time. New York: Vintage.

Gans, H. (1979b). Symbolic ethnicity: The future of ethnic groups and cultures in America. *Ethnic and Racial Studies, 2,* 1–19.

Ganz, M. (2004). Why David sometimes wins: Strategic capacity in social movements. In J. Goodwin & J. M. Jasper (Eds.), *Rethinking social movements: Structure, meaning, and emotion* (pp. 177–98). Lanham, MD: Rowman & Littlefield.

Gardner, H. (1983). *Frames of mind: The theory of multiple intelligences.* New York: Basic Books.

Garfinkel, H. (1967). *Studies in ethnomethodology.* Englewood Cliffs, NJ: Prentice-Hall.

Garrett, J. T. (1994). Health. In M. B. Davis (Ed.), *Native Americans in the 20th century* (pp. 233–37). New York: Garland.

Garrow, D. (1968). *Bearing the cross: Martin Luther King, Jr. and the Southern Christian Leadership Conference—A personal portrait.* New York: Morrow.

Gately, G. (2005, December 31). A town in the spotlight wants out of it. *The New York Times.* Retrieved from www.nytimes.com/2005/12/21/education/a-town-in-the-spotlight-wants-out-of-it.html.

Gaustad, E. (1962). *Historical atlas of American religion.* New York: Harper & Row.

Gaventa, J. (1980). *Power and powerlessness: Quiescence and rebellion in an Appalachian valley.* Urbana: University of Illinois Press.

Geertz, C. (1973). *The interpretation of cultures.* New York: Basic Books.

General Social Survey. (2012). General Social Survey 2012 cross-section and panel combined. Association of Religion Data Archives. Retrieved from www.thearda.com/Archive/Files/Codebooks/GSS12PAN_CB.asp.

Gerson, K. (1985). *Hard choices: How women decide about work, career, and motherhood.* Berkeley: University of California Press.

Gerson, K. (1993). *No man's land: Men's changing commitments to family and work.* New York: Basic Books.

Gibson, C., & Jung, K. (2006). Historical census statistics on the foreign-born population of the United States: 1850 to 2000. US Census Bureau. Retrieved from www.census.gov/population/www/documentation/twps0081/twps0081.pdf.

Gieryn, T. F. (1999). *Cultural boundaries of science: Credibility on the line.* Chicago: University of Chicago Press.

Gilbert, D. (1998). *The American class structure in an age of growing inequality.* Belmont, CA: Wadsworth.

Gilligan, C. (1982). *In a different voice: Psychological theory and women's development.* Cambridge: Harvard University Press.

Glaeser, E. L., & Ward, B. A. (2006). *Myths and realities of American political geography* (Harvard Institute of Economic Research Discussion Paper No. 2100). Retrieved from http://ssrn.com/abstract=874977.

Glass, J., & Jacobs, J. (2005). Childhood religious conservatism and adult attainment among black and white women. *Social Forces, 84,* 555–79.

Glassner, B. (1999). *The culture of fear: Why Americans are afraid of the wrong things.* New York: Basic Books.

Glaze, L. E., & Kaeble, D. (2014). Correctional populations in the United States, 2013. Bureau of Justice Statistics. Retrieved from www.bjs.gov/index.cfm?ty=pbdetail&iid=5177.

Glazer, N., & Moynihan, D. P. (1963). *Beyond the melting pot: The Negroes, Puerto Ricans, Jews, Italians, and Irish of New York City.* Cambridge: MIT Press.

Glenn, E. N. (1986). *Issei, Nisei, war bride: Three generations of Japanese American women in domestic service.* Philadelphia: Temple University Press.

Global Slavery Index. (2016). Global slavery index 2016 report. 3rd ed. Retrieved from www.globalslaveryindex.org/download.

Glock, C., & Stark, R. (1965). *Religion and society in tension.* Chicago: Rand McNally.

Goffman, E. (1959). *The presentation of self in everyday life.* New York: Doubleday.

Goffman, E. (1961). *Asylums: Essays on the social situation of mental patients and other inmates.* New York: Doubleday Anchor.

Goffman, E. (1963). *Stigma: Notes on the management of spoiled identity.* Englewood Cliffs, NJ: Prentice-Hall.

Goldberger, P. (1995, April 20). The gospel of church architecture, revised. *The New York Times,* pp. C1, 5. Retrieved from www.nytimes.com/1995/04/20/garden/the-gospel-of-church-architecture-revised.html?sq=+Megachurches%3A+User-friendly+architecture+&scp=1&st=cse.

Goldman, A. L., et al. (2018). Out-of-pocket spending and premium contributions after implementation of the Affordable Care Act. *JAMA Internal Medicine, 178*(3), 347–55.

Goodman, B. (2006a, April 17). People stand, the spirit walks: Easter at the Georgia Dome. *The New York Times.* Retrieved from www.nytimes.com/2006/04/17/us/17easter.html.

Goodman, B. (2006b, March 29). Teaching the Bible in Georgia's public schools. *The New York Times.* Retrieved from www.nytimes.com/2006/03/29/education/29bible.html.

Goodstein, L. (2005, December 4). Intelligent design might be meeting its maker. *The New York Times.* Retrieved from www.nytimes.com/2005/12/04/weekinreview/intelligent-design-might-be-meeting-its-maker.html.

Goodstein, L. A. (2007, February 25). A divide, and maybe a divorce. *The New York Times,* p. 1. Retrieved from www.nytimes.com/2007/02/25/weekinreview/25goodstein.html.

Goodwin, J. (2006). *No other way out: States and revolutionary movements.* New York: Cambridge University Press.

Gordon, M. M. (1964). *Assimilation in American life: The role of race, religion, and national origins.* New York: Oxford University Press.

Gould, E. (2014). Why America's workers need faster wage growth—And what we can do about it. Economic Policy Institute. Retrieved from www.epi.org/publication/why-americas-workers-need-faster-wage-growth.

Gourevitch, P. (1998). *We wish to inform you that tomorrow we will be killed with our families: Stories from Rwanda.* New York: Farrar, Straus & Giroux.

Gramsci, A. (1971). *Selections from the prison notebooks.* London: Lawrence & Wishart.

Grand View Research. (2016, October 10). Medical billing outsourcing market worth $16.9 billion by 2024. *PR Newswire*. Retrieved from www.prnewswire.com /news-releases/medical-billing-outsourcing-market -worth-169-billion-by-2024-grand-view-research-inc -596501001.html.

Grandey, A., et al. (2005). Must "service with a smile" be stressful? The moderating role of personal control for American and French employees. *Journal of Applied Psychology, 90*(5), 893–904.

Granovetter, M. (1973). The strength of weak ties. *American Journal of Sociology, 78*, 1360–80.

Granovetter, M. (1974). *Getting a job: A study of contacts and careers*. Chicago: University of Chicago Press.

Granovetter, M. (1978). Threshold models of collective behavior. *American Journal of Sociology, 83*(6), 1420–43.

Grant, M. (1936). *The passing of the great race* (4th rev. ed.). New York: Scribner. (Original work published 1916.) Retrieved from www.archive .org/stream/passingofgreatra00granuoft /passingofgreatra00granuoft_djvu.txt.

Gray, K. F., & Cunnyngham, K. (2017). Trends in Supplemental Nutritional Assistance Program participation rates: Fiscal year 2010 to fiscal year 2015. Retrieved from https://fns-prod.azureedge.net/sites /default/files/ops/Trends2010-2015.pdf.

Gray, N., & Nye, P. S. (2001). American Indian and Alaska Native substance abuse: Co-morbidity and cultural issues. *American Indian and Alaska Native Mental Health Research, 10*(2), 67–84. Retrieved from www.ucdenver .edu/academics/colleges/PublicHealth/research/centers /CAIANH/journal/Documents/Volume%2010/10(2) _Gray_Substance_Abuse_67-84.pdf.

Green, E. L. (2017, December 15). DeVos delays rule on racial disparities in special education. *The New York Times*. Retrieved from www.nytimes.com/2017/12/15/us/politics /devos-obama-special-education-racial-disparities.html.

Greenhouse, S. (2013, July 27). Fighting back against wretched wages. *The New York Times*. Retrieved from www.nytimes.com/2013/07/28/sunday-review/fighting -back-against-wretched-wages.html.

Greenhouse, S. (2014 April 24). In Florida tomato fields, a penny buys progress. *The New York Times*. Retrieved from www.nytimes.com/2014/04/25/business/in-florida -tomato-fields-a-penny-buys-progress.html.

Greer, C. (2006, April). *Black ethnicity: Political attitudes, identity, and participation in New York City*. Institute for Social and Economic Research and Policy, Columbia University. Retrieved from www.iserp.columbia.edu /news/articles/black_ethnicity.html.

Grose, T., & Kallerman, P. (2015). The 1099 economy— Elusive, but diverse, and growing. *Bay Area Economy*. Retrieved from https://medium.com/@BayAreaEconomy /the-1099-economy-elusive-but-diverse-and-growing -bcc6f65694fe#.t39kzun5w.

Grosz, E. A. (1994). *Volatile bodies: Toward a corporeal feminism*. Bloomington: Indiana University Press.

Guardian, The. (2014, June 9). Computer simulating 13-year-old boy becomes first to pass Turing test. Retrieved from www.theguardian.com/technology/2014/jun/08/super -computer-simulates-13-year-old-boy-passes-turing-test.

Gurian, M., & Stevens, K. (2005). *The minds of boys: Saving our sons from falling behind in school and life*. San Francisco: Jossey-Bass.

Gusfield, J. (1986). *Symbolic crusade: Status politics and the American temperance movement* (2nd ed.). Champaign: University of Illinois Press.

Haber, A. (1966). Poverty budgets: How much is enough? *Asia Pacific Journal of Human Resources, 1*(3), 5–22.

Hacker, J. (2006). *The great risk shift: The assault on American jobs, families, health care and retirement and how you can fight back*. New York: Oxford University Press.

Hackett, C., & McClendon, D. (2017, April 5). Christians remain world's largest religious group, but they are declining in Europe. Retrieved from www .pewresearch.org/fact-tank/2017/04/05/christians- remain-worlds-largest-religious-group-but -they-are-declining-in-europe.

Hacking, I. (1999). *The social construction of what?* Cambridge: Harvard University Press.

Hadaway, K. C., et al. (1993). What the polls don't show: A close look at U.S. church attendance. *American Sociological Review, 58*, 741–52.

Haley, A. (1976). *Roots: The saga of an American family*. New York: Doubleday.

Han, J. (2006). "We are Americans too": A comparative study of the effects of 9/11 on South Asian communities. Cambridge: Discrimination and National Security Initiative, Harvard University. Retrieved from www .pluralism.org/affiliates/kaur_sidhu/We_Are_Americans _Too.pdf.

Haney López, I. F. (1995). White by law. In R. Delgado (Ed.), *Critical race theory: The cutting edge* (pp. 542–50). Philadelphia: Temple University Press.

Hankins, J. D. (2014). *Working skin: Making leather, making a multicultural Japan*. Berkeley: University of California Press.

Hannaford, I. (1996). *Race: The history of an idea in the West*. Washington, DC: Woodrow Wilson Center Press.

Hanson, T. L., et al. (1997). Economic resources, parental practices and children's well-being. In G. Duncan & J. Brooks-Gunn (Eds.), *Consequences of growing up poor* (pp. 190–238). New York: Russell Sage Foundation.

Hanushek, E. A., & Wößmann, L. (2006). Does educational tracking affect performance and inequality? Differences-indifferences evidence across countries. *The Economic Journal, 116*(510), C63–76.

Hanushek, E. A., et al. (1998). *Teachers, schools, and academic achievement* (NBER Working Paper No. 6691). National Bureau of Economic Research.

Hanushek, E. A., et al. (2005). *The market for teacher quality* (NBER Working Paper No. 11154). National Bureau of Economic Research.

Harding, S. (1987). *Feminism and methodology: Social science issues*. Bloomington: Indiana University Press.

Harris, A. R., et al. (2002). Murder and medicine: The lethality of criminal assault, 1960–1999. *Homicide Studies, 6*, 128.

Harris, D. R., & Sim, J. J. (2002, August). Who is multiracial? Assessing the complexity of lived race. *American Sociological Review, 67*(4), 614–27.

Harris, S. (2006). *Letter to a Christian nation*. New York: Vintage.

Hartmann, H. (1976). Capitalism, patriarchy, and job segregation by sex. *Signs, 1*(2), 137–69.

Hartmann, H. (1981). The unhappy marriage of Marxism and feminism. In L. Sargent (Ed.), *Women and revolution: A discussion of the unhappy marriage between Marxism and feminism* (pp. 1–43). Boston: South End Press.

Hasenfeld, Y., et al. (1987). The welfare state, citizenship, and bureaucratic encounters. *Annual Review of Sociology, 13,* 387–415.

Hashima, P. Y., & Amato, P. R. (1994). Poverty, social support and parental behavior. *Child Development, 65,* 394–403.

Haskell, K. (2004, May 16). Revelation plus make-up advice. *The New York Times.* Retrieved from www.nytimes.com /2004/05/16/weekinreview/ideas-trends-revelation -plus-makeup-advice.html.

Hawley, A. (1968). Human ecology. In D. Sills (Ed.), *International encyclopedia of the social sciences* (pp. 327–37). New York: Macmillan.

Hayford, S. R., & Morgan, P. S. (2008). Religiosity and fertility in the United States: The role of fertility intentions. *Sociological Forces, 86*(1), 1163–88.

Hays, S. (2003). *Flat broke with children: Women in the age of welfare reform.* New York: Oxford University Press.

Heckman, J. J., et al. (2016). Returns to education: The causal effects of education on earnings, health and smoking (NBER Working Paper 2291). National Bureau of Economic Research. Retrieved from www.nber.org/papers/w22291.

Hegewisch, A., & Williams-Baron, E. (2013, September 13). The gender wage gap: 2016; earnings differences by gender, race, and ethnicity. Institute for Women's Policy Research. Retrieved from https://iwpr.org/publications /gender-wage-gap-2016-earnings-differences-gender -race-ethnicity.

Hegewisch, A., & Williams-Baron, E. (2017, September 13). The gender wage gap 2016: Earnings differences by gender, race, and ethnicity. Institute for Women's Policy Research. Retrieved from https://iwpr.org/publications /gender-wage-gap-2016-earnings-differences-gender -race-ethnicity.

Heilbroner, R. L. (1999). *The worldly philosophers.* New York: Simon & Schuster.

Hendricks, M. (1993, November 10). Is it a boy or a girl? *Johns Hopkins Magazine.*

Henley, J. (2018, January 12). Money for nothing: Is Finland's universal basic income trial too good to be true? *The Guardian.* Retrieved from www.theguardian.com /inequality/2018/jan/12/money-for-nothing-is-finlands -universal-basic-income-trial-too-good-to-be-true.

Henwood, B. F., et al. (2012). Substance abuse recovery after experiencing homelessness and mental illness *Journal of Dual Diagnosis, 8*(3), 238–46.

Henwood, B. F., et al. (2013, December). Permanent supportive housing: Addressing homelessness and health disparities? *American Journal of Public Health, 103*(S2), S188–92.

Herbert, W. (2011, June 7). Virginity and promiscuity: Evidence for the very first time. *Association for Psychological Science.* Retrieved from www .psychologicalscience.org/index.php/news/full-frontal -psychology/virginity-and-promiscuity-evidence-for -the-very-first-time.html.

Herdt, G. H. (1981). *Guardians of the flutes: Idioms of masculinity.* New York: McGraw-Hill.

Hernandez, D. J., et al. (2007, April). *Children in immigrant families in the U.S. and 50 states: National origins, language, and early education* (Research Brief Series Publication #2007–11). Albany, NY: Child Trends and the Center for Social and Demographic Analysis, SUNY.

Herrnstein, R. J., & Murray, C. (1994). *The bell curve: Intelligence and class structure in American life.* New York: Free Press.

Hetherington, E. M., & Kelly, J. (2002). *For better or for worse: Divorce reconsidered.* New York: Norton.

Heyrman, C. L. (1997). *Southern cross: The beginnings of the Bible belt.* New York: Knopf.

Hitchens, C. (2007). *God is not great: How religion poisons everything.* New York: Twelve Books.

Ho, M. K. (1987). *Family therapy with ethnic minorities.* Newbury Park, CA: Sage.

Ho, V. (2009, March 12). Native American death rates soar as most people are living longer. *Seattlepi.com.* Seattle. Retrieved from www.seattlepi.com/local/403196_tribes12 .html.

Hobbes, T. (1981). *Leviathan.* New York: Penguin Books. (Original work published 1651.)

Hochschild, A. R. (1983). *The managed heart: Commercialization of human feeling.* Berkeley: University of California Press.

Hochschild, A. R. (1989). *The second shift: Working parents and the revolution at home.* New York: Viking.

Hochschild, A. R. (1997). *The time bind: When work becomes home and home becomes work.* New York: Metropolitan Books.

Hochschild, A. R. (2003). *The commercialization of intimate life: Notes from home and work.* Berkeley: University of California Press.

hooks, b. (1984). Black women: Shaping feminist theory and feminism: A movement to end sexist oppression and the significance of the feminist movement. In *Feminist theory from margin to center* (pp. 1–17). Boston: South End Press.

Horowitz, J., et al. (2017, March 23). Americans widely support paid family and medical leave, but differ over specific policies. Pew Research Center. Retrieved from www.pewsocialtrends.org/2017/03/23/americans-widely -support-paid-family-and-medical-leave-but-differ -over-specific-policies.

Horwitz, A. V., & Wakefield, J. C. (2007). *The loss of sadness: How psychiatry transformed normal sorrow into depressive disorder.* New York: Oxford University Press.

Hosseini, B. (2018, February 3). Ethiopia bans foreign adoption. CNN. Retrieved from www.cnn.com/2018/01 /11/africa/ethiopia-foreign-adoption-ban/index.html.

Hout, M. (1983). *Mobility tables: Quantitative applications in the social sciences.* Beverly Hills, CA: Sage.

Howe, L. K. (1977). *Pink collar workers: Inside the world of women's work.* New York: Putnam.

Howes, C., & Olenick, M. (1986). Family and child influences on toddlers' compliance. *Child Development, 26,* 292–303.

Howes, C., & Stewart, P. (1987). Child's play with adults, toys, and peers: An examination of family and child care influences. *Developmental Psychology, 23,* 423–30.

Hoxby, C. (2000, August). *Peer effects in the classroom: Learning from gender and race variation* (NBER Working Paper No. 7867). National Bureau of Economic Research.

Human Rights Watch. (2006). Building towers, cheating workers. Retrieved from www.hrw.org/sites/default/files/reports/uae1106webwcover.pdf.

Hurley, D. (2005, April 19). The divorce rate: It's not as high as you think. *The New York Times*. Retrieved from www.nytimes.com/2005/04/19/health/19divo.html?scp=1&sq=&st=cse.

Hurtado, A. (1995). Variations, combinations, and evolutions: Latino families in the United States. In R. E. Zambrana (Ed.), *Understanding Latino families: Scholarship, policy, and practice* (pp. 40–61). Thousand Oaks, CA: Sage.

Iannaccone, L. R. (1997). Skewness explained: A rational choice model of religious giving. *Journal of the Scientific Study of Religion, 36*, 141–57.

Iannoccone, L. R., et al. (1995). Religious resources and church growth. *Social Forces, 74*, 705–31.

Ichihara, M. (2006). Making the case for soft power. *SAIS Review, 26*(1), 197–200.

Ikeda, H. (2001). Buraku students and cultural identity: The case of a Japanese minority. In K. Shimahara (Ed.), *Ethnicity, race, and nationality in education: A global perspective* (pp. 81–100). Mahwah, NJ: Erlbaum.

Imbens, G. W., et al. (2001). Estimating the effect of unearned income on labor earnings, savings, and consumption: Evidence from a survey of lottery players. *American Economic Review, 91*(4), 778–94.

Inglehart, R., & Baker, W. (2000, February). Modernization, cultural change and the persistence of traditional values. *American Sociological Review, 65*(1), 19–51.

International Bottled Water Association. (2013, April 25). US consumption of bottled water shows continued growth, increasing 6.2 percent in 2012; sales up 6.7 percent. Retrieved from www.bottledwater.org/us-consumption-bottled-water-shows-continued-growth-increasing-62-percent-2012-sales-67-percent.

International Bottled Water Association. (2018). Bottled water market. Retrieved from www.bottledwater.org/economics/bottled-water-market.

Intersex Society of North America. (2006). *What's wrong with the way intersex has traditionally been treated?* Retrieved from www.isna.org/faq/concealment.

Isaacs, H. (1975). *Idols of the tribe: Group identity and political change.* New York: Harper & Row.

Jackson, E. (2017, September 30). Re: Have you ever reconsidered being transgender? [Post to Quora.] Retrieved from www.quora.com/Have-you-ever-reconsidered-being-transgender.

Jackson, P. (1968). *Life in classrooms.* New York: Holt, Reinhart, & Winston.

Jacobi, T., & Schweers, D. (2017, April 11). Female Supreme Court justices are interrupted more by male justices and advocates. *Harvard Business Review.* Retrieved from https://hbr.org/2017/04/female-supreme-court-justices-are-interrupted-more-by-male-justices-and-advocates.

Jacobs, J. (1961). *The death and life of great American cities.* New York: Random House.

Jacobs, J., & Gerson, K. (2004). *The time divide: Work, family, and gender inequality.* Cambridge: Harvard University Press.

Jacobson, M. F. (1998). Anglo-Saxons and others, 1840–1924. In *Whiteness of a different color: European immigrants and the alchemy of race* (pp. 39–90). Cambridge: Harvard University Press.

Jacquet, J. (2016). *Is shame necessary: New uses for an old tool.* New York: Vintage Books.

James, G. (2003, June 29). Exurbia and God: Megachurches in New Jersey. *The New York Times*, pp. 1, 17. Retrieved from www.nytimes.com/2003/06/29/nyregion/exurbia-and-god-megachurches-in-new-jersey.html.

James, S. A., et al. (1987). Socioeconomic status, John Henryism, and hypertension in blacks and whites. *American Journal of Epidemiology, 126*, 664–73.

James, W. (1982). *The varieties of religious experience.* New York: Penguin. (Original work published 1903.)

Jencks, C., & Phillips, M. (Eds.). (1998). *The black-white test score gap.* Washington, DC: Brookings Institution Press.

Jencks, C., et al. (1972). *Inequality: A reassessment of the effect of family and schooling in America.* New York: Basic Books.

Jenkins, D., et al. (2018, April). What we are learning about guided pathways. Community College Research Center. Retrieved from https://ccrc.tc.columbia.edu/publications/what-we-are-learning-guided-pathways.html.

Jenson, A. R. (1969). How much can we boost IQ and scholastic achievement? *Harvard Educational Review, 39*, 1–123.

Johnson, E. R., et al. (2016). Extreme weight-control behaviors and suicide risk among high school students. *Journal of School Health, 86*(4), 281–87.

Johnson, L. (1998). Proposal for a nationwide war on the sources of poverty. In P. Halsall (Ed.), *Internet modern history sourcebook.* Retrieved from www.fordham.edu/halsall/mod/1964johnson-warpoverty.htm. (Original speech presented March 16, 1964.)

Kaeble, D., & Glaze, L. (2016). Correctional populations in the United States, 2015. Washington, DC: US Department of Justice, Bureau of Justice Statistics. Retrieved from www.bjs.gov/content/pub/pdf/cpus15.pdf.

Kandel, W. A. (2018, February 9). U.S. family-based immigration policy. Congressional Research Service. Retrieved from https://fas.org/sgp/crs/homesec/R43145.pdf.

Kane, T. J. (1998). Racial and ethnic preferences in college admissions. In C. Jencks & M. Phillips (Eds.), *The black-white test score gap* (pp. 431–56). Washington, DC: Brookings Institution Press.

Kanno-Youngs, Z. V. (2015, August 5). NCAA members slow to adopt transgender athlete guidelines. *USA Today.* Retrieved from www.usatoday.com/story/sports/college/2015/08/03/ncaa-transgender-athlete-guidelines-keelin-godsey-caitlyn-jenner/31055873.

Kanter, R. M. (1977). *Men and women of the corporation.* New York: Basic Books.

Kaplan, E. A. (2003, October 3–9). Black like I thought I was: Race, DNA, and a man who knows too much. *L. A. Weekly.* Retrieved from www.alternet.org/story.html?StoryID=16917.

Karen, D. (2002, July). Changes in access to higher education in the United States: 1980–1992. *Sociology of Education, 75*(3), 191–210.

Kashef, Z. (2003, February). Persistent peril: Why African American babies have the highest infant mortality rate in the developed world. *RaceWire.* Retrieved from www.arc.org/racewire/030210z_kashef.html.

Kasselstrand, I., et al. (2017). Institutional confidence in the United States: Attitudes of secular Americans. *Secularism and Nonreligion, 6*, p. 6.

Kelley, D. M. (1972). *Why conservative churches are growing: A study in sociology of religion*. San Francisco: Harper & Row.

Kelling, G. L., & Sousa, W. H., Jr. (2001). Do police matter? An analysis of the impact of New York City's police reforms. *Civic Report, 22.*

Kersley, R., & Stierli, M. (2015, October 13). Global wealth in 2015: Underlying trends look good. Credit Suisse. Retrieved from www.credit-suisse.com/us/en/about -us/responsibility/news-stories/articles/news-and -expertise/2015/10/en/global-wealth-in-2015-underlying -trends-remain-positive.html.

Kessler-Harris, A. (1990). *A woman's wage: Historical meanings and social consequences*. Lexington: University Press of Kentucky.

Khan, S. R. (2010). *Privilege: The making of an adolescent elite at St. Paul's School*. Princeton, NJ: Princeton University Press.

Kidder, T. (2003). *Mountains beyond mountains: Healing the world: The quest of Dr. Paul Farmer*. New York: Random House.

Kilbourne, J. (1979). *Killing us softly: Advertising's image of women* [Motion picture]. Belmont, MA: Cambridge Documentary Films.

Killewald, A. (2013). Return to *Being black, living in the red: A race gap in wealth that goes beyond social origins*. *Demography, 50*(4), 1177–95. Retrieved from http://link .springer.com/article/10.1007/s13524-012-0190-0.

Kim, J. (2001). Hegemony and cultural resistance. In N. J. Smelser & P. B. Baltes (Eds.), *International encyclopedia of the social & behavioral sciences*. New York: Elsevier.

Kimmel, M. S. (1996). *Manhood in America: A cultural history*. New York: Free Press.

Kimmel, M. S. (2000). *The gendered society*. New York: Oxford University Press.

Kinsey, A. C. (1948). *Sexual behavior in the human male*. Philadelphia: Saunders.

Kirkpatrick, D. (2002, June 8). Evangelical sales are converting publishers. *The New York Times*. Retrieved from www.nytimes.com/2002/06/08/books/evangelical -sales-are-converting-publishers.html.

Kitzmiller v. Dover Area School District, 400 F. Supp. 2d 707.

Klebanov, P. K., et al. (1994). Classroom behavior of very low birth weight elementary school children. *Pediatrics, 94*(5), 700–708.

Klein, N. (2000). *No logo: Taking aim at the brand bullies*. New York: Picador.

Klinenberg, E. (2002). *Heat wave: A social autopsy of disaster in Chicago*. Chicago: University of Chicago Press.

Klinenberg, E. (2013). *Going solo: The extraordinary rise and surprising appeal of living alone*. New York: Penguin.

Knafo, S. (2005, December 25). Praise the Lord and raise the curtain. *The New York Times*. Retrieved from www .nytimes.com/2005/12/25/nyregion/thecity/praise-the -lord-and-raise-the-curtain.html.

Knobel, D. T. (1986). *Paddy and the republic: Ethnicity and nationality in antebellum America*. Middletown, CT: Wesleyan University Press.

Kocchar, R., & Cilluffo, A. (2017). How wealth inequality has changed in the U.S. since the great recession, by race, ethnicity, and income. Pew Research Center. Retrieved from www.pewresearch.org/fact-tank/2017/11/01/how -wealth-inequality-has-changed-in-the-u-s-since-the -great-recession-by-race-ethnicity-and-income.

Koenig, H. G., et al. (1994). Religious practices and alcoholism in a southern adult population. *Hospital Communication Psychiatry, 45*, 225–31.

Kohn, M. L., & Schooler, C. (1983). *Work and personality: An inquiry into the impact of social stratification*. Norwood, NJ: Ablex.

Kozol, J. (1991). *Savage inequalities: Children in America's schools*. New York: Crown.

Kraybill, D. B. (Ed.). (1993). *The Amish and the state*. Baltimore, MD: Johns Hopkins University Press.

Kraybill, D. B., & Nolt, S. (1995). *Amish enterprise: From plows to profits*. Baltimore, MD: Johns Hopkins University Press.

Kreider, R., & Ellis, R. (2011). Number, timing, and duration of marriages and divorces: 2009. Current population reports, US Census Bureau. Retrieved from www.census .gov/prod/2011pubs/p70-125.pdf.

Kremer, M., & Levy, D. (2008). Peer effects and alcohol use among college students. *Journal of Economic Perspectives, 22*(3), 189–206.

Krueger, A. B., & Whitmore, D. M. (2001). The effect of attending a small class in the early grades on college test-taking and middle school test results: Evidence from Project STAR. *Economic Journal, 111*(468), 1–28.

Krueger, A. B., & Zhu, P. (2002). *Another look at the New York City school voucher experiment* (NBER Working Paper No. 9418). National Bureau of Economic Research.

Krugman, P. (2005, September 16). Not the new deal. *The New York Times*. Retrieved from www.nytimes.com/2005 /09/16/opinion/not-the-new-deal.html.

Kuhn, T. (1962). *The structure of scientific revolutions*. Chicago: University of Chicago Press.

Kulick, D. (1998). *Travesti: Sex, gender, and culture among Brazilian transgendered prostitutes*. Chicago: University of Chicago Press.

Kuruvil, M. C. (2006, September 3). 9/11: Five years later. Typecasting Muslims as a race. *San Francisco Chronicle*. Retrieved from www.sfgate.com/cgi-bin/article.cgi?f=/c /a/2006/09/03/MNG4FKUMR71.DTL.

Kurzman, C. (2002, December). Bin Laden and other thoroughly modern Muslims. *Contexts Magazine, 1*(4), 13–20.

Lamont, M. (1992). *Money, morals, and manners*. Chicago: University of Chicago Press.

Landau, E. (2012, July 30). Why polio hasn't gone away yet. CNN. Retrieved from www.cnn.com/2012/07/27/health /polio-eradication-efforts/index.html.

Landsbergis, P. A., et al. (2014). Work organization, job insecurity, and occupational health disparities. *American Journal of Industrial Medicine, 57*(5), 495–515.

Langlois, D. E., & Zales, C. R. (1992). Anatomy of a top teacher. *Education Digest, 57*(5), 31–34.

Laqueur, T. (1990). *Making sex: Body and gender from the Greeks to Freud*. Cambridge: Harvard University Press.

Lareau, A. (1987). Social class differences in family-school relationships: The importance of cultural capital. *Sociology of Education, 60*, 33–85.

Lareau, A. (2002). Invisible inequality: Social class and child-rearing in black families and white families. *American Sociological Review, 67*, 747–76.

Lareau, A. (2003). *Unequal childhoods: Class, race, and family life*. Berkeley: University of California Press.

Lareau, A., et al. (2006, July). *Social class and children's time use* (working paper). College Park: University of Maryland, Department of Sociology.

Lasch, C. (1977). *Haven in a heartless world: The family besieged.* New York: Basic Books.

Lasch, C. (1991). *The true and only heaven: Progress and its critics.* New York: Norton.

Latour, B., & Woolgar, S. (1979). *Laboratory life: The social construction of scientific facts.* Los Angeles: Sage.

Lauten, E. (2018, February 19). State Rep. Steve Hurst introduces bill to teach biblical creationism alongside evolution. *Alabama Today.* Retrieved from http://altoday .com/archives/21486-state-rep-steve-hurst-introduces -bill-to-teach-biblical-creationism-alongside-evolution.

LeBon, G. (2002). *La psychologie des foules.* (*The crowd: A study of the popular mind.*) Mineola, NY: Dover. (Original work published 1895.)

Lee, F. R. (2006, March 28). "Big Love": Real polygamists look at HBO polygamists and find sex. *The New York Times.* Retrieved from www.nytimes.com/2006/03/28 /arts/television/28poly.html.

Leiter, J. (1983). Classroom composition and achievement gains. *Sociology of Education, 56*(3), 126–32.

Leland, J. (2004, May 16). Alt-worship; Christian cool and the new generation gap. *The New York Times.* Retrieved from www.nytimes.com/2004/05/16/weekinreview /ideas-trends-alt-worship-christian-cool-and-the-new -generation-gap.html.

Lempers, J. D., et al. (1989). Economic hardship, parenting and distress in adolescence. *Child Development, 60,* 25–39.

Lenski, G. (1961). *The religious factor: A sociological study of religion's impact on politics, economics, and family life.* New York: Doubleday.

Leswing, Kif. (2015, June 29). Google is making your search results worse. *International Business Times.* Retreived from www.ibtimes.com/google-making-your-search-results -worse-study-1988372.

Levitt, S. D., & Dubner, S. J. (2005). *Freakonomics: A rogue economist explores the hidden side of everything.* New York: Morrow.

Lewis, A. E. (2004). *Race in the schoolyard: Negotiating the color line in classrooms and communities.* New Brunswick, NJ: Rutgers University Press.

Lewis, C., et al. (1992). Sex stereotyping of infants: A re-examination. *Journal of Reproductive and Infant Psychology, 10,* 53–61.

Lewis, O. (1966). The culture of poverty. *Scientific American, 215*(4), 19–25.

Lieberson, S. (1961). A societal theory of race and ethnic relations. *American Sociological Review, 26,* 902–10.

Lieberson, S., & Mikelson, K. S. (1995). Distinctive African American names: An experimental, historical, and linguistic analysis of innovation. *American Sociological Review, 60,* 928–46.

Lieberson, S., et al. (2000). The instability of androgynous names: The symbolic maintenance of gender boundaries. *American Journal of Sociology, 105*(5), 1249–87.

Lien, T. (2018, February 19). Uber class-action lawsuit over how drivers were paid gets green light from judge. *The Los Angeles Times.* Retrieved from www.latimes .com/business/technology/la-fi-tn-uber-class-action -20180219-story.html.

Lim, A. (2018, January 6). The alt-right's Asian fetish. *The New York Times.* Retrieved from https://www.nytimes .com/2018/01/06/opinion/sunday/alt-right-asian-fetish .html.

Lipka, M. (2015, May 12). 5 key findings about the changing U.S. religious landscape. Pew Research Center. Retrieved from www.pewresearch.org/fact-tank/2015/05/12 /5-key-findings-u-s-religious-landscape.

Liu, L., et al. (2015). Global, regional, and national causes of child mortality in 2000-13, with projections to inform post-2015 priorities: An updated systematic analysis. *Lancet, 385,* 430–40, figure 2.

Liu, Z. (2015). China's carbon emissions report 2015. Harvard University Belfer Center. Retrieved from http://belfercenter.ksg.harvard.edu/publication/25417 /chinas_carbon_emissions_report_2015.html.

Livingston, G. (2015, May 7). Childlessness falls, family size grows among highly educated women. Pew Research Center. Retrieved from www.pewsocialtrends.org/2015 /05/07/childlessness.

Lleras-Muney, A. (2005). The relationship between education and adult mortality in the United States. *Review of Economic Studies, 72,* 189–221.

Locke, J. (1980). *Second treatise of government.* Indianapolis, IN: Hackett. (Original work published 1690.)

López, G., & Bialik, K. (2017, May 3). Key findings about U.S. immigrants. Pew Research Center. Retrieved from www.pewresearch.org/fact-tank/2017/05/03 /key-findings-about-u-s-immigrants.

Lorber, J. (1994). *Paradoxes of gender.* New Haven, CT: Yale University Press.

Lucas, S. R. (1999). *Tracking inequality: Stratification and mobility in American high schools.* New York: Teachers College Press.

Lucas, S. R., & Good, A. D. (2001). Race, class, and tournament track mobility. *Sociology of Education, 74,* 139–56.

Lui, M. (2004). Doubly divided: The racial wealth gap. In C. Collins et al. (Eds.), *The wealth inequality reader.* Cambridge, MA: Economic Affairs Bureau.

Lukes, S., & British Sociological Association. (2005). *Power: A radical view.* New York: Palgrave Macmillan. (Original work published 1974.)

Luo, M. (2006, April 16). Evangelicals debate the meaning of "evangelical." *The New York Times,* pp. 4, 5. Retrieved from www.nytimes.com/2006/04/16/weekinreview /evangelicals-debate-the-meaning-of-evangelical.html.

Mack, J., & Lansley, S. (1985). *Poor Britain.* London: Allen & Unwin.

MacKinnon, C. (1983). Feminism, Marxism, method and the state: Toward feminist jurisprudence. *Signs: Journal of Women in Culture and Society, 8,* 635.

MacLeod, J. (1995). *Ain't no makin' it: Aspirations and attainment in a low-income neighborhood.* Boulder, CO: Westview Press. (Original work published 1987.)

MacMaster, N. (2001). *Racism in Europe 1870–2000.* Basingstoke, Hampshire, UK: Palgrave.

Malinowski, B. (1913). *The family among the Australian Aborigines.* London: University of London Press.

Maltz, M. D. (2001). *Recidivism.* Orlando, FL: Academic Press. (Original work published 1984.) Retrieved from www.uic .edu/depts/lib/forr/pdf/crimjust/recidivism.pdf.

Manza, J., & Uggen, C. (2006). *Locked out: Felon disenfranchisement and American democracy.* New York: Oxford University Press.

Maraniss, D. (1995). *First in his class: The biography of Bill Clinton.* New York: Simon & Schuster.

Marmot, M., & Wilkinson, R. G. (Eds). (1999). *Social determinants of health.* New York: Oxford University Press.

Martin, J. A., et al. (2010). Births: Final data for 2008. *National Vital Statistics Report, 59*(1). Retrieved from www.cdc.gov/nchs/data/nvsr/nvsr59/nvsr59_01.pdf.

Martin, J. A., et al. (2015). Births: Final data for 2013. *National Vital Statistics Reports, 64*(1). Retrieved from www.cdc.gov/nchs/data/nvsr/nvsr64/nvsr64_01.pdf.

Martin, J. A., et al. (2017). Births: Final data for 2015. *National Vital Statistics Reports, 66*(1). Retrieved from www.cdc.gov/nchs/data/nvsr/nvsr66/nvsr66_01.pdf.

Martin, J. A., et al. (2018). Births: Final data for 2016. *National Vital Statistics Reports, 67*(1). Retrieved from www.cdc.gov/nchs/data/nvsr/nvsr67/nvsr67_01.pdf.

Martin, J. T., & Nguyen, D. H. (2004). Anthropometric analysis of homosexuals and heterosexuals: Implications for early hormone exposure. *Hormones and Behavior, 45*(1), 31–39.

Martineau, H. (1837). *Theory and practice of society in America.* London: Saunders & Otley.

Martineau, H. (1838). *How to observe morals and manners.* London: Charles Knight.

Marx, A. W. (1998). To bind up the nation's wounds: The United States after the Civil War. In *Making race and nation: A comparison of South Africa, the United States, and Brazil* (pp. 120–57). Cambridge: Cambridge University Press.

Marx, K. (1932). Estranged labor. In *Economic and Philosophical Manuscripts of 1844.* Retrieved from www.marxists.org/archive/marx/works/1844/manuscripts/labour.htm. (Original work published 1844.)

Marx, K. (1978). Contribution to the critique of Hegel's philosophy of right: Introduction. In R. Tucker (Ed.), *The Marx-Engels reader* (pp. 53–65). New York: Norton. (Original work pulished 1844.)

Marx, K. (1999). Critique of the Gotha programme. *Marxists internet archive.* Retrieved from www.marxists.org/archive/marx/works/1875/gotha/index.htm. (Original work published 1890–91.)

Marx, K., & Engels, F. (1998). *The communist manifesto.* New York: Penguin Group. (Original work published 1848.)

Massey, D. S. (1995). The new immigration and ethnicity in the United States. *Population and Development Review, 21,* 631–52.

Massey, D. S., & Denton, N. A. (1993). *American apartheid: Segregation and the making of the underclass.* Cambridge: Harvard University Press.

Maus, J. (2012). It's the 20th anniversary of critical mass: What it meant to Portland then and now. Bikeportland.org. Retrieved from http://bikeportland.org/2012/09/28/its-the-20th-anniversary-of-critical-mass-what-it-meant-then-and-now-in-portland-78129.

Mayer, S. (1997). *What money can't buy: Family income and children's life chances.* Cambridge: Harvard University Press.

McAdam, D. (1982). *Political process and the development of black insurgency, 1930–1970.* Chicago: University of Chicago Press.

McCarthy, J. D., & Zald, M. N. (1973). *The trend of social movements in America: Professionalization and resource mobilization.* Morristown, NJ: General Learning Press.

McCarthy, J. D., & Zald, M. N. (1977). Resource mobilization and social movements: A partial theory. *American Journal of Sociology, 82*(6), 1212–41.

McCullough, M., & Smith, T. (2003). Religion and health: Depressive symptoms and mortality as case studies. In M. Dillon (Ed.), *Handbook of sociology of religion* (pp. 190–206). Cambridge: Cambridge University Press.

McDonald, M. P. (2018). National turnout rates. Retrieved from www.electproject.org/home/voter-turnout/voter-turnout-data.

McDonald, M. P., & Popkin, S. L. (2001). The myth of the vanishing voter. *American Political Science Review, 95*(4), 963–74.

McGregor, P. P. L., & Borooah, V. K. (1992). Is low spending or low income a better indicator of whether or not a household is poor: Some results from the 1985 Family Expenditure Survey. *Journal of Social Policy, 21*(1), 53–69.

McGuire, M. (2007). Embodied practices: Negotiation and resistance. In N. Ammerman (Ed.), *Everyday religion: Observing modern religious lives* (pp. 187–200). New York: Oxford University Press. Retrieved from www.oxfordscholarship.com/oso/public/content/religion/9780195305418/toc.html?q=Embodied`practices:`Negotiation`resistance.

McIntosh, P. (1988). *White privilege and male privilege: A personal account of coming to see correspondences through work in women's studies.* Wellesley, MA: Wellesley College Center for Research on Women.

McKenna, P. (2016, December 27). 2016: How Dakota pipeline protest became a Native American cry for justice. *InsideClimate News.* Retrieved from https://insideclimatenews.org/news/22122016/standing-rock-dakota-access-pipeline-native-american-protest-environmental-justice.

McLeod, J. D., & Shanahan, M. J. (1993). Poverty, parenting and children's mental health. *American Sociological Review, 58,* 351–66.

McMillan, C. (2011, July 14). State budget shortfall forces UC fee increase for 2011 [press release]. University of California Office of the President. Retrieved from http://newsroom.ucla.edu/stories/uc-regents-vote-tuition-increase-210662.

Mead, G. H. (1934). *Mind, self, and society from the standpoint of a social behaviorist.* C. W. Morris (Ed.). Chicago: University of Chicago Press.

Mead, M. (1928). *Coming of age in Samoa.* New York: Morrow.

Mechanic, D., & Meyer, S. (2000). Concepts of trust among patients with serious illness. *Social Science and Medicine, 51*(5), 657–68.

Meckler, L. (2002, July 24). Divorce, American style. Study: Certain couples more likely to split up than others. *CBS News.* Retrieved from www.cbsnews.com/stories/2002/07/24/national/main516165.shtml.

Meier, R. F. (1982). Perspectives on the concept of social control. *Annual Review of Sociology, 8,* 35–55.

Meighan, R. (1981). *A sociology of educating.* New York: Holt, Reinhart & Winston.

Melucci, A. (1989). Nomads of the present: Social movements and individual needs in contemporary society. New York: Vintage.

Merton, R. (1938). Social structure and anomie. *American Sociological Review, 3*, 672–82.

Merton, R. (1949a). Discrimination and the American creed. In R. M. MacIver (Ed.), *Discrimination and national welfare; a series of addresses and discussion* (pp. 99–126). New York: Institute for Religious and Social Studies.

Merton, R. (1949b). *Social theory and social structure*. Glencoe, IL: Free Press.

Meyer, D. S. (2007). *The politics of protest: Social movements in America*. New York: Oxford University Press.

Michels, R. (1915). *Political parties: A sociological study of the oligarchical tendencies of modern democracy* (E. Paul & C. Paul, Trans.). New York: Hearst's International Library.

Miele, F. (1995). For whom the bell curve tolls: Interview with Charles Murray. *Skeptic, 3*(2), 34–41. Retrieved from www.prometheism.net/articles/interview01.html.

Milkman, R. (2006). *LA Story: Immigrant workers and the future of the US labor movement*. New York: Russell Sage Foundation.

Milkman, R. (2007). Unions fight for work and family policies. In D. S. Cobble (Ed.), *The sex of class: Women transforming American labor* (pp. 63–80). Ithaca, NY: Cornell University Press.

Miller, C. C. (2015, May 26). When family-friendly policies backfire. *The New York Times*. Retrieved from www.nytimes.com/2015/05/26/upshot/when-family-friendly-policies-backfire.html.

Milloy, C. (2017, August 22). Want to see proof of institutional racism? Let weed open your eyes. *The Washington Post*. Retrieved from www.washingtonpost.com/local/want-to-see-proof-of-institutional-racism-let-weed-open-your-eyes/2017/08/22/099b7740-8751-11e7-a94f-3139abce39f5_story.html?utm_term=.abaaae83946b.

Mills, C. W. (1959). *The sociological imagination*. New York: Oxford University Press.

Mills, C. W. (2000). *The power elite*. New York: Oxford University Press. (Original work published 1956.)

Mishel, L., & Schieder, J. (2017). CEO pay remains high relative to the pay of typical workers and high-wage earners. Economic Policy Institute. Retrieved from www.epi.org/files/pdf/130354.pdf.

Mishel, L., & Shierholz, H. (2013, August 21). *A decade of flat wages: The key barrier to shared prosperity and a rising middle class*. Economic Policy Institute. Retrieved from www.epi.org/publication/a-decade-of-flat-wages-the-key-barrier-to-shared-prosperity-and-a-rising-middle-class.

Missouri Economic Research and Information Center. (2018). Cost of living data series: 2017 annual average. Retrieved from www.missourieconomy.org/indicators/cost_of_living.

Mohamed, B. (2018, January 3). New estimates show U.S. Muslim population continues to grow. Pew Research Center. Retrieved from www.pewresearch.org/fact-tank/2018/01/03/new-estimates-show-u-s-muslim-population-continues-to-grow.

Mohamed, B., & Sciupac, E. P. (2018, January 26). The share of Americans who leave Islam is offset by those who become Muslim. Pew Research Center. Retrieved from www.pewresearch.org/fact-tank/2018/01/26/the-share-of-americans-who-leave-islam-is-offset-by-those-who-become-muslim.

Monaghan, T. (1990, August). The thrill of poverty. *Harpers Magazine*, p. 22.

Montesquieu, C., Baron de. (1899). *The spirit of the laws*. (Vols. 1–2; T. Nugent et al., Trans.). New York: Colonial Press. (Original work published 1748/1750.)

Moore, B. (1993). *Social origins of dictatorship and democracy: Lord and peasant in the making of the modern world*. Boston: Beacon Press. (Original work published 1966.)

Morgan, P. L., et al. (2017). Replicated evidence of racial and ethnic disparities in disability identification in U.S. schools. *Educational Researcher, 46*(6): 305–22.

Morning, A. (2004). *The nature of race: Teaching and learning about human difference*. (Unpublished doctoral dissertation). Department of Sociology, Princeton University, Princeton, NJ.

Morris, A. (1984). *The origins of the civil rights movement*. New York: Free Press.

Morris, A. (2004). Reflections on social movement theory: Criticisms and proposals. In J. Goodwin & J. M. Jasper (Eds.), *Rethinking social movements: Structure, meaning, and emotion* (pp. 233–46). Lanham, MD: Rowman & Littlefield.

Morrison, D. R., & Coiro, M. J. (1999). Parental conflict and marital disruption: Do children benefit when high-conflict marriages are dissolved? *Journal of Marriage and the Family, 61*, 626–37.

Moynihan, D. P. (1965). *The Negro family: The case for national action*. Washington, DC: Office of Policy Planning and Research, US Department of Labor.

Mueller, A. S., et al. (2015). Suicide ideation and bullying among US adolescents: Examining the intersections of sexual orientation, gender and race/ethnicity. *American Journal of Public Health, 105*(5), 980–85. Retrieved from https://ajph.aphapublications.org/doi/pdf/10.2105/AJPH.2014.302391.

Muraki, I., et al. (2013, August 29). Fruit consumption and the risk of type 2 diabetes: Results from three prospective longitudinal cohort studies. *British Medical Journal, 347*.

Murnane, R. J., et al. (2005). Learning why more learning takes place in some classrooms than others. *German Economic Review, 6*(3), 309–30.

Murray, C. (1984). *Losing ground: American social policy, 1950–1980*. New York: Basic Books.

Murray, C. (2016). *In our hands: A plan to replace the welfare state*. Lanham, MD: Rowman & Littlefield.

Mustanski, B. S., et al. (2005, March). A genomewide scan of male sexual orientation. *Human Genetics, 116*(4), 272–78. Retrieved from http://mypage.iu.edu/~bmustans/Mustanski_etal_2005.pdf.

Nanda, S. (1990). *Neither man nor woman: The hijras of India*. New York: Wadsworth.

NASA. (2018). Global temperature: Global land-ocean temperature index, temperature anomalies. Retrieved from https://climate.nasa.gov/vital-signs/global-temperature.

Nash, J. M. (2000, July 31). This rice could save a million kids a year. *Time, 156*(5). Retrieved from http://content.time.com/time/magazine/article/0,9171,997586-6,00.html.

Nason-Clark, N. (2000). Making the sacred safe: Woman abuse and communities of faith. *Sociology of Religion, 61*(4), 349–68.

Natanson, H. (2017, June 5). Harvard rescinds acceptances for at least ten students for obscene memes. *The Harvard Crimson*. Retrieved from www.thecrimson.com/article /2017/6/5/2021-offers-rescinded-memes.

National Center for Education Statistics. (2014). Total fall enrollment in degree-granting postsecondary institutions, by attendance status, sex of student, and control of institution: Selected years, 1947 through 2024. Retrieved from https://nces.ed.gov/programs/digest/d14 /tables/dt14_303.10.10.asp?current=yes.

National Center for Education Statistics. (2017a). Graduation rate from first institution . . . cohort entry years, 1996 through 2009. Table 326.10. Retrieved from https://nces .ed.gov/programs/digest/d16/tables/dt16_326.10.asp?c.

National Center for Education Statistics. (2017b). Total undergraduate fall enrollment in degree-granting postsecondary institutions, by attendance status, sex of student, and control and level of institution: Selected years, 1970 through 2026. Retrieved from https://nces .ed.gov/programs/digest/d16/tables/dt16_303.70 .asp?current=yes.

National Center for Education Statistics. (2018a). The nation's report card: 2015 mathematics & reading assessments. Retrieved from www.nationsreportcard.gov /reading_math_2015/#mathematics?grade=4.

National Center for Education Statistics. (2018b, May). Characteristics of children's families. Retrieved from https://nces.ed.gov/programs/coe/indicator_cce.asp.

National Center for Health Statistics. (2017). National marriage and divorce rates 2000–16. Centers for Disease Control and Prevention, National Vital Statistics System. Retrieved from www.cdc.gov/nchs/data/dvs/national _marriage_divorce_rates_00-16.pdf.

National Institute on Drug Abuse. (2017, September). Overdose death rates. Retrieved from www.drugabuse .gov/related-topics/trends-statistics/overdose-death -rates.

National Philanthropic Trust. (2018). Charitable giving statistics. Retrieved from www.nptrust.org /philanthropic-resources/charitable-giving-statistics.

Nelson, M. K. (1990). *Negotiated care: The experience of family day care providers*. Philadelphia: Temple University Press.

New York State Division of Parole (2007, September). *New York State parole handbook* (NYSPH). Retrieved from http://parole.state.ny.us/Handbook.pdf.

Nicholas, R. W. (1973). Social and political movements. *Annual Review of Anthropology, 2*, 63–84.

Niebuhr, G. (1995, April 18). The gospels of management. The minister as marketer: Learning from business. *The New York Times*, p. 1. Retrieved from www.nytimes.com/1995 /04/18/us/megachurches-second-article-series-gospels -management-minister-marketer-learning.html.

Niebuhr, H. R. (1929). *The social sources of denominationalism*. New York: Holt.

Nissle, S., & Bshor, T. (2002). Winning the jackpot and depression: Money cannot buy happiness. *International Journal of Psychiatry in Clinical Practice, 6*(3), 183–86.

Nussbaum, P. (2006, January 21). A global ministry of "muscular Christianity." *The Washington Post*. Retrieved from www.washingtonpost.com/wp-dyn/content/article /2006/01/21/AR2006012100284_pf.html.

Nye, B. A., et al. (1994). Small is far better. *Research in the Schools, 1*(1), 9–20.

Nye, J. (1990). Soft power. *Foreign Policy, 80*, 153–71.

Oakes, J. (1985). *Keeping track: How schools structure inequality*. New Haven, CT: Yale University Press.

Oakley, A. (1972). *Sex, gender, and society*. London: Maurice Temple Smith.

O'Barr, W. (1995). *Linguistic evidence: Language, power and strategy in the courtroom*. San Diego, CA: Academic Press.

O'Connor, L. (2013, September 21). California universities come up with crazy crowdfunding scholarship idea. *Huffington Post*. Retrieved from www.huffingtonpost .com/2013/09/20/uc-promise-for-education-_n_3965021 .html.

O'Connor, N. (2009, April). Hispanic origin, socio-economic status, and community college enrollment. *Journal of Higher Education, 80*(2), 121–45.

Okahana, H., & Zhou, E. (2017). Graduate enrollment and degrees: 2006 to 2016. Council of Graduate Schools. Retrieved from http://cgsnet.org/ckfinder/userfiles/files /CGS_GED16_Report_Final.pdf.

Oliver, J. E. (1993). Intergenerational transmission of child abuse: Rates, research and clinical implications. *American Journal of Psychiatry, 150*, 1315–24.

Oliver, M., & Shapiro, T. (1995). *Black wealth, white wealth: A new perspective on racial inequality*. London: Routledge.

Olson, M. (1965). *The logic of collective action*. Cambridge: Harvard University Press.

Orfield, G. (1996). Turning back to segregation. In G. Orfield & S. E. Eaton (Eds.), *Dismantling desegregation: The quiet reversal of Brown v. Board of Education* (pp. 1–22). New York: New Press.

Orfield, G., et al. (2016, May 16). *Brown at 62: School segregation by race, poverty and state*. Civil Rights Project/Proyecto Derechos Civiles. Retrieved from https://civilrightsproject.ucla.edu/research/k-12 -education/integration-and-diversity/brown-at-62 -school-segregation-by-race-poverty-and-state/Brown -at-62-final-corrected-2.pdf.

Organisation for Economic Co-operation and Development. (2017, October 26). Table PF2.1A. Summary of paid leave entitlements available to mothers. Retrieved from www .oecd.org/els/soc/PF2_1_Parental_leave_systems.pdf.

Organisation for Economic Co-operation and Development. (2018a). Poverty rate (indicator). Retrieved from https://data.oecd.org/inequality/poverty-rate.htm.

Organisation for Economic Co-operation and Development. (2018b). Collective bargaining coverage. Retrieved from https://stats.oecd.org/Index.aspx?DataSetCode=CBC.

Orr, A. H. (2007). A mission to convert. *The New York Review of Books, 54*(1). Retrieved from www.nybooks.com /articles/19775.

Orshansky, M. (1963). *Children of the poor* (Social Security Bulletin). Washington, DC: US Department of Labor.

Ortner, S. (1974). Is female to male as nature to culture? In M. Z. Rosaldo & L. Lamphere (Eds.), *Woman, culture, and society* (pp. 67–88). Stanford, CA: Stanford University Press.

Oyěwùmí, O. (1997). *The invention of women: Making an African sense of Western gender discourses*. Minneapolis: University of Minnesota Press.

Padgett, D. K. (2007, May). There's no place like (a) home: ontological security among persons with serious mental illness in the United States. *Social Science & Medicine, 64*(9), 1925–36. Retrieved from www.ncbi.nlm.nih.gov /pubmed/17355900.

Pager, D. (2003). Blacks and ex-cons need not apply. *Contexts, 2*(4), 58–59.

Paik, A., et al. (2016). Broken promises: Abstinence pledging and sexual and reproductive health. *Journal of Marriage and Family, 78*(2), 546–61.

Pareto, V. (1983). *The mind and society* (A. Livingston, Ed.; A. Bongiorno & A. Livingston, Trans.). New York: AMS Press. (Original work published 1935.)

Pareto, V., & Finer, S. E. (1966). *Sociological writings.* New York: Praeger Sociology.

Pariser, E. (2011). *The filter bubble: How the new personalized web is changing what we read and how we think.* New York: Penguin Press.

Parker, I. (2007, July 30). Swingers. *The New Yorker.* Retrieved from www.newyorker.com/reporting/2007/07/30 /070730fa_fact_parker.

Parker, K., & Livingston, G. (2018, June 13). 7 facts about American fathers. Pew Research Center. Retrieved from www.pewresearch.org/fact-tank/2017/06/15/fathers -day-facts.

Parkin, F. (1982). *Max Weber.* London: Tavistock.

Parsons, E. F. (2005). Midnight rangers: Costume and performance in the reconstruction-era Ku Klux Klan. *Journal of American History, 92*(3), 811.

Parsons, T. (1951). *The social system.* Glencoe, IL: Free Press.

Pascoe, C. J. (2007). *Dude, you're a fag: Masculinity and sexuality in high school.* Berkeley: University of California Press.

Patillo, M. (2007). *Black on the block: The politics of race and class in the city.* Chicago: University of Chicago Press.

Payne, C. (2013, August 1). Fertility forecast: Baby bust is over, births will rise. *USA Today.* Retrieved from www .usatoday.com/story/news/nation/2013/08/01/usa-total -fertility-rate/2590781.

PBS. (2000, February 22). Lost tribes of Israel. *Nova.* Retrieved from www.pbs.org/wgbh/nova/transcripts /2706israel.html.

Pearson, B. Z. (1993). Predictive validity of the scholastic aptitude test (SAT) for Hispanic bilingual students. *Hispanic Journal of Behavioral Sciences, 15*(3), 342–56.

Penner, J. E., et al. (Eds.). (1999). *Aviation and the global atmosphere.* Cambridge: Cambridge University Press. Retrieved from www.ipcc.ch/ipccreports/sres/aviation /index.php?idp=64.

Perlman, J., & Parvensky, J. (2006). *Denver Housing First Collaborative: Cost benefit analysis and program outcomes report.* Denver: Colorado Coalition for the Homeless. Retrieved from www.denversroadhome.org/files /FinalDHFCCostStudy_1.pdf.

Perrow, C. (2007). *The next catastrophe: Reducing our vulnerabilities to natural, industrial, and terrorist disasters.* Princeton, NJ: Princeton University Press.

Petrosky, E., et al. (2017). Racial and ethnic differences in homicides of adult women and the role of intimate partner violence—United States, 2003–2014. *Morbidity and Mortality Weekly Report, 66*(28), 741–46.

Pew Forum on Religion and Public Life. (2008, February). US religious landscape survey: Religious affiliation: Diverse and dynamic.

Pew Forum on Religion and Public Life. (2011). Global Christianity—A report on the size and distribution of the world's Christian population. Retrieved from www .pewforum.org/2011/12/19/global-christianity-exec.

Pew Forum on Religion and Public Life. (2015). U.S. public becoming less religious. Retrieved from www.pewforum .org/2015/11/03/u-s-public-becoming-less-religious.

Pew Research Center. (2011a). Muslim American survey. Retrieved from www.people-press.org/2011/08/30 /section-1-a-demographic-portrait-of-muslim-americans.

Pew Research Center. (2011b, July 26). Pew Research Center's social & demographic trends. Wealth gaps rise to record highs between whites, blacks and Hispanics. Retrieved from http://pewresearch.org/pubs/2069 /housing-bubble-subprime-mortgages-hispanics-blacks -household-wealth-disparity.

Pew Research Center. (2011c, April 5). Multi-race Americans and the 2010 Census. Retrieved from www .pewsocialtrends.org/2011/04/05/multi-race-americans -and-the-2010-census.

Pew Research Center. (2013). Modern parenthood study: Parental time use table. Retrieved from www .pewresearch.org/data-trend/society-and-demographics /parental-time-use.

Pew Research Center. (2014). Less than half of U.S. kids today live in a "traditional" family. Retrieved from www .pewresearch.org/fact-tank/2014/12/22/less-than-half -of-u-s-kids-live-in-a-traditional-family.

Pew Research Center. (2015a). Christians decline as share of US population; other faiths and the unaffiliated are growing. Retrieved from www.pewforum.org/2015/05 /12/americas-changing-religious-landscape/pr_15-05-12 _rls-00.

Pew Research Center. (2015b). US public becoming less religious. Retrieved from www.pewforum.org/2015/11/03 /u-s-public-becoming-less-religious.

Pew Research Center. (2015c). Future immigration will change the face of America by 2065. Retrieved from http://www.pewresearch.org/fact-tank/2015/10 /05/future-immigration-will-change-the-face-of -america-by-2065/.

Pew Research Center. (2017a, February 5). Internet /broadband fact sheet. Retrieved from www.pewinternet .org/fact-sheet/internet-broadband.

Pew Research Center. (2017b). Most people say they have achieved the American dream, or are on their way to achieving it. Retrieved from www.pewresearch.org /fact-tank/2017/10/31/most-think-the-american-dream -is-within-reach-for-them/ft_17-10-31_americandream _demographic.

Pew Research Center. (2017c, June 26). Support for same-sex marriage grows, even among groups that had been skeptical. Retrieved from www.people-press.org/2017/06 /26/support-for-same-sex-marriage-grows-even-among -groups-that-had-been-skeptical.

Phelan, C., & Link, B. G. (2005). Controlling disease and creating disparities: A fundamental cause perspective. *Journals of Gerontology, 60B,* 30, 31.

Philips, M. E. et al. (2015, November 22). Transgender locker room policy eludes school district facing government sanctions under Title IX. Retrieved from www.collegeandprosportslaw.com/uncategorized /transgender-locker-room-policy-eludes-school-district -facing-government-sanctions-under-title-ix.

Phillips, D., et al. (1987). Child care quality and children's social development. *Developmental Psychology, 23,* 537–43.

Pickert, K. (2014). University of California approves steep tuition hike. *Time.* Retrieved from http://time.com /3598249/university-of-california-tuition-hike.

Pierce, J. L. (1995). *Gender trials: Emotional lives in contemporary law firms.* Berkeley: University of California Press.

Pierné, G. (2013). Hiring discrimination based on national origin and religious closeness: Results from a field experiment in the Paris area. *IZA Journal of Labor Economics, 2*(4). Retrieved from https://izajole .springeropen.com/articles/10.1186/2193-8997-2-4.

Piketty, T., et al. (2018). Distributional national accounts: Methods and estimates for the United States. *Quarterly Journal of Economics, 133*(2), 553–609.

Pilkington, D. (1996). *Follow the rabbit-proof fence.* St. Lucia, Australia: University of Queensland Press.

Pisetsky, E. M., et al. (2008). Disordered eating and substance abuse in high-school students: Results from the Youth Risk Behavior Surveillance System. *International Journal of Eating Disorders, 41*(5), 464–70.

Piven, F. F., & Cloward, R. A. (1988). *Why Americans don't vote: And why politicians want it that way.* Boston: Beacon Press.

Plath, S. (1971). *The bell jar.* New York: Harper & Row. (Original work published 1963.)

Pollan, M. (2006). *The omnivore's dilemma: A natural history of four meals.* New York: Penguin Press.

Poorman, E. (2018, January 14). Why does America still have so few female doctors? *The Guardian.* Retrieved from www.theguardian.com/commentisfree/2018/jan/14 /why-are-there-still-so-few-female-doctors.

Porter, J., & Jick, H. (1980). Addiction rare in patients treated with narcotics. *New England Journal of Medicine,* 302(2), 123.

Portes, A. (1969). Dilemmas of a golden exile: Integration of Cuban refugee families in Milwaukee. *American Sociological Review, 34,* 505–18.

Portes, A., & MacLeod, D. (1996). Educational progress of children of immigrants: The roles of class, ethnicity and school context. *Sociology of Education, 69*(4), 255–75.

Portes, A., & Zhou, M. (1993). The new second generation: Segmented assimilation and its variants. *Annals of the American Academy of Political and Social Science, 530,* 74–96.

Portes, A., et al. (1985). After Mariel: A survey of the resettlement experiences of 1980 Cuban refugees in Miami. *Estudios Cubanos* (Cuban Studies), *15,* 37–59.

Posner, R. A. (1992). *Sex and reason.* Cambridge: Harvard University Press.

Powell, B., & Steelman, L. C. (1990). Beyond sibship size: Sibling density, sex composition, and educational outcomes. *Social Forces, 69*(1), 181–206.

Pratt, L. A., et al. (2011, October). *Antidepressant use in persons aged 12 and over: United States, 2005–2008* (National Center for Health Statistics Data Brief No. 76). Retrieved from www.cdc.gov/nchs/data/databriefs/db76.html.

Public Policy Polling. (2011, March 17). Americans support unions over Walker; Sheen for president? Retrieved from www.publicpolicypolling.com/main/2011/03/americans -support-unions-over-walker-sheen-for-president.html.

Putnam, R. D. (2000). *Bowling alone: The collapse and revival of American community.* New York: Simon & Schuster.

Qaim, M., & Zilberman, D. (2003). Yield effects of genetically modified crops in developing countries. *Science, 299,* 900–902.

Qian, Z. (2013, September 11). Divergent paths of American families. US2010 Project. Retrieved from https://s4.ad.brown.edu/Projects/Diversity/Data/Report /report09112013.pdf.

Quadagno, J. (1987). Theories of the welfare state. *Annual Review of Sociology, 13,* 109–28.

Quadagno, J. (1996). *The color of welfare: How racism undermined the war on poverty.* New York: Oxford University Press.

Radway, J. (1987). *Reading the romance: Women, patriarchy, and popular literature.* Chapel Hill: University of North Carolina Press.

Rahman, Q., et al. (2003). Sexual orientation–related differences in prepulse inhibition of the human startle response. *Behavioral Neuroscience, 117*(5), 1096–102.

Rainwater, L. (1974). *What money buys: Inequality and the social meanings of income.* New York: Basic Books.

Ralli, T. (2005, September 5). Who's a looter? In storm's aftermath, pictures kick up a different kind of tempest. *The New York Times.* Retrieved from www.nytimes.com /2005/09/05/business/05caption.html.

Read, J. (2008, February). Muslims in America. Retrieved from http://contexts.org/articles/fall-2008/muslims-in- america.

Reardon, S. F. (2013). The widening income achievement gap. *Educational Leadership, 70*(8), 10–16. Retrieved from http://nppsd.fesdev.org/vimages/shared/vnews/stories /525d81ba96ee9/SI%20-%20The%20Widening%20 Income%20Achievement%20Gap.pdf.

Reardon, S. F., et al. (2012, August 3). Race, income, and enrollment patterns in highly selective colleges, 1982–2004. Center for Education Policy Analysis, Stanford University. Retrieved from http://cepa.stanford .edu/sites/default/files/race income %26 selective college enrollment august 3 2012.pdf.

Reddy, G. (2005). *With respect to sex: Negotiating hijra identity in South India.* Chicago: University of Chicago Press.

Reimer, S. (2016). It's just a very male industry: Gender and work in UK design agencies. *Gender, Place & Culture, 23*(7), 1033–46.

Reskin, B., & Roos, P. (1990). *Job queues, gender queues.* Philadelphia: Temple University Press.

Rich, A. (1980). Compulsory heterosexuality and lesbian existence. *Signs: Journal of Women in Culture and Society, 5,* 631–60.

Ringen, S. (1987). *The possibility of politics.* New York: Oxford University Press.

Rios, V. (2011). *Punished: Policing the lives of black and Latino boys.* New York: NYU Press.

Risman, B. J. (1998). *Gender vertigo: American families in transition.* New Haven, CT: Yale University Press.

Roberts, S. (2010, January 5). No longer majority black, Harlem is in transition. *The New York Times.* NY/Region section. Retrieved from www.nytimes.com/2010/01/06/nyregion/06harlem.html.

Rockoff, J. E. (2004). The impact of individual teachers on student achievement: Evidence from panel data. *The American Economic Review, 94*(2), 247–52.

Roof, W. C. (1989). Multiple religious switching. *Journal for the Scientific Study of Religion, 28*, 530–35.

Rosaldo, M. Z. (1974). Woman, culture and society: A theoretical overview. In M. Z. Rosaldo & L. Lamphere (Eds.), *Woman, culture, and society* (pp. 17–42). Stanford, CA: Stanford University Press.

Rosenbaum, J. E. (1980). Track misperceptions and frustrated college plans: An analysis of the effects of track perception in the National Longitudinal Survey. *Sociology of Education, 53*(2), 74–88.

Rosenbaum, J. E., with L. S. Rubinowitz. (2000). *Crossing the class and color lines: From public housing to white suburbia.* Chicago: University of Chicago Press.

Rosenfeld, J., & Kleykamp, M. (2009). Hispanics and organized labor. *American Sociological Review, 74*(4), 916–37.

Rosenhan, D. L. (1973). On being sane in insane places. *Science, 179*, 25–58.

Rosenthal, R., & Jacobson, L. (1968). *Pygmalion in the classroom: Teacher expectation and pupils' intellectual development.* New York: Rinehart & Winston.

Rosenzweig, M. R. (1982). Educational subsidy, agricultural development and fertility change. *Quarterly Journal of Economics, 97*(1), 67–88.

Rothstein, J. M. (2004). College performance predictions and the SAT. *Journal of Econometrics, 121*(1–2), 297–317.

Rousseau, J.-J. (2004). A dissertation on the origin and foundation of the inequality of mankind. In *Discourse on inequality.* Whitefish, MT: Kessinger Publishing. (Original work published 1754.)

Rowntree, B. S. (1910). *Poverty, A study of town life.* London: Macmillan.

Rubin, G. (1975). The traffic in women: Notes on the "political economy" of sex. In R. R. Reiter (Ed.), *Toward an anthropology of women* (pp. 157–210). New York: Monthly Review Press.

Ruggles, P. (1990). *Drawing the line: Alternative poverty measures and their implications for public policy.* Washington, DC: Urban Institute Press.

Ryan, C. L., & Bauman, K. (2016, March). Educational attainment in the United States: 2015. Population characteristics. US Census Bureau. Retrieved from www.census.gov/content/dam/Census/library/publications/2016/demo/p20-578.pdf.

Sacerdote, B. (2001). Peer effects with random assignment: Results for Dartmouth roommates. *Quarterly Journal of Economics, 116.*

Sachs, J. D. (2001). *Macroeconomics and health: Investing in health for economic development.* Report of the Commission on Macroeconomics and Health. Geneva: World Health Organization.

Sadker, M., & Sadker, D. (1994). *Failing at fairness: How America's schools cheat girls.* New York: Touchstone.

Saez, E. (2013). Striking it richer: The evolution of top incomes in the United States. University of California, Berkeley. Retrieved from http://elsa.berkeley.edu/~saez/saez-UStopincomes-2012.pdf.

Saez, E., & Zucman, G. (2016). Wealth inequality in the United States since 1913: Evidence from capitalized income tax data. *Quarterly Journal of Economics, 131*(2), 519–78.

Salai-i-Martin, X. (2002). *The disturbing "rise" of global income inequality* (NBER Working Paper No. 8904). National Bureau of Economic Research. Retrieved from www.nber.org/papers/w8904.

Salganik, M. J., et al. (2006). Experimental study of inequality and unpredictability in an artificial cultural market. *Science, 311*(5762), 854–56.

Samuels, A. (2015). Rachel Dolezal's true lies. *Vanity Fair.* Retrieved from www.vanityfair.com/news/2015/07.

Sanday, P. R. (1990). *Fraternity gang rape: Sex, brotherhood, and privilege on campus.* New York: New York University Press.

Sander, T. H., & Putnam, R. D. (2005, September 10). Sept. 11th as a civics lesson. *The Washington Post*, p. A23. Retrieved from www.washingtonpost.com/wp-dyn/content/article/2005/09/09/AR2005090901821.html.

Sander, T. H., & Putnam, R. D. (2010). Still bowling alone? The post-9/11 split. *Journal of Democracy, 21*(1), 9–16.

Santa Ana, O. (2004). Is there such a thing as Latino identity? *American Family: Journey of Dreams.* Retrieved from www.pbs.org/americanfamily/latino2.html.

Saroyan, S. (2006, April 16). Christianity, the brand. *The New York Times.* Retrieved from www.nytimes.com/2006/04/16/books/christianity-the-brand.html.

Sawyer, K., et al. (2016). Queering the gender binary: Understanding transgender workplace experiences. In T. Kollen (Ed.), *Sexual orientation and transgender issues in organizations* (pp. 21–42). Switzerland: Springer.

Sawyer, L. (2017, January 28). Waconia woman crawls back from the "dead." *Star Tribune.* Retrieved from www.startribune.com/waconia-woman-crawls-back-from-the-dead/412049283.

Scalise, I. M. (2014, January 26). Lo stato riconosca a casalinghe un salario e una vera pensione. *La Repubblica.* Retrieved from www.repubblica.it/economia/2014/01/26/news/bongiorno_lo_stato_riconosca_salario_casalinghe-76954421.

Schackner, B. (2002, September 27). College course focuses on the wealthy minority. *Post-Gazette.* Retrieved from www.post-gazette.com/localnews/20020929wealthy6.asp.

Schacter, J., & Thum, Y. M. (2004). Paying for high- and low-quality teaching. *Economics of Education Review, 23*(4), 411–30.

Scharf, S. A., & Flom, B. A. (2010, October). Report of the fifth annual national survey on retention and promotion of women in law firms. *National Association of Women Lawyers.* Retrieved from www.aauw.org/learn/research/upload/NewVoicesPayEquity_NAWL.pdf.

Scheitle, C., & Dougherty, K. (2010). Race, diversity, and membership duration in religious congregations. *Sociological Inquiry, 80*(3), 405–23.

Scherrer, K. S., & Pfeffer, C. A. (2017). None of the above: Toward identity and community-based understandings of (a)sexualities. *Archives of Sexual Behavior, 46*(3), 643–46.

Schippers, M. (2016). *Beyond monogamy: Polyamory and the future of polyqueer sexualities.* New York: NYU Press.

Schlosser, E. (2001). *Fast food nation: The dark side of the all-American meal*. Boston: Houghton Mifflin.

Sciolino, E. (2004, October 22). France turns to tough policy on students' religious garb. *The New York Times*. Retrieved from www.nytimes.com/2004/10/22/world/europe/france-turns-to-tough-policy-on-students-religious-garb.html.

Seelye, K. Q. (2005, May 16). *Newsweek* apologizes for report of Koran insult. *The New York Times*, p. A1. Retrieved from www.nytimes.com/2005/05/16/world/asia/newsweek-apologizes-for-report-of-koran-insult.html.

Semega, J. L., et al. (2017). Income and poverty in the United States: 2016. Current Population Reports, US Census Bureau. Retrieved from www.census.gov/content/dam/Census/library/publications/2017/demo/P60-259.pdf.

Setoodeh, R. (2006, March 20). Troubles by the score. *Newsweek*. Retrieved from www.msnbc.msn.com/id/11788171/site/newsweek.

Shah, D. (2004). *Defense of Marriage Act: An update*. Washington, DC: US General Accounting Office.

Shaheen, J. G. (1984). *The TV Arab*. Bowling Green, OH: Bowling Green State University Popular Press.

Shapiro, D., et al. (2016). Time to degree: A national view of the time enrolled and elapsed for associate and bachelor's degree earners [signature report 11]. Herndon, VA: National Student Clearinghouse Research Center. Retrieved from https://nscresearchcenter.org/wp-content/uploads/SignatureReport11.pdf.

Sheff, E. (2014). The polyamorists next door: Inside multiple-partner relationships and families. Maryland, MD: Rowman & Littlefield.

Sheler, J. (2001, October 29). Muslim in America. *U.S. News & World Report*, pp. 50–52.

Sherkat, D. E. (1991). Leaving the faith: Testing theories of religious switching using survival models. *Social Science Research, 20,* 171–87.

Sherkat, D. E. (1998). Counterculture or continuity? Competing influences on baby boomers. Religious orientations and participation. *Social Forces, 76,* 1087–115.

Sherkat, D. E., & Ellison, C. G. (1999). Recent developments and current controversies in the sociology of religion. *Annual Review of Sociology, 25,* 363–94.

Shiva, V. (1992a). Recovering the real meaning of sustainability. In D. E. Cooper & J. A. Palmer (Eds.), *The environment in question: Ethics in global issues* (pp. 187–93). London: Routledge.

Shiva, V. (1992b). Women's indigenous knowledge and biodiversity conservation. In G. Sen (Ed.), *Indigenous vision* (pp. 205–14). New Delhi: Sage.

Shiva, V. (2002). *Water wars: Privatization, pollution and profit*. Cambridge, MA: South End Press.

Shiva, V. (2004). Turning scarcity into abundance. *Yes! Magazine*. Retrieved from www.yesmagazine.org/article.asp?ID=698.

Shore, B. (1998). *Culture in mind: Cognition, culture, and the problem of meaning*. New York: Oxford University Press.

Shore, L. (2017, January 3). Gal interrupted, why men interrupt women and how to avert this in the workplace. *Forbes*. Retrieved from www.forbes.com/sites/womensmedia/2017/01/03/gal-interrupted-why-men-interrupt-women-and-how-to-avert-this-in-the-workplace/#437dd01217c3.

Shulevitz, J. (2016, January 8). It's payback time for women. *The New York Times*. Retrieved from www.nytimes.com/2016/01/10/opinion/sunday/payback-time-for-women.html.

Shulman, J., & Bowen, W. G. (2002). *The game of life: College sports and educational values*. Princeton, NJ: Princeton University Press.

Silva, T. J. (2017). Bud-sex: Constructing normative masculinity among rural straight men that have sex with men. *Gender & Society, 31*(1), 51–73.

Singer, N. (2017, May 13). How Google took over the classroom. *New York Times*. Retrieved from www.nytimes.com/2017/05/13/technology/google-education-chromebooks-schools.html?mcubz=0&_r=0.

Simmel, G. (1900). *Philosophie der Geldes* (The philosophy of money). Leipzig: Duncker & Humblot.

Simmel, G. (1950). Quantitative aspects of the group. In Kurt Wolff (Comp. and Trans.), *The sociology of Georg Simmel* (pp. 87–180). Glencoe, IL: Free Press.

Skocpol, T. (2004). Civic transformation and inequality in the contemporary United States. In K. Neckerman (Ed.), *Social inequality* (pp. 731–69). New York: Russell Sage Foundation.

Skocpol, T., & Amenta, E. (1986). States and social policies. *Annual Review of Sociology, 12,* 131–57.

Skogrand, L., et al. (2004, July). Understanding Latino families: Implications for family education. *Family Resources*. Retrieved from http://extension.usu.edu/diversity/files/uploads/Latino02-7-05.pdf.

Smedley, A. (1999). *Race in North America: Origin and evolution of a worldview*. Boulder, CO: Westview Press.

Smelser, N. (1962). *Theory of collective behavior*. New York: Free Press.

Smith, A. (2003). *The wealth of nations*. New York: Penguin. (Original work published 1776.)

Smith, B. H. (2009). *Natural reflections: Human cognition at the nexus of science and religion*. New Haven, CT: Yale University Press.

Smith, G. A., & Cooperman, A. (2016, September 14). The factors driving the growth of religious "nones" in the U.S. Pew Research Center. Retrieved from www.pewresearch.org/fact-tank/2016/09/14/the-factors-driving-the-growth-of-religious-nones-in-the-u-s.

Smith, J. R., et al. (1997). Consequences of living in poverty for young children's cognitive and verbal ability and early school achievement. In J. G. Duncan & J. Brooks-Gunn (Eds.), *Consequences of growing up poor* (pp. 132–89). New York: Russell Sage Foundation.

Smith, T. (2012). Beliefs about God across time and countries. National Opinion Research Center (NORC), University of Chicago. Retrieved from www.norc.org/pdfs/beliefs_about_god_report.pdf.

Snow, D. A., et al. (1986). Frame alignment processes, micromobilization, and movement participation. *American Sociological Review, 51*(4), 464–81.

Snowden, F. M. (1983). *Beyond color prejudice: The ancient view of blacks*. Cambridge: Harvard University Press.

Snyder, S., & Evans, W. (2002). *The impact of income on mortality: Evidence from the social security notch* (NBER Working Paper No. W9197). National Bureau of Economic Research. Retrieved from www.nber.org/papers/w9197.

Snyder, T. D., & Tan, A. G. (2005). *Digest of education statistics, 2004* (NCES 2006-005). US Department of Education, National Center for Education Statistics. Washington, DC: US Government Printing Office.

Sokal, A., & Bricmont, J. (1999). *Fashionable nonsense: Postmodern intellectuals' abuse of science.* London: Picador.

Sommers, C. H. (2000). *The war against boys: How misguided feminism is harming our young men.* New York: Simon & Schuster.

Sorokin, P. (1959). *Social and cultural mobility.* New York: Free Press. (Original work published 1927.)

Sorokin, P., & Lunden, W. A. (1959). *Power and morality: Who shall guard the guardians?* Boston: Porter Sargent.

Spencer, M. B., et al. (1987). Double stratification and psychological risk: Adaptational processes and school achievement of black children. *Journal of Negro Education, 56*(1), 77–87.

Spiegel, A. (Producer). (2006, December 15). Shouting across the divide, act one: Which one of them is not like the other? [Radio broadcast]. *This American Life.* Retrieved from www.thisamericanlife.org/Radio_Episode.aspx?episode=322.

Spock, B. (1998). *Baby and child care.* New York: Pocket Books. (Original work published 1946.)

Spohn, C., & Holleran, D. (2002). The effects of imprisonment on recidivism rates of felony offenders: A focus on drug offenders. *Criminology, 40*(2), 329–58.

Spurlock, M. (Director). (2004). *Super size me* [Motion picture]. United States: Sony Pictures.

Spurlock, M. (Host). (2005, June 29). Muslims and America [Television series episode]. In J. Chinn et al. (Producers), *30 days.* Los Angeles, Bluebush Productions LLC.

Stacey, J. (1987). Sexism by a subtler name? Postindustrial conditions and postfeminist consciousness in the Silicon Valley. *Socialist Review, 96,* 7–28.

Stacey, J. (1996). *In the name of the family: Rethinking family values in the postmodern age.* Boston: Beacon Press.

Stacey, J. (1997). Neo-family-values campaign. In R. N. Lancaster & M. di Lionardo (Eds.), *The gender/sexuality reader: Culture, history, political economy* (pp. 453–72). New York: Routledge.

Stacey, J. (2004). Cruising to familyland: Gay hypergamy and rainbow kinship. *Current Sociology, 52*(2), 181–97.

Stacey, J., & Thorne, B. (1985). The missing feminist revolution in sociology. *Social Problems, 32,* 301–16.

Stack, C. B. (1974). *All our kin: Strategies for survival in a black community.* New York: Harper & Row.

Stack, C. B., & Burton, L. M. (1994). Kinscripts: Reflections on family, generation and culture. In E. N. Glenn et al. (Eds.), *Mothering: Ideology, experience, and agency* (pp. 33–44). New York: Routledge.

Stark, R., & Bainbridge, W. S. (1985). *The future of religion: Secularization, revival, and cult formation.* Berkeley: University of California Press.

Stark, R., & Bainbridge, W. S. (1987). *A theory of religion.* Toronto: Lang.

Stark, R., & Finke, R. (2000). *Acts of faith: Explaining the human side of religion.* Berkeley: University of California Press.

Steele, C. M., & Aronson, J. (1998). Stereotype threat and the test performance of academically successful African Americans. In C. Jencks and M. Phillips (Eds.), *The black-white test score gap* (pp. 401–27). Washington, DC: Brookings Institution Press.

Steelman, L. C., & Powell, B. (1985). The social and academic consequences of birth order: Real, artifactual or both? *Journal of Marriage and the Family, 47*(1), 117–24.

Steelman, L. C., & Powell, B. (1989). Acquiring capital for college: The constraints of family configuration. *American Sociological Review, 54*(5), 844–55.

Steinberg, S. (1974). *The academic melting pot.* New York: McGraw-Hill.

Stevens, J. (2006). Pregnancy envy and the politics of compensatory masculinities. *Gender and Politics, 1,* 265–94.

Stewart, J. B. (2015, July 16). Convictions prove elusive in "London Whale" trading case. *The New York Times.* Retrieved from www.nytimes.com/2015/07/17/business/figures-in-london-whale-trading-case-escape-the-authorities-nets.html.

Stokes, B. (2017). Public divided on prospects for the next generation. Pew Research Center. Retrieved from www.pewglobal.org/2017/06/05/2-public-divided-on-prospects-for-the-next-generation.

Stone, D. A. (1994). Making the poor count. *American Prospect, 17,* 84–88.

Sunderam, S., et al. (2018). Assisted reproductive technology surveillance—United States, 2015. *Morbidity and Mortality Weekly Report, 67*(3), 1–28.

Super, D. A. (2018, January 15). "Work requirements" for public benefits are really just time limits. *The Los Angeles Times.* Retrieved from www.latimes.com/opinion/op-ed/la-oe-super-work-requirements-20180115-story.html.

Tavernise, S. (2016, April 22). U.S. Suicide Rate Surges to a 30-Year High. *The New York Times.* Retrieved from www.nytimes.com/2016/04/22/health/us-suicide-rate-surges-to-a-30-year-high.html.

Taylor, K. (2014, October 27). How buying beer paid me $4,000 a month for college. *Daily Finance.* Retrieved from www.dailyfinance.com/2014/10/27/how-buying-beer-paid-me-4-000-a-month-for-college.

Taylor, P., et al. (2011, July 26). Wealth gaps rise to record highs between whites, blacks, Hispanics twenty-to-one. Pew Research Center. Retrieved from www.pewsocialtrends.org/2011/07/26/wealth-gaps-rise-to-record-highs-between-whites-blacks-hispanics.

Tett, Gillian. (2010). Road map that opens up shadow banking. *Financial Times.* Retrieved from www.ft.com/intl/cms/s/0/1a222bf4-f33d-11df-a4fa-00144feab49a.html#axzz3x3F5SrvG.

Thomas, E. (2005, May 23). How a fire broke out. *Newsweek.* Retrieved from www.msnbc.msn.com/id/7857407/site/newsweek.

Thomas, W. I., & Thomas, D. S. (1928). *The child in America: Behavior problems and programs.* New York: Knopf.

Thomas Nelson, Inc. (2010). *Refuel: The complete New Testament* (2nd ed.), NCV [Catalog blurb]. Retrieved from www.thomasnelson.com/consumer/-product_detail.asp?dept_id=190900&sku=0718013026.

Thurow, L. (1970). *Investment in human capital.* Belmont, CA: Wadsworth.

Time. (1941, December 22). How to tell your friends from the Japs. Retrieved from http://content.time.com/time/magazine/article/0,9171,932034,00.html.

Tinkler, J. E. (2013). How do sexual harassment policies shape gender beliefs? An exploration of the moderating effects of norm adherence and gender. *Social Science Research, 42*(5), 1269–83.

Tocqueville, A. de. (1835). *Democracy in America*. Retrieved from http://xroads.virginia.edu/~HYPER/DETOC/toc_indx.html.

Torres, N. G. (2017, December 19). More than 37,000 Cubans face deportation orders. *The Miami Herald*. Retrieved from www.miamiherald.com/news/nation-world/world/americas/cuba/article190571369.html.

Tourism Economics. (2016). The economic impact of tourism in Lancaster County. Retrieved from www.discoverlancaster.com/Uploads/files/TE-DLEconImpact2015-2016-05-24.pdf.

Triester, R. (2005, July 22). "Real beauty"—or real smart marketing? *Salon.com*. Retrieved from www.salon.com/2005/07/22/dove_2.

Truth, S. (1851). Ain't I a woman? Speech delivered at the Women's Convention in Akron, Ohio. *Feminist.com*. Retrieved from www.nps.gov/articles/sojourner-truth.htm.

Turkewitz, J. (2018, January 4). For Native Americans, a "historic moment" on the path to power at the ballot box. *The New York Times*. Retrieved from www.nytimes.com/2018/01/04/native-american-voting-rights.html.

Turner, R., & Killian, L. M. (1987). *Collective behavior* (3rd ed.). Englewood Cliffs, NJ: Prentice-Hall.

Tyson, K., et al. (2005). It's not "a black thing": Understanding the burden of acting white and other dilemmas of high achievement. *American Sociological Review, 70*(4), 582–605.

Unah, I., & Boger, J. C. (2001, April 16). *Race and the death penalty in North Carolina*. Retrieved from www.common-sense.org/pdfs/NCDeathPenaltyReport2001.pdf.

UNAIDS. (2013). *Global report: UNAIDS report on the global AIDS epidemic 2013* (pp. A1–A15). Retrieved from www.unaids.org/en/media/unaids/contentassets/documents/epidemiology/2013/gr2013/UNAIDS_Global_Report_2013_en.pdf.

UNICEF. (2017). Levels and trends in child mortality. Retrieved from www.unicef.org/publications/files/Child_Mortality_Report_2017.pdf.

United Methodist Church. (2011). 2011 state of the church: The people. Retrieved from www.umc.org/site/c.lwL4KnN1LtH/b.6765229/k.D5ED/2011_State_of_the_Church_The_People.htm.

United Methodist Church. (2016). United Methodists at-a-glance. Retrieved from www.umc.org/who-we-are/united-methodists-at-a-glance.

United Methodist Church. (2018). United Methodists at-a-glance. Retrieved from www.umc.org/news-and-media/united-methodists-at-a-glance.

United Nations. (2001, May 17). Poverty biggest enemy of health in developing world, secretary-general tells World Health Assembly [Press release]. Retrieved from www.un.org/News/Press/docs/2001/sgsm7808.doc.htm.

US Census Bureau. (1993). We the first Americans. US Department of Commerce Economics and Statistics Administration. Retrieved from www.census.gov/apsd/wepeople/we-5.pdf.

US Census Bureau. (2010a, October 19). Nation's foreign-born population nears 37 million. Retrieved from www.census.gov/newsroom/releases/archives/foreignborn_population/cb10-159.html.

US Census Bureau. (2010b). Data profile highlights, population finder fact sheet. 2006–2008 American Community Survey 3-year estimates. Retrieved from http://factfinder.census.gov/servlet/ACSSAFFFacts?_sse=on.

US Census Bureau. (2010c). 2012 statistical abstract: Income, poverty, & wealth. Retrieved from www.census.gov/compendia/statab/cats/income_expenditures_poverty_wealth.html.

US Census Bureau. (2011, May). The Hispanic population: 2010. 2010 census briefs. Retrieved from www.census.gov/prod/cen2010/briefs/c2010br-04.pdf.

US Census Bureau. (2012a). The American Indian and Alaska Native population. Retrieved from www.census.gov/prod/cen2010/briefs/c2010br-10.pdf.

US Census Bureau. (2012b). Selected population profile in the United States. 2012 American Community Survey 1-year estimates. Retrieved from http://factfinder2.census.gov/faces/tableservices/jsf/pages/productview.xhtml?pid=ACS_12_1YR_S0201&prodType=table.

US Census Bureau. (2012c). Hispanic or Latino origin by specific origin. 2012 American Community Survey 1-year estimates. Retrieved from http://factfinder2.census.gov/faces/tableservices/jsf/pages/productview.xhtml?pid=ACS_12_1YR_B03001&prodType=table.

US Census Bureau. (2012d). Table 4: Percent distribution of household net worth, by amount of net worth and selected characteristics: 2011. Retrieved from www.census.gov/people/wealth.

US Census Bureau. (2013). Census Bureau reports 21 percent married households have at least one foreign-born spouse. Retrieved from www.census.gov/newsroom/releases/archives/foreignborn_population/cb13-157.html.

US Census Bureau. (2014). Asian/Pacific American Heritage Month: May 2014. Facts for features. Retrieved from www.census.gov/newsroom/facts-for-features/2014/cb14-ff13.html.

US Census Bureau. (2016a, November 17). The majority of children live with two parents, Census Bureau reports. Retrieved from www.census.gov/newsroom/press-releases/2016/cb16-192.html.

US Census Bureau. (2016b). Household type by household size. 2011–2015 American Community Survey 5-year estimates. Retrieved from https://factfinder.census.gov/faces/tableservices/jsf/pages/productview.xhtml?src=bkmk

US Census Bureau. (2016c). Age and sex. 2012–2016 American Community Survey 5-year estimates. Retrieved from https://factfinder.census.gov/bkmk/table/1.0/en/ACS/16_5YR/S0101/0100000US.

US Census Bureau. (2016d). Median age of the total population—United States—states; and Puerto Rico: Total population. 2012–2016 American Community Survey 5-year estimates. Retrieved from https://factfinder.census.gov/bkmk/table/1.0/en/ACS/16_5YR/GCT0101.US01PR/0100000US.

US Census Bureau. (2016e) Hispanic or Latino origin by specific origin. 2016 American Community Survey 1-year estimates. Retrieved from https://factfinder.census.gov/bkmk/table/1.0/en/ACS/16_1YR/B03001/0100000US.

US Census Bureau. (2016f). Educational attainment. 2016 American Community Survey 1-year estimates. Retrieved from https://factfinder.census.gov/faces/tableservices/jsf/pages/productview.xhtml?pid=ACS_16_1YR_S1501&prodType=table.

US Census Bureau. (2016g). Average household size of occupied housing units by tenure. 2016 American Community Survey 1-year estimates. Retrieved from https://factfinder.census.gov/faces/tableservices/jsf/pages/productview.xhtml?pid=ACS_16_1YR_B25010&prodType=table.

US Census Bureau. (2017a). American Indian and Alaska native alone or in combination with one or more other races. 2016 American Community Survey 1-year estimates. Retrieved from https://factfinder.census.gov/faces/tableservices/jsf/pages/productview.xhtml?pid=ACS_16_1YR_B02010&prodType=table.

US Census Bureau. (2017b). Race. 2016 American Community Survey 1-year estimates. Retrieved from https://factfinder.census.gov/faces/tableservices/jsf/pages/productview.xhtml?pid=ACS_16_1YR_B02001&prodType=table.

US Census Bureau. (2017c). Median household income in the past 12 months (in 2016 inflation-adjusted dollars) (white alone householder). 2016 American Community Survey 1-year estimates. Retrieved from https://factfinder.census.gov/faces/tableservices/jsf/pages/productview.xhtml?pid=ACS_16_1YR_B19013A&prodType=table.

US Census Bureau. (2017d). Median household income in the past 12 months (in 2016 inflation-adjusted dollars) (black or African American alone householder). 2016 American Community Survey 1-year estimates. Retrieved from https://factfinder.census.gov/faces/tableservices/jsf/pages/productview.xhtml?pid=ACS_16_1YR_B19013B&prodType=table.

US Census Bureau. (2017e). Hispanic or Latino origin by specific origin. 2016 American Community Survey 1-year estimates. Retrieved from https://factfinder.census.gov/faces/tableservices/jsf/pages/productview.xhtml?pid=ACS_16_1YR_B03001&prodType=table.

US Census Bureau. (2017f). Median family income in the past 12 months (in 2016 inflation-adjusted dollars) (Asian alone householder). 2016 American Community Survey 1-year estimates. Retrieved from https://factfinder.census.gov/faces/tableservices/jsf/pages/productview.xhtml?pid=ACS_16_1YR_B19113D&prodType=table.

US Census Bureau. (2017g). Place of birth for the foreign-born population in the United States. 2016 American Community Survey 1-year estimates. Retrieved from https://factfinder.census.gov/bkmk/table/1.0/en/ACS/13_1YR/B05006/0100000US|0400000US12|0400000US36.

US Census Bureau. (2017h). Poverty status in the past 12 months by sex by age (American Indian and Alaska native alone). 2016 American Community Survey 1-year estimates. Retrieved from https://factfinder.census.gov/faces/tableservices/jsf/pages/productview.xhtml?pid=ACS_16_1YR_B17001C&prodType=table.

US Census Bureau. (2017i). Poverty status in the past 12 months by sex by age (Asian alone). 2016 American Community Survey 1-year estimates. Retrieved from https://factfinder.census.gov/faces/tableservices/jsf/pages/productview.xhtml?pid=ACS_16_1YR_B17001D&prodType=table.

US Census Bureau. (2017j). Tenure (Asian alone householder). 2016 American Community Survey 1-year estimates. Retrieved from https://factfinder.census.gov/faces/tableservices/jsf/pages/productview.xhtml?pid=ACS_16_1YR_B25003D&prodType=table.

US Census Bureau. (2017k). ACS demographic and housing estimates. 2012–2016 American Community Survey 5-year estimates. Retrieved from https://factfinder.census.gov/bkmk/table/1.0/en/ACS/16_5YR/DP05/0100000US.

US Census Bureau. (2017l, March 16). Supplemental poverty measure. Retrieved from www.census.gov/topics/income-poverty/supplemental-poverty-measure/about.html.

US Census Bureau. (2017m). Median age at first marriage. Universe: population 15 to 54 years. 2016 American Community Survey 1-year estimates. Retrieved from https:// factfinder.census.gov/faces/tableservices/jsf/pages/productview.xhtml?pid=ACS_16_1YR_B12007&prodType=table.

US Census Bureau. (2017n). Decennial censuses, 1890 to 1940, and Current Population Survey, annual social and economic supplements, 1947 to 2017. Retrieved from www.census.gov/content/dam/Census/library/visualizations/time-series/demo/families-and-households/ms-2.pdf.

US Census Bureau. (2017o). Historical living arrangements of children. Retrieved from www.census.gov/data/tables/time-series/demo/families/children.html.

US Census Bureau. (2017p). America's families and living arrangements: 2017. Retrieved from www.census.gov/data/tables/2017/demo/families/cps-2017.html.

US Census Bureau. (2017q). Educational attainment in the United States: 2016. Retrieved from www.census.gov/data/tables/2016/demo/education-attainment/cps-detailed-tables.html.

US Census Bureau. (2018). Poverty status in the past 12 months. 2016 American Community Survey 1-year estimates. Retrieved from https://factfinder.census.gov/bkmk/table/1.0/en/ACS/16_1YR/S1701/0400000US27.

US Department of Agriculture. (2018). Supplemental Nutrition Assistance Program: Participation and costs, 1969–2017. Retrieved from www.fns.usda.gov/pd/supplemental-nutrition-assistance-program-snap.

US Department of Commerce. (2009, May 15). Confirmation hearing for Robert Groves, chaired by Senator Thomas Carper (D-DE). Retrieved from www.census.gov/newsroom/releases/pdf/confirmation_transcript.pdf.

US Department of Health and Human Services (2015). Dietary guidelines for Americans 2015–2020. Retrieved from http://health.gov/dietaryguidelines/2015/guidelines.

US Department of Health and Human Services. (2016). Health, the United States, 2015: In brief. Retrieved from www.cdc.gov/nchs/data/hus/hus15_inbrief.pdf.

US Department of Health and Human Services. (2018) US Federal poverty guidelines used to determine financial eligibility for certain federal programs. Retrieved from https://aspe.hhs.gov/poverty-guidelines.

US Department of the Interior, Indian Affairs. (2018). Frequently asked questions: What is the Bureau of Indian Education? Retrieved from www.bia.gov/frequently-asked-questions.

US Department of Justice, Civil Rights Division. (2007). *Enforcement and outreach following the September 11 terrorist attacks.* Retrieved from www.usdoj.gov/crt/legalinfo/discrimupdate.html.

US Department of State, Bureau of Consular Affairs. (2018). Adoption statistics. Retrieved from https://travel.state.gov/content/travel/en/Intercountry-Adoption/adopt_ref/adoption-statistics.html.

US Senate, Subcommittee on Immigration and Naturalization of the Committee on the Judiciary. (1965, February 10). Washington, DC, pp. 1–3.

Usher, N. (2014). *Making news at the* New York Times. Ann Arbor: University of Michigan Press.

Varian, H. R. (2006, May 4). Red states, blue states: New labels for long-running differences. *The New York Times,* p. 3. Retrieved from www.nytimes.com/2006/05/04/business/04scene.html.

Varon, J. (2004). *Bringing the war home: The Weather Underground, the Red Army faction, and revolutionary violence in the sixties and seventies.* Berkeley: University of California Press.

Vars, F. E., & Bowen, W. G. (1998). Scholastic aptitude test scores, race, and academic performance in selective colleges and universities. In C. Jencks & M. Phillips (Eds.), *The black-white test score gap* (pp. 457–79). Washington, DC: Brookings Institution Press.

Ventola, C. L. (2011). Direct-to-consumer pharmaceutical advertising: Therapeutic or toxic? *Pharmacy and Therapeutics, 36*(10), 669–74, 681–84.

Verba, S., et al. (2004). Political inequality: What do we know about it? In K. M. Neckerman (Ed.), *Social inequality* (pp. 635–66). New York: Sage.

Verdier, H. (2016). DTR: Define the relationship—Swipe right for Tinder's first podcast. *The Guardian.* Retrieved from www.theguardian.com/tv-and-radio/2016/dec/15/dtr-define-the-relationship-swipe-right-for-tinders-first-podcast.

Vespa, J., et al. (2013). America's families and living arrangements: 2012. Population characteristics. US Census Bureau. Retrieved from http://www.census.gov/prod/2013pubs/p20-570.pdf.

Vestal, S. (2017, December 1). Rachel Dolezal remains unabashadly Rachel Dolezal. *The Spokesman-Review.* Retrieved from www.spokesman.com/stories/2017/dec/01/shawn-vestal-rachel-dolezal-remains-unabashedly-ra.

Virupaksha, H. G., Muralidhar, D., & Ramakrishna, J. (2016). Suicide and suicidal behavior among transgender persons. *Indian Journal of Psychological Medicine, 38*(6), 505–509.

Wade, L. (2017). *American hookup: A new culture of sex on campus.* New York: Norton.

Wall, A. (2012). Gubernatorial electons, campaign costs, and winning governors. Council of State Governments. Retrieved from http://knowledgecenter.csg.org/kc/content/gubernatorial-elections-campaign-costs-and-winning-governors.

Waite, L., & Gallagher, M. (2000). *The case for marriage: Why married people are happier, healthier, and better off financially.* New York: Doubleday.

Wallerstein, J., et al. (2000). *The unexpected legacy of divorce.* New York: Hyperion.

Wang, W. (2012, February 16). The rise of intermarriage. Pew Research Center, Social and Demographic Trends. Retrieved from www.pewsocialtrends.org/2012/02/16/the-rise-of-intermarriage.

Wang, W. (2015, June 12). Interracial marriage: Who is "marrying out"? Pew Research Center. Retrieved from www.pewresearch.org/fact-tank/2015/06/12/interracial-marriage-who-is-marrying-out.

Ward, J. (2015). *Not gay: Sex between straight white men.* New York: NYU Press.

Warren, E. (2007). Unsafe at any rate. *Democracy: A Journal of Ideas, 5.* Retrieved from www.democracyjournal.org/article2.php?ID=6528&limit=0&limit2=1500&page=1.

Warren, E., & Warren Tyagi, A. (2003). *The two-income trap: Why middle-class mothers and fathers are going broke.* New York: Basic Books.

Warren, J. W., & Twine, F. W. (1997). White Americans, the new minority? Non-blacks and the ever-expanding boundaries of whiteness. *Journal of Black Studies, 28,* 200–18.

Washington, E. (2008). Female socialization: How daughters affect their legislator fathers' voting on women's issues. *American Economic Review, 98*(1), 311–32.

Water.org. (2014, May 18). Water facts. Retrieved from http://water.org/water-crisis/water-facts/water.

Watts, D. (2003). *Six degrees: The science of a connected age.* New York: Norton.

Weber, M. (1946). *From Max Weber: Essays in sociology.* H. H. Gerth & C. W. Mills (Eds. & Trans.). New York: Oxford University Press.

Weber, M. (1968). *Economy and society: An outline of interpretive sociology.* G. Rothe & C. Wittich (Eds.). (E. Fischoff et al., Trans.). New York: Bedminster Press. (Original work published 1922.)

Weber, M. (2003). *The Protestant ethic and the spirit of capitalism.* Oxford: Blackwell. (Original work published 1904.)

Weber, M. (2004). *The vocation lectures: Science as a vocation, politics as a vocation.* D. S. Owen et al. (Eds.). Indianapolis, IN: Hackett.

Wegener, B. (1991). Job mobility and social ties: Social resources, prior job, and status attainment. *American Sociological Review, 56*(1), 60–71.

West, C., & Zimmerman, D. (1987). Doing gender. *Gender & Society, 1*(2), 123–31.

Whitbeck, L. B., et al. (1991). Family economic hardship, parental support, and adolescent self-esteem. *Social Psychology Quarterly, 54,* 353–63.

Wilcox, W. B. (1999). *Religion and paternal involvement: Product of religious commitment or American convention?* Paper presented at the Annual Meeting of the American Sociological Association, Chicago.

Wilcox, W. B. (2004). *Soft patriarchs, new men: How Christianity shapes fathers and husbands.* Chicago: University of Chicago Press.

Wilensky, H. L. (1974). *The welfare state and equality: Structural and ideological roots of public expenditures.* Berkeley: University of California Press.

Williams, C. (1995). *Still a man's world: Men who do women's work.* Berkeley: University of California Press.

Williams, C. (2013). The glass escalator, revisited. *Gender & Society, 27*(5), 609–29.

Williams, S. (2005). Million dollar blocks. Public service announcement [Digital presentation]. Columbia University, Spatial Information Design Lab. Retrieved from www.spatialinformationdesignlab.org/movie.php?url=MEDIA/PSA_01.avi.

Willis, P. (1981). *Learning to labor: How working class kids get working class jobs.* New York: Columbia University Press. (Original work published 1977.)

Wilson, J. Q., & Kelling, G. L. (1982). Broken windows: The police and neighborhood safety. *The Atlantic Monthly, 249*(3), 29–37.

Wilson, V., & Rodgers III, W. M. (2016, September 20). Black-white wage gaps expand with rising wage inequality. Economic Policy Institute. Retrieved from www.epi.org/publication/black-white-wage-gaps-expand-with-rising-wage-inequality.

Wilson, W. J. (1978). *The declining significance of race: Blacks and changing American institutions.* Chicago: University of Chicago Press.

Wilson, W. J. (1987). *The truly disadvantaged: The inner city, the underclass, and public policy.* Chicago: University of Chicago Press.

Wilson, W. J. (1996). *When work disappears: The world of the new urban poor.* New York: Knopf.

Winant, H. (2001). *The world is a ghetto: Race and democracy since World War II.* New York: Basic Books.

Wingfield, A. H. (2009). Racializing the glass escalator: Reconsidering men's experiences with women's work. *Gender & Society, 23*(1), 5–26.

Wirth, L. (1938). Urbanism as a way of life. *American Journal of Sociology, 44*(1), 1–24.

Witte, J. F. (1998). The Milwaukee voucher experiment. *Educational Evaluation and Policy Analysis, 20*(4), 229–51.

Wolff, E. N. (2017, November 29). Has middle class wealth recovered? Table 2. Retrieved from www.aeaweb.org/conference/2018/preliminary/paper/5ZFEEf69.

Wood, W., et al. (2002). Sources of information about dating and their perceived influence on adolescents. *Journal of Adolescent Research, 17*, 401–17.

World Bank. (2002). Global partnership to eliminate riverblindness. Retrieved from www.worldbank.org/afr/gper.

World Bank. (2018). CO_2 emissions (metric tons per capita). Retrieved from https://data.worldbank.org/indicator/EN.ATM.CO2E.PC?year_high_desc=false.

World Economic Forum. (2017). The global gender gap report 2017. Retrieved from www3.weforum.org/docs/WEF_GGGR_2017.pdf.

World Food Programme. (2018). Hunger facts. Retrieved from www.wfp.org/share-a-hunger-fact.

World Health Organization. (2017). Fact sheet: Poliomyelitis. Retrieved from www.who.int/mediacentre/factsheets/fs114/en.

World Health Organization. (2018a). Life expectancy at birth (years), 2000–2016: Both sexes: 2016. Global Health Observatory (GHO) data. Retrieved from www.who.int/gho/mortality_burden_disease/life_tables/situation_trends/en.

World Health Organization. (2018b, February 7). Fact sheet: Drinking-water. Retrieved from www.who.int/mediacentre/factsheets/fs391/en.

World Health Organization. (2018c). HIV/AIDS. Global Health Observatory (GHO) data. Retrieved from www.who.int/gho/hiv/en.

Wuthnow, R. (1998). *Loose connections: Joining together in America's fragmented communities.* Cambridge: Harvard University Press.

Wyler, G. (2014, May 6). NYC's newest megachurch is more popular than Jesus. *Vice.* Retrieved from www.vice.com/en_us/article/8gd8jp/hillsong-nyc-more-popular-than-jesus.

Xu, J., et al. (2009, August 19). Deaths: Preliminary data for 2007. Centers for Disease Control and Prevention, National Center for Health Statistics. *National Vital Statistics Reports, 58*(1). Retrieved from www.cdc.gov/nchs/data/nvsr/nvsr58/nvsr58_01.pdf.

Yavorsky, J. E., et al. (2015). The production of inequality: The gender division of labor across the transition to parenthood. *Journal of Marriage and the Family, 77*(3): 662–79.

Yeakley, R. (2011, February 15). Evangelical churches still growing, mainline Protestantism in decline. *Huffington Post.* Retrieved from www.huffingtonpost.com/2011/02/15/report-us-churches-contin_n_823701.html.

Young, L., & Hammock, E. (2007). On switches and knobs, microsatellites and monogamy. *Trends in Genetics, 23*(5), 209–12.

Young, L., & Wang, Z. (2004). The neurobiology of pair bonding. *Nature Neuroscience, 7*(10), 1048–54. Retrieved from www.ecfs.org/projects/pchurch/AT%20BIOLOGY/PAPERS%5CNeurobiology%20of%20Pair%20Bonding.pdf.

Young, M. P. (2002). Confessional protest: The religious birth of U.S. national social movements. *American Sociological Review, 67*, 660–88.

Zelizer, V. A. (2005). *The purchase of intimacy.* Princeton, NJ: Princeton University Press.

Zenith USA. (2018). Top 30 global media owners 2017. Retrieved from www.zenithusa.com/top-30-global-media-owners-2017.

Zernike, K. (2006, April 23). College, my way. *The New York Times.* Retrieved from www.nytimes.com/2006/04/23/education/edlife/zernike.html.

Zill, N. (1988). Behavior, achievement, and health problems among children in stepfamilies: Findings from a national survey of child health. In E. M. Hetherington & J. Arasteh (Eds.), *Impact of divorce, single parenting and stepparenting in children* (pp. 325–68). Hillsdale, NJ: Erlbaum.

Zill, N., et al. (1991). *The life circumstances and development of children in welfare families: A profile based on national survey data.* Washington, DC: Child Trends.

Zimbardo, P. (1971). *Stanford prison experiment: A simulation study of the psychology of imprisonment conducted at Stanford University.* Retrieved from www.prisonexp.org.

Zimbardo, P. (2007). *The Lucifer effect: Understanding how good people turn evil.* New York: Random House.

Zimbardo, P. (2018). Philip Zimbardo's response to recent criticisms of the Stanford Prison Experiment. Retrieved from www.prisonexp.org/response.

Zimmer, R. W., & Toma, E. F. (2000). Peer effects in private and public schools across countries. *Journal of Policy Analysis and Management, 19*(1), 75–92.

Zimmerman, D. H., & West, C. (1975). Sex roles, interruptions and silences in conversation. In B. Thorned & N. Henley (Eds.), *Language and sex: Difference and dominance* (pp. 105–29). Rowley, MA: Newbury House.

Zuckerman, P. (2009). Atheism, secularity, and well-being: How the findings of social science counter negative stereotypes and assumptions. *Sociology Compass, 3*(6), 949–71.

Zukin, S. (2003). *Point of purchase: How shopping changed American culture.* New York: Routledge.

Zwick, R. (2002). *Fair game? The use of standardized admissions tests in higher education.* New York: Routledge.

Zwick, R., & Sklar, J. C. (2005). Predicting college grades and degree completion using high school grades and SAT scores: The role of student ethnicity and first language. *American Educational Research Journal, 42*(3), 439–65.

Credits

FIGURES & TABLES

CHAPTER 2: Figure 2.3: From Babbie, *The Practice of Social Research*, 11E. © 2007 South-Western, a part of Cengage Learning, Inc. Reproduced by permission. www.cengage.com/permissions.

CHAPTER 5: Figure 5.3: Originally from "Studies of independence and conformity: A minority of one against a unanimous majority," by Asch, S. E. *Psychological Monographs* 70 (9), 1–70. doi:10.1037/h0093718. **Figure 5.5:** Figure 2, "Structure of Romantic and Sexual Contact at Jefferson" from "Chains of Affection: The Structure of Adolescent Romantic and Sexual Networks" by Peter S. Bearman, James Moody, and Katherine Stovel, *American Journal of Sociology*, Vol. 110, No. 1, July 2004, pp. 44–91. Copyright © 2004, The University of Chicago Press. Reprinted by permission of the publisher. **Figure 5.6:** From "Chains of Affection: The Structure of Adolescent Romantic and Sexual Networks" by Peter S. Bearman, James Moody, and Katherine Stovel, *American Journal of Sociology*, Vol. 110, No. 1, July 2004, pp. 44–91. Copyright © 2004, The University of Chicago Press. Reprinted by permission of the publisher.

CHAPTER 7: Table 7.2: Figure from *Mobility Tables: Quantitative Applications in the Social Sciences* by Michael Hout. p. 11. Copyright © 1983 by SAGE Publications, Inc. Reprinted by permission of SAGE Publications, Inc.

CHAPTER 8: Elliot Jackson, "'Have You Ever Reconsidered Being Transgender?' Answer by Elliot Jackson," Quora.com, September 30, 2017. Reprinted by permission of the author.

CHAPTER 9: Table 9.1: Figure, "Gordon's Stages of Assimilation," *Assimilation in American Life: The Role of Race, Religion, and National Origins* by Milton M. Gordon, p. 71. Copyright © 1964, Oxford University Press, Inc. Reprinted by permission of Oxford University Press, Inc.

CHAPTER 10: Figure 10.1: Figure, "Work-Family Living Arrangements of Children, 1960 & 2012" by Philip Cohen. Reprinted by permission of the author.

PHOTOS

FRONTMATTER: p. ix: AP Photo/Paul Sakuma; **p. x:** CBS/Photofest; **p. xi:** Emmanuel Dunand/AFP/Getty Images; **p. xii:** Lee Celano/The New York Times/Redux; **p. xiv:** Patrick T. Fallon/Bloomberg via Getty Images; **p. xv:** imageBROKER/Alamy Stock Photo; **p. xvi:** Bettmann/Getty Images; **p. xvi:** John Birdsall/Alamy Stock Photo.

CHAPTER 1: p. 5: Gjon Mili/The LIFE Picture Collection/Getty Images; **p. 7:** Miramax/Photofest; **p. 11 (left):** AP Photo/Paul Sakuma, file; **p. 11 (right):** Sunday Times/Moeletsi Mabe/Shutterstock; **p. 14:** Courtesy of Asha Rangappa; **p. 16 (left):** Spencer Platt/Getty Images; **p. 16 (right):** Mario Tama/Getty Images; **p. 18 (from left to right):** SuperStock; GRANGER—All rights reserved; Sarin Images/GRANGER—All rights reserved; **p. 19 (from left to right):** Private Collection/Bridgeman Images; GRANGER—All rights reserved; Temple de la Religion de l'Humanite, Paris, France/Bridgeman Images; **p. 20 (left):** George Rinhart/Corbis via Getty Images; **p. 20 (right):** Bettmann/Getty Images; **p. 21 (from left to right):**

Photo; **p. 175:** Courtesy of Erica Rothman, Nightlight Productions and Dalton Conley; **p. 176:** Lee Celano/The New York Times/Redux; **p. 178:** Richard Perry/The New York Times/Redux; **p. 179:** © Tampa Bay Times/Alessandra Da Pra/The Image Works; **p. 180:** AP Photo/The Ledger Independent, Terry Prather; **p. 181:** AP Photo/Patriot-News, Paul Chaplin; **p. 182:** Doug Pensinger/Getty Images for AFL; **p. 183:** Courtesy of Erica Rothman, Nightlight Productions and Dalton Conley; **p. 184:** Lauren Greenfield/INSTITUTE; **p. 192:** keith morris/Alamy Stock Photo.

CHAPTER 6: p. 198: Courtesy of Erica Rothman, Nightlight Productions and Dalton Conley; **p. 200 (left):** © British Library Board/Robana/Art Resource, NY; **p. 200 (right):** Bettmann/Getty Images; **p. 201:** AP Photo/David J. Phillip; **p. 202 (both):** Bettmann/Getty Images; **p. 203:** Three Lions/Getty Images; **p. 204:** AP Photo/Greg Wahl-Stephens; **p. 207:** 3LH/SuperStock; **p. 210:** Pacific Press Service/GRANGER—All rights reserved; **p. 213 (clockwise from top left):** Jack Hollingsworth/Corbis; Anthony-Masterson/Getty Images; Joseph Scherschel/The LIFE Picture Collection/Getty Images; Kathy deWitt/Alamy Stock Photo; Seth Wenig/Reuters/Newscom; **p. 216:** Ruth Fremson/The New York Times/Redux; **p. 217:** Allied Artists/Photofest; **p. 218:** Phillip G. Zimbardo, Inc.; **p. 219:** AP Photo; **p. 222:** Courtesy of Erica Rothman, Nightlight Productions and Dalton Conley; **p. 223:** Phillip G. Zimbardo, Inc.; **p. 225:** Timothy A. Clary/AFP/Getty Images; **p. 231:** Courtesy of Matt Ramirez; **p. 232:** Q. Sakamaki/Redux; **p. 234:** Chronicle/Alamy Stock Photo; **p. 236:** Bettmann/Getty Images; **p. 241:** Brant Ward/San Francisco Chronicle/Polaris.

CHAPTER 7: p. 250: Photo12/UIG via Getty Images; **p. 253:** AP Photo/Themba Hadebe; **p. 254:** AP Photo/Kevork Djansezian; **p. 256:** © Ted Streshinsky/CORBIS/Corbis via Getty Images; **p. 257:** Ben Stechschulte/Redux; **p. 259:** Will Burgess/Reuters/Newscom; **p. 261:** Farooq Naeem/AFP/Getty Images; **p. 262:** AP Photo/Steve Coleman; **p. 264:** B. O'Kane/Alamy Stock Photo; **p. 268:** Patrick T. Fallon/Bloomberg via Getty Images; **p. 274:** Courtesy of Erica Rothman, Nightlight Productions and Dalton Conley; **p. 277:** Courtesy of Erica Rothman, Nightlight Productions and Dalton Conley; **p. 278:** Q. Sakamaki/Redux; **p. 284:** Katherine Taylor/The New York Times/Redux; **p. 285:** GRANGER—All rights reserved.

CHAPTER 8: p. 292: Al Drago/CQ Roll Call via AP Images; **p. 294:** 'Phall-O-meter,' Intersex Society of North America. Credit: Wellcome Collection. CC BY 4.0; **p. 298:** FO Travel/Alamy Stock Photo; **p. 299:** Courtesy of Amos Mac; **p. 301:** Marvin Joseph/The Washington Post via Getty Images; **p. 302 (left):** Erin Baiano/The New York Times/Redux; **p. 302 (right):** PYMCA/UIG via Getty Images; **p. 304:** AP Photo; **p. 306:** Harold M. Lambert/Getty Images; **p. 308:** Library of Congress; **p. 310:** AP Photo/Greg Smith; **p. 315:** The Advertising Archives; **p. 317:** AP Photo/Noah Berger; **p. 319:** Courtesy of Erica Rothman, Nightlight Productions and Dalton Conley; **p. 320:** Ashmolean Museum, University of Oxford, UK/Bridgeman Images; **p. 323:** University Archives and Records Center, University of Pennsylvania; **p. 324:** Jason Szenes/UPI/Newscom; **p. 325:** imageBROKER/Alamy Stock Photo; **p. 327:** Courtesy of Erica Rothman, Nightlight Productions and Dalton Conley.

CHAPTER 9: p. 339: The Ohio State University Billy Ireland Cartoon Library & Museum; **p. 341:** Reproduced by courtesy of Mauricio Hatchwell Toledano, www.facsimile-editions.com; **p. 343:** SPL/Science Source; **p. 344:** © Hulton-Deutsch Collection/CORBIS/Corbis via Getty Images; **p. 345:** National Park Service, Statue of Liberty National Monument; **p. 346:** Courtesy of Dr. Bhagat Singh Thind Spiritual Science Foundation; **p. 348:** Keith Bedford/The New York Times/Redux; **p. 349:** Simon Rawles/Alamy Stock Photo; **p. 350:** Courtesy of Erica Rothman, Nightlight Productions and Dalton Conley; **p. 351:** Reuters/Kamal Kishore; **p. 353:** Brad Rickerby/Reuters/Newscom; **p. 355:** Jonathan Ernst/Reuters/Newscom; **p. 357:** Photo by Mario Villafuerte; **p. 358:** AP Photo/Steven Senne; **p. 363:** AP Photo/The Sun Herald, Sean Loftin; **p. 368:** National Archives; **p. 369:** Bettmann/Getty Images; **p. 370:** Bettmann/Getty Images; **p. 372:** Reuters/Corrine Dufka; **p. 377:** Ralph Orlowski/Getty Images; **p. 379:** Library of Congress; **p. 384:** Courtesy of Erica Rothman, Nightlight Productions and Dalton Conley.

CHAPTER 10: p. 392: ABC/Photofest; **p. 394:** Francis Miller/The LIFE Picture Collection/Getty

Index

educational attainment, 359, 377, 378, 382–83

Educational Testing Service, 17

egalitarianism, 414

egoistic suicide, 208–9

Egypt, ancient, 340

Ehrenreich, Barbara, 420–21

80/20 rule (Pareto Principle), 267

elder abuse, 409–10

The Elementary Forms of the Religious Life (Durkheim), 27

elite–mass dichotomy system, 266–69

embeddedness, 171–74

emerging adulthood, 142

empiricists *versus* theorists, 40

employment, maternal, 406–7, 410, 437

endogamy, 394

England, Paula, 184, 327

English language, 17

Enlightenment, 20, 250–53

entrepreneurship, 180–81

Environmental Protection Agency (EPA), 102

Epstein, Cynthia Fuchs, 302–3

equality of condition, 256–57

equality of opportunity, 255–56

equality of outcome, 257–58

equal opportunity, 93–94

equity inequality, 378, 380

Erikson, Kai, 226–27

essentialist arguments, 70, 296, 308, 313, 365

Essman, Elliot, 271

estate system, 259–60

ethereum, 159, 160

ethics of social research, 72–73

Ethiopia, 400

ethnicity: and cultural practices, 28; and primordialism, 365; *versus* race, 351–54; in the United States, 354–61, 382–83. *See also* race

ethnocentrism, 84–85, 342

ethnography, 48, 58–60

ethnomethodology, 148–50, 308

eugenics, 344–46, 347

Europe, 276

European colonists, 354–55

European Commission, 191–92

European immigration, 338–40, 366, 370–71

Evans, Phyllis, 66

evolution and social change, 32

evolutionary biology, 39

exchange mobility, 281

Executive Order 9066, 379

exogamy, 347, 394, 430

experimental methods, 70–71

experimenter effects, 57–60

extended family, 396, 397

the eyes and ears of the street, 206–7

face, 146–47

Facebook, 101, 110, 152, 153, 191

face-to-face interactions, 166–69, 178–79, 182

fag discourse, 140–41

fair families, 415

Fair Sentencing Act (2010), 380

fair-weather liberals, 375, 376

fake news, 101

family(ies), 390–437; and domestic abuse, 409–10; family forms and changes, 393–400; feminist rethinking of, 407–9; future of, 425–32; and gendered division of labor, 402–3, 410–15, 437; historical divide between public and private, 400–405, 415–16, 423; homosexual and transgender, 428–29; and household definitions, 38; and inequality, 415–22, 422–25; multiracial, 430–31; nuclear family model, 391–93, 395, 400–401, 404–5; as primary group, 168; of prisoners, 230; and socialization, 123, 124, 129–32, 407; and work and gender, 405–7. *See also* children; fatherhood; marriage; motherhood

farmworkers, 280, 281

Fascism, 35, 36, 99

fashion industry, 319

Fassbender, Michael, 96

fatalistic suicide, 211

fatherhood, 397, 410–15

Fausto-Sterling, Anne, 295

Federal Bureau of Investigation (FBI), 225, 226

Feeding the Family (Duvall), 407

Feinberg, Kenneth, 413–14

fellatio, 320–21

female genital mutilation, 90

The Feminine Mystique (Friedan), 309, 408

femininity, 299–300, 303, 306–7, 309

feminism: and compensation for housework, 413, 414; and conflict theories, 307; and daughters, 129; defined, 303; feminist theory, 32–33, 304; and gender inequality theories, 303–11; and Martineau, 22; Marxist, 321–22; and Mead, 90; and media critiques, 108–9; and rethinking of the family, 407–9; second wave of, 303, 309, 316; and sexuality, 321–22; and suicide, 211; third wave of, 309; and women of color, 309–10

feminist methodology, 60–61, 72

feminist theory, 32–33, 304

Ferber, Abby, 363

Ferguson, Adam, 250, 251, 255

fertility rates, 403, 405, 406

fertility treatments, 399–400

FHA mortgages, 378

Figlio, David, 116

Filipino immigrants, 359

filter bubble, 101

financial crisis (2007–2008), 71, 159, 226, 269. *See also* Great Recession

Finney, Hal, 159

Flat Broke with Children (Hays), 421–22

Fong, Jeanne, 433

food and social construction of reality, 142, 143

food stamps. *See* SNAP (Supplemental Nutrition Assistance Program)

For Better or for Worse (Hetherington and Kelly), 427

Ford Motor Company, 268

foreign-born populations in the United States, 366–67, 381

formal social sanctions, 206, 207, 245

formal sociology, 27

Foster, Don, 141

Foucault, Michel, 233–37, 322–24, 325

France, 377

freedom of the press, 109–10

free market, 174

free rider problem, 258

French Revolution, 18

Freud, Sigmund, 306–7

Friedan, Betty, 24, 309, 310, 408

Friendster, 48

front-stage and backstage arenas, 145–46, 147

Fryer, Roland, 284

Ft. Myer, Virginia, 369

functionalism: and approaches to deviance and social control, 200–214; and family models, 395–97; and Parsons, 30–31; structural, 305–6

The Future of Marriage (Bernard), 408

sociological theorist, 23–25; and conflict theory, 31–32; *Critique of the Gotha Programme*, 257–58; and equality of outcome, 258; and feminism, 307, 321–22; and hegemony, 99; and reflection theory, 95; on timeline, 19, 20, 21; Weber compared to, 25–26

masculinity: and doing gender, 309; in dual-income families, 408, 412; femininity compared to, 300–301, 303; and Freud, 306–7; and gender inequality in the workplace, 317, 318; hegemonic, 301–2; and intersectionality, 309–10

Massey, Douglas, 25, 370

mass incarceration, 237–38, 240–41. *See also* prison and prisoners

mass media. *See* media

master-slave dialectic, 253–54

master status, 138

material culture, 86

maternal employment, 406–7, 410, 437

maternity leave, 414

matriarchal thesis, 416–17

matrix of domination, 309–10

Mattis, James, 268

McCain, John, 425

McCandless, Christopher Johnson, 213

McCarty, Oseola, 262

McGlaun, Jenni, 420

McIntosh, Peggy, 362

McNamara, Robert, 268

McRobbie, Angela, 108

McVeigh, Timothy, 204

me (concept of), 126

Mead, George Herbert, 23, 28, 33, 126, 127, 128

Mead, Margaret, 37, 90

meaning(s): and actions, 26, 33; and concepts, 32; and culture, 87–88; cycle of, 33; and interpretive sociology, 26, 41; and social interaction, 28, 33, 141, 143

means-ends theory of deviance, 212

Mean Streets (film), 224

Mears, Ashley, 319

mechanical solidarity, 201–2, 203–4, 205, 234

media, 97–114; defined, 97; effects of, 101–3; history of, 97–99; life cycle, 100–101;

and organicism, 31; political economy of, 109–14; race and gender politics in, 104–5; savvyness, 81–82; social norms and, 80; and stereotypes, 103–6

mediating variables, 56

mediators, 163–64

Medicaid, 285

medical education, 11

Meetup.com, 182

Men and Women of the Corporation (Kanter), 317–18

mental health and illness, 220–21

mental health institutions, 231–33

meritocracy, 267

Merton, Robert: and midrange theory, 23, 34, 136, 156, 157; prejudice and discrimination chart, 375–76; role theory of, 136–37, 156–57; strain theory of, 212–14; on timeline, 23

#MeToo movement, 330

metrosexual males, 302

Mexican Americans, 357, 358, 418

Mexico, 83, 84, 418

microsociology, 27, 33, 34, 41

micro theories, 214

middle class, 271–75. *See also* working class

Middle Eastern Americans, 360–61

midrange theory, 34

Mikelson, Kelly S., 116

Milgram, Stanley, 174–75

military and altruistic suicide, 209–10

military order, 267–68

Mill, John Stuart, 20

Millar, John, 250, 251, 255

Mills, C. Wright: and elite-mass dichotomy system, 266, 267–68; functionalist theory criticized by, 31; and power elite, 190; *The Power Elite*, 267; and sociological imagination, 4–5; *The Sociological Imagination*, 5; on timeline, 23, 24

Mind, Self, and Society (Mead), 28

The Mind and Society (Pareto), 266–67

"Mind over Splatter" (Foster), 141

minimum wage, 420–21

minorities: and affirmative action policies, 283–85; and equity inequality, 378; and social deviance, 200; and street crime, 224; and suicide, 209,

211–12. *See also* race; racism; *specific minority group names*

minority-majority group relations, 364–73; assimilation, 364–65; pluralism, 365–68; racial conflict, 372–73; segregation and discrimination, 368–72

miscegenation, 347, 393–94, 430

MIT, 110

Mnuchin, Steven, 268

mobility tables, 280–81, 282

Möbius strip, 296

modeling industry, 319

model minority myth, 360, 366, 384

moderating variables, 56

modern sociological theories, 31–34

Molina, Florencia, 254

Money, Morals, and Manners (Lamont), 63–64

monogamy, 394, 395

monogenism, 344

monopolies, 174

Montesquieu, 276

Moretti, Franco, 40

More Work for Mother (Cowan), 413

Morgan, Stephen, 37

Mormon Church and Mormons, 394

Morning, Ann, 69–70

mortgages, 378

Moss, Kate, 113

"Most Expensivest Shit" (*GQ* series), 288

motherhood: and African American families, 416; as full-time occupation, 402; and household division of labor, 410–15; women delaying, 399–400; in Zambian culture, 397. *See also* stay-at-home mothers; working mothers

Moveon.org, 182

moving pictures, creation of, 97, 98

Moynihan, Daniel P., 365, 416

Mozart, Wolfgang Amadeus, 96

Ms. (magazine), 108

multiculturalism, 367

multilevel marketing, 174

multiple births, 399–400

multiracial families, 430–31

Multiracial March on Washington (1996), 382

Munduruku people, 397

murder rate, 228

Murray, Charles, 31, 421

Murree Brewery Company, 261

Americans from safety net, 68; and global inequality, 276, 277; institutional, 379–81; and Irish immigrants, 338; in Japan, 349; in the media, 103–8; and poverty, 198; prejudice and discrimination, 375–76; and pseudoscience, 342–43, 344; scientific, 342, 347, 372–73; and social Darwinism, 344; and social norms, 199, 200; white privilege compared to, 362. *See also* white nationalism; white supremacists

radical feminists, 307, 308
Radway, Janice, 100
railroad, 20, 35, 37
Rainbow Man, 81–82, 113, 114–16
Ramirez, Marc, 230–31
Rangappa, Asha, 12–14
rape, 320, 372, 409. *See also* sexual assault
rationality, 26, 28
Rauscher, Emily, 129
Read, Jen'nan, 349–50
Reading the Romance (Radway), 100
Reagan, Ronald, 369
reality, social construction of, 28–29, 141–51
Reardon, Sean, 284
rebels, 213–14
recidivism, 229, 231, 240
Reddy, Gayatri, 298
redlining, 372, 378, 389. *See also* residential segregation
reference groups, 170
reflection theory, 94–97
reflexivity, 57
reliability, 56–57
religion(s): and caste system, 260, 261; Comte on, 19–20; and deviance, 226–27; Durkheim on, 27; as ideological framework, 89; media promotion of, 81–112, 114–16; and medical practices, 90; and suicide, 209. *See also specific religion names*
Renoir, Pierre-Auguste, 206, 207
reproductive labor, 413
Republic (Plato), 85
Republican Party, 432
research, 46–79; ethics of, 72–73; flowchart, 62; methods and basics of, 50–71; and policy, 73–75
research cycle, 50
residential segregation, 369–72, 378, 389

resistance *versus* acceptance to racial oppression, 373, 374–75
Reskin, Barbara, 316
resocialization, 135, 136
restitutive social sanctions, 204–5
retreatists, 213
returns to schooling, 8, 9, 10
reverse causality, 54
revolution and social change, 32
reweighting, 74
Reynolds, William Bradford, 369
Rich, Adrienne, 322
Ridgefield, Washington, 429
right to be forgotten, 191–92
Rios, Victor, 197–99
riots, 374
Risman, Barbara, 407, 415
ritualists, 212, 213
R. J. Reynolds, 114
Roberts, John G., Jr., 33
Robeson, Paul, 368
Rockin' Rollen, 81–82, 113, 114–16
role conflict, 137, 156–57
roles, 137
role strain, 137, 156–57
role theory, 136–38
romantic leftovers, 187–88
Rome, ancient, 341
roommates, 152–53
Roos, Patricia, 316
Roosevelt, Franklin D., 68, 379
Roots (Haley), 353
Rosaldo, Michelle, 305
Rosenhan, David L., 217–18
Rousseau, Jean-Jacques, 20, 249–50, 255, 258
Rowling, J. K., 95–96
Rubin, Gayle, 304–5
Rwanda, 372–73
Sacerdote, Bruce, 152
Sachs, Jeffrey, 252–53, 276–77
safety net, 68
safety net, informal, 418
Salai-i-Martin, Xavier, 278–79
Salganik, Matthew, 71
Sambia people, 320–21
same-sex marriage, 324, 326, 428–29, 432, 433
Samoans, 90
samples and sampling, 66–67, 73–75
Sanders, Bernie, 179
San Quentin State Prison, 241
sanskritization, 261
Sapir-Whorf Hypothesis, 87
sati (suttee), 209
SAT test, 122, 257, 282
scared straight programs, 240, 241
Schippers, Mimi, 326

schools: advertising in, 111–12; desegregation of, 368–69; primary and secondary, 17; segregation of, 368–69; and socialization, 123, 132–34. *See also* college; education
Schüll, Natasha Dow, 37, 38
science, 89
scientific method, 49
scientific racism, 342, 347, 372–73
Scottish Enlightenment, 250–53, 258
secondary deviance, 220–21, 231
secondary groups, 169
Secondat, Charles de, 343
second shift, 410–15
The Second Shift (Hochschild), 436
secretarial work, 317–18
secular morality, 19
segmental solidarity, 201
segregation: and the Big Sort, 388–89; in criminal justice system, 372; and equal opportunity concept, 94; Jim Crow laws, 200–201, 256, 301, 347, 368, 373; and minority-majority relations, 364, 368–72; and poverty, 198; residential, 369–72, 378, 389; of schools, 368–69
self, 28, 125–27, 232–33
self concept, 28
Senior, Jennifer, 412
seppuku (hara-kiri), 209, 210
September 11, 2001 (9/11), 349–50, 361, 413–14
September 11 Victim Compensation Fund, 413–14
SES. *See* socioeconomic status (SES)
Sesame Street (TV show), 102
sex: defined, 292; and feminist theory, 32; relationship with gender, 293–96. *See also* gender; sexuality
Sex, Gender, and Society (Oakley), 32
sex/gender system, Rubin's, 304–5
sexism, 94, 108–9, 311–12
sex-reassignment surgeries, 294–95
sex role theory, 305–6
sexual assault, 330–31. *See also* domestic abuse and violence; intimate partner violence (IPV); rape
Sexual Behavior in the Human Male (Kinsey), 324
sexual harassment, 140, 141, 313, 314–15, 318, 330–31